Water Resources Management Models for Policy Assessment

Water Resources Management Models for Policy Assessment

Editor

Luis Garrote

MDPI • Basel • Beijing • Wuhan • Barcelona • Belgrade • Manchester • Tokyo • Cluj • Tianjin

Editor
Luis Garrote
Civil Engineering:
Hydraulics, Energy and Environment
Universidad Politécnica de Madrid
Madrid
Spain

Editorial Office
MDPI
St. Alban-Anlage 66
4052 Basel, Switzerland

This is a reprint of articles from the Special Issue published online in the open access journal *Water* (ISSN 2073-4441) (available at: www.mdpi.com/journal/water/special_issues/Management_ Models_Policy).

For citation purposes, cite each article independently as indicated on the article page online and as indicated below:

LastName, A.A.; LastName, B.B.; LastName, C.C. Article Title. *Journal Name* **Year**, *Volume Number*, Page Range.

ISBN 978-3-0365-4100-6 (Hbk)
ISBN 978-3-0365-4099-3 (PDF)

Contents

About the Editor

Luis Garrote

Luis Garrote is a full professor of hydraulic engineering and head of the research group on hydroinformatics and water management at the Universidad Politécnica de Madrid, Spain. He has developed water resource management models for policy assessment, working in collaboration with national and international administrations and the private sector.

Preface to "Water Resources Management Models for Policy Assessment"

Water resources management models support a variety of research applications, including the assessment of water availability, allocation of water among competing uses, evaluation of system performance, identification of optimal system expansion, or definition of suitable operating strategies. System analysis tools, such as simulation and optimization, have been enriched with novel modelling concepts drawn from social sciences, economic analysis, conflict resolution, agent-based systems, or game theory, among others. This field has evolved from the traditional emphasis on cost–benefit analysis in water resource project investments to a wider scope that includes environmental implications, stakeholder concerns, social welfare, and human dimensions. We now face the challenge of developing integrated modelling frameworks to provide quantitative evidence to policymakers on water management issues.

This book compiles original research papers that apply a variety of techniques to identify and evaluate water resource management policies. The compilation presented here covers a wide range of topics and methodologies applied across the world, from a local to a continental scope. Open challenges in water resource management, such as quantitative assessment of policy impacts, trade-off analyses, understanding the water-energy-food-environment nexus, collaborative model development, stakeholder engagement, formalizing social interactions, or improving the theoretical understanding of complex adaptive systems, are outlined. Therefore, this book covers research areas that have emerged from the origins of water resource systems analysis, seeking to improve the way in which water policy is formulated and implemented.

Luis Garrote
Editor

Editorial

Water Resources Management Models for Policy Assessment

Luis Garrote

Department of Civil Engineering, Hydraulics, Energy and Environment, Universidad Politécnica de Madrid, 28040 Madrid, Spain; l.garrote@upm.es

Citation: Garrote, L. Water Resources Management Models for Policy Assessment. *Water* **2021**, *13*, 1063. https://doi.org/10.3390/w13081063

Received: 7 April 2021
Accepted: 9 April 2021
Published: 13 April 2021

Publisher's Note: MDPI stays neutral with regard to jurisdictional claims in published maps and institutional affiliations.

Water resources management models support a variety of research applications, including the assessment of water availability [1], the allocation of water among competing uses [2], the evaluation of system performance [3,4], the identification of optimal system expansion [5], and the definition of suitable operating strategies [6]. System analysis tools, like simulation and optimization, have been enriched with novel modelling concepts drawn from social sciences [7], economic analysis [8], conflict resolution [9], agent-based systems [10], and game theory [11], among others. The field has evolved from a traditional emphasis on cost–benefit analysis in water resource project investments to a wider scope that includes environmental implications, stakeholder concerns, social welfare, and human dimensions [12].

This Special Issue of Water integrates a collection of research papers that develop or apply water resources management models for policy identification and assessment. Active research has been conducted to address the challenge of developing integrated modelling frameworks to provide quantitative evidence for policymakers on water management issues. The compilation presented here covers a wide range of topics and methodologies applied across the world, from a local to continental scope. It illustrates open challenges in water resources management, like quantitative assessment of policy impacts, trade-off analyses, understanding the water-energy-food-environment nexus, collaborative model development, stakeholder engagement, formalizing social interactions, or improving the theoretical understanding of complex adaptive systems. This issue is therefore a representation of research areas that have emerged from the origins of water resource systems analysis seeking to improve the way water policy is formulated and implemented.

The contributions to the Special Issue may be classified into four major topics: water availability and accessibility, management of water infrastructure, environmental concerns, and social and economic issues. Contributions in the first group focus on the estimation of water availability under different climate and policy scenarios. Two papers are focused on Europe and two are focused on China. The paper by Sordo-Ward et al. [13] presented a regional assessment of future water availability in Europe. They applied a high-resolution model to produce detailed maps of water availability in European rivers and evaluated model and scenario uncertainties under different climate projections. The work presented in [14] was specifically focused on the role of reservoir storage to enhance resilience to climate change. The authors studied 16 major river basins in Southern Europe and found that increased storage capacity attenuated the reduction of water availability and reduced its uncertainty under climate change projections. Li et al. [15] evaluated five spatial factors to obtain a water accessibility index in Southwest China. They produced a spatial pattern and compared water accessibility and water demand at the county level. As a result of their analysis, the authors provided policy recommendations to correct the imbalance. Finally, Wang et al. [16] studied the water-carrying capacity of the Chang-Ji region in Northeast China. They applied techniques such as the fuzzy comprehensive evaluation method, gray correlation analysis, and multiple linear regression models to evaluate water-carrying capacity under different social development plans, identified critical issues, and provided suggestions to allow for a sustainable development of the economy in the region.

The second topic deals with models intended to provide support for management policies for water infrastructure. The paper by Rubio-Martin et al. [17] presented an application of system dynamics for the strategic planning of drought management in a river basin located in Southeast Spain. The authors proposed a system state index that is used to trigger dynamic reservoir operating rules, policies, and drought management strategies. They argued that application of their decision support system may lead to a substantial reduction of the economic impact of droughts in the basin. Gabriel-Martin et al. [18] aimed at solving conflicts that arise in the operation of multipurpose reservoirs. Their technical contribution is a model that maximizes reservoir yield subject to constraints imposed by hydrological dam safety and downstream river safety. They produced a set of Pareto optimal configurations that may be used by policymakers to emphasize water availability or flood protection. Bejarano et al. [19] offered a computational tool intended to summarize data on sub-daily streamflow into manageable, comprehensive, and ecologically meaningful metrics, which can be used to qualify and quantify flow alteration. This tool may be used by policymakers to evaluate the potential ecological consequences of the hydrological alteration produced by water infrastructure. The contribution by Martin-Candilejo et al. [20] is focused on energy efficiency. They proposed a novel method to account for energy costs associated to water pumping in the design and operation of water supply systems.

Water quality is the major focus of the third topic, which deals with environmental concerns. Xie et al. [21] reported on the experience of implementing the nation-wide freshwater health evaluation in China. They proposed a new indicator framework combining ecosystem integrity with non-ecological performance with the objective of improving water governance. The result of their work is directly policy-relevant because it will be integrated into a new national standard. Salehi et al. [22] evaluated the pollutant discharge characteristics for 12 facilities in an industry sector in the United States. They applied principal component analysis to water quality parameters and developed water quality indexes to monitor water quality fluctuations. They characterized stormwater quality variations among studied facilities and seasons, concluding with suggestions for future changes for decision makers. The work by Duan et al. [23] focused on background pollutants and their influence on water quality management and assessment methods in China. The authors argue that it is unreasonable to use a uniform standard to evaluate water quality across the country. They defined a suitable pollutant yield coefficient by coupling an export coefficient model with a mechanistic model. Based on their results, they proposed a more reasonable sewage discharge limit and water quality evaluation method. Best management practices to control water pollution were analyzed in [24]. The authors evaluated the performance of three types of pollution control measures on dissolved nitrogen by coupling an improved watershed model with a multi-objective optimization algorithm. Their optimization model system could assist decision-makers in selecting the most appropriate measures for pollution control in a watershed. Wang et al. [25] proposed an index system to evaluate the degree of coordination between economic development and infrastructure construction in a sponge city in China. They studied the spatial statistical pattern of coordination and concluded that the problems due to inadequate coordination were prominent in the region. They suggested a stronger emphasis on the construction of green infrastructure.

The fourth topic is related to social and economic issues. Lima-Quispe et al. [26] discussed river basin planning in Bolivia from the wider perspective of regional planning. They tackled the problems of coordinating watershed planning with other planning units and integrating watershed management with water resources management. The authors proposed the novel technique of robust decision support to help stakeholders discern positive and negative interactions of interventions, use spatially explicit indicators, and identify adequate management strategies. Li et al. [27] explored the applicability of China's policy based on water saving contracts by risk assessment. Overall risk was found to be low, but they showed concern for some potential risk factors, such as audit, financing, and payment risk. Feria-Dominguez et al. [28] analyzed the impact of a severe drought on the Brazilian stock market. They found statistical evidence of financial impact caused by the declaration of drought among agri-food firms, particularly in those companies that shell perishable products. Shen et al. [29] studied the impact of tourism on the sustainable development of a reservoir in China. They applied different analytical techniques to process hundreds of questionnaires filled by the local population. In their conclusions, they found that stakeholders were very critical of the consequences of tourism development in the region and provided suggestions to mitigate the negative impacts. Santasusagna Riu et al. [30] also used questionnaires to analyze the management of urban public services in the internal border area between two Spanish regions. Based on their analysis of the replies, they concluded that there are deficiencies to correct and suggested enhanced cooperation across the border to improve priority urban public services.

This Special Issue is a compilation of 18 contributions that offer a wide perspective of the potential of water resources management models for policy assessment. The papers focus on a diversity of topics, geographical locations, spatial scales, and methodologies that illustrate successful case studies of science inspiring policy. This work is offered as an asset for researchers and policymakers.

Funding: This research received no external funding.

Acknowledgments: The Guest Editor wishes to thank the authors for their relevant contributions to the Special Issue and to the anonymous reviewers for their constructive comments and their dedication. Special thanks to the editorial managers who worked hard to speed up the handling of the manuscripts submitted to this Special Issue. Their help is gratefully acknowledged.

Conflicts of Interest: The author declares no conflict of interest.

References

1. Alcamo, J.; Doll, P.; Henrichs, T.; Kaspar, F.; Lehner, B.; Rösch, T.; Siebert, S. Development and testing of the WaterGAP 2 global model of water use and availability. *Hydrol. Sci.* **2003**, *48*, 317–337. [CrossRef]
2. Letcher, R.A.; Jakeman, A.J.; Croke, B.F.W. Model development for integrated assessment of water allocation options. *Water Resour. Res.* **2004**, *40*, W05502. [CrossRef]
3. Hashimoto, T.; Stedinger, J.R.; Loucks, D.P. Reliability, Resiliency, and Vulnerability Criteria for Water-Resource System Performance Evaluation. *Water Resour. Res.* **1982**, *18*, 14–20. [CrossRef]
4. Zou, H.; Liu, D.; Guo, S.; Xiong, L.; Liu, P.; Yin, J.; Zeng, Y.; Zhang, J. Shen, Y Quantitative assessment of adaptive measures on optimal water resources allocation by using reliability, resilience, vulnerability indicators. *Stoch. Environ. Res. Risk Assess.* **2020**, *34*, 103–119. [CrossRef]
5. Girard, C.; Rinaudo, J.; Pulido-Velazquez, M.; Caballero, Y. An interdisciplinary modelling framework for selecting adaptation measures at the river basin scale in a global change scenario. *Environ. Model. Softw.* **2015**, *69*, 42–54. [CrossRef]
6. Labadie, J.W. Optimal Operation of Multireservoir Systems: State-of-the-Art Review. *J. Water Resour. Plan. Manag.* **2004**, *130*, 93–111. [CrossRef]
7. Purkey, D.R.; Huber-Lee, A.; Yates, D.N.; Hanemann, M.; Harrod-Julius, S. Integrating a Climate Change Assessment Tool into Stakeholder-Driven Water Management Decision-Making Processes in California. *Water Resour. Manag.* **2007**, *21*, 315–329. [CrossRef]
8. Harou, J.J.; Pulido-Velazquez, M.; Rosenberg, D.E.; Mediellín-Azuara, J.; Lund, J.K.; Howitt, R.E. Hydro-economic models: Concepts, design, applications, and future prospects. *J. Hydrol.* **2009**, *375*, 627–643. [CrossRef]
9. Jury, W.A.; Vaux, H.J. The emerging global water crisis: Managing scarcity and conflict between water users. *Adv. Agron.* **2007**, *95*, 1–76. [CrossRef]
10. Berger, T.; Birner, R.; Díaz, J.; McCarthy, N.; Wittmer, N. Capturing the complexity of water uses and water users within a multi-agent framework. *Water Resour. Manag.* **2007**, *21*, 129. [CrossRef]
11. Madani, K.; Lund, J.R. A Monte-Carlo game theoretic approach for multi-criteria decision making under uncertainty. *Adv. Water Resour.* **2011**, *34*, 607–616. [CrossRef]
12. Brown, C.M.; Lund, J.R.; Cai, X.; Reed, P.M.; Zagona, E.A.; Ostfeld, A.; Hall, J.; Characklis, G.W.; Yu, Y.; Brekke, L. The future of water resources systems analysis: Toward a scientific framework for sustainable water management. *Water Resour. Res.* **2015**, *51*, 6110–6124. [CrossRef]
13. Sordo-Ward, A.; Granados, I.; Iglesias, A.; Garrote, L. Blue Water in Europe: Estimates of Current and Future Availability and Analysis of Uncertainty. *Water* **2019**, *11*, 420. [CrossRef]
14. Granados, A.; Sordo-Ward, A.; Paredes-Beltrán, B.; Garrote, L. Exploring the Role of Reservoir Storage in Enhancing Resilience to Climate Change in Southern Europe. *Water* **2021**, *13*, 85. [CrossRef]
15. Li, T.; Qiu, S.; Mao, S.; Bao, R.; Deng, H. Evaluating Water Resource Accessibility in Southwest China. *Water* **2019**, *11*, 1708. [CrossRef]

16. Wang, G.; Xiao, C.; Qi, Z.; Liang, X.; Meng, F.; Sun, Y. Water Resource Carrying Capacity Based on Water Demand Prediction in Chang-Ji Economic Circle. *Water* **2021**, *13*, 16. [CrossRef]
17. Rubio-Martin, A.; Pulido-Velazquez, M.; Macian-Sorribes, H.; Garcia-Prats, A. System Dynamics Modeling for Supporting Drought-Oriented Management of the Jucar River System, Spain. *Water* **2020**, *12*, 1407. [CrossRef]
18. Gabriel-Martin, I.; Sordo-Ward, A.; Santillán, D.; Garrote, L. Flood Control Versus Water Conservation in Reservoirs: A New Policy to Allocate Available Storage. *Water* **2020**, *12*, 994. [CrossRef]
19. Bejarano, M.D.; García-Palacios, J.H.; Sordo-Ward, A.; Garrote, L.; Nilsson, C. A New Tool for Assessing Environmental Impacts of Altering Short-Term Flow and Water Level Regimes. *Water* **2020**, *12*, 2913. [CrossRef]
20. Martin-Candilejo, A.; Santillán, D.; Garrote, L. Pump Efficiency Analysis for Proper Energy Assessment in Optimization of Water Supply Systems. *Water* **2020**, *12*, 132. [CrossRef]
21. Xie, C.; Yang, Y.; Liu, Y.; Liu, G.; Fan, Z.; Li, Y. A Nation-Wide Framework for Evaluating Freshwater Health in China: Background, Administration, and Indicators. *Water* **2020**, *12*, 2596. [CrossRef]
22. Salehi, M.; Aghilinasrollahabadi, K.; Salehi Esfandarani, M. An Investigation of Stormwater Quality Variation within an Industry Sector Using the Self-Reported Data Collected under the Stormwater Monitoring Program. *Water* **2020**, *12*, 3185. [CrossRef]
23. Duan, M.; Du, X.; Peng, W.; Zhang, S.; Yan, L. Necessity of Acknowledging Background Pollutants in Management and Assessment of Unique Basins. *Water* **2019**, *11*, 1103. [CrossRef]
24. Qi, Z.; Kang, G.; Wu, X.; Sun, Y.; Wang, Y. Multi-Objective Optimization for Selecting and Siting the Cost-Effective BMPs by Coupling Revised GWLF Model and NSGAII Algorithm. *Water* **2020**, *12*, 235. [CrossRef]
25. Wang, K.; Zhang, L.; Zhang, L.; Cheng, S. Coupling Coordination Assessment on Sponge City Construction and Its Spatial Pattern in Henan Province, China. *Water* **2020**, *12*, 3482. [CrossRef]
26. Lima-Quispe, N.; Coleoni, C.; Rincón, W.; Gutierrez, Z.; Zubieta, F.; Nuñez, S.; Iriarte, J.; Saldías, C.; Purkey, D.; Escobar, M.; et al. Delving into the Divisive Waters of River Basin Planning in Bolivia: A Case Study in the Cochabamba Valley. *Water* **2021**, *13*, 190. [CrossRef]
27. Li, Q.; Shangguan, Z.; Wang, M.Y.; Yan, D.; Zhai, R.; Wen, C. Risk Assessment of China's Water-Saving Contract Projects. *Water* **2020**, *12*, 2689. [CrossRef]
28. Feria-Domínguez, J.M.; Paneque, P.; de la Piedra, F. Are the Financial Markets Sensitive to Hydrological Risk? Evidence from the Bovespa. *Water* **2020**, *12*, 3011. [CrossRef]
29. Shen, C.-C.; Liang, C.-F.; Hsu, C.-H.; Chien, J.-H.; Lin, H.-H. Research on the Impact of Tourism Development on the Sustainable Development of Reservoir Headwater Area Using China's Tingxi Reservoir as an Example. *Water* **2020**, *12*, 3311. [CrossRef]
30. Santasusagna Riu, A.; Galindo Caldés, R.; Tort Donada, J. Assessing Inter-Administrative Cooperation in Urban Public Services: A Case Study of River Municipalities in the Internal Border Area between Aragon and Catalonia (Spain). *Water* **2020**, *12*, 2505. [CrossRef]

Article

Blue Water in Europe: Estimates of Current and Future Availability and Analysis of Uncertainty

Alvaro Sordo-Ward [1], Isabel Granados [1], Ana Iglesias [2] and Luis Garrote [1,*]

[1] Department of Civil Engineering: Hydraulics, Energy and Environment, Universidad Politécnica de Madrid, Madrid 28040, Spain; alvaro.sordo.ward@upm.es (A.S.-W.); i.granados@upm.es (I.G.)

[2] Department of Agricultural Economics & CEIGRAM, Universidad Politécnica de Madrid, Madrid 28040, Spain; ana.iglesias@upm.es

* Correspondence: l.garrote@upm.es; Tel.: +34-913-366-672

Received: 29 December 2018; Accepted: 21 February 2019; Published: 26 February 2019

Abstract: This study presents a regional assessment of future blue water availability in Europe under different assumptions. The baseline period (1960 to 1999) is compared to the near future (2020 to 2059) and the long-term future (2060 to 2099). Blue water availability is estimated as the maximum amount of water supplied at a certain point of the river network that satisfies a defined demand, taking into account specified reliability requirements. Water availability is computed with the geospatial high-resolution Water Availability and Adaptation Policy Assessment (WAAPA) model. The WAAPA model definition for this study extends over 6 million km^2 in Europe and considers almost 4000 sub-basins in Europe. The model takes into account 2300 reservoirs larger than 5 hm^3, and the dataset of Hydro 1k with 1700 sub-basins. Hydrological scenarios for this study were taken from the Inter-Sectoral Impact Model Inter-Comparison Project and included simulations of five global climate models under different Representative Concentration Pathways scenarios. The choice of method is useful for evaluating large area regional studies that include high resolution on the systems´ characterization. The results highlight large uncertainties associated with a set of local water availability estimates across Europe. Climate model uncertainties for mean annual runoff and potential water availability were found to be higher than scenario uncertainties. Furthermore, the existing hydraulic infrastructure and its management have played an important role by decoupling water availability from hydrologic variability. This is observed for all climate models, the emissions scenarios considered, and for near and long-term future. The balance between water availability and withdrawals is threatened in some regions, such as the Mediterranean region. The results of this study contribute to defining potential challenges in water resource systems and regional risk areas.

Keywords: climate change; water resources; water availability; uncertainty; WAAPA model; Western Europe

---

1. Introduction

Water management is challenged by climate change. By the 2070s, the percentage of the surface area under conditions of severe water stress is expected to increase from the current 19% to 35% in central and southern Europe [1]. Populations living under water stress conditions in regions from 17 countries of Western Europe are projected to increase by between 16 and 44 million [2]. It is also predicted that the runoff of certain rivers may diminish by up to 80% during the summers. Reservoirs may lose resources due to a decrease in rainfall and the frequency of droughts will increase. The consensus is that the effect of climate change will also exacerbate precipitation extremes with more pronounced drought and flood periods [3–5]. At the same time, future water demand is increasing due to climate and social changes. Higher temperatures lead to increased water demand for irrigation and urban supply, hydroelectric potential of Europe may decrease 6% on average, and between 20 and 50% in the Mediterranean region. Advances in technology efficiency may only affect industrial demand [2]. In the Mediterranean region, impacts of climate change on water will certainly have a large influence on human water security and biodiversity [6]. There are several hundred local studies on the potential impacts of climate change on water resources in the Mediterranean, which apply many different approaches. Although the results are diverse and sometimes contradictory, a common element is that one of the primary impacts of climate change will be a reduction of water availability in the Mediterranean Region [1,2]. Furthermore, several authors showed that Global Climate Models (GCMs) were the main source of uncertainty when assessing the impacts of climate change on hydrologic processes [7,8]. Meanwhile, uncertainty associated with streamflow appeared to be more consistent with precipitation than temperature and showed higher sensitivity to the selection of GCMs than to the Regional Climate Models (RCPs) [9,10].

Water availability focuses on blue water, which is defined as water that runs off the landscape into streams, rivers, reservoirs, and groundwater [11]. However, the term "water availability" includes multiple aspects. A multitude of studies consider water availability to be directly linked to changes in average runoff, estimated as the net difference between precipitation and evapotranspiration [12,13]. In non-altered basins, water availability would be either null or extremely low because it would be determined by long term minimum values of flow. It is clear that hydraulic infrastructure plays an important role in making water available for users, mainly by the regulation and transportation of water resources. Even though the storage-based strategy proved to be very successful in the past [14], expanding infrastructure is not an option to increase availability in many regions due to social and environmental constraints [15]. As a result, increasing demand relies heavily on management. The emphasis is currently being placed on how to improve management of existing infrastructure and on socio-economic measures through demand management and water use efficiency [16,17]. The main factors to be considered in regulated water resource systems are the stream flow variability, storage capacity, and yield reliability. In this study, we define blue water availability as the amount of water that can be supplied at a certain point of the river network to satisfy a regular demand under specified reliability requirements [18,19]. Therefore, water availability is the combined result of natural processes, existing infrastructure, and policy. A wide range of techniques have been proposed to analyse water availability, from relatively simple stochastic processes relating these variables to highly complex models solving the water allocation problem [20–24], even including social and economic considerations [25]. In the water sector, institutions, users, technology, and the economy cooperate to achieve equilibrium between water supply and demand in water resource systems. In order to understand the process of reaching future goals for water under climate change, science has developed a set of tools to understand uncertainty [26–29], assess future impacts [30,31], and facilitate policy development [1,16,18,32]. However, most studies were developed using detailed water management and planning models, and were applied at the local scale. In systems and situations where limited information is available and regional or continental-scale studies are needed, it is generally better to obtain a global overview of the water supply systems' performance under different climate and policy scenarios, using simplified regional models rather than carrying out very detailed simulations with conventional models, which require very specific information on water demands and infrastructure [18,33,34]. These continental scale-models are conceived to estimate the maximum water availability and to provide technical and quantitative support to possible water policies in the short and long term. Then, these models and detailed water management and planning models should be considered as complementary tools.

Over forty percent of the total water withdrawal in Europe is used for agriculture. Southern countries use the largest percentages of abstracted water for agriculture. This generally accounts for more than two thirds of total abstraction. In northern member States, levels of water use in agriculture are much smaller, with irrigation being less important but still accounting for more than 30% in some areas [35]. Moreover, if the climate in a given region gets drier and warmer, water availability will decrease, and the issue will be exacerbated by increasing water demand [36]. For example, it is expected that areas of maize grain cultivation will expand up to 30–50% in Europe [37–40] with increases of up to 50% in net primary productivity in northern European ecosystems, as a result of a longer growing season and higher CO_2 concentrations [37]. As the projected impacts on productivity of crops and ecosystems included the direct effects of increased CO_2 concentration on photosynthesis, the variation in simulated results attributed to differences between the climate models were, in all cases, smaller than the variation attributed to emissions scenarios [37]. The objective of this study is to estimate future potential blue water availability in Europe and its associated uncertainty, which is induced by emissions scenarios and climate change models. This study first proposes a methodology to conduct climate change analyses in water resource systems, which is based on a high-resolution geospatial model and the use of information available in public databases. Second, the study evaluates distributed mean annual runoff and its uncertainty in main rivers within Western Europe in the baseline period and in two future periods. Third, the study analyses water availability changes and its uncertainty across Western Europe under different climate change scenarios and climate models. Finally, the study analyses the geographically distributed relationships at a continental-scale among the mean annual runoff, water availability, and water withdrawals under the baseline and future periods.

2. Materials and Methods

The methodological approach is detailed in Figure 1. The methodology is based on a high-resolution GIS-based model, named "Water Availability and Adaptation Policy Assessment (WAAPA)" which enables the estimation of water availability under many climate scenarios to produce a global picture of the situation [33]. The model assimilates climate and geospatial information seamlessly, accounts for reservoir storage (from an individual reservoir or from a system of reservoirs), and produces blue water availability estimates. The model computes net blue water availability for consumptive use of a river basin, taking into account the regulation capacity of its water supply system, and a set of management standards defined by water policy. The model estimates the water availability not only at the outlet of sub-basins (e.g., river intersections), but also at any desired point of the defined river network (e.g., each dam location), by accounting for the entire system of dams in the upstream basin. Basic components of WAAPA are reservoirs, inflows, and demands and they are linked to nodes of the river network. The joint reservoir operation model simulates the behaviour of a set of reservoirs that supply water for a set of prioritized demands, complying with specified ecological flows and accounting for evaporation losses.

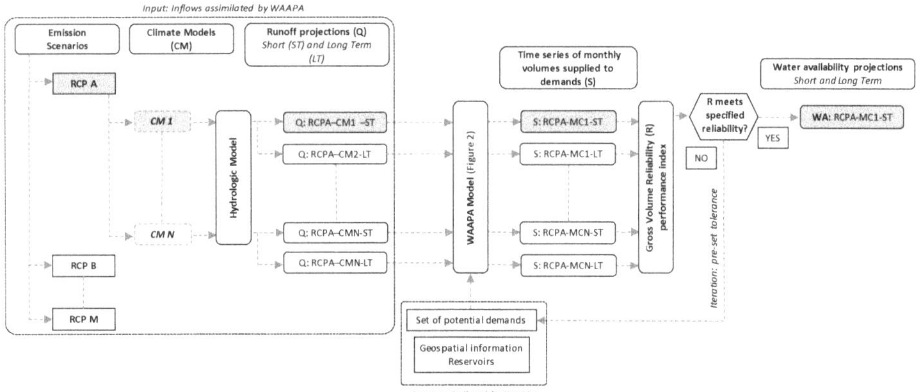

Figure 1. General scheme of the applied methodology. The displayed procedure was applied to each defined sub-basin. Grey areas indicate the first path carried out, from the selection of the emissions scenario to the estimation of the water availability.

In this study, we evaluated the water availability of the joint reservoir operation model following a high resolution and global management scheme (Figure 2). For each selected sub-basin (derived from dam locations and river confluences), this scheme considers each reservoir individually and all reservoirs are jointly operated to supply a set of prioritized demands. It is assumed that any demand at a given point in the stream network can be supplied by any reservoir located upstream from it. It corresponds to a situation where there is little development of system interconnections, but a large development of water distribution networks, which are managed globally to supply all demands present in the analysed system. Water is first released (to satisfy demands) from the reservoirs located at low areas of the basin. If these reservoirs are full and receive more contributions, uncontrolled spills are released and water falls out of the system. On the other hand, if upstream reservoirs are full and receive more inflows, the extra water is collected by the downstream reservoirs. This management criterion is not totally real, because real systems usually are managed taking into account more conditions and constraints. The joint reservoir operation model maximizes water availability because it minimizes the excess storage. In each time step, the model performs the following operations:

1. It satisfies the environmental flow requirement in every reservoir with the available inflow. Environmental flows are passed to downstream reservoirs and added to their inflows.
2. It computes evaporation in every reservoir and reduces storage accordingly.
3. It computes excess storage (storage above maximum capacity) in every reservoir (if there is an increment of storage with the remaining inflow).
4. It satisfies demands ordered by priority, if possible. It uses excess storage first, then available storage starting from higher priority reservoirs.
5. If excess storage remains in any reservoir, it computes uncontrolled spills.

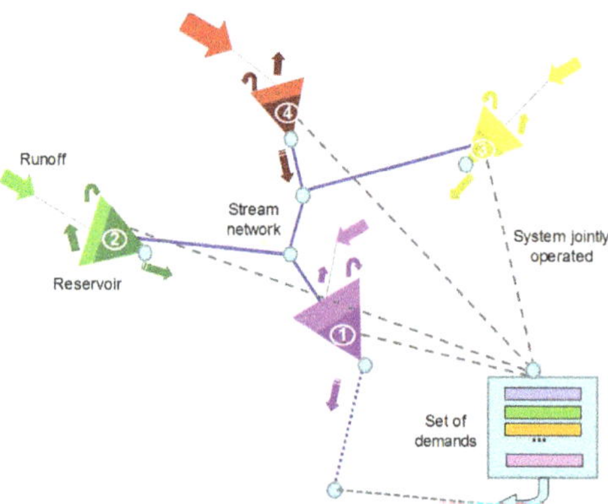

Figure 2. Operation scheme of the high-resolution Water Availability and Adaptation Policy Assessment (WAAPA) model for each given point of the stream network (blue lines). Triangles represent dams, big coloured arrows represent inflows, small arrows represent reservoir evaporation, uncontrolled spills, and environmental flows, and grey dashed lines represent supplies from each reservoir to the basin demands (rectangles).

The result of the joint reservoir operation model is a set of time series of monthly volumes supplied to each demand, monthly storage values, monthly values of spills, environmental flows, and evaporation losses in every reservoir. Finally, we calculated the system performance by applying the Gross Volume Reliability performance index. This index is the ratio of total volume supplied to demand in the system and the total volume demanded by the system, during the analysed period [33]. In this study, water availability is estimated by considering only one demand present in the system under the hypothesis of 90% reliability.

To define the maximum amount of water that can be supplied at a certain point of the river network to satisfy a regular demand, a bipartition method is applied: Excessive values of demands are set (for example, similar to mean monthly runoff) and the simulation is carried out. The deficits are obtained and specified reliability requirements are checked. If the specified reliability requirements are not fulfilled, the demand is reduced by half and simulated again. If the specified reliability requirements are satisfied, half of the difference is added and simulated again, and so on until the deficit (or gain) is smaller than a pre-set tolerance (e.g., 0.1 hm^3/year).

Case Study

The area under analysis is composed of the major river basins in Western Europe. WAAPA model data are geographically referenced (Figure 3). Following, we present the data used to build the WAAPA model. We determined the topology of the model by dividing the area under study into a number of units of analysis, which are homogeneous sub-basins from the water management perspective. The sub-basins are related through the "drain to" relationship, and the analysis is applied to all possible basins, from the small headwater sub-basins to the largest basin draining to the sea. In this work, we divided western Europe into sub-basins (3839), based on the Hydro1k data set (1.538 sub-basins [41]), and the derived-from dam locations (2.301 sub-basins), which belong to 621 large basins draining to the sea. The total area under study is over 6,000,000 km^2.

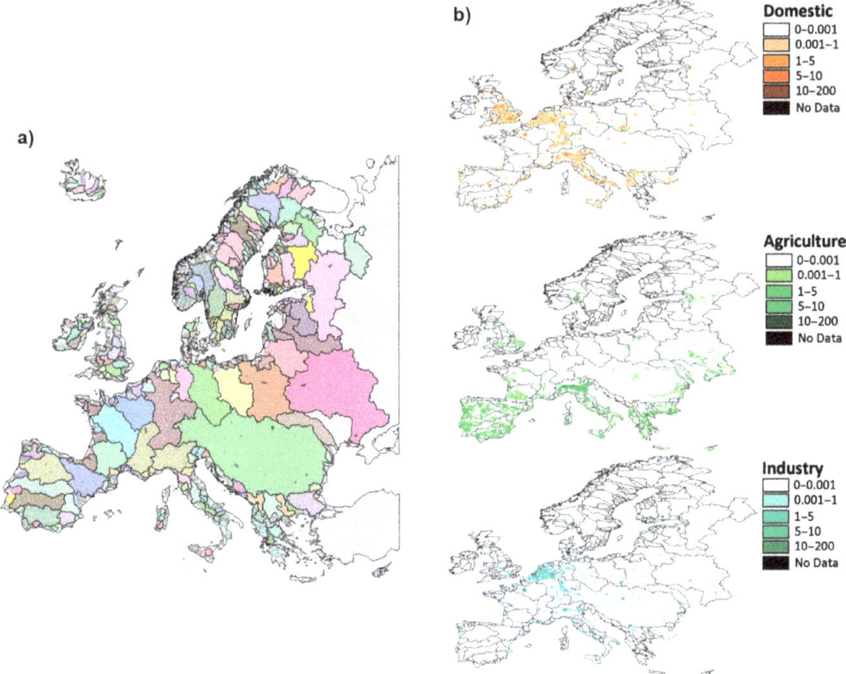

Figure 3. Case study: Western Europe. (**a**) Domain under analysis. Colours represent the 621 major river basins draining to the sea. (**b**) Information utilized for the estimation of withdrawals (domestic (hm^3/km^2), agriculture (hm^3/km^2) and industry (hm^3/km^2)) in present and future scenarios and for each analysed sub-basin.

Naturalized streamflow was obtained from the results of the application of the PCRGLOBWB model [42] to the Inter-Sectoral Impact Model Inter-Comparison Project [43]. The PCRGLOBWB model was run for the entire globe at 0.5° resolution, using forcing from five global climate models under historical conditions and climate change projections, corresponding to four Representative Concentration Pathways scenarios: RCP2.6, RCP4.5, RCP6.0, and RCP8.5. The following climate models were used as input: GFDL-ESM2NM (GFDL), HadGEM2-ES (HadGEM2), IPSL-CM5A-LR (IPSL), MIROC-ESM-CHEM (MIROC), and NorESM1-M (NorESM1).

Three time periods were considered: Reference (1960–1999), short term (ST, 2020–2059), and long term (LT, 2060–2099). Since runoff obtained from climate model input usually presents significant bias, average runoff values were corrected for bias using the UNH/GRDC (University of New Hampshire/Global Runoff Data Centre) composite runoff field, which combines observed river discharges with a water balance model [44], and is a reference of the current global surface runoff [34,44,45]. Following González-Zeas [45], we applied the bias-correction methodology based on the determination of a monthly correction factor. We calculated the monthly mean runoff series for the control scenario to obtain twelve representative statistical parameters: The ratios between the UNH/GRDC values (observed) and the simulated runoff. These multiplying factors were used to correct bias in the control and the projected series. The reservoir storage volume available for regulation in every sub-basin was obtained from the ICOLD World Register of Dams [46]. Dams in the register with more than 5 hm^3 of storage capacity were georeferenced and linked to the corresponding storage capacity and flooded area (2.301 dams). Environmental flows were computed through a hydrologic method. The minimum environmental flow was set to the 10% percentile of the marginal monthly distribution, according to Spanish legislation. In the absence of more advanced methods, the Spanish regulation for river basin plans establishes several hydrologic methods to define minimum environmental flows [31]. One of them is based on the percentile of the marginal distribution of monthly flows, defining a range between 5 and 15%.

In this study, we estimated current, short-, and long-term geographically distributed water withdrawals. Country-based data on current freshwater withdrawal were taken from the World Bank database. These data were spatially distributed using proxy variables: Population density for urban and industrial withdrawals and irrigated area for agricultural withdrawals. The population density was obtained from the Gridded Population of the World product of the Global Rural–Urban Mapping Project (GRUMP), available at the Centre for International Earth Science Information Network (Figure 3b) [47]. The area potentially under irrigation was estimated from the Global Map of Irrigated Area dataset [48]. Future withdrawals were estimated using the projections of population and gross domestic product (GDP) provided by IIASA. These projections were estimated following RCP scenario assumptions [38,39]. Projections of total freshwater withdrawal and industrial withdrawal were estimated from regressions based on World Bank data using per capita GDP projections [40].

3. Results

Figure 4 shows the comparison of streamflow change from reference (1960–1999) to climate change RCP4.5 scenarios, both for short (2020–2059) and long term (2060–2099), and over the five climate models. Figure 4 is dimensionless (percentage), and the values were obtained by applying Equation (1). The red shading represents a decrease (negative values) and green shading an increase (positive values) of the future mean annual runoff. The yellow shading represents no changes of mean annual runoff for future periods compared to the reference scenario.

$$\text{Mean annual runoff change} = \left(\frac{\text{Mean annual runoff at future scenario}}{\text{Mean annual runoff at reference scenario}} - 1 \right) \times 100 \qquad (1)$$

Overall, the models produce a smooth picture of mean annual runoff change in Europe, with decreases in the South. Severe negative changes are projected in the Iberian Peninsula, from the Black Sea in the South almost to the Baltic Sea in the North, and predominantly positive changes are projected in western to central Europe and in northern Europe. A mixed pattern with higher variability in mean annual runoff is shown across central Europe and the Carpathians. The climate models that produce more annual runoff reduction are HadGEM2 and NorEsM1. However, it can be seen that the values and spatial extent of the regions with reduced streamflow (in brownish colours) vary significantly from one climate model to another. This is remarkable considering that all simulations were performed with the same hydrologic model. As expected, in general, the changes are more intense in the long-term period. The region of neutral changes (represented in yellow) moves toward the north from low carbon (RCP2.6) to high carbon (RCP8.5) emissions scenarios (not shown).

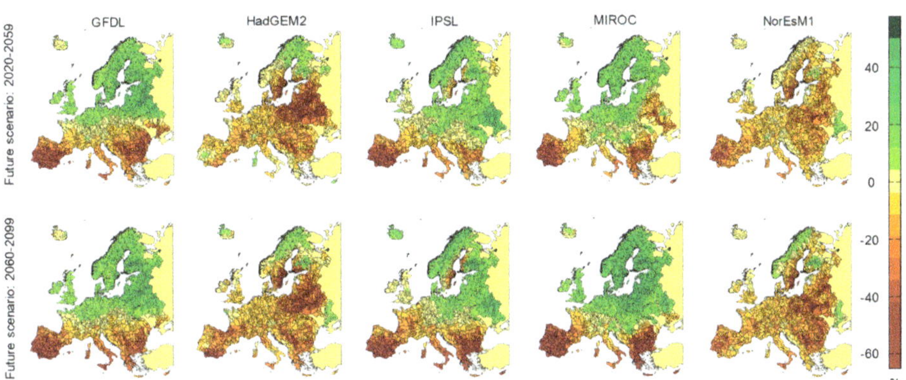

Figure 4. Changes (percentage) of mean annual runoff in future scenarios (2020–2059 and 2060–2099) compared with the reference scenario (1960–1999), according to different climate models and for the emissions scenario RCP4.5. Red shading represents a decrease of the mean annual runoff and green shading an increase.

The results of potential water availability in historical conditions (1960–1999) for all climate models are shown in Figure 5. It shows the values of potential water availability as a function of mean annual runoff in all the analysed sub-basins. Small, blue dots represent results in intermediate sub-basins, while larger, darker blue dots represent results in the global basins. All models show a similar picture, with a large variation of water availability among basins as a consequence of differences in hydrologic regime and reservoir storage.

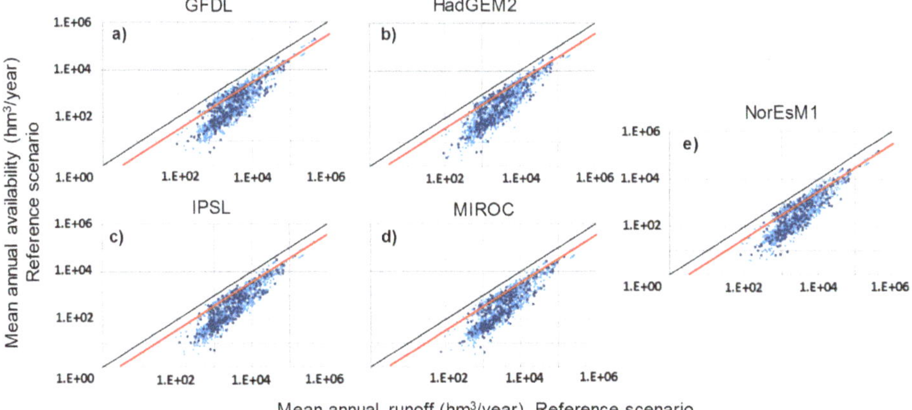

Figure 5. Mean annual water availability as a function of mean annual flow for the historical period (1960–1999) and for the different climate models. Small, blue dots represent results in intermediate sub-basins, while larger, darker blue dots represent results in the global river basins. Red line shows the value of 40% of mean annual runoff. (**a**) GFDL model, (**b**) HadGEM2, (**c**) IPSL, (**d**) MIROC and (**e**) NorEsM1.

The spatial distribution of changes (between the long term and reference periods) of potential water availability along the major rivers in Europe is presented in Figure 6, for all emissions scenarios and climate models analysed. Figure 6 is dimensionless, and the values were obtained by applying an equation similar to Equation 1, but using potential water availability instead of runoff. Red shading represents a decrease (negative values) of the future potential water availability and green shading an increase (positive values). The yellow shading represents no changes of potential water availability compared to the reference scenario. Although, in general, the climate models show a gradient of potential water availability changes with larger reductions in South Western Europe and larger increases in Northern Europe, values show important differences by comparing the results among climate models (same emissions scenario). By comparing the maps within each column (Figure 6), we visualize important differences in the results from one climate model to another, and by keeping each emissions scenario unaltered. The models that produce the most potential water availability reduction are HadGEM2 and NorEsM1, while IPSL and MIROC produce the least reductions. On the other hand, by comparing the maps within each row (Figure 6), we observe the different results obtained for the same climate model and different emissions scenarios. It can be seen that, in general, differences among the emissions scenarios (for each climate model) are smaller than those among different models (for each emissions scenario). The driest scenario is RCP8.5 for all analysed climate models.

Figure 7 shows, for each analysed sub-basin, the changes in the potential water availability in the long-term period with respect to the reference period (y axis), as a function of changes in the mean annual runoff in the long-term period with respect to the reference period (x axis), for all emissions scenarios and the climate model GFDL. The equations used to plot the results are similar to the proposed Equation 1 for the runoff variable (see Figure 4) and the proposed for the water availability variable (see Figure 6). Quadrant 1 (q.I) shows sub-basins where runoff decreases in the future and water availability increases. Both runoff and water availability increase in q.II, runoff increases and water availability decreases in q.III, and both runoff and water availability decrease in q.IV. In addition, basins with the same reduction of runoff experience different reductions in availability as a result of changes in the hydrologic variability and their different regulation capacity.

Figure 8 shows the spatial distribution of the ratio of the runoff, water availability, and water withdrawal for the model GFDL in emissions scenario RCP4.5 for the reference (1960–1999) and long-term period (2060–2099). The bottom row shows potential water availability as a fraction of runoff, the central one shows water withdrawal, also as a fraction of runoff, and the upper row shows the water withdrawal as a fraction of water availability.

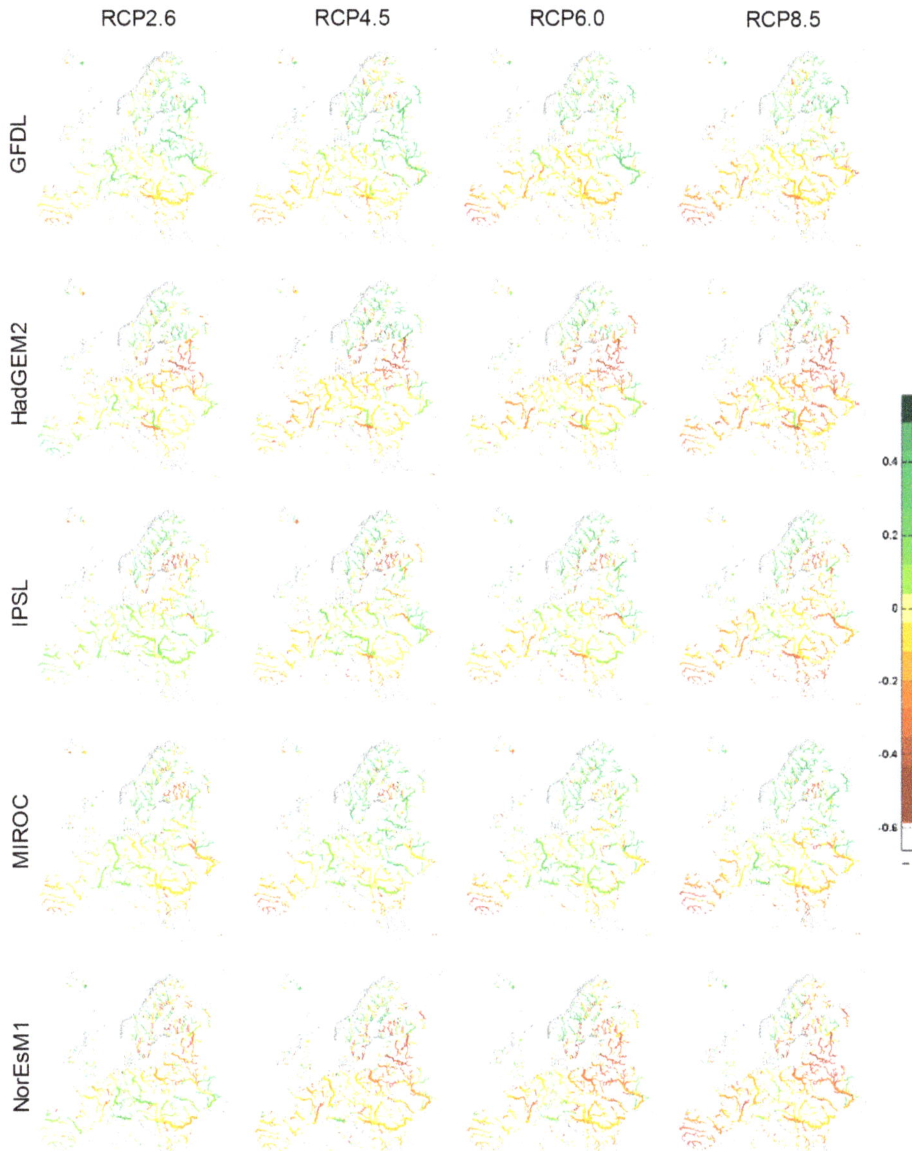

Figure 6. Changes in potential water availability for the long-term scenario (2060–2099) compared to the reference scenario (1960–1999), according to all climate models and emissions scenarios analysed. Red shading represents a decrease of the potential water availability and green shading an increase (individual maps at full resolution are available as supplementary files).

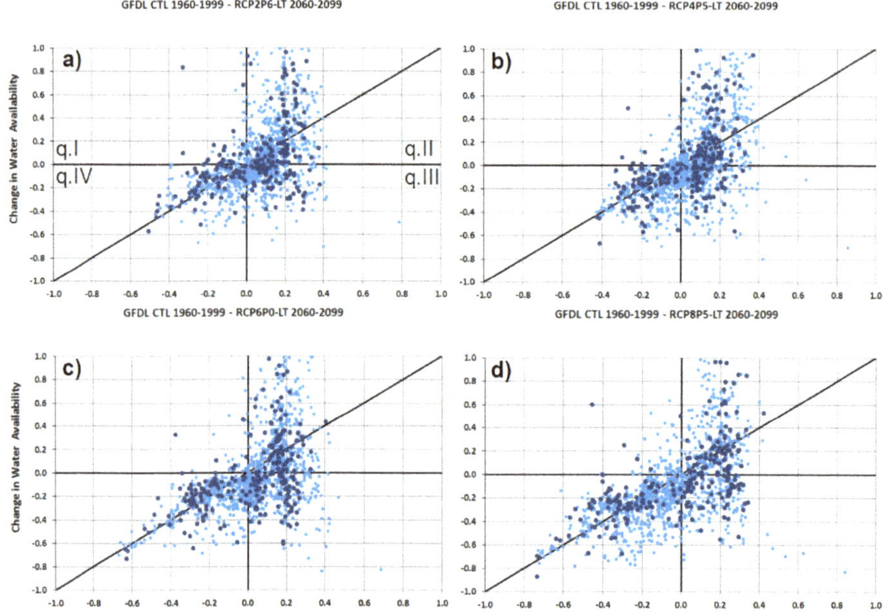

Figure 7. Changes of mean annual water availability from historical period (1960–1999) to long-term period (2060–2099) as a function of changes in runoff for model GFDL and emissions scenario RCP2.6, RCP4.5, RCP6.0, and RCP8.5. Small, blue dots represent results in intermediate sub-basins, while larger, darker blue dots represent results in the global basins. q.I, q.II, q.III, and q.IV point out each quadrant. (**a**) RCP2.6, (**b**) RCP4.5, (**c**) RCP6.0 and (**d**) RCP8.5.

Figure 9 shows the uncertainty associated with the climate models and emissions scenarios, both for mean annual runoff and mean water availability, by calculating the coefficient of variation (CV, standard deviation divided by mean) in each calculation point, for each climate model (five), for each emissions scenario (four) and for the short term (ST) and the long term (LT). We represented the probability distribution function (Pdf) of the CVs in each case. Continuous lines represent the uncertainty for each climate model, obtained by comparing the four emissions scenarios for each climate model. The dashed lines represent uncertainty for each emissions scenario, obtained by comparing the CV of the five models for each emissions scenario. Figure 9a,c shows the uncertainty associated with runoff for the ST and LT, respectively. Comparatively, Figure 9b,d shows the uncertainty associated with availability for the ST and LT, respectively.

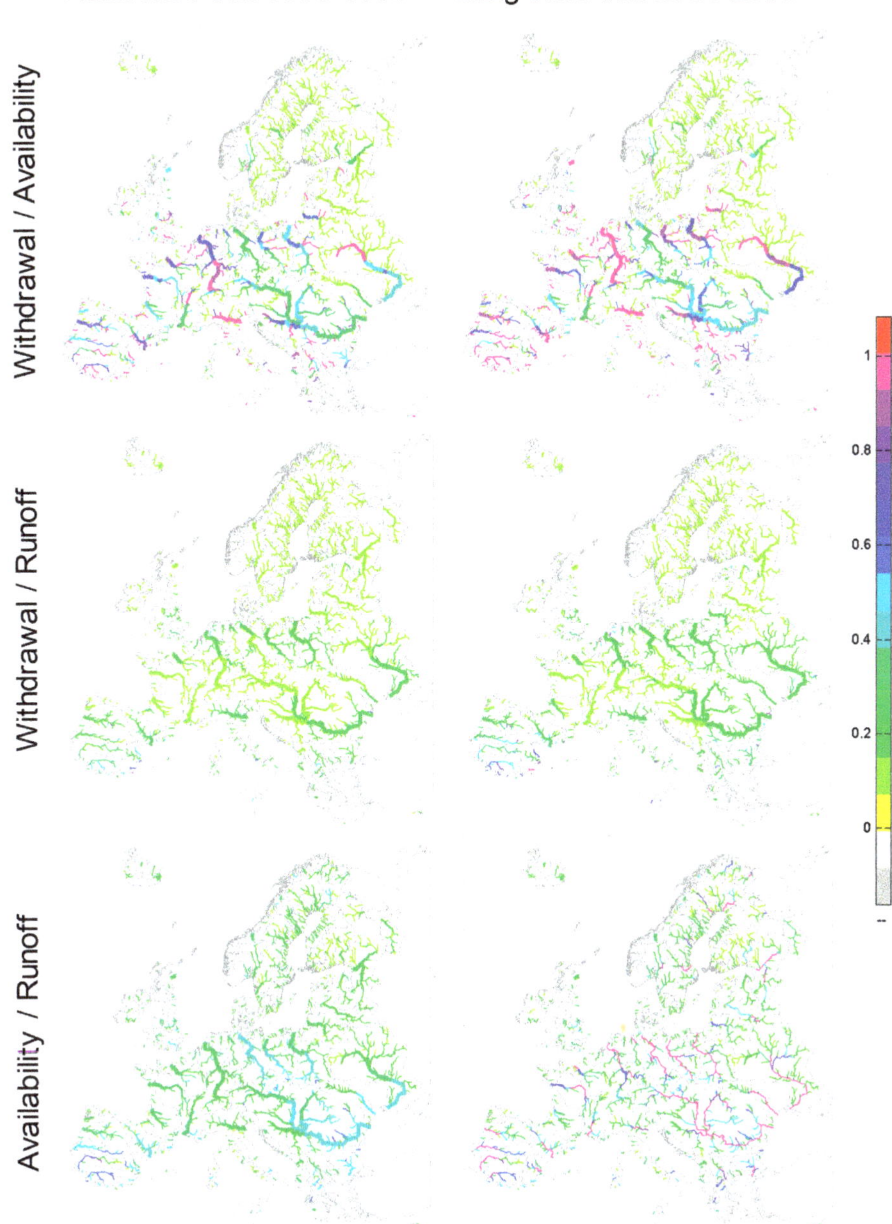

Figure 8. Spatial distribution of the per unit change of potential water availability between the historical period (1960–1999) and the long-term period (2070–2099) for the model GFDL, under the emissions scenario RCP4.5. (**Top row**) Withdrawal as a fraction of availability. (**Centre row**) Water withdrawal as a fraction of runoff. (**Bottom row**) Potential water availability as a fraction of runoff.

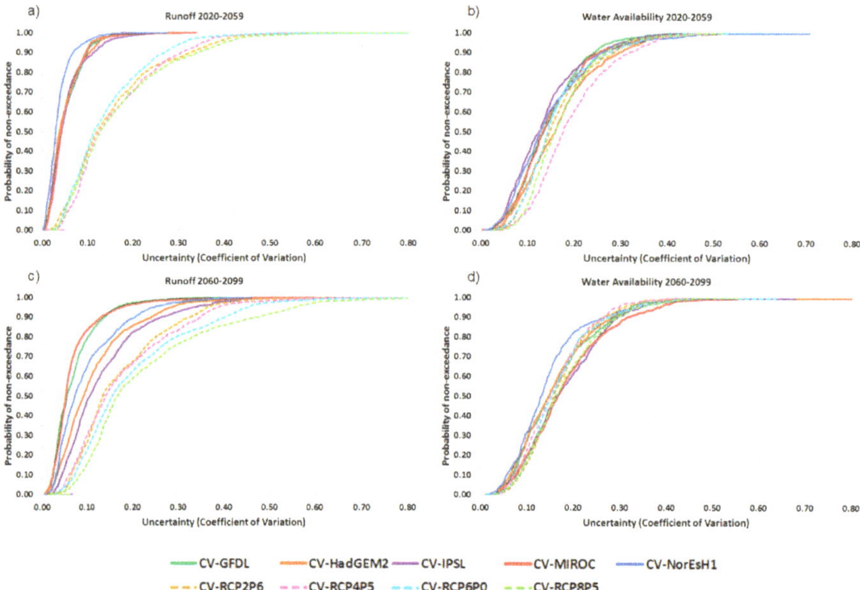

Figure 9. Climate model and emissions scenario uncertainties. Each continuous line represents the probability distribution function (Pdf) of the coefficient of variation (CV) corresponding to mean annual runoff and mean annual water availability in each calculation point, for each climate model (GFDL, green; HadGEM2, brown; IPSL, purple; MIROC, red; and NorEsH1, blue) and four emissions scenarios (RCP2.6, RCP4.5, RCP6.0, and RCP 8.5). Dashed line represents the Pdf of CV for each emissions scenario and the five climate models. (**a**) Runoff and short-term period (ST) and (**c**) long-term period (LT); (**b**) Water availability and ST and (**d**) LT.

4. Discussion

Both for short- and long-term periods, the models show similar spatial patterns of mean annual runoff changes in Europe, with decreases in the south, especially in the south-west and increases in the north. Our results agree with global and continental-scale studies that reported mean annual runoff projections [1,49,50]. These studies provide a coherent pattern of change in annual runoff, predicting with a high degree of confidence severe decreases (up to 40%) of surface runoff in areas already affected by water scarcity, like the Mediterranean region, and are consistent with the projected runoff increases in northern Europe (5–30%). However, it can be seen that the values and spatial extent of the regions with reduced streamflow vary significantly from one climate model to another. It suggests that there is an important climate model uncertainty, being the changes of mean annual runoff among emissions scenarios (and the same climate model) smaller than those among the different climate models (same emissions scenario).

In the short-term runoff, Figure 9a clearly shows that higher CV values are more frequent (amplitude of each Pdf curve) by comparing the results among models (and the same emissions scenario, dashed lines) than among emissions scenarios (and the same climate model, continuous lines). In addition, the uncertainties associated with the emissions scenarios are also similar among them (differences among continuous lines for the y axis). Although climate change models are the most robust tools available to generate consistent climate change projections, they are still a source of considerable uncertainties [10,51]. In this regard, Garrote [18] highlighted that the uncertainty has not been reduced with the progressive improvement of modelling tools; on the contrary, it seems to be increasing as a result of the evolving approach to generating emissions scenarios. On the other hand, results suggest that, because of the number of variables and complexity involved in the estimation of the future climate, its estimation has an implicit uncertainty that should be acknowledged for the development of climate adaptation plans. In the long-term runoff (Figure 9c), the uncertainty increases (increasing the amplitude of the Pdf curves) for both climate models and emissions scenarios, although climate models remain more uncertain than emissions scenarios. Also, greater dispersion of uncertainty is found among models than among emissions scenarios. It could be partially explained by the increase of the differences between emissions scenarios for the long-term analysis. These results are consistent with several inter-comparison studies that also show considerable variability in the magnitude and timing of the projected runoff [9,49,50,52,53]. At this point, it is remarkable that all simulations in this study were performed with the same hydrologic model. Databases of climate scenarios are available from different research projects [54,55], including surface runoff among their output variables. As the characterization of the water cycle in the models used in these types of studies usually is very simple and results provide a low signal-to-noise ratio (especially in arid and semi-arid regions), varying the large-scale hydrological models incorporates an additional source of uncertainty [18,50,52]. Some authors state that hydrologic model uncertainties are less significant than those originating from climate change models [9,56].

Changes in potential water availability in short- and long-term scenarios according to all climate models and emissions scenarios were analysed. High resolution results showed similar future spatial patterns to mean annual runoff, with the differences among the emissions scenarios (for each climate model) being smaller than those among different models (for each emissions scenario). Figure 9b shows that the uncertainty associated with the emissions scenarios increases and their values draw near to the climate model uncertainties. Furthermore, the Pdfs of the uncertainty associated with the climate models for water availability remain similar to that for runoff. Similar behaviour is observed for the long-term period (Figure 9d). It suggests that the management of hydraulic infrastructures (mainly reservoirs in this study) plays an important role by decoupling water availability from hydrologic variability. This is observed for all climate models and emissions scenarios considered. Svensson et al. [57] reinforced the importance of the installation of reservoirs in several river basins in Europe in the last century, by attenuating the basins' drought conditions. For quantifying and summarizing purposes, Table 1 shows the emissions scenarios' and climate models' uncertainty for the 50% probability of exceeding CV values. Several local and regional studies agree that the propagation of the uncertainties affects water resource system performances [26,58–60]. Thus, the assessment (or projection) of the performance of a water resources system should be evaluated with extreme care. As previously stated, the reservoir operation model applied in WAAPA is highly simplified and was designed to maximize water availability. Thus, the reality of reservoir operation is much more complex. Usually, not all reservoirs in the basin are jointly managed to supply all demands. They are either managed individually to supply local demands or grouped in systems that are managed independently. Availability of storage volume for water conservation management is also variable according to local conditions, due to the need to allocate storage volume to flood control. Therefore, it is unlikely that upstream reservoirs are kept full to release space in downstream reservoirs. Normal operation would tend to balance storage in all reservoirs to prevent uncontrolled spills. In practice, the spatial pattern of water availability will differ from that obtained in WAAPA. WAAPA results should only be considered as an upper bound of the actual water availability that could be obtained in practice.

Results from the comparisons of the changes in potential water availability with changes in runoff clearly show how changes in the former are not proportional to changes in the latter, suggesting the inadequacy of methodologies that estimate availability as a fraction of mean annual runoff. As an example, in Figure 5, the red line shows the traditional value of 40% of the mean annual runoff adopted for water availability when no simulation of reservoir regulation is performed [61]. It can be seen that adopting this constant value as a proxy of water availability can be strongly misleading, since only those basins with very regular flow or very large reservoir storage can reach this value. In most basins, water availability is a smaller fraction of the mean annual runoff.

Table 1. Summary of the emissions scenarios' and climate models' uncertainty for the 50% probability of exceeding CV values.

Climate Models	Scenario Uncertainty			
	Runof ST	**Runof LT**	**Availability ST**	**Avaliability LT**
CV-GFDL	0.05	0.07	0.15	0.16
CV-HadGEM2	0.05	0.12	0.17	0.16
CV-IPSL	0.05	0.13	0.14	0.18
CV-MIROC	0.05	0.07	0.15	0.18
CV-NorEsH1	0.04	0.10	0.15	0.15
Average	0.05	0.10	0.15	0.17
Emission Scenarios	**Model Uncertainty**			
CV-RCP2P6	0.16	0.17	0.16	0.17
CV-RCP4P5	0.16	0.18	0.19	0.16
CV-RCP6P0	0.14	0.20	0.16	0.17
CV-RCP8P5	0.17	0.23	0.17	0.18
Average	0.16	0.19	0.17	0.17

As shown, availability and withdrawal are only a small fraction of runoff in most of Europe and their projected changes are small, except for the south-east and the south-west. However, the representation of the water withdrawal as a fraction of water availability (Figure 8, upper row) shows that these two variables have similar values in many regions of Europe, and that they are getting closer in the long-term scenario. It means that in many regions, water shortage struggles to satisfy the demand with a specific reliability could emerge or increase, both for the present and future periods. It can also be seen that the relationship between these variables is complex, and that it varies significantly among regions, depending on hydrologic regime, climate, reservoir storage, and socioeconomic factors.

Green water (not analysed in this study), similarly to blue water, is also expected to decrease in most of western Europe except for northern countries. However, changes in green water result from complex interplay of impacts on precipitation, temperature, and CO_2 concentration, which ultimately affects potential evapotranspiration, soil moisture conditions, and growing periods. Thus, patterns of expected changes differ for green and blue water [62]. Irrigation demands will also be affected, due to modified seasonal patterns and evapotranspiration demands [36,63].

Finally, some limitations of this study should be noted. We estimated the potential water availability (upper theory limit) by considering only one demand present in the system. System performance was evaluated as gross volume reliability. Potential water availability was obtained under the hypothesis of 90% reliability. The data used in this study were obtained from specific climate models and emissions scenarios, thus, the conclusions derived from this study are inextricably affected by the models' uncertainty. Additionally, we made a series of simplifying assumptions. We assumed variable geographic and temporal water withdrawals, both in the present and future climate, from indirect methods (GDP and population). We assumed that the reservoirs, whose sole purpose was hydropower generation, were not included in the systems to manage the water resources. We considered that the hydraulic infrastructure corresponding to each analysed sub-basin (determined from a given point in the stream network) was being jointly managed to supply global demands, while in some real cases it could have been divided in to several rather independent subsystems. Furthermore, in our model, there were no system interconnections nor a large-scale water distribution infrastructure. We did not consider other sources of uncertainty as, for instance, the observed climate data source or the hydrologic model applied and the inclusion of regional climate models (RCMs). It is expected that RCMs have less associated uncertainty than GCMs when a particular region is analysed, as they account for more detailed and specific regional characteristics.

5. Conclusions

This study presents the potential water availability changes under alternative climate change scenarios in western Europe. Results are geographically referenced at high resolution across the major European river basins. The study includes the estimation of the associated uncertainties, resulting from differences among climate change scenarios and climate models. The authors are not aware of similar studies conducted at such a high-resolution continental scale. In this study, we applied the WAAPA model on a high-resolution dataset to analyse water availability changes across western Europe. The proposed model and the applied methodology demonstrated their ability to perform regional studies covering extensive domains, while maintaining high resolution on the characterization of the systems. The climate models that produced the most reduction of mean annual runoff and potential water availability were HadGEM2 and NorEsM1, while IPSL and MIROC produced the least reduction. Overall, for both mean annual runoff and potential water availability, a gradually varying picture of change in Europe was observed, with a decrease in the south (especially in the south-west) and an increase in the north. Moreover, the region of neutral changes moves to the north, from low carbon (RCP2.6) to high carbon (RCP8.5) emissions scenarios. Climate model uncertainties for mean annual runoff and potential water availability were found to be higher than scenario uncertainties. This conclusion was derived by comparing the variability of the results obtained, while the PCRGLOBWB model was forced with different climate models under the same emissions scenario to that of the results from different emissions scenarios for the same climate model forcing. Thus, although climate change models are the most robust tools available to generate consistent climate change projections, they are still a source of considerable uncertainties and their results should be carefully used for operative purposes.

While potential water availability and water withdrawal are only a small fraction of runoff in most of Europe for current and future scenarios (except in the south-east and the south-west of Europe), water withdrawal and water availability are similar in many regions of Europe, and they are getting closer in the long-term scenario (2060–2099). Thus, the balance between water availability and withdrawals is threatened in some regions. Furthermore, social factors, like management of hydraulic infrastructure, play an important role by decoupling water availability from hydrologic variability. This is observed for all climate models and emissions scenarios considered. Finally, although this study presents significant progress in terms of spatial scale and detail compared to previous studies, it is still only indicative of the importance of regional change, due to the assumptions and uncertainties discussed. Nevertheless, the results are useful for envisioning potential water resource system conflicts and contributing to the identification of regions where an in-depth analysis may be necessary.

Supplementary Materials: The following are available online at http://www.mdpi.com/2073-4441/11/3/420/s1.

Author Contributions: Conceptualization, L.G. and A.I.; methodology, A.S.-W. and A.I.; software, L.G.; investigation and formal analysis, A.S.-W. and I.G.; resources and data curation, I.G.; writing—original draft preparation, I.G.; writing—review and editing, A.S.-W.; visualization and supervision, L.G.; funding acquisition, A.S.-W. and A.I.

Funding: This research was partially funded by Universidad Politécnica de Madrid through the "Programa propio: ayudas a proyectos de I+D de investigadores posdoctorales" and the "ADAPTA" project. We also acknowledge the financial support of the European Commission BASE project (grant agreement no.: ENV-308337) of the 7th Framework Program (http://base-adaptation.eu).

Conflicts of Interest: The authors declare no conflict of interest.

References

1. IPCC 2014. *Climate Change 2014: Impacts, Adaptation, and Vulnerability Part A: Global and Sectoral Aspects; Contribution of Working Group II to the Fifth Assessment Report of the Intergovernmental Panel on Climate Change 2014;* Field, C.B., Barros, V.R., Dokken, D.J., Mach, K.J., Mastrandrea, M.D., Bilir, T.E., Chatterjee, M., Ebi, K.L.,

Estrada, Y.O., Genova, R.C., et al., Eds.; Cambridge University Press: Cambridge, UK; New York, NY, USA, 2017; pp. 1–32.

2. European Environment Agency (EEA). *Climate Change, Impacts and Vulnerability in Europe 2016. An Indicator-Based Report*; EEA Report No 1/2017; European Environment Agency: Copenhagen, Denmark, 2017.

3. Arnell, N.W.; Van Vuuren, D.P.; Isaac, M. The implications of climate policy for the impacts of climate change on global water resources. *Glob. Environ. Chang.* **2011**, *21*, 592. [CrossRef]

4. Alcamo, J.; Floerke, M.; Maerker, M. Future long-term changes in global water resources driven by socioeconomic and climatic changes. *Hydrol. Sci.* **2007**, *52*, 247–275. [CrossRef]

5. Easterling, D.R.; Meehl, J.; Parmesan, C.; Changnon, S.A.; Karl, T.R.; Mearns, L.O. Climate extremes: Observations, modeling, and impacts. *Science* **2000**, *289*, 2068–2074. [CrossRef] [PubMed]

6. Vörösmarty, C.J.; McIntyre, P.B.; Gessner, M.O.; Dudgeon, D.; Prusevich, A.; Green, P.; Glidden, S.; Bunn, S.E.; Sullivan, C.A.; Liermann, C.R.; et al. Global threats to human water security and river biodiversity. *Nature* **2010**, *467*, 555. [CrossRef] [PubMed]

7. Chawla, I.; Mujumdar, P.P. Partitioning uncertainty in streamflow projections under nonstationary model conditions. *Adv. Water Resour.* **2108**, *112*, 266–282. [CrossRef]

8. Yen, H.; Wang, X.; Fontane, D.G.; Harmel, R.D.; Arabi, M. A framework for propagation of uncertainty contributed by parameterization, input data, model structure, and calibration/validation data in watershed modeling. *Environ. Model. Softw.* **2014**, *54*, 211–221. [CrossRef]

9. Gao, J.; Sheshukovb, A.Y.; Yena, H.; Douglas-Mankin, K.R.; White, M.J.; Arnold, J.G. Uncertainty of hydrologic processes caused by bias-corrected CMIP5 climate change projections with alternative historical data sources. *J. Hydrol.* **2019**, *568*, 551–561. [CrossRef]

10. Déqué, M.; Somot, S.; Sanchez-Gomez, E.; Goodess, C.M.; Jacob, D.; Lenderink, G.; Christensen, O.B. The spread amongst ENSEMBLES regional scenarios: Regional climate models, driving general circulation models and interannual variability. *Clim. Dyn.* **2012**, *38*, 951–964. [CrossRef]

11. Falkenmark, M.; Rockström, J. The New Blue and Green Water Paradigm: Breaking New Ground for Water Resources Planning and Management. *Water Resour. Plan. Manag.* **2006**, *132*, 129–132. [CrossRef]

12. García-Ruiz, J.M.; López-Moreno, J.I.; Vicente-Serrano, S.M.; Lasanta–Martínez, T.; Beguería, S. Mediterranean water resources in a global change scenario. *Earth Sci. Rev.* **2011**, *105*, 121–139. [CrossRef]

13. Gardner, L.R. Assessing the effect of climate change on mean annual runoff. *J. Hydrol.* **2009**, *379*, 351–359. [CrossRef]

14. Simonovic, S.P. *Managing Water Resources: Methods and Tools for a Systems Approach*; UNESCO Publishing: Paris, France, 2009.

15. Nilsson, C.; Reidy, C.A.; Dynesius, M.; Revenga, C. Fragmentation and flow regulation of the World's large river systems. *Science* **2005**, *308*, 405–408. [CrossRef] [PubMed]

16. Iglesias, A.; Santillán, D.; Garrote, L. On the Barriers to Adaption to Less Water under Climate Change: Policy Choices in Mediterranean Countries. *Water Resour. Manag.* **2018**. [CrossRef]

17. Iglesias, A.; Garrote, L. Adaptation strategies for agricultural water management under climate change in Europe. *Agric. Water Manag.* **2015**, *155*, 113–124. [CrossRef]

18. Garrote, L. Managing Water Resources to Adapt to Climate Change: Facing Uncertainty and Scarcity in a Changing Context. *Water Resour. Manag.* **2017**, *31*, 2951–2963. [CrossRef]

19. Vogel, R.M.; Sieber, J.; Archfield, S.A.; Smith, M.P.; Apse, C.D.; Huber-Lee, A. Relations among storage, yield, and instream flow. *Water Resour. Res.* **2007**, *43*, W05403. [CrossRef]

20. Arnold, J.; Bieger, K.; White, M.; Srinivasan, R.; Dunbar, J.; Allen, P. Use of Decision Tables to Simulate Management in SWAT+. *Water* **2018**, *10*, 713. [CrossRef]

21. White, M.J.; Gambone, M.; Yen, H.; Arnold, J.; Harmel, D.; Santhi, C.; Haney, R. Regional Blue and Green Water Balances and Use by Selected Crops in the U.S. *J. Am. Water Resour. Assoc.* **2015**, *51*, 1626–1642. [CrossRef]

22. Wurbs, R.A.; Muttiah, R.S.; Felden, F. Incorporation of climate change in water availability modeling. *J. Hydrol. Eng.* **2005**, *10*, 375. [CrossRef]

23. Yates, D.; Sieber, J.; Purkey, D.; Huber-Lee, A. WEAP21—A demand-, priority-, and preference-driven water planning model. Part 1: Model characteristics. *Water Int.* **2005**, *30*, 487–500. [CrossRef]

24. Andreu, J.; Capilla, J.; Sanchís, E. AQUATOOL, a generalized decision-support system for water-resources planning and operational management. *J. Hydrol.* **1996**, *177*, 269–291. [CrossRef]

25. Harou, J.J.; Pulido-Velazquez, M.; Rosenberg, D.E.; Medellín-Azuara, J.; Lund, J.R.; Howitt, R.E. Hydro-economic models: Concepts, design, applications, and future prospects. *J. Hydrol.* **2009**, *375*, 627–643. [CrossRef]

26. Sordo-Ward, A.; Granados, I.; Martín-Carrasco, F.; Garrote, L. Impact of Hydrological Uncertainty on Water Management Decisions. *Water Resour. Manag.* **2016**, *30*, 5535–5551. [CrossRef]

27. Chávez-Jimenez, A.; Lama, B.; Garrote, L.; Martin-Carrasco, F.; Sordo-Ward, A.; Mediero, L. Characterisation of the Sensitivity of Water Resources Systems to Climate Change. *Water Resour. Manag.* **2013**, *27*, 4237–4258. [CrossRef]

28. World Water Assessment Programme (WWAP). *The United Nations World Water Development Report 4: Managing Water under Uncertainty and Risk*; UNESCO: Paris, France, 2012.

29. Moss, R.H.; Edmonds, J.A.; Hibbard, K.A.; Manning, M.R.; Rose, S.K.; Van Vuuren, D.P.; Carter, T.R.; Emori, S.; Kainuma, M.; Kram, T.; et al. The next generation of scenarios for climate change research and assessment. *Nature* **2010**, *463*, 747–756. [CrossRef] [PubMed]

30. Estrela, T.; Perez-Martin, M.A.; Vargas, E. Impacts of climate change on water resources in Spain. *Hydrol. Sci. J.* **2012**, *57*, 1154–1167. [CrossRef]

31. Pulido-Velazquez, D.; Garrote, L.; Andreu, J.; Martin-Carrasco, F.J.; Iglesias, A. A methodology to diagnose the effect of climate change and to identify adaptive strategies to reduce its impacts in conjunctive-use systems at basin scale. *J. Hydrol.* **2011**, *405*, 110–122. [CrossRef]

32. Iglesias, A.; Garrote, L.; Diz, A.; Schlickenrieder, J.; Martin-Carrasco, F. Rethinking water policy priorities in the Mediterranean region in view of climate change. *Environ. Sci. Policy* **2011**, *14*, 744–757. [CrossRef]

33. Garrote, L.; Iglesias, A.; Granados, A.; Mediero, L.; Martin-Carrasco, F. Quantitative assessment of climate change vulnerability of irrigation demands in Mediterranean Europe. *Water Resour. Manag.* **2015**, *29*, 325–338. [CrossRef]

34. Abbaspoura, K.C.; Rouholahnejada, E.; Vaghefia, S.; Srinivasan, R.; Yang, H.; Kløve, B. A continental-scale hydrology and water quality model for Europe: Calibration and uncertainty of a high-resolution large-scale SWAT model. *J. Hydrol.* **2015**, *524*, 733–752. [CrossRef]

35. European Commission: Agriculture and Rural Development. Agriculture and Environment. Agriculture and Water. Available online: http://ec.europa.eu/agriculture/envir/water/ (accessed on 7 February 2019).

36. Döll, P. Impact of climate change and variability on irrigation requirements: A global perspective. *Clim. Chang.* **2002**, *54*, 269–293. [CrossRef]

37. Olesen, J.E.; Carter, T.R.; Díaz-Ambrona, C.H.; Fronzek, S.; Heidmann, T.; Hickler, T.; Holt, T.; Minguez, M.I.; Morales, P.; Palutikof, J.P.; et al. Uncertainties in projected impacts of climate change on European agriculture and terrestrial ecosystems based on scenarios from regional climate models. *Clim. Chang.* **2007**, *81*, 123–143. [CrossRef]

38. Arnell, N.W. Climate change and global water resources: SRES emissions and socio-economic scenarios. *Glob. Environ. Chang.* **2004**, *14*, 31–52. [CrossRef]

39. SRES Final Data (Version 1.1, July 2000). Center for International Earth Science Information Network. Available online: http://sres.ciesin.columbia.edu/final_data.html (accessed on 23 December 2018).

40. The World Bank Data. Available online: https://data.worldbank.org/ (accessed on 23 December 2018).

41. EROS, USGS. *HYDRO1k Elevation Derivative Database*; Tech. Rept.; U.S. Geological Survey Centre for Earth Resources Observation and Science (EROS): Garretson, SD, USA, 2008.

42. Van Beek, L.P.H.; Bierkens, M.F.P. *The Global Hydrological Model PCR-GLOBWB: Conceptualization, Parameterization and Verification*; Utrecht University, Faculty of Earth Sciences, Department of Physical Geography: Utrecht, The Netherlands, 2009.

43. Warszawski, L.; Frieler, K.; Huber, V.; Piontek, F.; Serdeczny, O.; Schewe, J. The Inter-Sectoral Impact Model Intercomparison Project (ISI–MIP): Project framework. *Proc. Natl. Acad. Sci. USA* **2014**, *111*, 3228–3232. [CrossRef] [PubMed]

44. Fekete, B.M.; Vörösmarty, C.J.; Grabs, W. High-resolution fields of global runoff combining observed river discharge and simulated water balances. *Glob. Biogeochem. Cycles* **2002**, *16*, 1–6. [CrossRef]

45. Gonzalez-Zeas, L.; Garrote, A.; Iglesias, A.; Sordo-Ward, A. Improving runoff estimates from regional climate models: A performance analysis in Spain. *Hydrol. Earth Syst. Sci.* **2012**, *16*, 1709–1723. [CrossRef]

46. World Register of Dams/Registre Mondial des Barrages (WRD). Available online: http://www.icold-cigb. net/GB/world_register/world_register_of_dams.asp (accessed on 23 December 2018).

47. CIESIN. Global Urban-Rural Mapping Project (GRUMP). Socioeconomic Data and Applications Centre (SEDAC), Columbia University: Palisades, NY, USA. Available online: http://sedac.ciesin.columbia.edu/gpw/ (accessed on 23 December 2018).
48. Döll, P.; Siebert, S. A digital global map of irrigated areas. *ICID J.* **2000**, *49*, 55–66.
49. Forzieri, G.; Feyen, L.; Rojas, R.; Flörke, M.; Wimmer, F.; Bianchi, A. Ensemble projections of future streamflow droughts in Europe. *Hydrol. Earth Syst. Sci.* **2014**, *18*, 85–108. [CrossRef]
50. Stahl, K.; Tallaksen, L.M.; Hannaford, J.; Van Lanen, H.A.J. Filling the white space on maps of European runoff trends: Estimates from a multi-model ensemble. *Hydrol. Earth Syst. Sci.* **2012**, *16*, 2035–2047. [CrossRef]
51. Murphy, J.M.; Sexton, D.M.H.; Barnett, D.H.; Jones, G.S.; Webb, M.J.; Collins, M.; Stainforth, D.A. Quantification of modelling uncertainties in a large ensemble of climate change simulations. *Nature* **2004**, *430*, 768–772. [CrossRef] [PubMed]
52. Gudmundsson, L.; Tallaksen, L.M.; Stahl, K.; Clark, D.B.; Dumont, E.; Hagemann, S.; Bertrand, N.; Gerten, D.; Heinke, J.; Hanasaki, N.; et al. Comparing Large-scale Hydrological Model Simulations to Observed Runoff Percentiles in Europe. *J. Hydrometeorol.* **2012**, *13*, 604–620. [CrossRef]
53. Prudhomme, C.; Parry, S.; Hannaford, J.; Clark, D.B.; Hagemann, S.; Voss, F. How well do large-scale models reproduce regional hydrological extremes in Europe? *J. Hydrometeorol.* **2011**, *12*, 1181–1204. [CrossRef]
54. CORDEX, Coordinated Regional Climate Downscaling Experiment. Available online: http://www.cordex.org/ (accessed on 6 February 2019).
55. Christensen, J.H.; Christensen, O.B. A summary of the PRUDENCE model projections of changes in European climate by the end of this century. *Clim. Chang.* **2007**, *81*, 7–30. [CrossRef]
56. Chen, C.; Haerter, J.O.; Hagemann, S.; Piani, C. On the contribution of statistical bias correction to the uncertainty in the projected hydrological cycle. *Geophys. Res. Lett.* **2011**, *38*, L20403. [CrossRef]
57. Svensson, C.; Kundzewicz, Z.W.; Maurer, T. Trend detection in river flow series: 2. Flood and low-flow index series. *Hydrol. Sci. J.* **2005**, *50*, 811–824. [CrossRef]
58. Nazemi, A.; Wheater, H.S. How can the uncertainty in the natural inflow regime propagate into the assessment of water resource systems? *Adv. Water Resour.* **2014**, *63*, 131–142. [CrossRef]
59. Steinschneider, S.; Wi, S.; Brown, C. The integrated effects of climate and hydrologic uncertainty on future flood risk assessments. *Hydrol. Process.* **2014**, *29*, 2823–2839. [CrossRef]
60. Fowler, H.J.; Blenkinsop, S.; Tebaldi, C. Linking climate change modelling to impacts studies: Recent advances in downscaling techniques for hydrological modeling. *Int. J. Clim.* **2007**, *27*, 1547–1578. [CrossRef]
61. Falkenmark, M. Environment and development: Urgent need for a water perspective. *Water Int.* **1991**, *16*, 229–240. [CrossRef]
62. Gerten, D.; Heinke, J.; Hoff, H.; Biemans, H.; Fader, M.; Waha, K. Global Water Availability and Requirements for Future Food Production. *Am. Meteorol. Soc.* **2011**, *12*, 885–899. [CrossRef]
63. Wisser, D.; Frolking, S.; Douglas, E.M.; Fekete, B.M.; Vörösmarty, C.J.; Schumann, A.H. Global irrigation water demand: Variability and uncertainties arising from agricultural and climate data sets. *Geophys. Res. Lett.* **2008**, *34*, L24408. [CrossRef]

water

MDPI

Article

Exploring the Role of Reservoir Storage in Enhancing Resilience to Climate Change in Southern Europe

Alfredo Granados [1],*, Alvaro Sordo-Ward [1], Bolívar Paredes-Beltrán [1,2] and Luis Garrote [1]

1   Departamento de Ingeniería Civil, Hidráulica, Energía y Medio Ambiente, Universidad Politécnica de Madrid, 28040 Madrid, Spain; alvaro.sordo.ward@upm.es (A.S.-W.); be.paredes@alumnos.upm.es (B.P.-B.); l.garrote@upm.es (L.G.)

2   Carrera de Ingeniería Civil, Facultad de Ingeniería Civil y Mecánica, Universidad Técnica de Ambato, Ambato 180206, Ecuador

*   Correspondence: a.granados@upm.es

Abstract: Recent trends suggest that streamflow discharge is diminishing in many rivers of Southern Europe and that interannual variability is increasing. This threatens to aggravate water scarcity problems that periodically arise in this region, because both effects will deteriorate the performance of reservoirs, decreasing their reliable yield. Reservoir storage is the key infrastructure to overcome variability and to enhance water availability in semiarid climates. This paper presents an analysis of the role of reservoir storage in preserving water availability under climate change scenarios. The study is focused on 16 major Southern European basins. Potential water availability was calculated in these basins under current condition and for 35 different climatic projections for the period 2070–2100. The results show that the expected reduction of water availability is comparable to the decrease of the mean annual flow in basins with large storage capacity. For basins with small storage, the expected reduction of water availability is larger than the reduction of mean annual flow. Additionally, a sensitivity analysis was carried out by replicating the analysis assuming variable reservoir volumes from 25% to 175% of current storage. The results show that increasing storage capacity attenuates the reduction of water availability and reduces its uncertainty under climate change projections. This feature would allow water managers to develop suitable policies to mitigate the impacts of climate change, thus enhancing the resilience of the system.

Keywords: climate change; reservoir performance; water availability; water resources

Citation: Granados, A.; Sordo-Ward, A.; Paredes-Beltrán, B.; Garrote, L. Exploring the Role of Reservoir Storage in Enhancing Resilience to Climate Change in Southern Europe. *Water* **2021**, *13*, 85. https://doi.org/10.3390/w13010085

Received: 10 November 2020
Accepted: 29 December 2020
Published: 1 January 2021

Publisher's Note: MDPI stays neutral with regard to jurisdictional claims in published maps and institutional affiliations.

1. Introduction

Climate change, associated with the recorded rise of average temperatures, which are expected to continue increasing to a greater or lesser extent, may also influence other climatic variables such as precipitation, frost, or evapotranspiration [1]. All these changes may affect, in turn, the hydrological processes and consequently net water resources. This threatens the performance of water resource systems and their capability to supply demand and ecological need as presently planned. Therefore, it is necessary to assess both the impact on water resources and the behavior of water systems under such a scenario [2].

Many authors have devoted significant efforts to evaluate net water resources in climate change projections, on all scales from global to basin [3–9]. Their results show that climate change will affect, in varying ways and to different extents, each region of the planet. As a global result, it could be synthesized that there will be a reduction of water resources of between 10% and 30% [1]. This is an indicative value, useful for developing macro-policies and for raising awareness in the population.

With regard to Southern Europe, despite the dispersion of the various models, the general trend indicates that net resources will decrease, and that the variability of their distribution will increase, as shown in the results of the Prediction of regional scenarios and uncertainties for defining European climate change risks and effects (PRUDENCE) [10] and Climate change and its impacts at seasonal, decadal and centennial timescales (EN-SEMBLES) [11] projects. In Southern Europe, the prognosis is that traditional water scarcity problems will be aggravated. In many basins in this region, the available resources can hardly meet the existing water demands [12]. These are areas with a benign climate, which favors the implementation of agriculture and the development of tourism and services. All these activities require substantial amounts of water. Although these regions have scarce water resources, they are also resilient as they have long experience in dealing with water scarcity and are well adapted to its management [13].

The present study focuses on understanding the effect of reservoir storage capacity on water availability in Southern European basins under climate change. As above mentioned, these basins are typically characterized by scarce and highly variable water resources. The adopted strategy for water resources development in the last century relied on reservoir storage, as it is necessary to store water during the wet periods for its use in the dry ones. In Spain, for example, the existing 1350 large dams helped to increase water availability from 10% to between 40% and 50% of mean natural flow during the last century [14]. As storage capacity grew in parallel with water use, this water availability is used strictly enough to serve current water needs. Alternative adaptation and mitigation measures are being developed in the current century: controlling irrigation water rights, increasing water use efficiency through localized and drip irrigation, developing non-conventional resources, such as water reutilization and desalination, among others [15,16]. Despite these efforts, projections of climate change suggest less water resources with higher variability, which will negatively affect system performance, so water availability is expected to reverse its growing trend [17].

The analysis of reservoir storage capacity and its relationship with safe yield has been a topic of study since the beginning of the development of large hydraulic systems [18]. Initially, graphical methods were developed to determine the reservoir capacity needed to satisfy a given demand with required reliability, and their use was restricted to single reservoir models. Later methods introduced uncertainty of future inflows and attempted to estimate required reservoir size through statistical analysis of inflows, leading to the concepts of risk of failure and reliability. The development of computing allowed the stochastic generation of synthetic series and the disaggregated analysis of multiple reservoirs in a system [19]. Löf and Hardison [20] provided storage-reliability-yield (SRY) relations for assessing the required storage capacity in the USA. The study was later revisited by Vogel et al. [21], who concluded that areas with lower variability tend to be equipped with within-year storage systems while those with large variability required larger over-year storage facilities. Further developments introduced the concepts of resilience and robustness [22] to complete the reliability-yield analysis [23]. An alternative approach is the simulation of the water resources system behavior, which in conjunction with the power of computers allows the development of complex models that reproduce a simulated operation of the system [24–26]. These models are useful as decision-support tools for allocating water among users and assessing the effectiveness of structural and managerial actions [27,28] and their capabilities are even extended to groundwater resources and social and economic considerations [29].

Focus is slowly being placed on the impact of climate change on water availability and the role of reservoir storage to increase resilience. Wurbs et al. [30] highlighted the need to introduce climate change in the analysis of water availability and proposed a methodology to couple climatic and system behavior models. Garrote et al. [31] developed a simulation model specifically suited to account for the role of reservoirs in providing water availability in the context of climate change. Several authors [32–34] have argued in favor of adaptive reservoir management as an effective mitigation measure during climate change. Adaptive management requires a good knowledge of the interplay between reservoir storage and the reliability, resilience, and vulnerability of a water supply system subject to uncertain input [35]. Water availability deriving from reservoir systems may become increasingly unstable under climate change [36] and knowledge on how regulated water supply systems react to flow alterations is essential for system managers to design climate adaptation policies.

This paper looks beyond the impact of climate change on water availability to provide insight into the performance of reservoir storage systems and their effectiveness of adaptation and mitigation measures. With this purpose we include a regional analysis of the performance of reservoir storage and a sensitivity analysis of the reservoir-yield relations under less abundant resources and larger variability conditions in 16 representative European basins. The objective of the research is to check if reservoir storage enhances resilience to climate change. Given the uncertainty of climate projections, the adopted approach is to evaluate basin response under a large ensemble of plausible future scenarios and to evaluate if reservoir storage plays a role in determining the response to changes in hydrologic forcing. System response is quantified in terms of the elasticity of water availability to climate change, comparing changes in potential water availability with changes in mean annual flow. Elasticity is evaluated with the help of two new indices proposed in this work, which characterize the attenuation of changes and the reduction of uncertainty provided by reservoir storage.

2. Data and Methods

2.1. Area under Analysis

We present results for 16 major river basins in Europe, which are shown in Figure 1. Basin selection was based on a regional focus on Southern Europe, but including different climates, hydrologic regimes, and storage capacities to allow for a more effective comparison. The selected basins cover a large fraction of the Atlantic and Mediterranean divides of Southern Europe and are representative of the variety of conditions that can be found across the region. The main characteristics of the basins considered in this study are shown in Table 1. Basin areas range from 17,550 km^2 (Segura) to 115,910 km^2 (Loire). Reservoir Storage Volume V includes all reservoirs, except those managed exclusively for hydropower. The basin with largest storage volume is Guadiana, which includes two of the largest reservoirs in Southern Europe: Alqueva (4.15 km^3) and La Serena (3.21 km^3). Basin hydrology is very variable, with Specific Runoff ranging from 11 mm/year in the Segura basin to 563 mm/year in the Po basin. The most relevant characteristic for this study is specific storage, defined as the ratio of Storage Volume V in km^3 divided by Mean Annual Flow F in km^3/year for the period 1960–1999. This ratio is usually called Residence Time (in years) and represents the regulation capacity of reservoirs in the basin. In the basins under study, it ranges across three orders of magnitude, from 0.01 years (Arno) to nearly 6 years (Segura).

The spatial support for the analysis is taken from the "Hydro1k" data set [37], derived from the Global 30 Arc-Second Elevation (GTOPO) 30 arc-second digital elevation model of the world. The dataset provides a digital elevation map and a set of topographically derived rasters at 1 km resolution, including streams and drainage basins divided into catchments. The original drainage basins in "Hydro1k" were processed to eliminate catchments which were too small (less than 1000 km^2), which were merged to neighboring catchments. The merging was always done with downstream areas and avoiding catchments including reservoirs. The reservoir storage volume in every catchment was obtained from the International Commission on Large Dams (ICOLD) World Register of Dams [38]. We selected dams in the register with more than 0.005 km^3 of storage capacity, excluding dams managed only for hydropower. The reservoirs were georeferenced and linked to the corresponding Hydro1k streams. All dams located in the same Hydro1k subbasin were grouped in an equivalent reservoir adding the storage volume and flooded area (to account for reservoir evaporation losses).

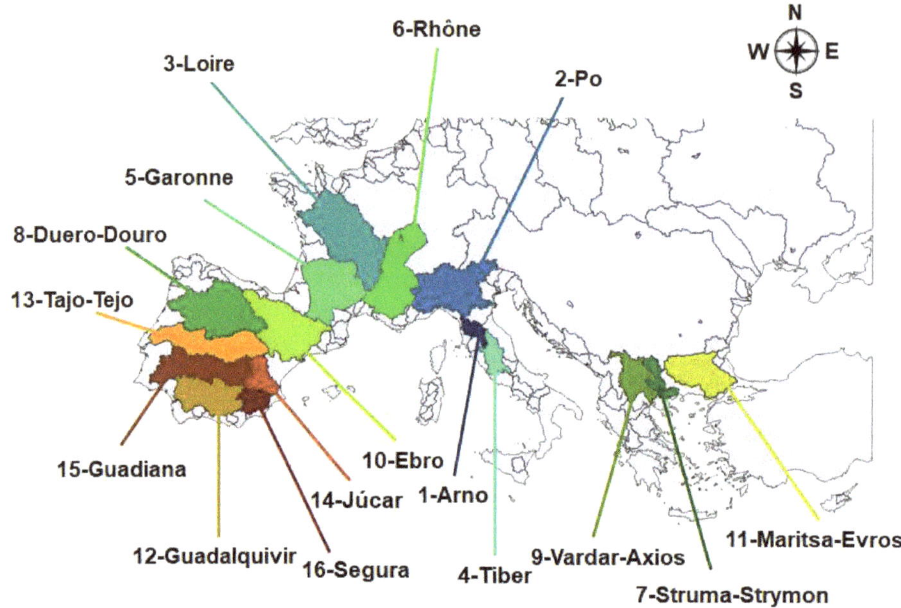

Figure 1. Southern European basins considered in this study.

Table 1. Basic characteristics of the basins analyzed in this study.

Basin	Basin Area (A) (10^3 km^2)	Mean Annual Flow (F) (km^3/year)	Storage Volume (V) (km^3)	Specific Runoff (F/A) (mm/year)	Residence Time (V/F) year
1-Arno	10.30	4.75	0.07	462	0.01
2-Po	84.73	47.68	0.93	563	0.02
3-Loire	115.91	28.82	0.72	249	0.02
4-Tiber	17.31	7.95	0.36	459	0.04
5-Garonne	79.67	26.21	1.81	329	0.07
6-Rhône	88.43	43.79	3.72	495	0.08
7-Struma-Strymon	16.81	2.24	0.23	133	0.10
8-Duero-Douro	96.24	19.91	3.48	207	0.17
9-Vardar-Axios	22.73	4.56	1.17	201	0.26
10-Ebro	84.90	15.33	4.63	181	0.30
11-Maritsa-Evros	52.60	7.70	3.57	146	0.46
12-Guadalquivir	54.96	8.66	6.27	158	0.72
13-Tajo-Tejo	69.73	11.99	8.88	172	0.74
14-Júcar	21.83	0.89	2.58	41	2.91
15-Guadiana	60.85	4.23	14.19	70	3.35
16-Segura	17.55	0.20	1.17	11	5.83

2.2. Methodological Overview

The methodological approach is presented in Figure 2. The analysis is structured in three steps: analysis of the forcing scenarios for water resources systems, analysis of system response in terms of potential water availability, and analysis of the sensitivity of system response to reservoir storage. The analysis of system forcing consists of the compilation of a large name of model runs producing monthly streamflow series in the basins under analysis, both for a historic control period and for a projected future period. Streamflow series for the control period were corrected for bias. Streamflow series for the future period were obtained under different climate scenarios. The scenarios were characterized in terms of the expected changes of mean, standard deviation, and coefficient of variation of annual streamflow. The analysis of system response is focused on the estimation of the potential water availability allowed by current reservoir storage in the basins under analysis. Uncertainty of potential water availability is first characterized for the control and the future periods. Then, the elasticity of water availability to climate changes is explored by comparing changes in potential water availability to changes in mean annual flow, both for individual projections and for the distribution of all projections in each basin. The focus of the analysis is to explore how this elasticity is affected by reservoir storage and streamflow variability. The third step is focused on exploring the sensitivity to reservoir storage. The analyses of the previous step are repeated considering variable storage in each basin. The performance of the system is characterized by two new indices proposed in this study: the attenuation index and the uncertainty index. These indices describe how the performance of the water supply system is affected by changes in streamflow. The main conclusions of the study are obtained by comparing how these indices change as a function of reservoir storage for all basins.

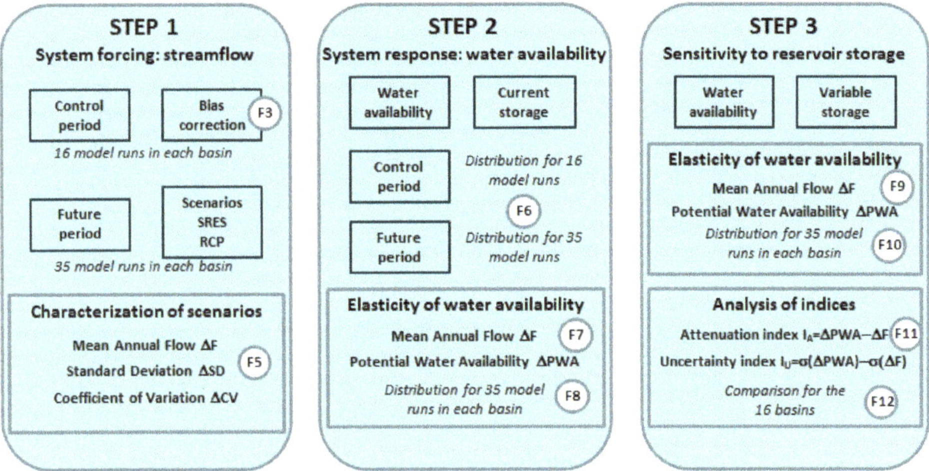

Figure 2. Main methodological steps followed in the analysis. White circles show the figures that illustrate results from each step.

2.3. Current and Future Runoff Scenarios

The focus of the present study is the analysis of the role of reservoir storage to determine how water resources systems react to changes in hydrologic forcing. An effort was made to obtain a wide ensemble of scenarios that would represent the uncertainty linked to climate projections. Therefore, we chose to combine model results obtained under two sets of emission scenarios, the Special Report on Emission Scenarios (SRES) and Representative Concentration Pathways (RCP), in order to increase the size of the ensemble. Current and future runoff scenarios were compiled from three previous studies that include Southern Europe [39–41]. These studies were based on results from different climate models developed over the last 15 years under two sets of emission scenarios: SRES and RCP. They jointly describe the uncertainty that is currently challenging water managers.

The first set of scenarios was taken from the output of regional climate models from the PRUDENCE project [10]. The study by González-Zeas et al. [39] was based on the projections of surface runoff made by eight RCMs at 50 km resolution nested in a single global model, referred to as HadAM3H, in emission scenarios A2 and B2. They analyzed current (1960–1990) and future (2070–2100) time slices. The second set of scenarios was based on the results of the Regional Climate Models (RCMs) of the ENSEMBLES project [11]. The project produced many transient model runs for the time period from 1960 to 2100 using RCMs to characterize model uncertainty. The study by Garrote et al. [40] selected runoff output from four ENSEMBLES models at 25 km resolution under emission scenario A1B to study the major Mediterranean river basins of Europe. They worked with windows of analysis on the transient model runs in three time slices: historical (1960–1990), short term (2020–2050) and long term (2070–2100). In the first two sets of scenarios, monthly runoff time series were directly obtained from the "Total runoff" variable (*mrro*) produced by RCMs. The values of surface runoff flux available at the nodes of the native grid of the RCMs (50 km resolution in PRUDENCE and 25 km resolution in ENSEMBLES) were used to produce monthly runoff maps by interpolation at the finer grid provided by the Hydro1k dataset (1 km). The center of the RCM grid was taken as a point equal to the average for that cell. Interpolation was based on a weighted mean using the inverse of the distance squared as weight. These runoff maps were combined with the subbasin definitions of Hydro1k to obtain monthly streamflow values for each subbasin. The third set of scenarios was based on the results of the global hydrological model PCRaster GLOBal Water Balance (PCRGLOBWB) model [42] in the Inter-Sectorial Impact Model Intercomparison Project (ISIMIP) [43]. In ISIMIP, the PCRGLOBWB model was forced with five global climate models under historical conditions and climate change projections corresponding to four Representative Concentration Pathways scenarios: RCP-2., RCP-4., RCP-6. and RCP-8., corresponding to radiative forcing in the year 2100 of 2.6, 4.5, 6.0, and 8.5 W/m^2, respectively. The study by Sordo-Ward et al. [41] used naturalized streamflow from PCRGLOBWB at 50 km resolution to analyze 1261 subbasins covering the entire territory of Western Europe. They considered two time slices in their analysis: historical (1960–1999) and long-term projection (2060–2099). The monthly streamflow time series in the subbasins were also obtained from monthly runoff maps derived from the runoff produced from the PCRGLOBWB model through interpolation at the Hydro1k 1 km grid.

A total of 16 model runs were compiled for the historical period (eight model runs from the PRUDENCE project, three model runs from the ENSEMBLES projects and five model runs from the PCRGLOBWB model). The windows of analysis in this period overlap for years 1960–1990. All these model runs produced different results in the basins under analysis. To assess the quality of these hydrological projections, the results obtained at the working scale of each model run were compared to a reference estimate of mean annual runoff under current conditions. The selected reference was the annual surface runoff layer (Global Composite Runoff Fields) of the University of New Hampshire Global Runoff Data Centre (GRDC) [44]. This data layer was produced by combining a database of observed river discharge information in more than 9900 gauging stations with a climate-driven water balance model to develop consistent runoff fields. The combination of direct readings from gauging stations with the water balance model preserves the spatial distribution of runoff generation and provides the best estimate of observed runoff over large domains. The mean values of the time series compiled for the historical period were compared with mean annual runoff produced by GRDC. The results are presented in Figure 3, which shows the scatterplot resulting from comparing catchment mean annual runoff produced from GRDC with that produced by model runs for the historical period. Model runs corresponding to the Special Report on Emissions Scenarios (SRES) (PRUDENCE and ENSEMBLES projects) show poor agreement. The models that performed best were Universidad de Castilla La Mancha (UCM) and Eidgenössische Technische Hochschule Zürich (ETHZ2), with coefficients of determination slightly lower than 0.6. This poor performance can be explained because runoff was obtained directly from RCM output. Model runs for the RCP scenarios (ISIMIP project) were produced by the hydrological model PCRGLOBWB. They show better performance, with coefficients of determination close to 0.7, but they reveal significant bias for low runoff. The discrepancies obtained in the comparison suggest that bias correction is necessary to overcome this very large model uncertainty. Using the monthly series of individual models without bias correction would imply significant distortion in the regulation provided by reservoirs in each basin. The ratio between reservoir storage capacity and mean annual flow would change for each model run, affecting the evaluation of the regulation capacity provided by the reservoirs. For this reason, runoff derived from RCM results and from PCRGLOBWB was corrected for bias in each location. The chosen method for bias correction was linear scaling [45]. This method is justified by data availability, because GRDC only provides monthly long-term means of runoff. Therefore, all model projections for the historical period have the same mean.

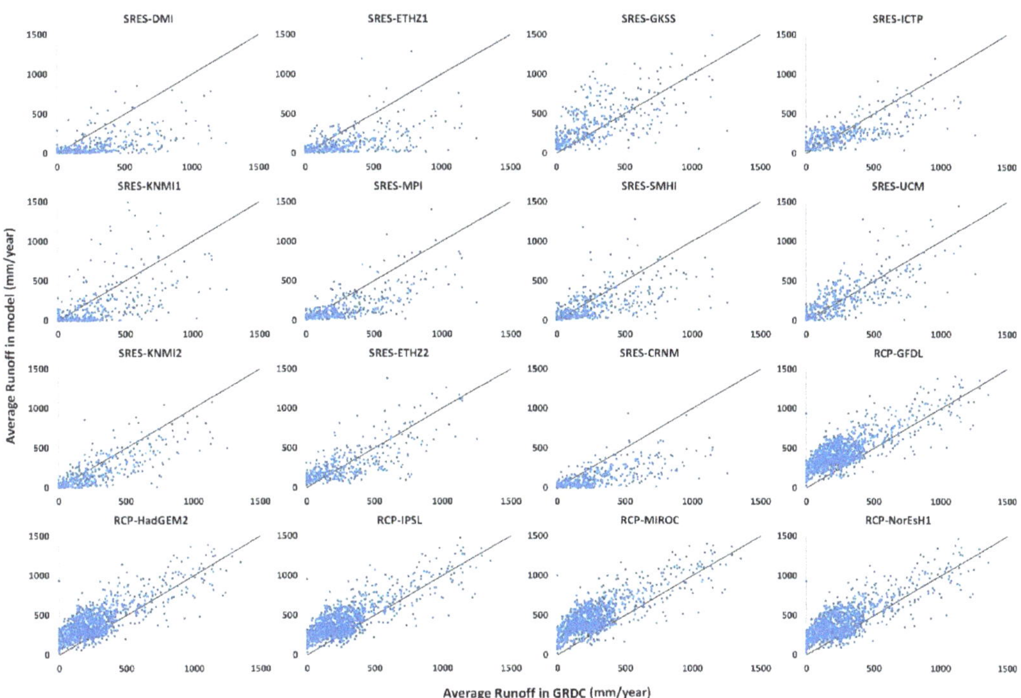

Figure 3. Comparison of mean annual runoff in catchments produced from the Global Runoff Data Set (GRDC) dataset with that produced by model runs for the historical period.

The number of model runs compiled for the long-term climate change projection was 35: eight model runs corresponding to the A2 scenario, four model runs corresponding to the B2 scenario, three model runs for the A1B scenario, five model runs for RCP-2 scenario, five model runs for RCP-4 scenario, five model runs for RCP-6 scenario and five model runs for RCP-8 scenario. The windows of analysis in the long-term projection overlap for the years 2070–2099. These projections were corrected for model bias by applying the same correction as in the corresponding model in the historical period. This ensemble of climate projections was put together from different projects developed over a 15-year period, running a range of global climate models under two sets of emission scenarios, and applying different methodologies. It can thus be considered a representative description of the range of scenarios that climate change science is projecting for the region. However, it should be noted that runoff projections derived from climate models are uncertain. Climate models provide a good overall representation of climate, but their performance degrades at the scale of individual grid boxes, indicating that they are not skillful at their smallest scale. The performance of RCMs generally improves after suitably removing bias. However, model errors still remain large, particularly for climatic variables relevant for hydrology, like precipitation or runoff [46]. Given this inherent uncertainty, a basic hypothesis of this work is that water management decisions based on the global analysis of a wide range of projections produces better results than decisions based on a very detailed analysis of a reduced number of projections.

The average annual runoff obtained from GRDC in the period 1960–2000 was also used to characterize the basins under analysis. The relationship between Specific Runoff and reservoir Residence Time is plotted in Figure 4 for all Hydro1k basins in Southern Europe, highlighting the 16 basins under analysis. As can be seen in Figure 4, there is a clear relation between both variables, with larger values of storage corresponding to basins with lower values of specific runoff. The selected basins produce a good coverage of the possible range of behaviors found in the region, from basins with large specific water resources and low storage capacity like Arno, Po or Loire, to others in the opposite situation with very low water resources and large storage volumes as Júcar, Guadiana or Segura.

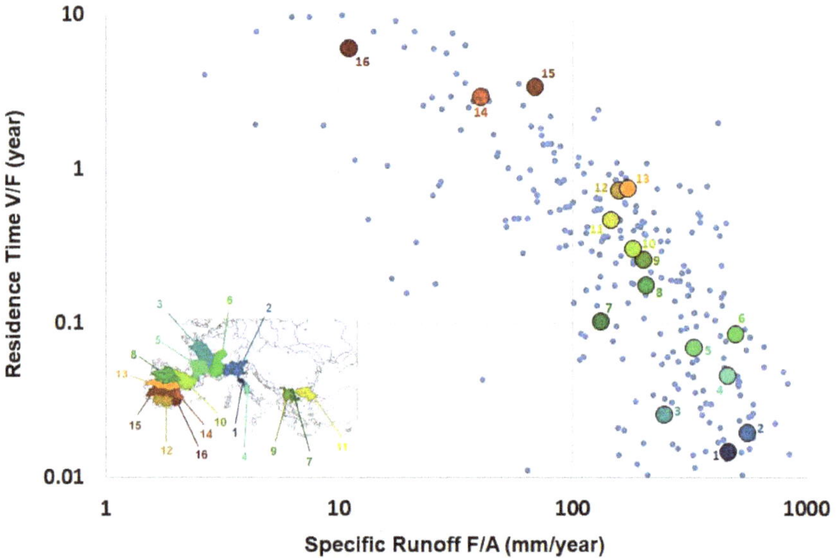

Figure 4. Relation between Specific Runoff (F/A) and reservoir Residence Time (V/F) for Hydro1K basins in Southern Europe. The 16 basins under study are highlighted using the same color coding and numbering as in Figure 1.

2.4. Water Availability Analysis

The study is based on the analysis of how climate change affects water availability in the different basins, and how this affect is modified by available reservoir storage. Potential Water Availability (PWA) is defined as the annual water demand that can be satisfied in a point of the drainage network with a given reliability. PWA depends on the mean and variability of the streamflow series, the storage available for flow regulation, the monthly distribution of the demand and the reliability indicator adopted in the analysis. In this study, PWA was estimated with the Water Availability and Adaptation Policy Analysis (WAAPA) model [31,47]. WAAPA simulates the operation of a complex water resources system with many reservoirs. The basic topological unit of WAAPA is the river network. The main components are inflows, reservoirs and demands, all linked to nodes in the network. WAAPA computes the amount of water supplied to demands from a system of reservoirs accounting for ecological flows and evaporation losses. Input data for WAAPA are monthly inflows in relevant points of the river network, monthly demand values, and reservoir data. Reservoirs are described by monthly maximum and minimum capacity, storage-area relationship, monthly rates of evaporation, and monthly required environmental flow. WAAPA applies an algorithm with simple operating rules, where all reservoirs in the basin are jointly managed to satisfy the set of demands, drawing water preferably from reservoirs located upstream. This algorithm is applied to potential demands located in every node in the river network, and therefore water availability is obtained for the entire river network. The main results of WAAPA are time series of monthly volumes supplied to each demand, monthly storage values and monthly values of spills, environmental flows, and evaporation losses in every reservoir. From this output, demand reliability can be computed for the criterion of choice (volume reliability, time reliability at the monthly or annual scale, or more complex criteria).

WAAPA can obtain PWA for a given demand reliability criterion through an iterative scheme that changes local demand values until the reliability criterion is met with a given precision. In this study, PWA is estimated by considering only one type of demand in the system, with constant monthly distribution. This choice was made because the true monthly distribution of demands in each model node is unknown. Results therefore should be considered only approximate and could be fine-tuned if the ratio between urban and irrigation demand was known in every model node. Ecological flows were specified as the 10% percentile of the monthly marginal distribution of natural flows. System performance is evaluated as gross volume reliability. PWA is obtained for 92% volume reliability. This reliability level was chosen as an intermediate value between reliabilities required from urban demands (usually close to 100%) and those required from irrigation demand (usually close to 90%), assuming an approximate distribution of 20% urban demand and 80% irrigation demand, which is typical of Portugal, Spain, and Greece [48].

3. Results and Discussion

The WAAPA model was run for the European Mediterranean region for the 16 hydrologic scenarios corresponding to the historical period (1960–2000) and for the 35 hydrologic scenarios corresponding to climate projections for the long-term time horizon 2070–2100. The long-term time horizon was chosen for two reasons. Firstly, the results from PRUDENCE project were only available for this time horizon. Secondly, the changes in the long-term time horizon are usually more accentuated than in the mid-term time horizon and the effects are more apparent. Results were obtained for all catchments in the Hydro1k dataset, but, for the sake of simplicity, we only present global results for the 16 basins under analysis. We first analyze the climate projections, then we present the results obtained for PWA in the basins. Average values of these results are summarized in Table 2 and presented and discussed in detail in the following section. Finally, the role of storage is studied through a sensitivity analysis.

3.1. Climate Projections

We first present the characterization of climate projections for the basins under study. Climate projections were taken from the runoff variable of RCM models in the PRUDENCE and ENSEMBLES projects (under SRES emission scenarios) and of the PCRGBLOBWB hydrologic model (under RCP emission scenarios). Mean and coefficient of variation of annual flows were computed for each basin during the historical period and during the long-term projection. Changes in the long-term projection were estimated taking the control period as a reference, applying the following expressions:

$$\Delta F = \frac{F_{PROJ} - F_{HIST}}{F_{HIST}} \; ; \; \Delta SD = \frac{SD_{PROJ} - SD_{HIST}}{SD_{HIST}} \; ; \; \Delta CV = \frac{CV_{PROJ} - CV_{HIST}}{CV_{HIST}} \tag{1}$$

where F is Mean Annual Flow, SD is the Standard Deviation of the annual time series of streamflow, and CV is the Coefficient of Variation of the annual time series of streamflow (standard deviation of the annual time series divided by mean annual flow). The subindices HIST and PROJ refer to the historical period and to the long-term projection.

Table 2. Summary of the results of the analysis of changes in streamflow ΔF, ΔSD and ΔCV and Potential Water Availability, ΔPWA, in the basins analyzed in this study (Ave: average of values for the 35 projections; Std: standard deviation of the values for the 35 projections).

Basin	ΔF		ΔSD		ΔCV		ΔPWA	
	Ave	Std	Ave	Std	Ave	Std	Ave	Std
1-Arno	−0.10	0.19	0.04	0.22	0.23	0.36	0.00	0.36
2-Po	−0.04	0.18	0.09	0.21	0.20	0.29	−0.23	0.27
3-Loire	−0.09	0.19	0.05	0.28	0.43	0.45	−0.21	0.32
4-Tiber	−0.12	0.19	0.03	0.26	0.53	0.53	−0.17	0.32
5-Garonne	−0.14	0.19	0.02	0.26	0.46	0.43	−0.18	0.27
6-Rhône	−0.06	0.17	0.11	0.26	0.28	0.33	−0.16	0.26
7-Struma-Strymon	−0.26	0.20	−0.13	0.22	0.93	1.00	−0.17	0.26
8-Duero-Douro	−0.22	0.22	−0.01	0.34	0.67	0.64	−0.28	0.20
9-Vardar-Axios	−0.23	0.19	−0.06	0.22	1.01	1.06	−0.18	0.23
10-Ebro	−0.20	0.20	−0.07	0.22	0.50	0.53	−0.19	0.17
11-Maritsa-Evros	−0.20	0.21	−0.04	0.32	0.21	0.78	−0.15	0.28
12-Guadalquivir	−0.43	0.31	−0.33	0.32	0.50	0.53	−0.35	0.28
13-Tajo-Tejo	−0.29	0.25	−0.12	0.31	0.75	0.93	−0.27	0.21
14-Júcar	−0.27	0.27	−0.11	0.38	1.38	1.63	−0.27	0.24
15-Guadiana	−0.35	0.40	−0.21	0.43	1.46	1.42	−0.35	0.29
16-Segura	−0.29	0.33	−0.15	0.47	0.74	1.40	−0.27	0.28

The results are depicted in Figure 5, which compares the relative changes in Standard Deviation (ΔSD) and Coefficient of Variation (ΔCV) of annual flows versus changes in Mean Annual Flow (ΔF) for the 35 available projections in the 16 basins under study. All projections are shown together in the left plots of Figure 5, showing for basins the same color codes as in Table 1 and Figure 1. The plots on the right show the mean value for each basin. A plot of each basin is available in the Supplementary Materials, showing individual projections. Projections under SRES emission scenarios are represented as plus signs and projections under RCP scenarios are represented as circles. The analysis of chart (a) of Figure 5 shows positive correlation between changes in Mean Annual Flow ΔF and Standard Deviation ΔSD. If the changes of F and SD were similar, the scatter plot of Figure 5a would be centered around the main diagonal (highlighted in grey). The mean values of changes are above the main diagonal for all basins, suggesting a relative increase of variability in future projections. The joint analysis of all projections for all basins in chart (c) of Figure 5 shows negative correlation between changes in Mean Annual Flow ΔF and Coefficient of Variation ΔCV: reduction of F and increase of CV. The general shape of the scatter plot is similar in all basins in Southern Europe. This has clear implications for water management since both factors will negatively impact water availability. This tendency is stronger for basins with larger residence times that, as seen in Figure 4, are located in water scarce regions, already facing strong hydrologic irregularities. The dispersion of results is stronger for basins with larger residence times, presenting an additional challenge for water management. The ensemble of projections, jointly considered, suggests that water managers should be ready to cope with less abundant and more variable water resources in the future. Given the large dispersion of results, water managers should also be ready to deal with greater year-on-year variability or extreme events than in the past. Figure 5 also shows that expected changes in CV are much larger than changes in F, with many basins reaching extreme values close to 2 (a 100% increase). The basins showing more extreme projections are Guadalquivir, Júcar and Guadiana.

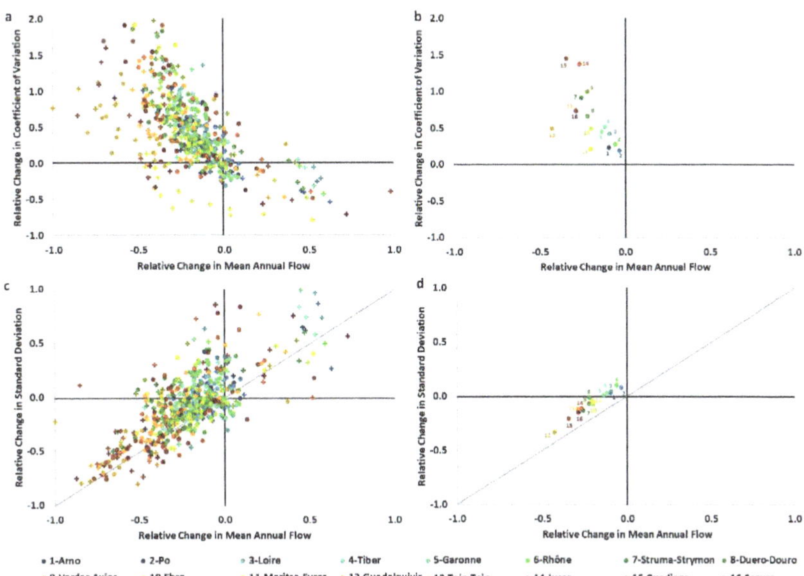

Figure 5. Relative changes in Standard Deviation (ΔSD) and Coefficient of Variation (ΔCV) of annual flow versus changes in Mean Annual Flow (ΔF) for the 16 basins under study. (**a**) ΔF vs. ΔSD, all projections; (**b**) ΔF vs. ΔSD, mean values; (**c**) ΔF vs. ΔCV, all projections; (**d**) ΔF vs. ΔCV, mean values.

3.2. Water Availability

The WAAPA model was used to compute Potential Water Availability (PWA) for the historical period and for the long-term projection in the 16 basins under analysis. The results are shown in Figure 6, which presents the value of PWA obtained in each basin as a function of the relative rank of the corresponding projection. All 35 projections were used to prepare this figure, thus mixing projections under SRES and RCP emission scenarios. An individual plot of each basin is included in the Supplementary Materials, where the joint distribution is compared to the distributions of both sets of emission scenarios. The corresponding emission scenario is identified for each model run available in the long-term projection. These plots show that there is no clear correlation between the emission scenario and the projected PWA. Values corresponding to different emission scenarios are mixed and the most extreme scenarios (A2 and RCP-8) do not always produce the minimum values for PWA.

PWA is expressed as a fraction of Mean Annual Flow (F) in the historical period. Results for the historical period are shown in the upper chart (a) and results for the long-term projection are shown in the lower chart (b). If all model runs were assumed equiprobable, this plot would correspond to the empirical estimation of the probability distribution function of PWA expected in each basin. The results show that the relative value of PWA to F tends to be larger for basins with larger storage capacity, both in the historical and in the projection periods. This fact clearly illustrates the effectiveness of reservoir storage to increase water availability. The plots also show large uncertainty in the estimation of PWA. For the historical period, this result is remarkable because historical time series were corrected for bias with respect to the GRDC estimation of F and therefore all had the same Mean Annual Flow. The uncertainty in PWA reflects model uncertainty because the differences in PWA can only be attributed to the differences in the seasonal and interannual variability of the time series produced by each model run. Unfortunately, the skill of the models to reproduce current hydrological irregularity cannot be evaluated because there are no available regional data sets for Southern Europe on interannual naturalized streamflow variability.

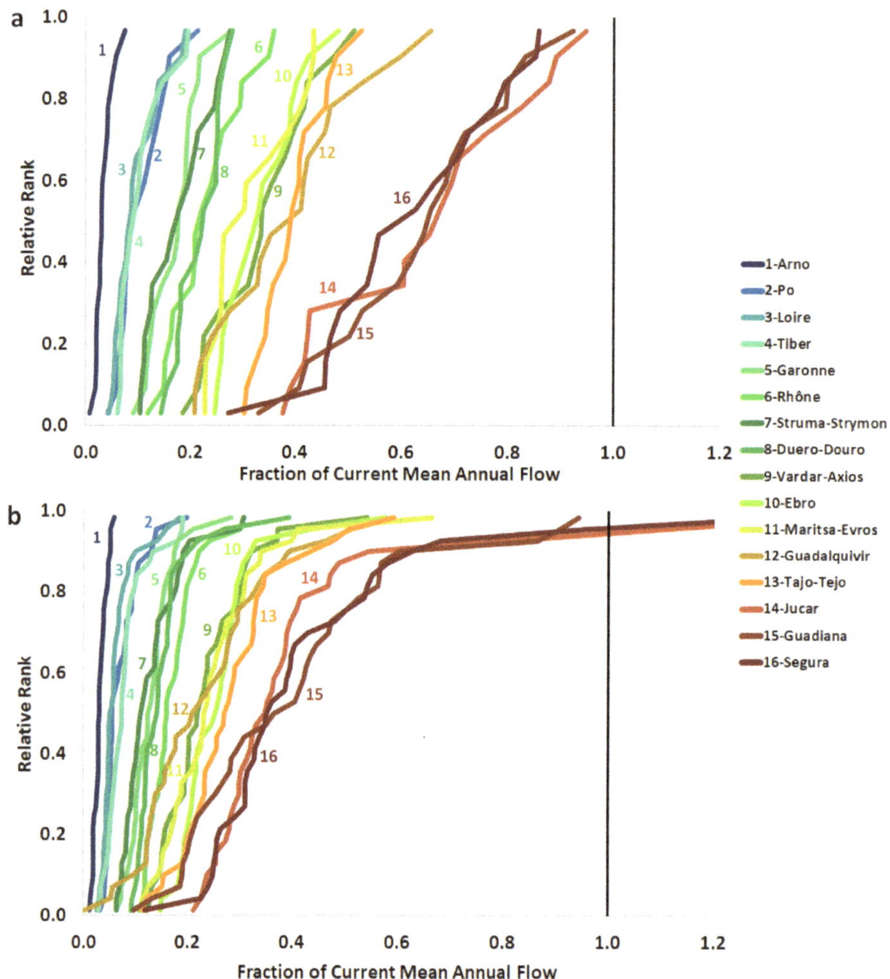

Figure 6. Estimated cumulative probability distribution function of Potential Water Availability (PWA) expressed as a fraction of current Mean Annual Flow (F) in the 16 basins under study. (**a**) historical period; (**b**) long-term projection.

Except in the Arno basin, PWA is expected to decrease significantly in the long-term projection with respect to the historical period, with average reductions between 15% and 35%. These reductions are the consequence of reduced F and increased CV. The most significant reductions are projected for the basins of South Western Europe: Guadiana and Guadalquivir (35% on average) and Duero (28% on average). The uncertainty of PWA in the long-term projection is larger than that in the historical period due to the additional variability introduced by emission scenarios. However, the large model uncertainty hinders the interpretation of results obtained for different emission scenarios.

The estimated changes in PWA are compared to estimated changes in F in Figures 7 and 8. Figure 7 shows the scatter plot of changes in both variables for the set of emission scenarios analyzed in all basins. A plot of each basin is available in the Supplementary Materials, showing individual projections. Projections under SRES emission scenarios are represented as plus signs and projections under RCP scenarios are represented as circles. Figure 8 shows the comparison of the estimated probability distributions of F and PWA. All 35 projections were used to prepare this figure, thus mixing projections under SRES and

RCP emission scenarios. An individual plot of each basin is included in the Supplementary Materials, where the joint distribution of PWA is compared to the distributions of both sets of emission scenarios.

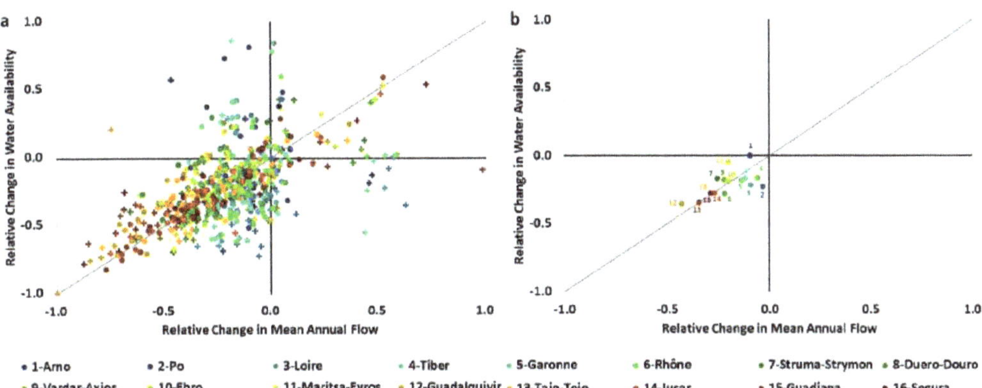

Figure 7. Comparison of the estimated changes in Mean Annual Flow (ΔF) and the estimated changes in Potential Water Availability (ΔPWA) for the 35 available projections in the 16 basins under study. (**a**) ΔF vs. ΔPWA, all projections; (**b**) ΔF vs. ΔPWA, mean values.

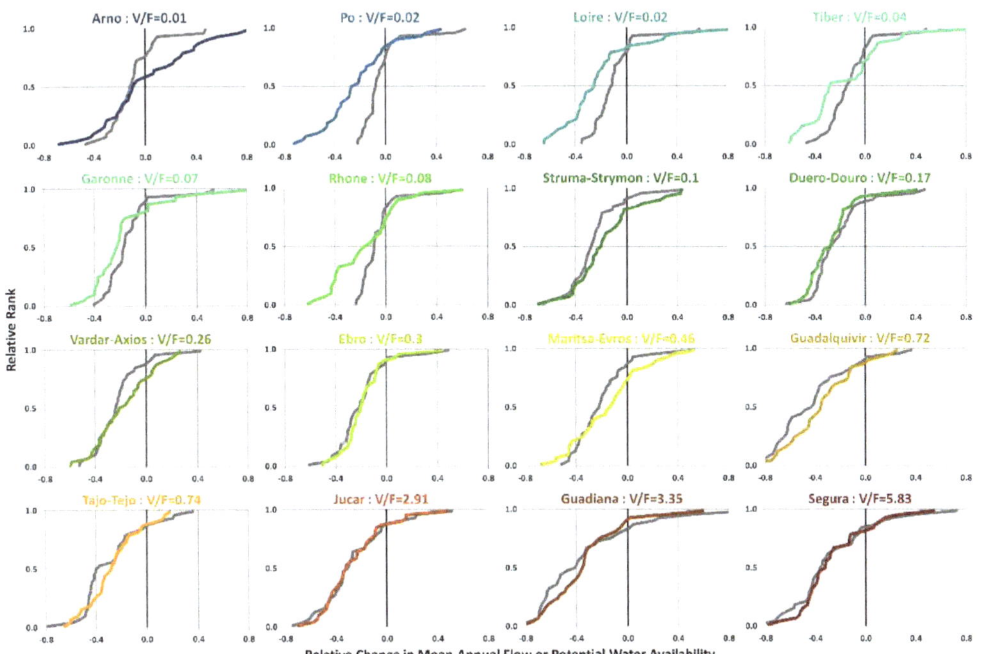

Figure 8. Estimated cumulative probability distribution function of changes in Mean Annual Flow (ΔF, in gray) and changes in Potential Water Availability (ΔPWA, in the color code for each basin) for the 16 basins under analysis.

In Figures 7 and 8, the changes of F are estimated from the first expression shown in Equation (1). The changes of PWA are similarly estimated from the comparison of values obtained in the long-term projection and the control period for the same model, applying the following expression:

$$\Delta PWA = \frac{PWA_{PROJ} - PWA_{HIST}}{PWA_{HIST}} \tag{2}$$

where PWA is Potential Water Availability and the sub-indices HIST and PROJ refer to the historical period and to the long-term projection.

The mean values plotted in chart (b) of Figure 7 reveal that basins with small storage capacity (Po, Loire, Tiber, Garonne and Rhône) show a larger reduction of PWA than the reduction of F. The Arno basin is the exception, with no reduction of PWA despite a small average reduction of F. The basins with larger storage capacity, in general, show a smaller reduction of PWA than the reduction of F. Struma-Strymon, Vardar-Axios, Ebro, Maritsa-Evros, Guadalquivir, Tajo-Tejo, Júcar, Guadiana and Segura belong to this group. Duero-Douro is an exception, with larger reduction of PWA than of F. This may be explained because Duero-Douro shows the largest difference between change in F and change in SD. The wide scatter of changes in F and PWA in chart (a) of Figure 7 shows that there is no exact relation between changes in Mean Annual Flow (ΔF) and changes in Potential Water Availability (ΔPWA). For individual projections, changes in PWA may be larger, equal or smaller than changes in F. This is, in part, a consequence of changes in hydrologic variability, which may explain why negative changes in F produce positive changes in PWA and vice versa. However, the comparison of Figures 5 and 7 shows that changes in hydrologic variability alone cannot explain the diversity of behaviors seen in Figure 7. Hydrologic variability is measured in terms of Coefficient of Variation of annual flows and is therefore referred to interannual variability. Basins with small storage capacity show a behavior more exposed to changes in CV because, for them, water availability is almost directly determined by short-duration dry periods of the streamflow series. These dry periods show a large variability among model runs, which explains the variability observed in values of PWA. As basin storage grows larger, the reservoirs attenuate the effect of short-duration dry periods and the interannual variability becomes less important. Basins with storage capacity larger than mean annual flow show a much less sensitivity to changes in the coefficient of variation of mean annual flows.

Figure 8 is useful to analyze the effect of the uncertainty on emission scenarios. The estimated probability distributions shown in Figure 8 reveal a wide range of behaviors. The basins were classified in five groups (A1, A2, A3, B1 and B2), according to the relative value of the distributions of changes in F and PWA. Group A1 is integrated by basins where the distribution of expected reductions in PWA is to the left of the distribution of expected reductions in F, suggesting that the availability of reservoir storage tends to dampen the effect of climate change. Struma-Strymon, Vardar-Axios, Guadalquivir and Tajo belong to this group. In the second group, A2, the distributions of expected changes in F and PWA are very similar. This group is formed by Ebro, Júcar and Segura. The only basin in Group A3 (Douro-Duero) presents larger expected reductions in PWA than in F. In group B, the probability distributions of F and PWA cross each other. In group B1, the distribution of changes in PWA is to the left of the distribution of changes in F for low probability values. For high probability values, the distribution of changes in PWA is to the right of the distribution of changes in F. This results in larger uncertainty for changes in PWA than in F. This effect may be due to increased exposure to changes in variability due to lack of regulation storage. Arno, Po, Loire, Tiber, Garonne, Rhône and Maritsa-Evros belong to group B1. The only basin in group B2 is Guadiana, where the distribution of changes in PWA is to the right of the distribution of changes in F for low probability values and to the left for high probability values. Guadiana shows less uncertainty in changes of PWA than in changes in F, due to its large reservoir storage.

A remarkable effect shown in Figure 8 is that the uncertainty regarding changes in F (gray line) seems to grow as specific storage grows. The larger spreads of the estimated probability distributions appear in basins with larger specific storage, in the bottom row.

Basins with comparatively smaller storage capacity, in the first row, show much less uncertainty on changes in F. This shows that reservoir storage was developed where it was required: in basins with large hydrologic variability. Furthermore, the difference in uncertainty between changes in F and changes in PWA, which is large in basins with small storage, is progressively reduced as specific storage increases.

3.3. The Influence of Storage

The results obtained in Section 2.2 suggest that reservoir storage plays a relevant role in controlling how projected changes in Mean Annual Flow may be translated into changes in Potential Water Availability. However, the large variability of local conditions in the studied basins introduces uncertainties in the analysis. In this section we further explore the influence of reservoir storage on changes in water availability through a sensitivity analysis that discounts for local conditions. We repeated the water availability analysis but considering different storage volumes in each basin. Potential Water Availability was computed in current and future scenarios in the 16 basins, assuming changing reservoir volumes of 25%, 50%, 75%, 100%, 125%, 150% and 175% of current storage. Storage was proportionally reduced or increased in the same location of existing reservoirs. This choice was made for convenience, without any implications for projected future evolution of storage in the region. In fact, the most likely scenario in the future for European basins is a progressive reduction of available storage due to reservoir sedimentation, with very little additional storage being built. Figure 9 shows the scatter plots of changes in F versus changes in PWA for four basins covering a wide range of values of reservoir storage: Loire (V/F = 0.03), Ebro (V/F = 0.30), Guadalquivir (V/F = 0.73) and Guadiana (V/F = 3.35). Individual plots for all basins are included in the Supplementary Materials. Projections under SRES emission scenarios are represented as diamonds and projections under RCP scenarios are represented as circles. Results shown in Figure 9 reveal that the dispersion of the scatter plot gets reduced as the storage capacity is increased. This effect is more marked for the Guadiana basin, which has the largest reservoir storage.

In order to assess the global behavior, the values obtained for changes in F and PWA were classified according to the relative rank of the corresponding projection, obtaining an empirical estimate of their probability distributions, under the assumption that all scenarios analyzed are equally likely. The results are shown in Figure 10, which presents the estimate of the probability distribution of changes in Mean Annual Flow (ΔF, blue line) and changes in Potential Water Availability (ΔPWA) for different storage values in colored lines from brown (25% of current storage volume) to green (175% of current reservoir storage). The basins analyzed showed variable sensitivity to storage. Some basins, like Arno, Ebro, Maritsa-Evros or Guadiana, showed very little sensitivity to storage capacity because the distributions of expected changes of PWA are very similar. Other basins, like Po, Tiber, Rhône or Struma-Strymon, presented significant differences in behavior depending on the storage volume assumed. In some of these basins, the estimated probability distributions of changes in PWA for high storage values (green color) were located to the right, indicating less reductions of PWA. Po, Loire, Rhône, Duero-Douro, Ebro, Júcar, Guadiana and Segura show this behavior. For other basins, however, the probability distributions for high storage values were located to the left. Tiber, Struma-Strymon and Guadalquivir belong to this group.

This range of behaviors illustrates the complex relations between hydrologic variability and reservoir storage in determining water availability in climate change scenarios, suggesting that a specific analysis for local conditions is required to translate projections of changes in mean annual flow into projections of changes in water availability in basins with significant storage capacity.

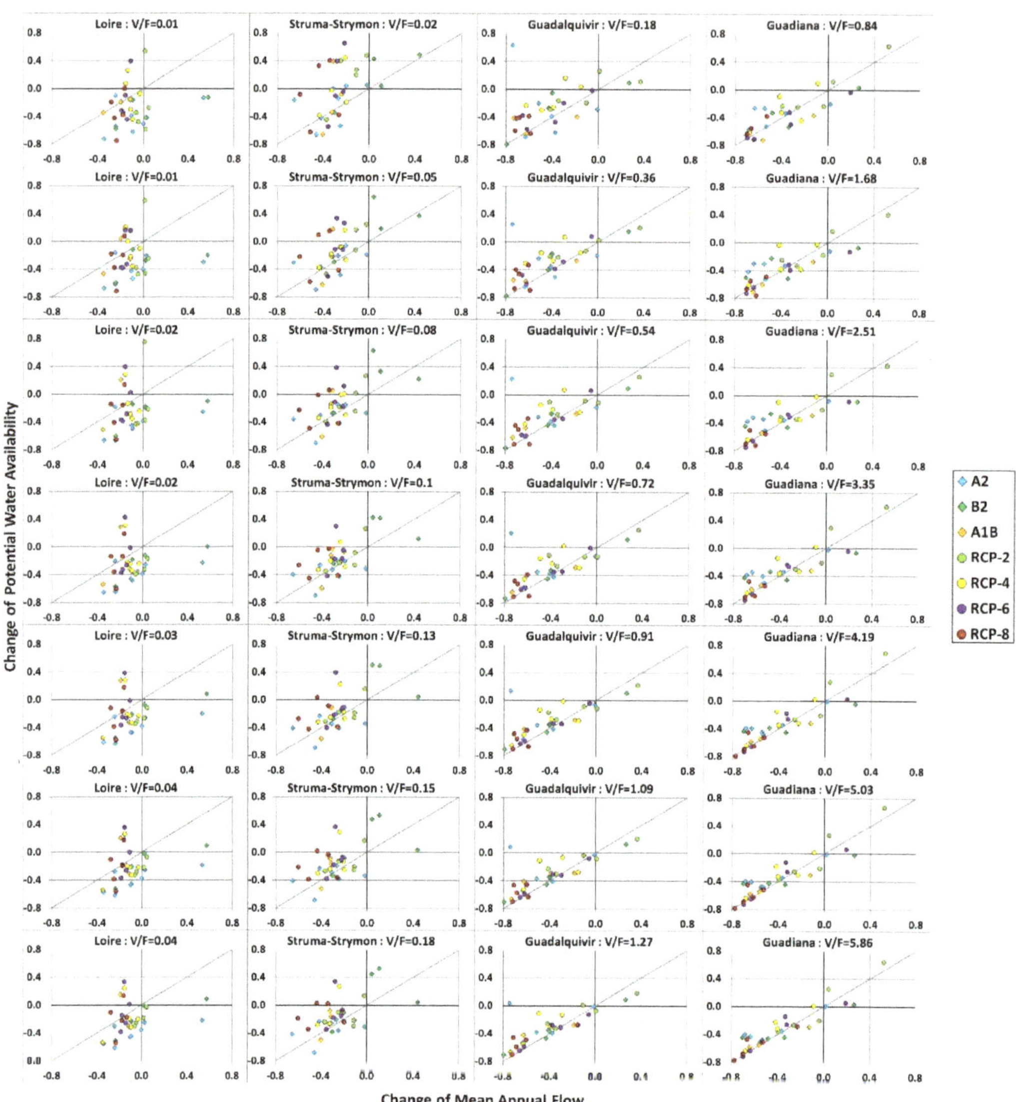

Figure 9. Comparison of the estimated changes in Mean Annual Flow (ΔF) and the estimated changes in Potential Water Availability (ΔPWA) for 25%, 50%, 75%, 100%, 125%, 150% and 175% (in rows, ordered from top to bottom) for four representative basins of the study: Loire (left column), Struma-Strymon (center-left column), Guadalquivir (center-right column) and Guadiana (right column).

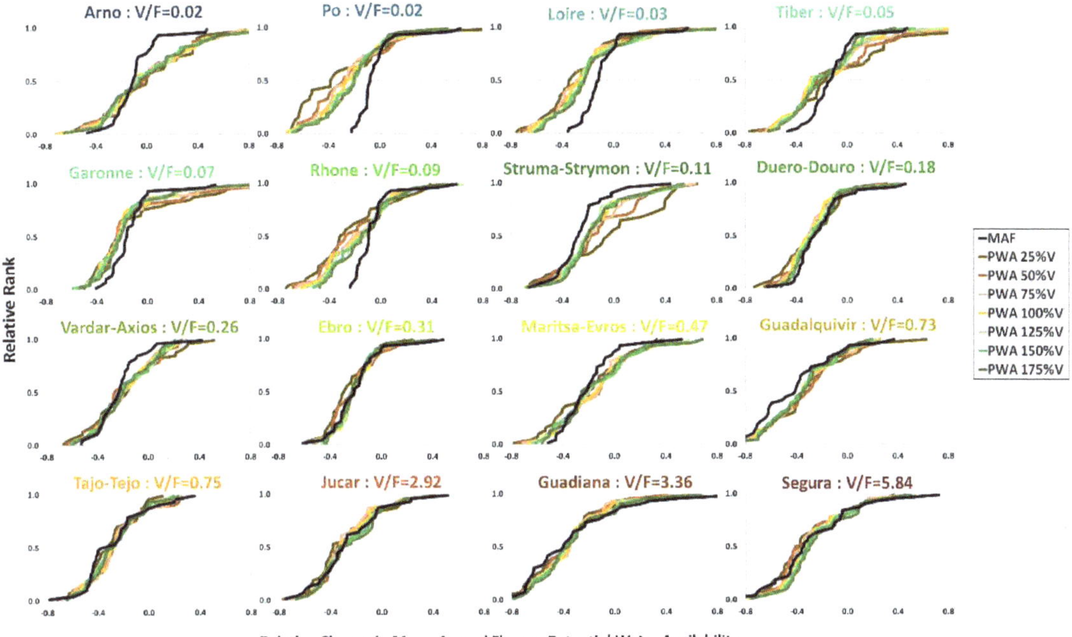

Figure 10. Probability distribution of changes in Mean Annual Flow (ΔF, in blue) and changes in Potential Water Availability (ΔPWA) for different storage values (color-coded, from 25% to 175% of current storage volume) in the 16 studied basins.

The influence of reservoir storage on the elasticity of water availability was analyzed by computing an attenuation index I_A, defined in the following expression:

$$I_A = \Delta PWA - \Delta F \tag{3}$$

where ΔPWA is the change in Potential Water Availability and ΔF is the change in Mean Annual Flow. As seen in the previous sections, most changes in PWA and F are reductions and therefore ΔPWA and ΔF are negative. A positive value of this index indicates an attenuation of the effect of climate change: the absolute value of ΔPWA is smaller than the absolute value ΔF.

We explore how the attenuation index I_A changes with reservoir storage. The results are shown in Figure 11, which presents the value of the attenuation index as a function of reservoir storage in each basin for all available projections (thin grey lines), the average values (solid lines in the color code corresponding to the basin) and average values plus and minus one standard deviation (dotted lines in the color code corresponding to the basin). The results show a large variability for individual projections, which translates into large uncertainty for water managers. The variability of the attenuation index appears to be progressively reduced as specific storage grows across basin locations (from top row to bottom row). This suggests that reservoir storage plays a relevant role in reducing uncertainty on the effects of climate change projections on water availability.

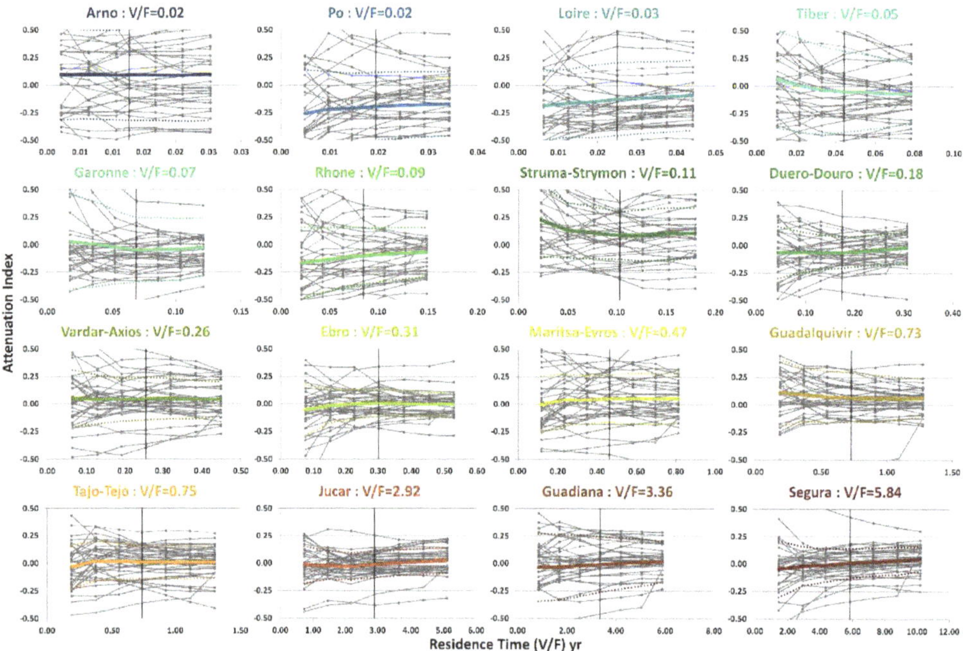

Figure 11. Values of the attenuation index I_A for different relative storage volumes in the 16 studied basins. Current relative storage is marked with a vertical black line.

The effect of reservoir storage on the variability of the attenuation index I_A is further explored by analyzing the uncertainty index I_U, defined as:

$$I_U = \sigma(\Delta PWA) - \sigma(\Delta F) \tag{4}$$

where $\sigma(\Delta PWA)$ is the standard deviation of the changes in Potential Water Availability for all projections and $\sigma(\Delta F)$ is the standard deviation of the changes in Mean Annual Flow for all projections. I_U index compares the variability of the projections of changes in water availability to that of the projections of mean flow. A negative value of this index indicates a reduction of the uncertainty of climate change projections: the variability of ΔPWA is smaller than the variability ΔF.

The summary of results found in the analysis of the I_A and I_U indices is shown in Figure 12. Chart (a) of Figure 12 compares the average of the values of the I_A index obtained in the sensitivity analyses of reservoir storage for all basins. Chart (b) of Figure 12 represents the corresponding values of the I_U index.

The plots shown in Figure 12 indicate that increased reservoir storage results in larger values of the attenuation index I_A in most basins and smaller values of the uncertainty index I_U in all basins. The results for the attenuation index are less conclusive than those for the uncertainty index. Out of the 16 basins analyzed, I_A is observed to decrease with increasing reservoir storage in four basins: Tiber, Garonne, Struma-Strymon and Guadalquivir. In the case of Garonne and Struma-Strymon, the decrease only covers the range from 25% to 100% of current reservoir storage. For storage volumes between 100% and 175% of current reservoir storage I_A is increasing with increasing reservoir storage. The case of Tiber basin may be explained because reservoirs only cover 12% of the contributing area and therefore 88% of the flow is unregulated. Guadalquivir basin is exposed to the most extreme reduction of Mean Annual Flow (43% on average) and this may have an influence on the observed behavior. In the case of the uncertainty index, the reduction with increasing storage is observed for all basins. These results are valid for the range of storage volumes explored in the sensitivity analysis in each basin individually and for all basins as a whole, regardless of basin size and location in Southern Europe, and therefore show a clear picture of the role played by reservoir storage in attenuating the impact of reduced streamflow on water availability and on reducing the uncertainty of climate change projections. It is unlikely that reservoir storage will be further increased in Southern Europe. Most basins already have an adequate amount of storage and additional storage capacity would not increase water availability in a scenario of decreasing resources. However, water managers should be aware that proper management of currently available storage will be helpful to address the challenges posed by climate change.

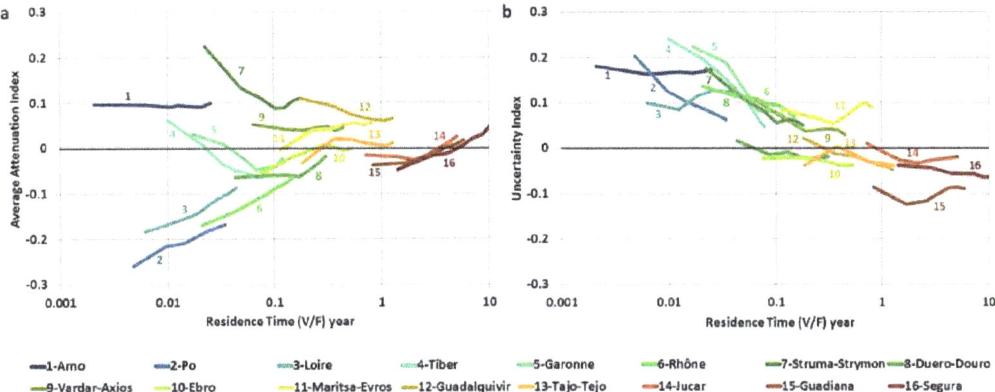

Figure 12. Comparison of the values of average I_A index and I_U index obtained in the sensitivity analyses of reservoir storage in the 16 basins under study. (**a**) Average value of the attenuation index (I_A) for the 35 available projections as a function of reservoir storage. (**b**) Value of the uncertainty index (I_U) as a function of reservoir storage.

4. Conclusions

Projected changes in hydrologic regime and water availability were analyzed in 16 basins in Southern Europe applying the WAAPA model to streamflow time series obtained from 35 climate projections under 7 emission scenarios. The analysis of climate projections concluded that a significant reduction of mean annual flow can be expected in most basins. The reduction in the mean is supplemented by a strong increase in the coefficient of variation, due to an increase of the variability of the projected series. This analysis is uncertain due to the very large variability introduced by the different models and emission scenarios examined. The overall result implies a corresponding reduction in potential water availability, with variable results across basins depending on hydrologic regime and reservoir storage. Basins with large storage values showed reductions of water availability comparable to the reductions of mean annual flow. Basins with small storage capacity showed a larger reduction of water availability than the reduction of mean annual flow. Although model and emission scenario uncertainties are larger than the expected reduction of water availability, a consistent picture emerges from the joint analysis of all projections, requiring significant adaptation measures to compensate for the projected reduction of water availability.

The influence of reservoir storage on basin response to climate change was studied through a sensitivity analysis where current reservoir storage was modified to examine its effects on water availability with values ranging from 25% to 175% of current storage values. The results showed very large variability, which illustrates the complex interplay between hydrologic regime, reservoir storage and water availability. Two indices were introduced to clarify the overall behavior: the attenuation index and the uncertainty index. The attenuation index compares the changes in water availability to the changes in mean annual flow. Positive values of this index indicate an attenuation of the impact of climate change projection on water availability. The uncertainty index compares the variability of changes in water availability and in mean annual flow. Positive values of this index indicate a reduction of the uncertainty of climate change projections on water availability. The results of the sensitivity analysis showed that increasing reservoir storage attenuates the reduction of water availability and reduces the uncertainty of climate projection. The results are valid for each individual basin within the range of storage volumes examined and for the set of 16 Southern European basins analyzed in this work. The effect observed for reservoir storage is a positive factor for system managers since decisions become harder as uncertainty grows. This feature would allow water managers to develop suitable policies to mitigate the impacts of climate change, thus enhancing the resilience of the system.

Supplementary Materials: The following are available online at https://www.mdpi.com/2073-4441/13/1/85/s1: High resolution images of Figures 1–12 and individual plots for each basin in Figures 5–11.

Author Contributions: Conceptualization, L.G., A.G. and A.S.-W.; methodology, A.G.; data processing, B.P.-B.; software, L.G. and B.P.-B.; validation, A.S.-W., A.G. and L.G.; writing—original draft preparation, A.G.; writing—review and editing, L.G.; visualization, A.S.-W.; supervision, L.G. All authors have read and agreed to the published version of the manuscript.

Funding: This research was funded by the Spanish Ministry of Science and Innovation, grant number PID2019-105852RA-I00: "Simulation of climate scenarios and adaptation in water resources systems (SECA-SRH)". B.P.-B. would like to acknowledge Universidad Técnica de Ambato for the financial support through its doctoral student mobility program (award No. 1886-CU-P-2018 Resolución HCU).

Acknowledgments: Data from the PRUDENCE and ENSEMBLES projects were used in this work. PRUDENCE was funded by the EU FP5 (EVK2-CT2001-00132) and ENSEMBLES was funded by the EU FP6 (contract No. 505539). Support from both projects is gratefully acknowledged. Moreover, the authors acknowledge the World Climate Research Programme's Working Group on Regional Climate, and the Working Group on Coupled Modelling, former coordinating body of CORDEX and responsible panel for CMIP5; and also thank the climate modelling groups for producing and making available their model output. Finally, the authors also acknowledge the Earth System Grid Federation infrastructure an international effort led by the U.S. Department of Energy's Program for Climate Model Diagnosis and Intercomparison, the European Network for Earth System Modelling and other partners in the Global Organisation for Earth System Science Portals (GO-ESSP).

Conflicts of Interest: The authors declare no conflict of interest. The funders had no role in the design of the study; in the collection, analyses, or interpretation of data; in the writing of the manuscript, or in the decision to publish the results.

References

1. IPCC. Climate change 2014: Impacts, adaptation and vulnerability. Part A: Global and sectoral aspects. In *Contribution of Working Group II to the Fifth Assessment Report of the Intergovernmental Panel on Climate Change*; Field, C.B., Barros, V.R., Dokken, D.J., Mach, K.J., Mastrandrea, M.D., Bilir, T.E., Chatterjee, M., Ebi, K.L., Estrada, Y.O., Genova, R.C., et al., Eds.; Cambridge University Press: Cambridge, UK; New York, NY, USA, 2014; p. 1132.
2. Garrote, L. Managing Water Resources to Adapt to Climate Change: Facing Uncertainty and Scarcity in a Changing Context. *Water Resour. Manag.* **2017**, *31*, 2951–2963. [CrossRef]
3. Arnell, N.W. The effect of climate change on hydrological regimes in Europe: A continental perspective. *Glob. Environ. Chang.* **1999**, *9*, 5–23. [CrossRef]
4. Arnell, N.W. Effects of IPCC SRES* emissions scenarios on river runoff: A global perspective. *Hydrol. Earth Syst. Sci.* **2003**, *7*, 619–641. [CrossRef]
5. Vörösmarty, C.J.; McIntyre, P.B.; Gessner, M.O.; Dudgeon, D.; Prusevich, A.; Green, P.; Glidden, S.; Bunn, S.; Sullivan, C.A.; Liermann, C.R.; et al. Global threats to human water security and river biodiversity. *Nat. Cell Biol.* **2010**, *467*, 555–561. [CrossRef]
6. Krysanova, V.; Dickens, C.; Timmerman, J.G.; Varela-Ortega, C.; Schlüter, M.; Roest, K.; Huntjens, P.; Jaspers, F.; Buiteveld, H.; Moreno, E.; et al. Cross-Comparison of Climate Change Adaptation Strategies Across Large River Basins in Europe, Africa and Asia. *Water Resour. Manag.* **2010**, *24*, 4121–4160. [CrossRef]
7. García-Ruiz, J.M.; López-Moreno, J.I.; Vicente-Serrano, S.M.; Lasanta–Martínez, T.; Beguería, S. Mediterranean water resources in a global change scenario. *Earth-Sci. Rev.* **2011**, *105*, 121–139. [CrossRef]

8. Forzieri, G.; Feyen, L.; Rojas, R.G.; Flörke, M.; Wimmer, F.; A Bianchi, A. Ensemble projections of future streamflow droughts in Europe. *Hydrol. Earth Syst. Sci.* **2014**, *18*, 85–108. [CrossRef]

9. Suárez-Almiñana, S.; Solera, A.; Andreu, J.; García-Romero, L. Análisis de incertidumbre de las proyecciones climáticas en relación a las aportaciones históricas en la Cuenca del Júcar. *Ing. Agua* **2020**, *24*, 89–99. [CrossRef]

10. Christensen, J.H.; Carter, T.R.; Rummukainen, M.; Amanatidis, G. Evaluating the performance and utility of regional climate models: The PRUDENCE project. *Clim. Chang.* **2007**, *81*, 1–6. [CrossRef]

11. Hewitt, C.D.; Griggs, D.J. Ensembles-based predictions of climate changes and their impacts. *Eos* **2004**, *85*, 566. [CrossRef]

12. Sordo-Ward, A.; Granados, A.; Iglesias, A.; Garrote, L.; Bejarano, M.D. Adaptation Effort and Performance of Water Management Strategies to Face Climate Change Impacts in Six Representative Basins of Southern Europe. *Water* **2019**, *11*, 1078. [CrossRef]

13. Iglesias, A.; Garrote, L.; Flores, F.; Moneo, M. Challenges to Manage the Risk of Water Scarcity and Climate Change in the Mediterranean. *Water Resour. Manag.* **2007**, *21*, 775–788. [CrossRef]

14. MMA. White Paper on Water in Spain. In *Spanish Ministry of the Environment*; MMA: Madrid, Spain, 2000.

15. Berbel, J.; Expósito, A.; Gutiérrez-Martín, C.; Mateos, L. Effects of the Irrigation Modernization in Spain 2002–2015. *Water Resour. Manag.* **2019**, *33*, 1835–1849. [CrossRef]

16. Gil-Meseguer, E.; Bernabé-Crespo, M.B.; Gómez-Espín, J.M. Recycled Sewage—A Water Resource for Dry Regions of Southeastern Spain. *Water Resour. Manag.* **2019**, *33*, 725–737. [CrossRef]

17. Garrote, L.; Granados, A.; Iglesias, A. Assessing water availability in Europe: A comparative study. In Proceedings of the EWRA Conference 2015, Istanbul, Turkey, 10–13 June 2015.

18. Rippl, W. The Capacity of Storage-Reservoirs for Water-Slpply (Including Plate). *Minutes Proc. Inst. Civ. Eng.* **2007**, *71*, 270–278. [CrossRef]

19. Klemeš, V. One hundred years of applied storage reservoir theory. *Water Resour. Manag.* **1987**, *1*, 159–175. [CrossRef]

20. Löf, G.O.G.; Hardison, C.H. Storage requirements for water in the United States. *Water Resour. Res.* **2008**, *2*, 323–354. [CrossRef]

21. Vogel, R.M.; Lane, M.; Ravindiran, R.S.; Kirshen, P. Storage Reservoir Behavior in the United States. *J. Water Resour. Plan. Manag.* **1999**, *125*, 245–254. [CrossRef]

22. Hashimoto, T.; Stedinger, J.R.; Loucks, D.P. Reliability, resiliency, and vulnerability criteria for water resource system performance evaluation. *Water Resour. Res.* **1982**, *18*, 14–20. [CrossRef]

23. Vogel, R.M.; Bolognese, R.A. Storage-Reliability-Resilience-Yield Relations for Over-Year Water Supply Systems. *Water Resour. Res.* **1995**, *31*, 645–654. [CrossRef]

24. Andreu, J.; Capilla, J.; Sanchís, E. AQUATOOL, a generalized decision-support system for water-resources planning and operational management. *J. Hydrol.* **1996**, *177*, 269–291. [CrossRef]

25. Wurbs, R.A. Assessing Water Availability under a Water Rights Priority System. *J. Water Resour. Plan. Manag.* **2001**, *127*, 235–243. [CrossRef]

26. Yates, D.; Sieber, J.; Purkey, D.; Huber-Lee, A. WEAP21—A Demand-, Priority-, and Preference-Driven Water Planning Model. *Water Int.* **2005**, *30*, 487–500. [CrossRef]

27. McMahon, T.A.; Adeloye, A.J. *Water Resources Yield*; Water Resources Publications: Highlands Ranch, CO, USA, 2005.

28. McMahon, T.A.; Adeloye, A.J.; Zhou, S.-L. Understanding performance measures of reservoirs. *J. Hydrol.* **2006**, *324*, 359–382. [CrossRef]

29. Pulido-Velazquez, M.; Andreu, J.; Sahuquillo, A.; Pulido-Velazquez, D. Hydro-economic river basin modelling: The application of a holistic surface–groundwater model to assess opportunity costs of water use in Spain. *Ecol. Econ.* **2008**, *66*, 51–65. [CrossRef]

30. Wurbs, R.A.; Muttiah, R.S.; Felden, F. Incorporation of Climate Change in Water Availability Modeling. *J. Hydrol. Eng.* **2005**, *10*, 375–385. [CrossRef]

31. Garrote, L.; Iglesias, A.; Martín-Carrasco, F.J.; Mediero, L. WAAPA: A model for water availability and climate change adaptation policy analysis. In Proceedings of the EWRA International Symposium—Water Engineering and Management in a Changing Environment, Catania, Italy, 29 June–2 July 2011.

32. Georgakakos, A.; Yao, H.; Kistenmacher, M.; Graham, N.; Cheng, F.-Y.; Spencer, C.; Shamir, E. Value of adaptive water resources management in Northern California under climatic variability and change: Reservoir management. *J. Hydrol.* **2012**, *412–413*, 34–46. [CrossRef]

33. Alimohammadi, H.; Bavani, A.M.; Roozbahani, A. Mitigating the Impacts of Climate Change on the Performance of Multi-Purpose Reservoirs by Changing the Operation Policy from SOP to MLDR. *Water Resour. Manag.* **2020**, *34*, 1495–1516. [CrossRef]

34. He, S.; Guo, S.; Yang, G.; Chen, K.; Liu, D.; Zhou, Y. Optimizing Operation Rules of Cascade Reservoirs for Adapting Climate Change. *Water Resour. Manag.* **2019**, *34*, 101–120. [CrossRef]

35. Biglarbeigi, P.; Strong, W.A.; Finlay, D.; McDermott, R.; Griffiths, P. A Hybrid Model-Based Adaptive Framework for the Analysis of Climate Change Impact on Reservoir Performance. *Water Resour. Manag.* **2020**, *34*, 4053–4066. [CrossRef]

36. Gorguner, M.; Kavvas, M.L. Modeling impacts of future climate change on reservoir storages and irrigation water demands in a Mediterranean basin. *Sci. Total Environ.* **2020**, *748*, 141246. [CrossRef] [PubMed]

37. EROS. HYDRO1k Elevation Derivative Database. Technical Report. Center for Earth Resources Observation and Science (EROS), U.S. Geological Survey (USGS). 2008. Available online: http://eros.usgs.gov/products/elevation/hydro1k.html-97 (accessed on 27 October 2020).

38. Delliou. The History of the World Register of Dams. International Commission on Large Dams. 2020. Available online: https://www.icold-cigb.org/userfiles/files/CIGB/History%20of%20the%20WRD-PleDelliou.pdf (accessed on 27 October 2020).
39. González-Zeas, D.; Garrote, L.; Iglesias, A.; Granados, A.; Chavez-Jimenez, A. Hydrologic Determinants of Climate Change Impacts on Regulated Water Resources Systems. *Water Resour. Manag.* **2015**, *29*, 1933–1947. [CrossRef]
40. Garrote, L.; Iglesias, A.; Granados, A.; Mediero, L.; Martín-Carrasco, F. Quantitative Assessment of Climate Change Vulnerability of Irrigation Demands in Mediterranean Europe. *Water Resour. Manag.* **2014**, *29*, 325–338. [CrossRef]
41. Bejarano, M.D.; Granados, I.; Iglesias, A.; Garrote, L. Blue Water in Europe: Estimates of Current and Future Availability and Analysis of Uncertainty. *Water* **2019**, *11*, 420. [CrossRef]
42. Van Beek, L.P.H.; Bierkens, M.F.P. *The Global Hydrological Model PCR-GLOBWB: Conceptualization, Parameterization and Verification*; Department of Physical Geography, Utrecht University: Utrecht, The Netherlands, 2008; Available online: http://vanbeek.geo.uu.nl/suppinfo/vanbeekbierkens2009.pdf (accessed on 27 October 2020).
43. Warszawski, L.; Frieler, K.; Huber, V.; Piontek, F.; Serdeczny, O.; Schewe, J. The Inter-Sectoral Impact Model Intercomparison Project (ISI–MIP): Project framework. *Proc. Natl. Acad. Sci. USA* **2014**, *111*, 3228–3232. [CrossRef]
44. Fekete, B.M.; Vörösmarty, C.J.; Grabs, W. High-resolution fields of global runoff combining observed river discharge and simulated water balances. *Glob. Biogeochem. Cycles* **2002**, *16*, 15-1–15-10. [CrossRef]
45. Tan, Y.; Guzman, S.M.; Dong, Z.; Tan, L. Selection of Effective GCM Bias Correction Methods and Evaluation of Hydrological Response under Future Climate Scenarios. *Climate* **2020**, *8*, 108. [CrossRef]
46. Teutschbein, C.; Seibert, J. Bias correction of regional climate model simulations for hydrological climate-change impact studies: Review and evaluation of different methods. *J. Hydrol.* **2012**, 12–29. [CrossRef]
47. Sordo-Ward, A.; Bejarano, M.D.; Granados, I.; Garrote, L. Facing Future Water Scarcity in the Duero-Douro Basin: Comparative Effect of Policy Measures on Irrigation Water Availability. *J. Water Resour. Plan. Manag.* **2020**, *146*, 04020011. [CrossRef]
48. Estrela, T.; Marcuello, C.; Dimas, M. Las Aguas Continentales en los Países Mediterráneos de la Unión Europea. CEDEX Report. 2000. Available online: http://hispagua.cedex.es/sites/default/files/aguas_continentales_union_europea.pdf (accessed on 22 December 2020).

Article

Evaluating Water Resource Accessibility in Southwest China

Tao Li [1,2], **Sha Qiu** [1,2], **Shuxin Mao** [3], **Rui Bao** [1,2] and **Hongbing Deng** [1,*]

[1] State Key Laboratory of Urban and Regional Ecology, Research Center for Eco-Environmental Sciences, Chinese Academy of Sciences, Beijing 100085, China

[2] University of Chinese Academy of Sciences, Beijing 100049, China

[3] School of Economics and Management, Shanxi University of Science & Technology, Xi'an 710021, China

* Correspondence: denghb@rcees.ac.cn; Tel.: +86-010-6284-9112

Received: 18 July 2019; Accepted: 14 August 2019; Published: 16 August 2019

Abstract: The accessibility, quantity, and quality of water resources are the basic requirements for guaranteeing water resource security. Research into regional water resource accessibility will contribute to improving regional water resource security and effective water resource management. In this study, we used a water resource accessibility index model considering five spatial factors to evaluate the grid-scale water resource accessibility and constructed the spatial pattern of water resource accessibility in Southwest China. Then, we analyzed the coupling coordination degree between county-level water resource accessibility and eco-socio-economic water demand elements. The water resource accessibility showed obvious regional differences, and the overall trend gradually decreased from Southeast to Northwest. The coupling coordination degree between county-level water resource accessibility and eco-socio-economic water demand elements was between 0.26 and 0.84, and was relatively low overall, whereas the counties (districts) with high coordination, moderate coordination, low coordination, reluctant coordination, and incoordination accounted for 0.92%, 5.31%, 21.06%, 59.71%, and 13.00% of total counties (districts), respectively. Therefore, the Southwest region needs to further strengthen the construction of its agricultural irrigation facilities, protect the water resources, and coordinate the relationship between water resource management and water demand elements to comprehensively guarantee regional sustainable development.

Keywords: water resource accessibility; spatial pattern; coupling coordination degree; water resource management; Southwest China

---

1. Introduction

Water resources are essential for maintaining the sustainable development of eco-socio-economic systems [1,2]. However, due to climate change, economic growth, population increase, and improper water resource management, many environmental problems, such as serious water pollution, the deterioration of the water environment, and an increased contradiction between water supply and demand, have become increasingly severe, creating strategic problems worldwide [3–7]. The United Nations (UN) estimates that more than one-third of the population on the planet will face a freshwater crisis by 2030 [8], and residents in regions with relatively abundant water resources will still have to spend time and energy to obtain water resources, which not only increases their living costs but also impacts health [9,10]. Therefore, whether people living in different regions can easily and fairly obtain sufficient water resources to meet their water demand has become a hot topic for academics and policymakers.

Accessibility is a broadly accepted concept in various scientific fields such as urban planning, transportation planning, and geography [11], and was initially used to measure potential interaction opportunities in transportation networks [12]. As research progressed, accessibility was defined as the difficulty in reaching a destination from a given location [13]. There are two main manifestations of accessibility: The number of opportunities or benefits that can be obtained within a given time or distance, and the amount of spatial resistance that needs to be overcome to reach a destination [14,15]. Therefore, the measurement of accessibility depends primarily on the spatial distribution of potential destinations and the spatial resistance that needs to be overcome to reach each destination [16,17]. The former is mainly measured by quantity or quality, reflecting the attractiveness of the destination, and the latter is mainly measured by indicators such as time, distance, or cumulative cost, reflecting the convenience of reaching the destination [17,18].

The quantity, quality, and accessibility of water resources are the basic requirements for ensuring water resource security [19,20]. However, the existing water resource evaluations mostly focus on the assessment of water quality and water quantity [21–25], but less attention has been paid to water resource accessibility [10,26–29]. Water resource accessibility refers to the difficulty of obtaining water resources from water sources [30], which is the fundamental factor determining the quantity, quality, and the efficiency of the water supply. Water resource accessibility is essential for human well-being, economic development, and ecological maintenance [31] and includes both spatial accessibility and time accessibility [10,28], often measured by indicators such as water intake distance (Euclidean distance, cost distance, path distance, etc.) and water collection time (shortest distance time, shortest path time, self-report time, etc.) [32–34]. However, these indicators ignore the impact of water quantity and various spatial resistance factors. Therefore, some development space remains within the existing quantitative research for the examination of water resource accessibility, and the quantitative methods need to be further improved.

Southwest China is the source and upstream of many rivers, and is also an important ecological barrier zone. It plays a key role in maintaining the ecological and socio-economic security of East China, South China, and even Southeast Asia [35]. The region has abundant rainfall and a large amount of water resources. However, due to the uneven water distribution in time and space, coupled with the limited infrastructure and the influence of complex topography, the use of water resources is difficult and costly, resulting in serious seasonal, regional, and engineering water shortages [36]. With the rapid population and economy growth, the demand for water resources in Southwest China continues to increase, and the misalignment between supply and demand is becoming increasingly acute. Therefore, we used a water resource accessibility index model considering five factors—runoff, slope, relative height difference, water intake distance and land use resistance—to evaluate the grid-scale water resource accessibility in Southwest China, using the ArcGIS platform (Environmental Systems Research Institute, Redlands, California, America) to construct the spatial pattern of water resource accessibility. Then, we analyzed the coupling coordination degree between water resource accessibility and eco-socio-economic water demand elements. The aims are to improve the water resource accessibility evaluation method, identify the areas with relatively low water resource accessibility and the key regions that the coupling coordination degree between water resource accessibility and eco-socio-economic water demand elements is relatively low, which is important for improving the determination of regional water resource security levels, strengthening regional water resource management allocation, and effectively implementing water conservancy facilities planning and urban development planning.

2. Materials and Methods

2.1. Study Area

The study area is located in Southwest China, including Chongqing Municipality, Sichuan Province, Guizhou Province, Yunnan Province, and the Guangxi Zhuang Autonomous Region, between 97°21′–112°3′ E and 20°53′–34°18′ N and covers a total area of 1.362 million km^2 (Figure 1). Southwest China is one of the three karst-concentrated contiguous areas in the world; the terrain in the area is complex and diverse, and the landforms are mainly plateaus and mountains in which basins and hills are widely distributed. The altitude difference is large, and the average elevation is as high as 1700 m. At the end of 2015, the region had a total resident population of 242.89 million, with an urbanization rate of 47.5%, the gross domestic product (GDP) was 8669.52 billion yuan, and a farmland irrigation area was 7.86 million ha [37–41].

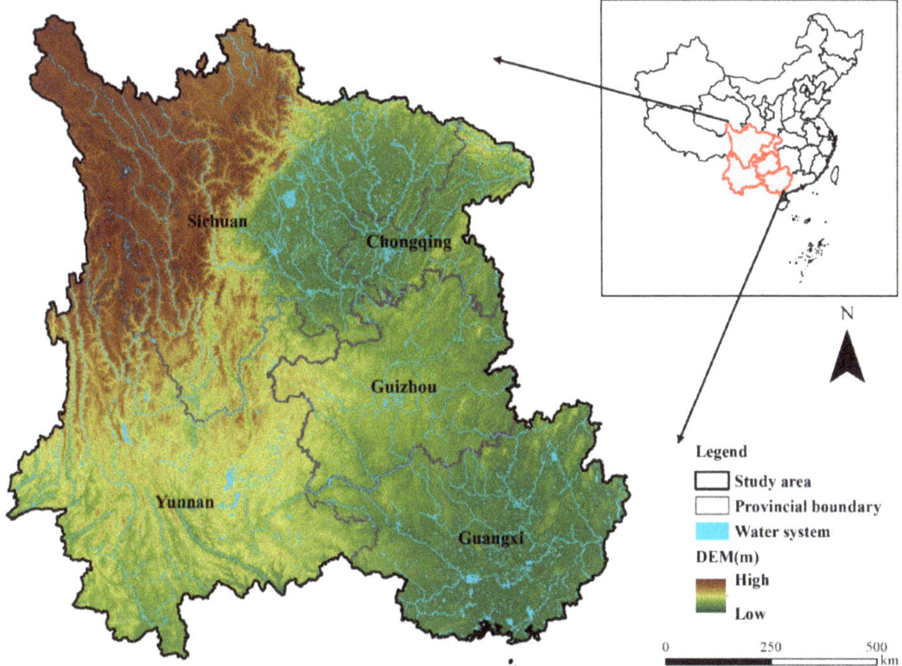

Figure 1. Location of study area in Southwest China (Note: The digital elevation model is abbreviated as DEM).

Southwest China is located in a tropical and subtropical humid region and is dominated by tropical and subtropical monsoon climates, with sufficient heat and abundant rainfall; however, rainfall is unevenly distributed in space and time. The region has developed water systems, including the Yangtze River, Yellow River, Irrawaddy River, Nujiang River, Lancang River, Yuanjiang River and Pearl River, which are the most important water resource enrichment areas in China [32]. In 2015, the average precipitation in the whole study area was 1211 mm, the total water resources amounted to 813.59 billion m^3, and the per capita water resource was 3350 m^3. The annual water supply was 89.13 billion m^3; and the agricultural, industrial, domestic, and ecological water consumptions were 54.31 billion m^3, 19.19 billion m^3, 14.48 billion m^3, and 1.15 billion m^3, respectively; but the water resource development use rate was only 11.0% [42–46], which is related to the difficulty in using water resources in the region [36].

2.2. Data Sources

Four main types of data sources were used in this study. Administrative boundary vector data, water system vector data, and digital elevation model (30 × 30 m) were downloaded from Institute of Remote Sensing and Digital Earth, Chinese Academy of Sciences [47]. The land use type (30 × 30 m) and the normalized vegetation index (1000 × 1000 m) were derived from Resource and Environment Data Cloud Platform, Chinese Academy of Sciences [48]. The meteorological and hydrological data were sourced from National Meteorological Information Center [49], in which the runoff coefficient was derived from the water resource bulletin of each province (municipality and autonomous region) [42–46]. Socioeconomic data (GDP, population, food production) were derived from the statistical yearbooks of provinces (municipality and autonomous region) [37–41]. The spatial data coordinate system had a unified projection of WGS_1984_Albers, and the resolution after data resampling was 90 × 90 m.

2.3. Methods

2.3.1. Basic Theory and Hypothesis

Water resource accessibility is closely related to a series of natural and human factors. These natural factors mainly include the water source, distance, relative height difference, slope, and land use type, whereas human factors mainly include funds (income, water fee), infrastructure (water supply pipelines, waterworks), and technology (irrigation technology, water treatment technology). In this study, rivers, lakes, and reservoirs were selected as the main water sources. We assumed that water users collect water from the nearest water source, and the water resource accessibility is mainly affected by natural factors such as runoff, water intake distance, relative height difference, slope, and land use type, and ignoring human factors. Considering the amount of water is still a problem in Southwest China, and incorporating water quality elements would increase the calculation and interpretation complexity, the water quality was not measured.

Different spatial resistance factors have significant impacts on the amount of accessible water and the difficulty in obtaining water. The runoff is positively correlated with the water resource accessibility: The greater the runoff, the greater the available water, and the higher the water resource accessibility. The other factors (slope, water intake distance, relative height difference, and land use resistance) are negatively correlated with the water resource accessibility: The larger the factors, the greater the water intake difficulty, and the lower the water resource accessibility. The calculation methods of each spatial element index in this study are as follows: (1) Runoff. We first selected the rainfall data of the main meteorological stations in the study area to interpolate the rainfall. Then, we calculated the runoff by combining the spatial distribution of the average runoff coefficient of each administrative unit and the rainfall. (2) Water intake distance. We used the Euclidean tool in ArcGIS software (Environmental Systems Research Institute, Redlands, California, America) to calculate the distance from each grid unit to the water source. (3) Slope. We used the Slope tool in ArcGIS software (Environmental Systems Research Institute, Redlands, California, America) to calculate the slope of each grid unit. (4) Relative height difference. We first used the Mask tool in the ArcGIS software (Environmental Systems Research Institute, Redlands, California, America) to extract the elevation of the water system in the study area. Then, we used the Euclidean Allocation tool to assign the water system elevation to the nearest grid cell. Finally, the Raster Calculator tool was used to calculate the relative height difference between the elevation value of each grid cell and the water system elevation. The spatial factor resistance was referenced from the literature [50] (Table 1).

Different eco-socio-economic factors have different spatial impacts on water demand. Four factors at the county level—normalized difference vegetation index (NDVI), per capita GDP, population density, and grain yield per unit area—were selected to reflect the spatial characteristics of water demand, which were used to characterize ecological water demand, industrial water demand, domestic water demand, and agricultural water demand, respectively. Among them, the NDVI was obtained by mask extraction and zonal statistics of national data, per capita GDP, population density, and grain yield per unit area were calculated based on statistical data. All data for each indicator in this study were converted to dimensionless using the maximum difference normalization method before calculation.

Table 1. Resistance classification and assignment of different spatial elements.

Spatial Element	Grade	Resistance Value
	5050 to 0	1
	0–287	3
Relative height difference (m)	287–746	5
	746–1420	7
	1420–7122	9
	−90 to 0 [1]	1
	0–5	3
Slope (°)	5–15	5
	15–25	7
	25–90	9
	Rivers, lakes, reservoirs	1
	Transportation land, grassland, green land, farmland	3
Land use type	Woodlands, garden	5
	Residential land, industrial and mining land	7
	Swamp, glaciers, bare land	9

[1] When the relative height difference is negative, the slope value is also negative, and the spatial resistance is smaller.

2.3.2. Water Resource Accessibility Index Model

We selected the runoff in each grid unit to represent the attractiveness of a water source to water users, and selected the cost distance constrained by the three resistance factors of slope, relative height difference, and land use to reflect the spatial resistance. Among them, runoff is positively correlated with water resource accessibility: The greater the runoff, the higher the accessibility. The cost distance is negatively correlated with the water resource accessibility: The larger the cost distance, the greater the accessibility. The cost distance can be determined using the cost distance model in ArcGIS software (Environmental Systems Research Institute, Redlands, California, America), which requires the input of two raster layers: The target layer and the resistance layer. In this study, the target layer was the water source (water system), whereas the resistance layer was the resistance matrix of three spatial factors: Slope, relative height difference, and land use type. According to the research [18,51], the water resource accessibility index model is as follows:

$$A_i = W_j \times fmin\left(\sum_{j=i,m}^{i=1,n} D_{ij} \times R_i\right) \tag{1}$$

where A_i refers to the water resource accessibility index, W_j refers to the water source attraction capacity (runoff), f is a positive correlation function that reflects the relationship between the minimum cumulative resistance and the spatial resistance from the water users to the water source, and D_{ij} and R_i refer to the distance and space resistance from the water users to the water source, respectively.

2.3.3. Coupling Coordination Degree Model

Coupling refers to the phenomenon by which two or more systems interact with each other to achieve synergy, and the coupling coordination degree refers to the degree of coordinated development between two or more systems [52]. Water resource accessibility and eco-socio-economic water demand elements are two closely related systems that restrict and promote each other. Thus, we used a coupling coordination degree model to express the degree of coordinated development between the two systems. The equations are as follows [53–55]:

$$C = 2\sqrt{f(a) \times g(b)/[f(a) + g(b)]} \tag{2}$$

$$T = \alpha f(a) + \beta g(b) \tag{3}$$

$$D = \sqrt{C \times T} \tag{4}$$

where C refers to the coupling degree, with a value in the interval [0, 1]; $f(a)$ refers to the water resource accessibility; $g(b)$ refers to eco-socio-economic water demand elements; D refers to the coupling coordination degree, with a value in the interval [0, 1] where the greater the D value, the higher the coupling coordination degree of the two systems, and vice versa; T refers to the comprehensive coordination index; α and β refer to the contribution of water resource accessibility and eco-socio-economic water demand elements to the coupling coordination degree, respectively. According to related research [53–55], we selected $\alpha = \beta = 0.5$ and divided the coupling coordination degree into 10 stages (Table 2).

Table 2. Classification standard and the types of coupling coordination degree (D).

Category	D Value	Subclass
Coordination category	0.9–1	Extreme coordination
	0.8–0.9	High coordination
	0.7–0.8	Moderate coordination
	0.6–0.7	Low coordination
Transition category	0.5–0.6	Reluctant coordination
	0.4–0.5	Near incoordination
Incoordination category	0.3–0.4	Slight incoordination
	0.2–0.3	Moderate incoordination
	0.1–0.2	High incoordination
	0–0.1	Extreme incoordination

3. Results

3.1. Spatial Pattern of Water Resource Accessibility

The spatial distribution characteristics of five factors—relative height difference, slope, land use resistance, water intake distance, and runoff—were analyzed. The relative height difference varies obviously. Extremely high mountains, such as Minshan, Nushan, and Hengduan Mountains, are concentrated in West Sichuan (in some areas, due to the existence of plateau lakes, the relative height difference is a large negative value) and West Yunnan, and the relative height difference is large, whereas the relative height differences in the Sichuan Basin, Guangxi, and Guizhou are relatively small (Figure 2a). The steep slope areas in the study area are relatively large, mainly distributed in West Sichuan and Northwest Yunnan, whereas the Sichuan Basin, Southwest Guangxi, and East Yunnan have relatively flat terrain with relatively low spatial resistance (Figure 2b). The concentrated distribution of glaciers and marshes in West Sichuan leads to a relatively high resistance value of land use in the region. In the Sichuan Basin and Northwest Sichuan, grassland, green land, and farmland are widely distributed, so the spatial resistance is relatively low (Figure 2c). The water intake distance is closely related to the spatial distribution of the water system. The water systems in Southwest Yunnan, North-Central Sichuan, and Guizhou are sparse, and the water intake distance is relatively large (Figure 2d). Due to the differences in the precipitation, temperature, and underlying surface of the watershed, and the influence of human activities, the regional differences in runoff are significant. The runoff is relatively high in East Guangxi and relatively low in the Sichuan Basin, West Sichuan, and Central and North Yunnan, whereas the overall trend is decreasing from the Southeast to the Northwest (Figure 2e).

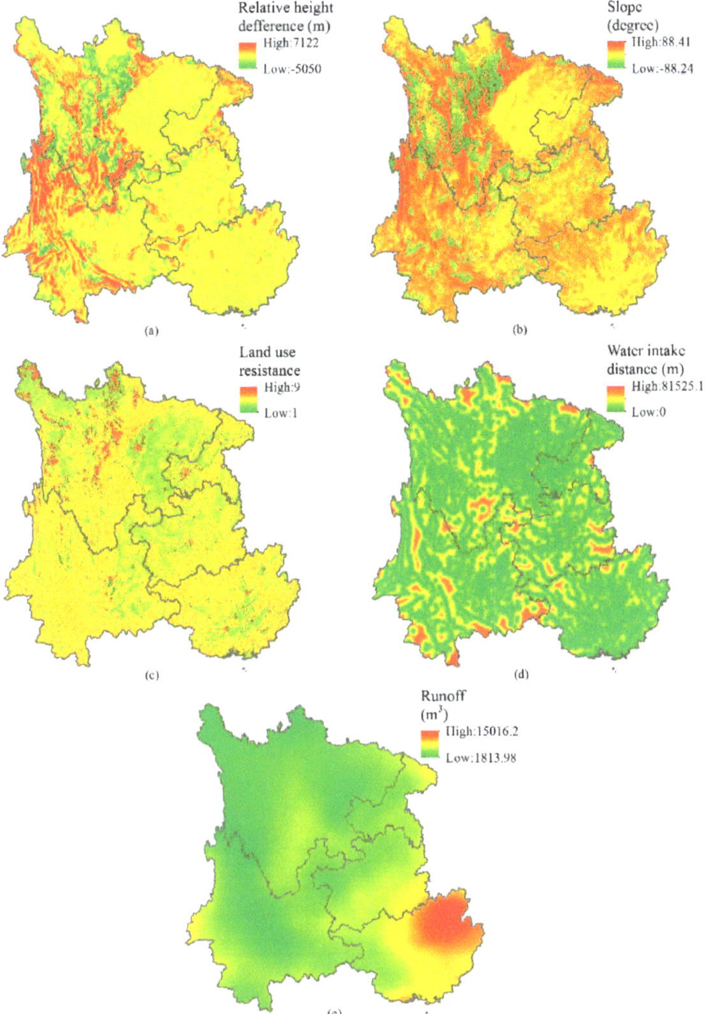

Figure 2. Spatial distribution of different factors affecting water resource accessibility: (**a**) Relative height difference, (**b**) slope, (**c**) land use, (**d**) water intake distance, and (**e**) runoff.

The grid-scale water resource accessibility has obvious regional differences, and the overall trend gradually decreases from Southeast to Northwest (Figure 3). The high-value area is mainly concentrated in Northeast Guangxi, whereas the low-value areas are mainly concentrated in West Sichuan and North-Central Yunnan, which is closely related to the spatial distributions of the water system, slope, elevation, runoff, and land use in Southwest China. In the Southeast, especially the Guangxi Zhuang Autonomous Region, the geomorphological type is a mountainous and hilly basin, and the terrain is relatively flat, whereas spatial resistance, such as the slope and relative height difference, is relatively low. The water system in the region is well-developed and the runoff is relatively high, with fewer constraints on access to water resources, so the water resource accessibility is relatively high. However, in the Northwest, especially in West Sichuan and Northwest Yunnan, the wide distribution of extremely high mountains, glaciers, and swamps results in a significant elevation difference, a large slope and land use resistance in the region, coupled with the relatively sparse water system and fewer water resources, which considerably increase the difficulty in obtaining water resources, resulting in relatively low water resource accessibility.

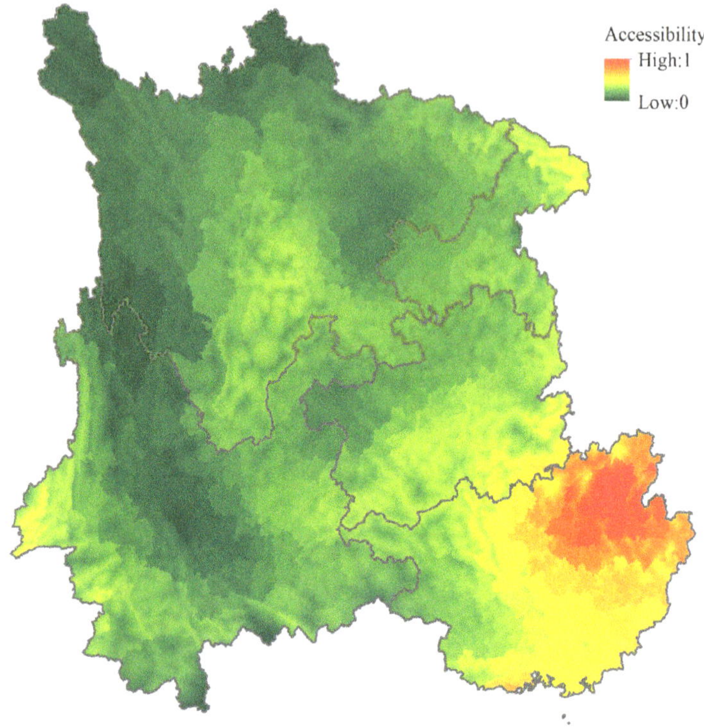

Figure 3. The spatial pattern of the water resource accessibility in Southwest China at the grid-scale.

Taking the county-level administrative district as the statistical unit, the water resource accessibility at the grid cell was calculated, and the spatial pattern of water resource accessibility at the county level was determined (Figure 4). The water resource accessibility varies considerably between different counties (districts) in Southwest China. The counties (districts) with relatively high water resource accessibility are mainly concentrated in Northeast Guangxi, whereas the water resource accessibility in some counties (districts) in the Sichuan Basin, West Sichuan, and Central and North Yunnan is relatively low, and the overall trend is a gradual decrease from Southeast to Northwest. The maximum water resource accessibility value was 0.996 in the Qixing District of Guangxi Zhuang Autonomous Region, and the lowest value was 0.113 in Derong County, Sichuan Province.

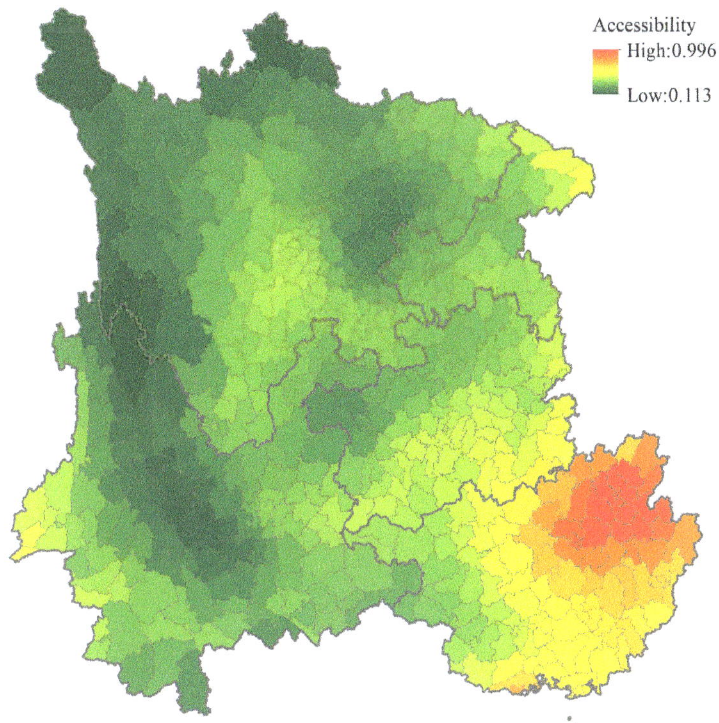

Figure 4. The spatial pattern of water resource accessibility in Southwest China at the county scale.

3.2. Spatial Distribution Characteristics of Different Water Demand Elements

In the study area, the superior hydrological and climatic conditions provide a suitable environment for the growth of vegetation, which increases the vegetation index overall. Only a few counties (districts) in West Sichuan and the urban center areas of each province (municipality and autonomous region) have relatively low NDVI scores (Figure 5a). As a typical ethnic minority settlement in China, Southwest China has a developing economy, and there is a significant difference in per capita GDP between different counties (districts) (Figure 5b). Among them, the per capita GDP in Central Sichuan, West Chongqing, and the urban center area of each province (municipality and autonomous region) is relatively high, and the per capita GDP in West and Northeast Sichuan, East Guizhou, Northwest Guangxi, and most parts of Yunnan is relatively low. The counties (districts) with high population density in the study area are mainly concentrated in the Sichuan Basin and the urban centers of each province (municipality and autonomous region) (Figure 5c). These regions have rapid economic development and a high level of urbanization, providing superior conditions for human survival and development. The low-value areas are mainly distributed in the ethnic minority areas of West Sichuan, West Yunnan, Southeast Guizhou, and Northwest Guangxi. East Sichuan, West Chongqing, Northeast Yunnan, and East Guangxi have flat terrain, superior climate and hydrological conditions, and the grain yield per unit area is relatively high (Figure 5d). In West Sichuan, Northwest Yunnan, and Guizhou, widespread mountainous areas, water shortages, and extensive desertification seriously affect food production.

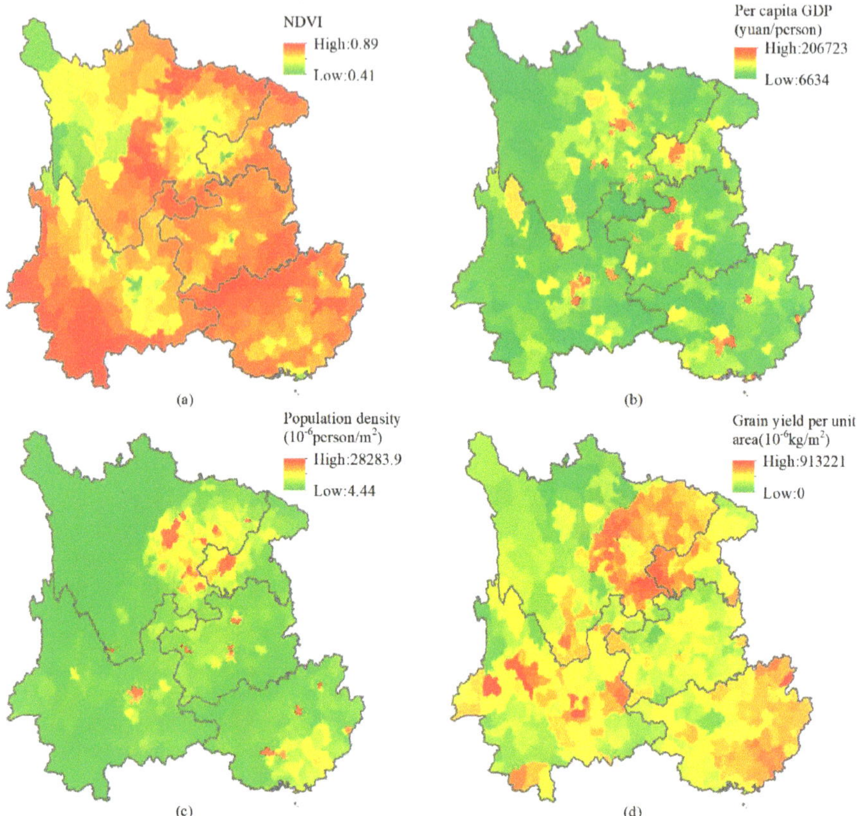

Figure 5. The spatial distribution of different eco-socio-economic water demand elements at the county level: (**a**) Normalized difference vegetation index (NDVI), (**b**) per capita GDP, (**c**) population density, (**d**) grain yield per unit area.

To more accurately reflect the water demand characteristics of each county (district) in Southwest China, the different water demand elements of each county-level administrative unit were weighted and summed according to the proportion of the water resource use structure for each province (municipality and autonomous region) in 2015; then, the spatial distribution of the comprehensive water demand elements in Southwest China was determined (Figure 6). The counties (districts) with high water resource demand are mainly concentrated in the Sichuan Basin, East Guangxi, and Central Yunnan, where the agricultural production level is relatively high, economic development is relatively fast, and the population is relatively concentrated. West Sichuan, Northwest Yunnan, and Guizhou, where agricultural production and the population density are low and economic development is relatively slow, have a lower water demand.

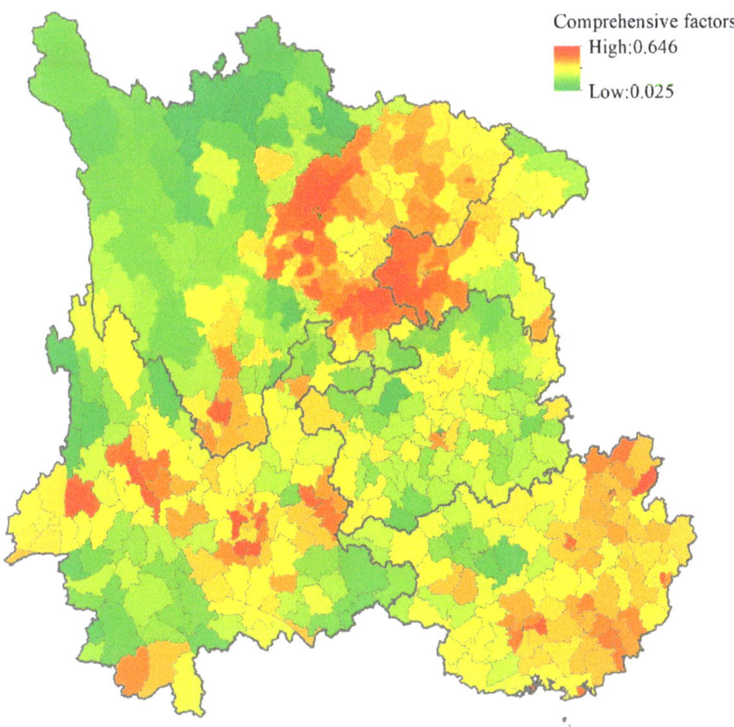

Figure 6. The spatial distribution of comprehensive water demand elements at the county level.

3.3. Coupling Coordination Degree of Water Resource Accessibility and Water Demand Elements

The coupling coordination degree of water resource accessibility and eco-socio-economic water demand elements in Southwest China was between 0.26 and 0.84, showing significant regional differences (Figure 7). The coupling coordination degree in Northeast Guangxi is relatively high overall, whereas the coupling coordination degree in West and North Sichuan and Northwest Yunnan, is relatively low. The overall trend is a decrease from Southeast to Northwest. The highest is Diecai District, Guangxi Zhuang Autonomous Region, and the lowest is Hongyuan County, Sichuan Province. According to the statistical results, the coupling coordination degree of water resource accessibility and eco-socio-economic water demand elements in the study area are mainly distributed in the coordination and transition stages, among which 5 counties (districts) show high coordination, accounting for 0.92%; 29 counties (districts) show moderate coordination, accounting for 5.31%; 115 counties (districts) show low coordination, accounting for 21.06%; 326 counties (districts) show reluctant coordination, accounting for 59.71%; and 68 counties (districts) show near incoordination, accounting for 12.45%; 2 counties show reluctant coordination, accounting for 0.37%; and 1 counties show near incoordination, accounting for 0.18%. The coupling coordination degree of water resource accessibility and eco-socio-economic water demand elements in Southwest China is relatively low overall.

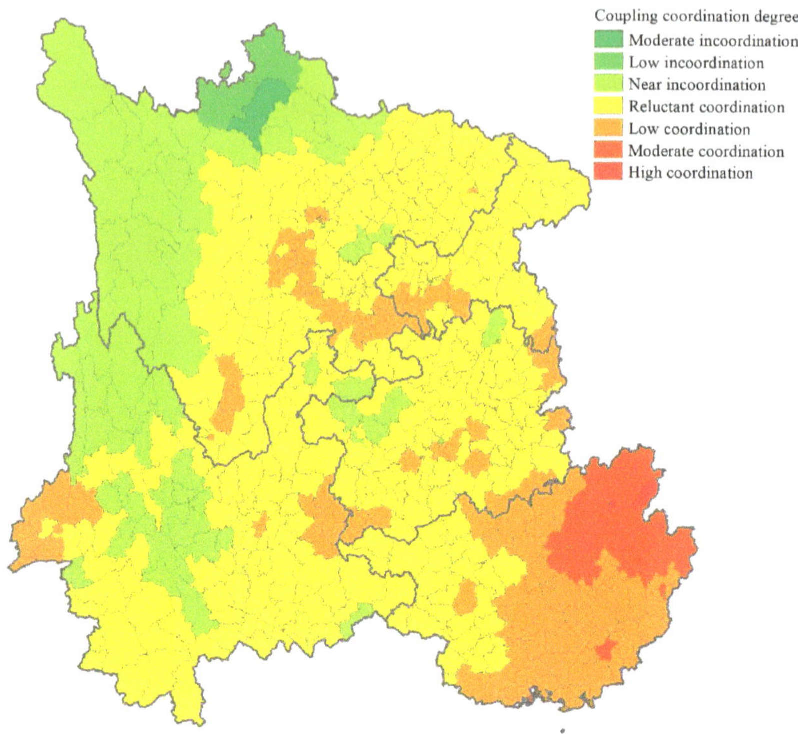

Figure 7. Spatial distribution of coupling coordination degree between water resource accessibility and water demand elements.

The regional difference in the coupling coordination degree between water resource accessibility and eco-socio-economic water demand elements in Southwest China are mainly due to the spatial differences in hydrological conditions, topography, and economic development level. East Guangxi has a relatively flat terrain, superior natural conditions, relatively low spatial resistance, and abundant precipitation, so the water resource accessibility is relatively high, which guarantees good growth and the efficient production of rice and other crops. East Guangxi is also the economic development center of Guangxi Zhuang Autonomous Region, and the population is highly concentrated, resulting in high water resource demand. Therefore, the water resource accessibility and the eco-socio-economic water demand elements show a relatively high coordinated development. The Sichuan Basin is one of the most important grain production bases in China. It has a relatively dense population and rapid economic development. However, due to the small amount of water resources and relatively low water resource accessibility, which lead to limited eco-socio-economic sustainable development to a certain extent, the coupling coordination degree between the two is relatively low. In Northwest Yunnan and West and North Sichuan, the wide distribution of extremely high mountains has caused large altitude differences and steep slopes, coupled with a relatively sparse water system and a low amount of water resources, so water resource accessibility in the region is extremely low, which means that it is difficult to meet the water resource demand of the eco-socio-economic elements. Therefore, the coupling coordination degree between the water resource accessibility and eco-socio-economic water demand elements shows near incoordination.

4. Discussion

4.1. Evaluation Method of Water Resource Accessibility

Water resource accessibility refers to the difficulty in obtaining water resources from water sources and is affected by multiple factors such as water quality, water quantity, distance, elevation, slope, land use, capital, infrastructure, and technology. Among them, the quantity and quality of water resources are decisive factors for the availability of water resources, whereas the distance, altitude, slope, land use, and others are factors affecting the convenience of obtaining water resources. The existing quantitative analyses of water resource accessibility often involved a single indicator or a few factors. For example, Jeff et al. [9] only considered the linear distance to the water source. Smiley [56] selected the four elements of water quality, water cost, water reliability, and water intake burden, and measured water resource accessibility through questionnaires and statistical analysis. Yu et al. [28] selected four factors including the slope, relative height difference, distance, and runoff to comprehensively analyze the accessibility of river water resources in the Hanjiang River Basin. Li et al. [29] constructed a grid-scale water accessibility evaluation model based on the length, runoff and viewshed value. Li et al. [57] constructed a water accessibility index by selecting indicators such as distance, altitude, ditch density, road density, and culvert number to study the water resource accessibility of freshwater wetland. In this study, we evaluated water resource accessibility by considering the five factors of runoff, slope, relative height difference, water intake distance, and land use resistance, and analyzed the spatial pattern of the water resource accessibility in different grid units. Among them, water quantity represents the attractiveness of the water source to the water users, and the other factors reflect the spatial resistance.

However, the water resource accessibility evaluation in this study still has some room for improvement. First, our evaluation only considered the impact of water quantity and ignored the water quality. Thus, in future research, the quality of water resources should be measured in terms of water quality requirements for different water demand elements. Second, we assumed that water users obtain water from the closest water source, but in practice, multiple water sources provide water for users. Therefore, it is necessary to weight the multiple water sources within a certain range in future research. Third, we ignored the impact of socio-economic factors, such as water supply facilities, water treatment technology, irrigation technology, and water fees, on water resource accessibility, which affects the accuracy of the evaluation results to a certain extent; thus, future research needs to comprehensively measure multiple factors.

4.2. Water Resource Accessibility and Regional Eco-Socio-Economic Development

Water resources are an important basis for supporting the development of eco-socio-economic systems, whereas social and economic development provides the necessary funds and conditions for ensuring the sustainable development and use of water resources [58], which affect and restrict each other. The limited water resource accessibility of spatial units not only threatens the supply of drinking water and irrigation water, but also threatens the sustainable and healthy development of the ecosystem [20,59,60]. Therefore, spatially accessible water resources are essential for an adequate freshwater supply [28].

In Southwest China, the topography and geomorphology are particularly complex, the spatial-temporal distribution of water resources is uneven, and the water supply facilities lack expansion and improvement potential, resulting in serious seasonal, regional, and engineering water shortages. The use of water resources is difficult and costly, which seriously restricts regionally sustainable eco-socio-economic development. Therefore, water resource management in Southwest China must fully consider the characteristics and formation of water resources in the region and should adopt different water resource development and use models. The analysis of the coupling coordination degree between water resource accessibility and eco-socio-economic water demand elements can be used to effectively identify the areas where the coupling coordination degree between the two is relatively low. This can provide a decision-making basis for strengthening regional water resource management and allocation, for effectively implementing water conservancy facilities planning, and for urban development planning, thus ensuring the coordinated and sustainable development of various systems.

In West Sichuan and Northwest Yunnan, there are many extremely high mountains, resulting in significant altitude differences and steep slopes, and the water sources are far away from water users, so the water resource accessibility is low and agricultural irrigation is difficult. Therefore, these areas require more investment to improve irrigation conditions and increase irrigation efficiency to meet crop water requirements. In the Sichuan Basin, Central Yunnan, Northwest Guizhou, and Southeast Guangxi, the population density is relatively high; it is necessary to continuously strengthen water resource protection and infrastructure construction in densely populated areas to ensure a safe and adequate supply of drinking water. The economic development level of most counties (districts) in Southwest China is low, somewhat lagging behind the water resource accessibility level, whereas the water resource accessibility in West Sichuan and Northwest Yunnan is relatively low, which restricts economic development to some extent. Therefore, according to the spatial pattern of water resource accessibility, rationally adjusting the industrial structure and developing a circular economy is necessary, which contribute to promoting rational and rapid economic development and ensuring the coordinated development of water resource accessibility and social and economic elements. In Northwest Yunnan and West and Northeast Sichuan, relatively poor natural conditions and scarce water resources result in relatively low water resource accessibility, which considerably limits vegetation growth. Therefore, these regions should be the focus for ecological conservation and restoration.

5. Conclusions

This paper applied a water resource accessibility index model, considering five spatial factors of runoff, slope, relative height difference, water intake distance and land use resistance, which enabled the quantitative analysis of the spatial distribution characteristics of water resource accessibility on a grid-scale in Southwest China. The results show that due to the large spatial distribution differences of different spatial elements, the spatial differences in water resource accessibility in Southwest China are relatively significant, and the overall trend is a decrease from Southeast to Northwest.

Due to the differences in hydrological conditions, topography, and economic development level, the coupling coordination degree between water resource accessibility and eco-socio-economic water demand elements in Southwest China has obvious regional differences, and the overall distribution characteristics are higher in the Southeast and lower in the Northwest. The proportion of counties (districts) with moderate coordination or higher was only 6.23%, mainly concentrated in the Northeast part of Guangxi. The counties (districts) with near incoordination, low incoordination, and moderate incoordination accounted for 13.00%, mainly concentrated in West Sichuan and Northwest Yunnan. The coupling coordination degree between the two is relatively low overall.

The water resource accessibility and the eco-socio-economic system in Southwest China have not achieved coordinated or sustainable development. The insufficient water resource support capacity in the region has restricted the development of the region to a certain extent, and the rapidly increasing population and economic development have increased water supply stress to a certain extent. Therefore, it is necessary to continuously coordinate the relationship between water resource management and regional development.

Author Contributions: Conceptualization, H.D.; Data curation, T.L., S.Q., S.M. and R.B.; Formal analysis, T.L.; Investigation, T.L., S.Q., S.M. and R.B.; Methodology, T.L., S.Q. and H.D.; Project administration, H.D.; Visualization, T.L.; Writing—original draft, T.L.; Writing—review & editing, T.L., S.Q., S.M., R.B. and H.D.

Funding: This research was funded by the National Key Research and Development Program of China (No. 2016YFC0502106).

Acknowledgments: We thank Yuebo Su for help and advice in data processing, article writing and modification.

Conflicts of Interest: The authors declare no conflict of interest.

References

1. Liquete, C.; Maes, J.; La Notte, A.; Bidoglio, G. Securing water as a resource for society: An ecosystem services perspective. *Ecohydrol. Hydrobiol.* **2011**, *11*, 247–259. [CrossRef]
2. Zhang, J.; Chen, G.; Xing, S.; Shan, Q.; Wang, Y.; Li, Z. Water shortages and countermeasures for sustainable utilisation in the context of climate change in the Yellow River Delta region, China. *Int. J. Sustain. Dev. World Ecol.* **2011**, *18*, 177–185. [CrossRef]
3. Vorosmarty, C.J.; McIntyre, P.B.; Gessner, M.O.; Dudgeon, D.; Prusevich, A.; Green, P.; Glidden, S.; Bunn, S.E.; Sullivan, C.A.; Liermann, C.R.; et al. Global threats to human water security and river biodiversity. *Nature* **2010**, *467*, 555–561. [CrossRef] [PubMed]
4. Lopez-Moreno, J.I.; Zabalza, J.; Vicente-Serrano, S.M.; Revuelto, J.; Gilaberte, M.; Azorin-Molina, C.; Moran-Tejeda, E.; Garcia-Ruiz, J.M.; Tague, C. Impact of climate and land use change on water availability and reservoir management: Scenarios in the Upper Aragon River, Spanish Pyrenees. *Sci. Total Environ.* **2014**, *493*, 1222–1231. [CrossRef] [PubMed]
5. Martinez-Santos, P.; Cervan, J.A.; Cano, B.; Diaz-Alcaide, S. Water versus Wireless Coverage in Rural Mali: Links and Paradoxes. *Water* **2017**, *9*, 375. [CrossRef]
6. Mokssit, A.; de Gouvello, B.; Chazerain, A.; Figuères, F.; Tassin, B. Building a Methodology for Assessing Service Quality under Intermittent Domestic Water Supply. *Water* **2018**, *10*, 1164. [CrossRef]
7. Brown, C.M.; Lund, J.R.; Cai, X.M.; Reed, P.M.; Zagona, E.A.; Ostfeld, A.; Hall, J.; Characklis, G.W.; Yu, W.; Brekke, L. The future of water resources systems analysis: Toward a scientific framework for sustainable water management. *Water Resour. Res.* **2015**, *51*, 6110–6124. [CrossRef]
8. Yang, Z.; Zhao, Y.; Cui, B.; Hu, T. Ecocity-oriented water resources supply-demand balance analysis. *China Environ. Sci.* **2004**, *24*, 636–640. (In Chinese) [CrossRef]
9. Ho, J.C.; Russel, K.C.; Davis, J. The challenge of global water access monitoring: Evaluating straight-line distance versus self-reported travel time among rural households in Mozambique. *J. Water Health* **2014**, *12*, 173–183. [CrossRef]
10. Purshouse, H.; Roxburgh, N.; Javorszky, M.; Sleigh, A.; Kimani, D.; Evans, B. Effects of water source accessibility and reliability improvements on water consumption in eastern Nairobi. *Waterlines* **2017**, *36*, 204–215. [CrossRef]
11. Hu, Y.; Domns, J. Measuring and visualizing place-based space-time job accessibility. *J. Transp. Geogr.* **2019**, *74*, 278–288. [CrossRef]
12. Hansen, W.G. How Accessibility Shapes Land Use. *J. Am. Inst. Plan.* **1959**, *25*, 73–76. [CrossRef]
13. Ala-Hulkko, T.; Kotavaara, O.; Alahuhta, J.; Helle, P.; Hjort, J. Introducing accessibility analysis in mapping cultural ecosystem services. *Ecol. Indic.* **2016**, *66*, 416–427. [CrossRef]
14. Cascetta, E.; Cartenì, A.; Montanino, M. A behavioral model of accessibility based on the number of available opportunities. *J. Transp. Geogr.* **2016**, *51*, 45–58. [CrossRef]
15. Chen, Y.; Ravulaparthy, S.; Deutsch, K.; Dalal, P.; Yoon, S.Y.; Lei, T.; Goulias, K.G.; Pendyala, R.M.; Bhat, C.R.; Hu, H.-H. Development of Indicators of Opportunity-Based Accessibility. *Transp. Res. Rec. J. Transp. Res. Board* **2011**, *2255*, 58–68. [CrossRef]
16. Páez, A.; Scott, D.M.; Morency, C. Measuring accessibility: Positive and normative implementations of various accessibility indicators. *J. Transp. Geogr.* **2012**, *25*, 141–153. [CrossRef]
17. Kwan, M.-P. Space-Time and Integral Measures of Individual Accessibility: A Comparative Analysis Using a Point-Based Framework. *Geogr. Anal.* **1998**, *30*, 191–216. [CrossRef]
18. Wang, Y.; Chen, B.; Yuan, H.; Wang, D.; Lam, W.K.H.; Li, Q. Measuring temporal variation of location-based ccessibility using spacetimeutility perspective. *J. Transp. Geogr.* **2018**, *73*, 13–24. [CrossRef]
19. Nastiti, A.; Sudradjat, A.; Geerling, G.W.; Smits, A.J.M.; Roosmini, D.; Muntalif, B.S. The effect of physical accessibility and service level of water supply on economic accessibility: A case study of Bandung City, Indonesia. *Water Int.* **2017**, *42*, 831–851. [CrossRef]
20. Biggs, E.M.; Duncan, J.M.A.; Atkinson, P.M.; Dash, J. Plenty of water, not enough strategy: How inadequate accessibility, poor governance and a volatile government can tip the balance against ensuring water security: The case of Nepal. *Environ. Sci. Policy* **2013**, *33*, 388–394. [CrossRef]

21. Manandhar, S.; Pandey, V.P.; Kazama, F. Application of Water Poverty Index (WPI) in Nepalese Context: A Case Study of Kali Gandaki River Basin (KGRB). *Water Resour. Manag.* **2011**, *26*, 89–107. [CrossRef]

22. Dong, G.; Shen, J.; Jia, Y.; Sun, F. Comprehensive Evaluation of Water Resource Security: Case Study from Luoyang City, China. *Water* **2018**, *10*, 1106. [CrossRef]

23. Yan, Y.; Qian, Y.; Wang, Z.Y.; Yang, X.Y.; Wang, H.W. Ecological risk assessment from the viewpoint of surface water pollution in Xiamen City, China. *Int. J. Sustain. Dev. World Ecol.* **2018**, *25*, 403–410. [CrossRef]

24. Cai, J.; Varis, O.; Yin, H. China's water resources vulnerability: A spatio-temporal analysis during 2003–2013. *J. Clean. Prod.* **2017**, *142*, 2901–2910. [CrossRef]

25. Rodrigues, D.B.B.; Gupta, H.V.; Mendiondo, E.M. A blue/green water-based accounting framework for assessment of water security. *Water Resour. Res.* **2014**, *50*, 7187–7205. [CrossRef]

26. Lester, S.; Rhiney, K. Going beyond basic access to improved water sources: Towards deriving a water accessibility index. *Habitat Int.* **2018**, *73*, 129–140. [CrossRef]

27. Kohli, A.; Komisar, S.J.; Montenegr, C.E. Maximizing domestic water accessibility: A statistical model. *Desalination* **2009**, *248*, 530–536. [CrossRef]

28. Yu, G.; Chen, X.; Tu, Z.; Yu, Q.; Liu, Y.A.; Yu, H. Modeling Water Accessibility of Natural River Networks Using the Fine-Grained Physical Watershed Characteristics at the Grid Scale. *Water Resour. Manag.* **2017**, *31*, 2271–2284. [CrossRef]

29. Li, F.; Liu, H.; Chen, X.; Yu, D. Trivariate Copula Based Evaluation Model of Water Accessibility. *Water Resour. Manag.* **2019**, *33*, 3211–3225. [CrossRef]

30. Crow, B.; Sultana, F. Gender, Class, and Access to Water: Three Cases in a Poor and Crowded Delta. *Soc. Nat. Resour.* **2002**, *15*, 709–724. [CrossRef]

31. Gedo, H.W.; Morshed, M.M. Inadequate accessibility as a cause of water inadequacy: A case study of Mpeketoni, Lamu, Kenya. *Water Policy* **2013**, *15*, 598–609. [CrossRef]

32. Smiley, S.L. Complexities of water access in Dar es Salaam, Tanzania. *Appl. Geogr.* **2013**, *41*, 132–138. [CrossRef]

33. Sorenson, S.B.; Morssink, C.; Campos, P.A. Safe access to safe water in low income countries: Water fetching in current times. *Soc. Sci. Med.* **2011**, *72*, 1522–1526. [CrossRef] [PubMed]

34. Li, S. A preliminary study on spatial accessibility of Nanchang urban rivers based on nearest distance analysis. *Jiangxi Hydraul. Sci. Technol.* **2017**, *43*, 254–257. (In Chinese) [CrossRef]

35. Liu, G. Formation and evolution mechanism of ecological security pattern in Southwest China. *Acta Ecol. Sin.* **2016**, *36*, 7088–7091. (In Chinese) [CrossRef]

36. Wu, C.; Dery, S.; Wu, W.; Liu, X.; Xiong, J.; Gao, W. A review of water resources utilization and protection in Southwest China. *Sci. Cold Arid Reg.* **2015**, *7*, 736–746. [CrossRef]

37. Chongqing Municipal Bureau of Statistics. *Chongqing statistical Yearbook 2016*; China Statistics Press: Beijing, China, 2016. (In Chinese)

38. Guangxi Bureau of Statistics. *Guangxi Statistical Yearbook 2016*; China Statistics Press: Beijing, China, 2016. (In Chinese)

39. Guizhou Bureau of Statistics. *Guizhou Statistical Yearbook 2016*; China Statistics Press: Beijing, China, 2016. (In Chinese)

40. Sichuan Bureau of Statistics. *Sichuan Statistical Yearbook 2016*; China Statistics Press: Beijing, China, 2016. (In Chinese)

41. Yunnan Bureau of Statistics. *Yunnan Statistical Yearbook 2016*; China Statistics Press: Beijing, China, 2016. (In Chinese)

42. Water Resources Department of Chongqing Municipal. *Chongqing Water Resources Bulletin 2015*; Water Resources Department of Chongqing Municipal: Chongqing, China, 2015. (In Chinese)

43. Water Resources Department of Guangxi Zhuang Autonomous Region. *Guangxi Water Resources Bulletin 2015*; Water Resources Department of Guangxi Zhuang Autonomous Region: Nanning, China, 2015. (In Chinese)

44. Water Resources Department of Guizhou Province. *Guizhou Water Resources Bulletin 2015*; Water Resources Department of Guizhou Province: Guiyang, China, 2015. (In Chinese)

45. Water Resources Department of Sichuan Province. *Sichuan Water Resources Bulletin 2015*; Water Resources Department of Sichuan Province: Chengdu, China, 2015. (In Chinese)

46. Water Resources Department of Yunnan Province. *Yunnan Water Resources Bulletin 2015*; Water Resources Department of Yunnan Province: Kunming, China, 2015. (In Chinese)

47. Institute of Remote Sensing and Digital Earth, Chinese Academy of Sciences. Available online: http://eds.ceode.ac.cn/sjglb/dataservice.htm (accessed on 10 August 2018).

48. Resource and Environmental Science Data Cloud Platform, Chinese Academy of Sciences. Available online: http://www.resdc.cn/ (accessed on 15 August 2018).

49. National Meteorological Information Center. Available online: http://data.cma.cn/data/detail/dataCode/A.0029.0001.html (accessed on 15 August 2018).

50. Ye, Y.; Su, Y.; Zhang, H.; Liu, K.; Wu, Q. Construction of an ecological resistance surface model and its application in urban expansion simulations. *J. Geogr. Sci.* **2015**, *25*, 211–224. [CrossRef]

51. Geurs, K.T.; van Wee, B.; Rietveld, P. Accessibility Appraisal of Integrated Land-Use—Transport Strategies: Methodology and Case Study for the Netherlands Randstad Area. *Environ. Plan. B Plan. Des.* **2016**, *33*, 639–660. [CrossRef]

52. Wang, S.; Lu, D.; Wang, Z. Analysis of Spatial Coupling of Urbanization and Ecological Environment in Central Plains Economic Region (CPER). *J. Pingdingshan Univ.* **2018**, *33*, 104–110. (In Chinese)

53. He, J.; Wang, S.; Liu, Y.; Ma, H.; Liu, Q. Examining the relationship between urbanization and the eco-environment using a coupling analysis: Case study of Shanghai, China. *Ecol. Indic.* **2017**, *77*, 185–193. [CrossRef]

54. Li, M.; Mao, C. Spatial-Temporal Variance of Coupling Relationship between Population Modernization and Eco-Environment in Beijing-Tianjin-Hebei. *Sustainability* **2019**, *11*, 991. [CrossRef]

55. Yao, L.; Li, X.; Li, Q.; Wang, J. Temporal and Spatial Changes in Coupling and Coordinating Degree of New Urbanization and Ecological-Environmental Stress in China. *Sustainability* **2019**, *11*, 1171. [CrossRef]

56. Smiley, S.L. Defining and measuring water access: Lessons from Tanzania for moving forward in the post-Millennium Development Goal era. *Afr. Geogr. Rev.* **2016**, *36*, 168–182. [CrossRef]

57. Li, H.; Fan, Y.; Gong, Z.; Zhou, D. Water accessibility assessment of freshwater wetlands in the Yellow River Delta National Nature Reserve, China. *Ecohydrol. Hydrobiol.* **2019**. [CrossRef]

58. Yu, X.; Zhang, L.; Chen, X.; Yang, K.; Huang, Y. Analysis of Coupling and Coordinated Development Between Water Resources and Social Economy in Hubei Province. *Resour. Environ. Yangtze Basin* **2018**, *27*, 809–817. (In Chinese) [CrossRef]

59. Yu, Q.; Tu, Z.; Yu, G.; Xu, L.; Yang, D.; Yang, Y. Modelling the crop water-satisfied degree on the grid scale: A CropWRA model and the case study of Hanjiang River Basin, China. *Agric. For. Meteorol.* **2018**, *262*, 215–226. [CrossRef]

60. Meunier, S.; Manning, D.T.; Quéval, L.; Cherni, J.A.; Dessante, P.; Zimmerle, D. Determinants of the marginal willingness to pay for improved domestic water and irrigation in partially electrified Rwandan villages. *Int. J. Sustain. Dev. World Ecol.* **2019**, 1–13. [CrossRef]

 **water**

Article

Water Resource Carrying Capacity Based on Water Demand Prediction in Chang-Ji Economic Circle

Ge Wang [1,2,3,4], Changlai Xiao [1,2,3,4], Zhiwei Qi [1,2,3,4], Xiujuan Liang [1,2,3,4,*], Fanao Meng [1,2,3,4] and Ying Sun [1,2,3,4]

[1] Key Laboratory of Groundwater Resources and Environment, Jilin University, Ministry of Education, No 2519, Jiefang Road, Changchun 130021, China; wangge18@mails.jlu.edu.cn (G.W.); jluxcl@126.com (C.X.); jluqzw@163.com (Z.Q.); mengfanao2014@163.com (F.M.); 17678318582@163.com (Y.S.)

[2] Jilin Provincial Key Laboratory of Water Resources and Environment, Jilin University, Changchun 130021, China

[3] National-Local Joint Engineering Laboratory of In-Situ Conversion, Drilling and Exploitation Technology for Oil Shale, Changchun 130021, China

[4] College of New Energy and Environment, Jilin University, No 2519, Jiefang Road, Changchun 130021, China

* Correspondence: jlulxj@126.com

Abstract: In view of the large spatial difference in water resources, the water shortage and deterioration of water quality in the Chang-Ji Economic Circle located in northeast China, the water resource carrying capacity (WRCC) from the perspective of time and space is evaluated. We combine the gray correlation analysis and multiple linear regression models to quantitatively predict water supply and demand in different planning years, which provide the basis for quantitative analysis of the WRCC. The selection of research indicators also considers the interaction of social economy, water resources, and water environment. Combined with the fuzzy comprehensive evaluation method, the gray correlation analysis and multiple linear regression models to quantitatively and qualitatively evaluate the WRCC under different social development plans. The developmental trends were obtained from 2017 to 2030 using four plans designed for distinct purposes. It can be seen that the utilization of water resource is unreasonable now and maintains a poor level under a business-as-usual Plan I. Plan II and Plan III show that resource-based water shortage is the most critical issue in this region, and poor water quality cannot be ignored either. Compared with Plan I, the average index of WRCC in Plan IV increased by 51.8% and over 84% of the regions maintain a good level. Strengthening sewage treatment and properly using transit water resources are more conducive to the rapid development of Chang-Ji Economic Circle.

Keywords: fuzzy comprehensive evaluation method; water resource carrying capacity; gray correlation analysis; multiple linear regression models; water environment capacity

Citation: Wang, G.; Xiao, C.; Qi, Z.; Liang, X.; Meng, F.; Sun, Y. Water Resource Carrying Capacity Based on Water Demand Prediction in Chang-Ji Economic Circle. *Water* **2021**, *13*, 16. https://dx.doi.org/10.3390/w13010016

Received: 5 December 2020
Accepted: 21 December 2020
Published: 24 December 2020

Publisher's Note: MDPI stays neutral with regard to jurisdictional claims in published maps and institutional affiliations.

1. Introduction

With the development of the urbanization process, the demand for water resources has increased significantly, but the pollution of water resources has caused serious problems. These changes pose a potential threat to water resource carrying capacity in many regions [1]. The water resources carrying capacity (WRCC) refers to the ability of water resources to withstand the largest population, socioeconomic, and ecological environment requirements under the premise of maintaining sustainable development. Studies of the WRCC can provide helpful information about how the socioeconomic system is both supported and restrained by the water resources system, such as intuitively measuring regional development potential [2], etc. Since the 21st century, people began to pay attention to the research of WRCC. Through continuous improvement of related influencing factors, a relatively mature evaluation system has initially formed [3–5]. Nowadays, the research on the WRCC has changed from a simple natural factor to a water-ecological-economic factor [6].

The research on the WRCC's theory in the international context is focused more on the relationship between carrying capacity and sustainable economic and social development [7]. However, the research on the limitation of water resources security to the WRCC is relatively late. According to the latest analysis of the obstacle degree for the WRCC system in Northeast China, the agricultural water pollution index emerged as the main factor that is restricting the steady rise of the WRCC since 2004. Following 2014, with the upsurge of industrialization, the percentage of industrial wastewater discharge has increased significantly [8]. There are many old industrial bases in the Chang-Ji Economic Circle, and the percentage of industrial wastewater discharge and the risk of water pollution has increased year by year. Furthermore, the data published in the Water Resources Bulletin in the study area over the years show that the surface water quality is bad. So, the water quality should be included in the analysis of WRCC in the Chang-Ji economic circle. In China, there are few studies on the water resources carrying capacity of the Chang-Ji economic circle [9]. Moreover, most of the studies in the Chang-Ji Economic Circle are carried out in administrative regions, and few studies take into account the constraints of water pollution on the WRCC, which cannot directly and accurately measure the development potential of the entire Chang-Ji Economic Circle [10,11]. Therefore, this study takes into account the quantity and quality of water resources and combines the comprehensive carrying capacity of social economic development and ecological environment and selects appropriate methods to evaluate the WRCC of the Chang-Ji Economic Circle.

At present, there are plenty of methods to evaluate the WRCC, such as the traditional trend analysis method [12], the principal component analysis method [13], the fuzzy comprehensive evaluation method (FCE) [14], the multiobjective analysis method [15], the artificial neural network method [16], and the system dynamics (SD) method [17,18]. Zhang et al. [13] applied the principal component analysis method to evaluate the temporal scale variation tendency of the WRCC. However, there was still some uncertainty when integrating the method WRCC index standardization, the method of principal component determination, and the weights of contribution rates. Multiobjective analysis is influenced by its limitations and is more suitable for smaller areas of research. The SD model can effectively simulate and predict through negative feedback adjustment [19]. However, this method requires a large number of parameter settings and data simulations [20], which cannot achieve rapid evaluation. In fact, various WRCC prediction methods are based on the further evaluation of their influencing factors. The FCE method which is widely used by scholars can analyze the WRCC from all aspects [21,22], making the research results more reliable. For example, Zhang et al. [23] used fuzzy set pair analysis theory to evaluate the WRCC in Dagong Yellow River Diversion Irrigation District from 2013 to 2017. The study qualitatively measured the water resources carrying capacity of the ecological irrigation area, and the evaluation results can provide a scientific basis for optimal allocation of water resources in the Dagong Yellow River diversion irrigation district. At the same time, through the comparative analysis of some indicators, the FCE method can solve the defects of the parameters that are difficult to grasp and easily lead to unreasonable conclusions. Moreover, while assessing the regional WRCC, it is necessary to predict future development trends based on the status quo. Accurate trend analysis has become an important part of reasonable evaluation. However, most scholars use the simple linear equations to predict related influencing factors. To make up for the shortcomings of large errors in traditional trend analysis, this study quantitatively predicts relevant influencing factors by the method of gray correlation analysis (GCA) combined with multiple linear regression (MLR) models. Then, on the basis of quantitatively predicting the development trend of social economy, water resources, and water environment evaluation indicators, the FCE method is used to make a reliable assessment of the WRCC.

2. Study Area and Data Sources

2.1. Study Area

The Chang-Ji Economic Circle includes the whole of Changchun City, Jilin City, Jiutai District, Shuangyang District, and Yongji County, as well as parts of Nongan County, Gongzhuling City, and Yitong County (Figure 1). The study area is located in the center of Jilin Province, accounting for 7.96% of the total province's area. The total population of the region is about 648 million and the urbanization rate is 59.64%. There are many water systems in the area, and river networks are dense, all belonging to the second Songhua River system. The large reservoirs include Fengman Reservoir, Shitoukoumen Reservoir, and Xinlicheng Reservoir. The total water storage capacity is as high as 134.1×10^8 m^3.

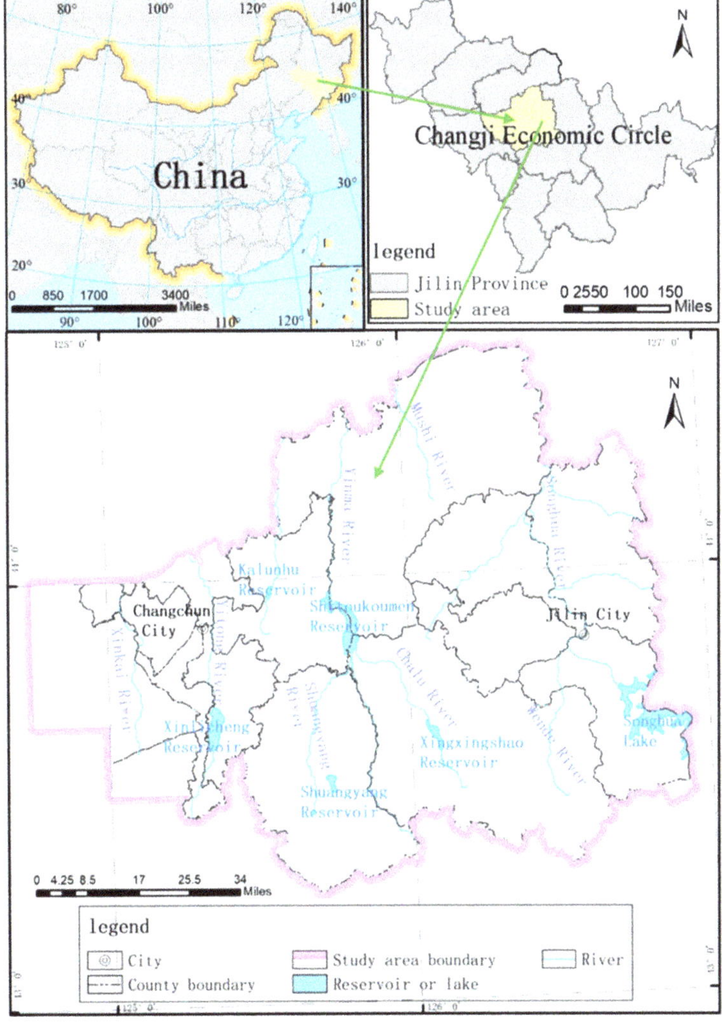

Figure 1. The location of the study area and the distribution of the river system and the water bodies.

2.2. Data Sources

The study area involves Jilin City and Changchun City, and the data used include socioeconomic data, water supply and demand data, and water environment data. The data sources are Jilin statistical yearbook (2010–2017), Changchun statistical yearbook (2010–2017), Jilin water resource bulletin (2010–2017), and Changchun water resource bulletin (2010–2017). In addition, water quota references the industry water quota standards of Jilin province and some environmental data comes from the website of Jilin province environmental protection bureau.

3. Establishment and Prediction of Water Demand Model

The analysis of the balance between supply and demand of water resources can reflect the overall level of water use in a region and is an important factor affecting the WRCC. In recent years, ecological and environmental problems have become increasingly prominent, and ecological water demand will also become an important issue in the analysis of the balance between water supply and demand. This study collects water supply and demand data for the past eight years in the study area and focuses on ecological water demand to predict the water demand in the next 13 years.

3.1. Model Related Methods

Since the study area involves 15 different administrative districts, if the forecast is made one by one according to the different water demand industries, the amount of data required is large and it is difficult to maintain consistency. The combination of the GCA and MLR model can make up for this deficiency. Using the GCA method, the correlation coefficients between the six water demands and the total water demand are calculated respectively. The six water requirements are ecological and environmental water demand, urban public water demand, domestic water demand, forestry, animal husbandry and fishery water demand, farmland irrigation water demand, and industrial water demand. One selects indicators with high correlation coefficients and establishes MLR models of each indicator for total water demand, then obtains total water demand at different planning levels.

3.1.1. The GCA Method

The GCA is to find the important factors that affect the target value by looking for the correlation between various factors between random sequences. This method is mainly to determine the correlation between the influencing factors and the target value, find the main characteristics of the problem, and intuitively quantify the complex influencing relationship. The main calculation method is to perform dimensionless processing on all data, set the processed target sequence to x_0, and set the factor sequence to x_i ($i = 1, 2, \ldots, n$); perform a difference between the two sequences. Calculate to find the maximum difference and minimum difference between the two poles. Among them, the difference sequence is expressed as:

$$\Delta_{0i}(k) = |x_0(k) - x_i(k)| \tag{1}$$

Find the correlation coefficient between each factor and the target value $\varepsilon_i(k)$:

$$\varepsilon_i(k) = \frac{\min\min|x_0(k) - x_i(k)| + \rho\max\max|x_0(k) - x_i(k)|}{|x_0(k) - x_i(k)| + \rho\max\max|x_0(k) - x_i(k)|} \tag{2}$$

Equation (1), $\Delta_{0i}(k)$ is the absolute value of the difference between two sequences.

Equation (2), ρ for the resolution coefficient, the empirical value is 0.5; $\max\max|x_0(k) - x_i(k)|$ is the maximum difference between the two levels; $\min\min|x_0(k) - x_i(k)|$ is the minimum difference between the two levels; the greater the calculated value of $\varepsilon_i(k)$, the greater the correlation between the factor and the target value.

Finally, according to the correlation coefficient, determine the degree of correlation between the influencing factors and the target sequence r_i:

$$r_i = \frac{1}{n}(\varepsilon_i(1) + \varepsilon_i(2) + \ldots + \varepsilon_i(n)) \tag{3}$$

Similarly, among all influencing factors, r_i with the larger value has a higher degree of correlation with the target sequence.

3.1.2. The MLR Model

The MLR model is a predictive method that participates in the analysis of the linear relationship between two or more independent and dependent variables. The method of establishing MLR model is as follows:

Suppose there is a linear relationship between the dependent variable y and the independent variable x_i ($i = 1, 2, \ldots, n$):

$$y_i = \beta_0 + \beta_1 x_1 + \beta_2 x_2 + \ldots\ldots + \beta_n x_n + \varepsilon \tag{4}$$

In Equation (4), β_i is the regression coefficient, $i = 1, 2, \ldots, n$; generally $n > 2$; ε is the random factor of error, and it follows the $N(0, \sigma^2)$ distribution.

$$\text{Let } Y = \begin{pmatrix} y_1 \\ y_2 \\ \vdots \\ y_m \end{pmatrix}, X = \begin{pmatrix} 1 & x_{11} & \cdots & x_{1n} \\ 1 & x_{21} & \cdots & x_{2n} \\ \vdots & \vdots & \cdots & \vdots \\ 1 & x_{m1} & \cdots & x_{mn} \end{pmatrix}, \beta = \begin{pmatrix} \beta_0 \\ \beta_1 \\ \vdots \\ \beta_n \end{pmatrix}, \varepsilon = \begin{pmatrix} \varepsilon_0 \\ \varepsilon_1 \\ \vdots \\ \varepsilon_n \end{pmatrix}, \text{ the MLR}$$

models in matrix form are available as follows:

$$Y = X\beta + \varepsilon \tag{5}$$

3.2. Model Establishment and Error Analysis

Combining the water supply and demand data from 2010 to 2017, the GCA method is used to obtain the correlation between the total water demand and the other basic water demand. The correlations obtained are ecological and environmental water demand (0.819), urban public water demand (0.652), domestic water demand (0.610), forestry, animal husbandry and fishery water demand (0.670), farmland irrigation water demand (0.709), and industrial water demand (0.744). Three indicators with a correlation degree greater than 0.7 are selected as the main influencing factors of total water demand.

With the help of SPSS statistical analysis software, a multivariate regression matrix with three factors of ecological environment (X_1), farmland irrigation (X_2) and industrial water demand (X_3) as independent variables and total water demand (Y) as dependent variable was constructed. Solve the regression model as follows:

$$Y = -2.408X_1 + 2.082X_2 + 1.087X_3 - 7.243 \tag{6}$$

The MLR model is tested for errors based on each water demand and total water demand from 2010 to 2017, as shown in Table 1. According to the analysis results, the error values are all within 1%, which meets the prediction accuracy requirements. Therefore, the established MLR model is suitable for the prediction of the total water demand in this study area.

Table 1. Fitting result of actual value and predicted value of total water demand.

	Ecological and Environmental Water Demand	Farmland Irrigation Water Demand	Industrial Water Demand	Actual Total Water Demand	Forecast Total Water Demand	Relative Error
2010	0.365	11.326	13.596	30.261	30.241	0.06%
2011	0.587	11.308	13.454	29.436	29.515	−0.27%
2012	0.603	11.492	13.193	29.655	29.576	0.27%
2013	0.598	11.755	12.550	29.348	29.438	−0.30%
2014	0.640	11.654	12.500	29.146	29.071	0.26%
2015	0.576	11.531	12.208	28.642	28.651	−0.03%
2016	0.637	11.601	12.444	29.014	28.907	0.37%
2017	0.591	11.834	12.528	29.393	29.592	−0.68%

4. Establishment of a Rapid Evaluation Model

In order to achieve efficient and accurate evaluation of WRCC, the study selects the FCE method to establish corresponding evaluation system. Measuring the size of WRCC by setting various plans provides a basis for regional water resources development and utilization.

4.1. Evaluation Method

First, establish the set of influencing factors of WRCC $U = (u_1, u_2, \ldots, u_n)$, and the corresponding comment set $V = (V_1, V_2, \ldots, V_n)$. Second, the membership degree r_{ij} of the comment set v_j is determined by a single factor fuzzy evaluation, which means to evaluate the single factor u, and the single factor evaluation set $r_i = (r_{i1}, r_{i2}, r_{i3})$ is obtained. Then, the evaluation index is quantified separately, and the corresponding membership degree r_{ij} is obtained, thus establishing the fuzzy relation matrix R with the amount of m evaluation factors. Finally, by analyzing weights $A = (\alpha_1, \alpha_2, \ldots, \alpha_n)$ of different influencing factors, the fuzzy operation $B = A \times R$ is used to synthesize the weight vector of the influencing factors and the fuzzy evaluation matrix to obtain the FCE's results.

The degree of membership usually indicates a certain degree of deviation, using two sets of formulas to calculate the membership function. x_1 and x_3 are the critical values of V_1, V_2, and V_2, V_3, respectively, and the grade of V_2 is the interval midpoint value of x_2 [14], where $x_2 = \frac{x_1 + x_3}{2}$. The first set of formulas is applicable to the case where the larger the index U_i is, the better the system is. The second set of formulas is applicable to the case where the smaller the index U_i is, the better the system is. The two sets of calculation formulas are as follows:

First set of formulas:

$$V_1(ui) = \begin{cases} 0.5\left(1 + \frac{u_i - x_1}{u_i - x_2}\right), & u_i \geq x_1 \\ 0.5\left(1 - \frac{u_i - x_1}{x_2 - x_1}\right), & x_2 < u_i < x_1 \\ 0, & u_i \leq x_2 \end{cases} \tag{7}$$

$$V_2(ui) = \begin{cases} 0.5\left(1 - \frac{u_i - x_1}{u_i - x_2}\right), & u_i \geq x_1 \\ 0.5\left(1 + \frac{x_1 - u_i}{x_1 - x_2}\right), & x_2 \leq u_i < x_1 \\ 0.5\left(1 + \frac{u_i - x_3}{x_2 - x_3}\right), & x_3 < u_i < x_2 \\ 0.5\left(1 - \frac{x_3 - u_i}{x_2 - u_i}\right), & u_i \leq x_3 \end{cases} \tag{8}$$

$$V_3(ui) = \begin{cases} 0, & u_i \geq x_2 \\ 0.5\left(1 - \frac{u_i - x_3}{x_2 - x_3}\right), & x_3 < u_i < x_2 \\ 0.5\left(1 + \frac{x_3 - u_i}{x_2 - u_i}\right), & u_i \leq x_3 \end{cases} \tag{9}$$

Second set of formulas:

$$V_1(ui) = \begin{cases} 0.5\left(1 + \frac{u_i - x_1}{u_i - x_2}\right), & u_i \leq x_1 \\ 0.5\left(1 - \frac{u_i - x_1}{x_2 - x_1}\right), & x_1 < u_i < x_2 \\ 0, & u_i \geq x_2 \end{cases} \tag{10}$$

$$V_2(ui) = \begin{cases} 0.5\left(1 - \frac{u_i - x_1}{u_i - x_2}\right), & u_i \leq x_1 \\ 0.5\left(1 + \frac{x_1 - u_i}{x_1 - x_2}\right), & x_1 < u_i \leq x_2 \\ 0.5\left(1 + \frac{u_i - x_3}{x_2 - x_3}\right), & x_2 < u_i < x_3 \\ 0.5\left(1 - \frac{x_3 - u_i}{x_2 - u_i}\right), & u_i \geq x_3 \end{cases} \tag{11}$$

$$V_3(ui) = \begin{cases} 0, \ u_i \leq x_2 \\ 0.5\left(1 - \frac{u_i - x_3}{x_2 - x_3}\right), \ x_2 < u_i < x_3 \\ 0.5\left(1 + \frac{x_3 - u_i}{x_2 - u_i}\right), \ u_i \geq x_3 \end{cases} \tag{12}$$

The corresponding membership degree r_{ij} is calculated separately, and matrix R corresponding to different horizontal years of each administrative region in the study area is obtained as follows:

$$R = \begin{bmatrix} r_{11} & r_{12} & \cdots & r_{1n} \\ r_{21} & r_{22} & \cdots & r_{2n} \\ \vdots & \vdots & \ddots & \vdots \\ r_{m1} & r_{m2} & \cdots & r_{mn} \end{bmatrix} \tag{13}$$

The vectorization evaluation result B is obtained from the membership matrix and the weight, and the formula is as follows:

$$B = A \times R = (\alpha_1 \alpha_2 \ldots \alpha_n) \cdot \begin{bmatrix} r_{11} & r_{12} & \cdots & r_{1n} \\ r_{21} & r_{22} & \cdots & r_{2n} \\ \vdots & \vdots & \ddots & \vdots \\ r_{m1} & r_{m2} & \cdots & r_{mn} \end{bmatrix} = (b_1 b_2 \ldots b_m) \tag{14}$$

To facilitate the comparative evaluation, the vectorization result is quantified by the following formula, and the corresponding comprehensive score value λ is obtained, where $k = 1$, and the higher λ indicates that the regional WRCC is higher.

$$\lambda = \frac{\sum_{i=1}^{n=3} b_i^k \times \lambda_i}{\sum_{i=1}^{n=3} b_i^k} \tag{15}$$

4.2. Evaluation of Index Selection

The key to correctly evaluate the regional WRCC is to properly select fuzzy evaluation indicators. To better reflect the status of regional WRCC, it is particularly important to select regionally representative evaluation indicators. Many factors influence the WRCC. Most studies only consider the impact of the amount of water resources, social economy, and ecological environment on WRCC [24–26] but ignore the impact of water quality on regional WRCC. This tends to make the calculation of the WRCC too large and fails to reflect the real situation in the study area. In fact, due to the different degrees of water pollution, the actual water supply and availability of water are smaller than themselves. Combined with the actual situation of the uneven distribution of water resources and the outstanding water quality problems in the study area, this study incorporates the water environment capacity (WEC) into the evaluation of WRCC. We considered the factors of water resources (including water quality and water quantity), as well as the socioeconomic and ecological environment to calculate WRCC. This lays a foundation for truly establishing a new realm of harmonious development of economic, social, and humanities based on the principle of sustainable development (Figure 2).

4.2.1. Evaluation Indicators U_1, U_2, and U_3

Based on the surface water and groundwater supply capacity, actual water supply, and water demand in this region, the WRCC is evaluated from the perspective of the quantity of water resources. Per capita available water resource (U_1) = available water resources/total population (m^3/person); per capita water supply quantity (U_2) = actual water supply/total population (m^3/person); water resource utilization rate (U_3) = water demand/available water resources (%).

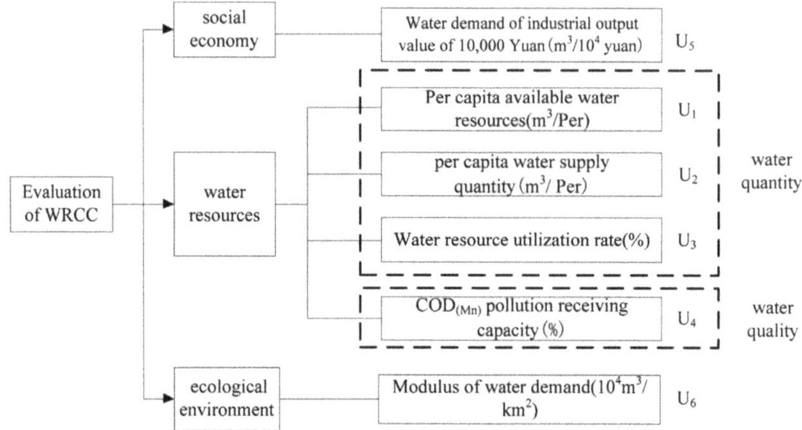

Figure 2. The research system of water resource carrying capacity (WRCC) index evaluation.

4.2.2. Evaluation Index U_4

Chemical oxygen demand (COD) pollution receiving capacity (U_4) = the water environment capacity of COD at this stage/the maximum water environment capacity of COD (%). The maximum water environment capacity of COD means the maximum pollution capacity that the region can received.

According to the results of the Water Resources Bulletin, Class IV, V, and above Class V water quality river sections are shown in the rivers in the study area. The quality of regional water resources is not optimistic [27,28]. In fact, the quality of water supply directly affects the availability of actual water supply, and the WEC can reflect the two main capabilities of water body dilution and natural purification. Therefore, based on the preliminary understanding of the water quality of the Chang-Ji Economic Circle, this study uses the water environmental capacity as the water quality evaluation index in the evaluation system. Since the flow rate of the river in the study area is stable as well as small, the calculation method of the overall standard is used.

The WEC is calculated as follows [29]:

$$W = 86.4Q_0(C_s - C_0) + 0.001KVC_s + 86.4qC_s \tag{16}$$

where W is the initial value for WEC of the water body, C_s is the standard water quality of the water body, Q_0 is the incoming water flow, C_0 is the upstream background concentration of the incoming water, K is the water quality degradation coefficient, Q is the side flow of the side stream, V is the flow rate, and q is the side inflow flow.

The overall standard calculation method usually does not consider the location of the pollution source, so the calculation results tend to be too large, which is nonconservative. Therefore, in order to conform to the reality, an uneven coefficient is introduced for correction [29]; the method is as follows:

$$W' = \alpha W \tag{17}$$

where W' represents the corrected WEC, and α denotes the uneven coefficient.

Because the rivers in the area belong to the small and the middle rivers [30], their flow is small and slow, and the uneven coefficient takes an empirical value of 0.8.

4.2.3. Evaluation Index U_5

The Chang-Ji Economic Circle is a gathering place for old industrial cities. The development of its industry can reflect the social and economic conditions of the region. So, we select evaluation index U_5 to effectively reflect this. Water demand of industrial output value of 10,000 Yuan (U_5) = industrial water demand/industrial output value (m^3/10,000 yuan). The industrial water demand and the industrial output value is obtained by the GCA and MLR model.

4.2.4. Evaluation Index U_6

Lou [31] and Wang [9] confirmed that the modulus of water demand can reflect the level of regional economic development. Thus, we select the Modulus of water demand as the evaluation index U_6 that effectively reflects the ecological environment of the study area. Modulus of water demand (U_6) = total water demand/land area (10,000 m^3/km^2). The total amount of water demand is obtained by the GCA and MLR model and the land area used the government published data. The required water modulus reflects the restriction of the ecological environment on the WRCC.

By selecting the above six indicators, it can effectively reflect the balance of supply and demand of water resources in the region, the amount of water resources, the quality of water resources, and the impact of socioeconomic conditions and ecological environment on the regional WRCC.

Based on the above-mentioned evaluation indicators U_1 to U_6, the impact degree of WRCC is analyzed, and its comment set $V = (V_1, V_2, V_3)$ is established (as shown in Table 2 below). The WRCC of V_1 to V_3 is gradually weakened. For this evaluation, the second set of formulas applies to U_1, U_2, and U_4; however, U_3, U_5, and U_6 apply to the first set of formulas. The weight is determined according to the influence degree of the influencing factors on the WRCC. According to expert analysis, the corresponding weights are obtained from empirical values, that is, $A = (\alpha_1, \alpha_2, \alpha_3, \alpha_4, \alpha_5, \alpha_6) = (0.2, 0.2, 0.3, 0.1, 0.1, 0.1)$. Then, according to the score value ($\lambda_1 = 0.95$, $\lambda_2 = 0.5$, and $\lambda_3 = 0.05$), the water carrying capacity of each area is analyzed and evaluated, and the score value directly reflects the WRCC in the area.

Table 2. Evaluation standard of WRCC grading indicators.

Evaluating Indicators Set U	Judgment Set V		
	V_1	V_2	V_3
U_1 Per capita available water resources (m^3/Per)	>1200	1200~400	<400
U_2 per capita water supply quantity (m^3/Per)	>1000	1000~500	<500
U_3 Water resource utilization rate (%)	<40	40~90	>90
U_4 COD$_{(Mn)}$ pollution receiving capacity (%)	78	50	22
U_5 Water demand of industrial output value of 10,000 Yuan (m^3/10^4 yuan)	<20	20~90	>90
U_6 Modulus of water demand (m^3/km^2)	<10	10~60	>60
Score value λ	0.95	0.5	0.05

4.3. Reasonably Evaluate Regional WRCC

The evaluation value of WRCC is statistically analyzed and divided into three levels: an evaluation value of WRCC greater than 0.6 is an area with good WRCC (I), an area evaluation value of water carrying capacity ranging between 0.3 and 0.6 is medium (II), and an area evaluation value of WRCC less than 0.3 is poor (III).

4.4. The Establishment of 4 Different Plans

4.4.1. Plan I

Under the current conditions, Plan I only considers the WRCC of self-produced water, does not increase the local water supply, or expand the capacity of water transfer outside the region, and predicts the carrying capacity of water resources in different years.

4.4.2. Plan II

On the basis of Plan I and considering the project "Carrying Water from Songhua River to Changchun" to increase the local water supply, we predict the water carrying capacity of different years. According to the government's economic development plan of the Chang-Ji Economic Circle, the cumulative water supply capacity of the design diversion project in Changchun City is 3.25×10^8 m^3. After the completion of the water supply project in the central city of Jilin Province, the cumulative water intake will be 5.83×10^8 m^3 in 2020. Furthermore, the cumulative water intake will be 6.92×10^8 m^3 in vision level year 2030.

4.4.3. Plan III

Based on Plan II, Plan III strengthens water governance and considers industrial, agricultural, and domestic water conservation. By reducing water usage quotas, increasing the reuse rate of reclaimed water and treating sewage as the most important measures.

According to the "Standards for Local Standard Water Use in Jilin Province" [32], the plan will appropriately reduce the industrial water quota, where each administration increases the amount of water reuse by 0.05 billion m^3/per year.

4.4.4. Plan IV

Based on Plan III, Plan IV increases an appropriate amount of transit water. The inflow water of Songhua River is 62.02×10^8 m^3 [27]. The increase in water supply is 40% of the inflow water of Songhua River, while the increase in actual water supply is 20%.

5. Results

Due to the inconsistent development speed of various regions, even within the same city, there are differences in regional WRCC. Most of the research only stays at the holistic research within the scope of the region, while ignoring the research of small administrative units [22,33]. Zhou et al. [11] compared with the temporal dynamic process of index change in the water environment carrying capacity and thought that it is urgent to carry on spatiotemporal dynamic change analysis in the WRCC considering spatial heterogeneity and spatial evolution. As a result, this article uses the smallest administrative unit to analyze change trend of WRCC from both time and space perspectives.

5.1. Results of Each Program from the Perspective of Time

5.1.1. Plan I

According to the results of Plan I (Figure 3), the study area is still generally in the middle area of WRCC (II). In 2020, the WRCC of Fengman District is the largest (0.606), while Chaoyang District is poor (0.287). By 2030, both Lvyuan District (0.298) and Chaoyang District (0.27) are in areas with poor WRCC (III). The comparison shows that the comprehensive score of WRCC in 2030 is decreasing compared with 2020, but the rate of decline is slow. It is comprehensively reflected that under the condition of not changing the status quo, the WRCC of the study area will continue to weaken. According to existing research, it likely due to the uneven distribution of the regional water resources [34], which makes the contradiction between water supply and demand increasingly prominent and the development potential decreases [35].

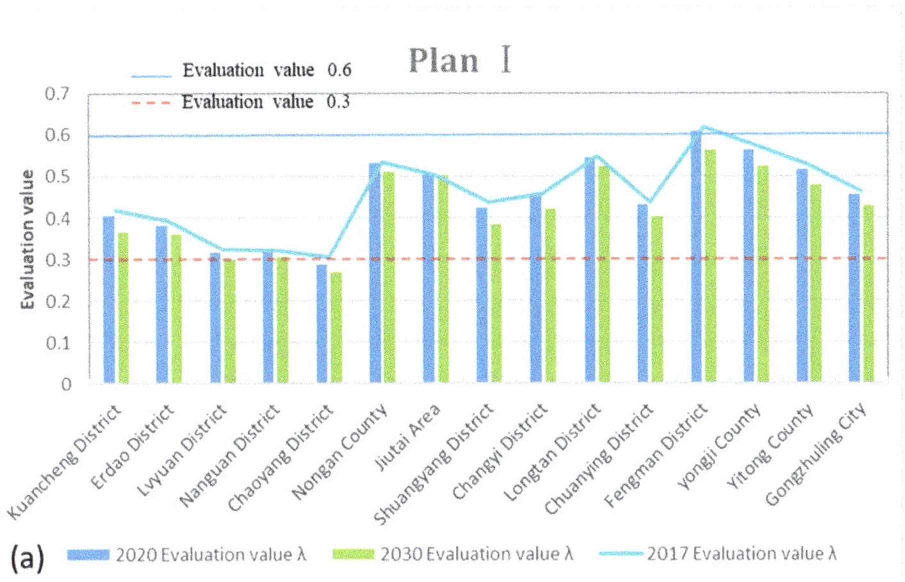

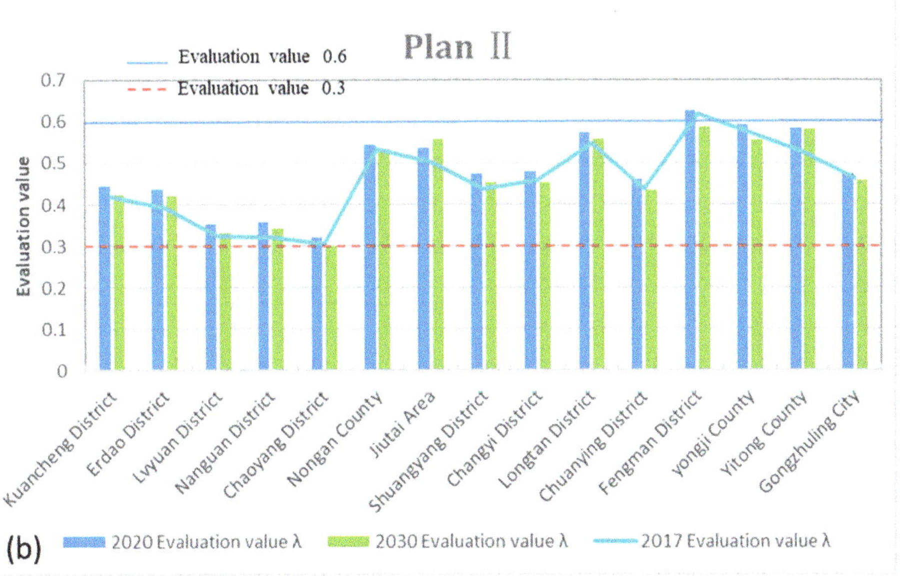

Figure 3. *Cont.*

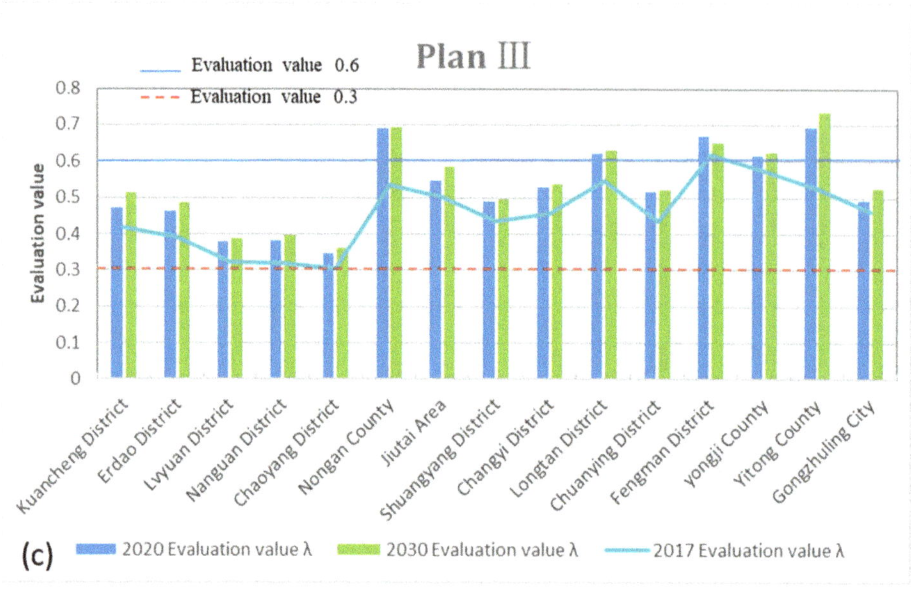

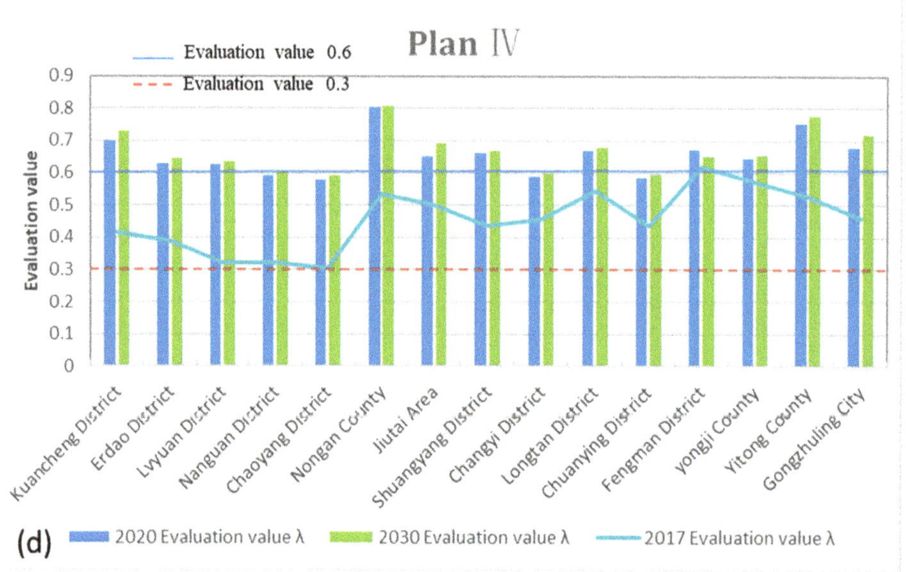

Figure 3. WRCC evaluation value of each region in different Plan. (**a**) Plan I (**b**) Plan II (**c**) Plan III (**d**) Plan IV.

5.1.2. Plan II

Under the influence of the open-source program, U_1 and U_2 increase in different degrees and U_3 decreases accordingly. The evaluation results have the highest degree of membership to V_2, and the WRCC in the area has been improved. There is no poor (III) in the near and long term (see from Figure 3). In the next 13 years, the WRCC will show a downward trend after a short period of improvement. It shows that the fewer water resources chiefly caused its long-term overloaded status [36] and the construction of water diversion project will promote the improvement of WRCC. However, relying solely on the construction of drinking water projects will not alleviate the problem of resource constraints for a long time in the future. Thus, water conservation should be promoted while transiting water [11].

5.1.3. Plan III

With the construction of the water diversion project and the popularization of water saving and sewage treatment policy, the WRCC of the study area will improve greatly compared with 2017. This plan can improve the utilization rate of water resources and Alleviate the tight of the water supply and demand. Unlike the previous plan, the WRCC of plan III will continue to grow in the future. It can be seen in Figure 3 that the WRCC of the whole study area has improved significantly compared to Plan I, and the average growth rate is 23%. In most areas of Jilin, there are areas with good (I) and medium (II) WRCC, and the difference in WRCC of each administrative region will gradually reduce. It shows that saving water and improving water quality are also important factors for enhancing the WRCC [9,37].

5.1.4. Plan IV

Plan IV not only maintains the water resource utilization rate at a high level but also greatly increases the water resource availability. So, the WRCC of the whole region has been significantly improved compared with Plan I. Moreover, the difference in WRCC of each administrative region has gradually decreased. The WRCC of most areas in the region is good (I), and the WRCC in Changchun and surrounding areas has increased significantly to a relatively high level (Figure 3 shows). As of 2030, the WRCC of the eight administrative regions will increase by more than 50% over 2017. It shows great potential for regional development and utilization. The shortage and uneven spatial and temporal distribution of water resources has seriously restricted the sustainable development of regional society and economy [38]. Plan IV is more in line with the principle of sustainable development of society and meets the development goals of combining water quality, water quantity, water ecology, and water environment, which can be used as a recommended plan.

5.2. Comparative Analysis of the Plan from the Perspective of Space

The predicted levels of WRCC in 2020 and 2030 at 15 observation locations in the Chang-Ji Economic Circle were analyzed in four plans, the development potential of water resources was evaluated (Figures 4 and 5), and the evaluation level was tested. According to Figures 4 and 5, the WRCC at each observation location exhibited a continuously increasing trend from Plan I to IV. These findings are consistent with the measures used in the design of the plan, which provides a certain level of reliability and reference to the present research.

5.2.1. WRCC Spatial Distribution in 2020

Figure 4 shows that the level of WRCC in the Fengman District of Jilin City, which has unique natural resource surrounding the Songhua River, will be good (I) during each plan in 2020. As the administrative center of Changchun City, Chaoyang District has a relative shortage of water resources and poor water quality. The level is predicted to improve from III to II, and the value to increase from 0.287 to 0.579. Nanguan, Changyi, and Chuanying District will retain the high level of II and will always have a certain development potential; the values for the three areas will change by 0.268, 0.136, and 0.153, respectively. The rest of the region will change from II to I and may adapt to social and economic development.

5.2.2. WRCC Spatial Distribution in 2030

Figure 5 shows that the initial level of WRCC in Chaoyang and Lvyuan District are both at III, and the potential of water resources exploitation is small. The WRCC level Chaoyang changes from III to II, and it is greatly improved from III to I for Lvyuan District. Additionally, their carrying capacity is greatly improved. Changyi and Chuanying District will maintain the high level of II, and the value will change by 0.176 and 0.192, respectively. The rest of the region will change from II to I, which will gradually increase the security effect of the economy and society. The different plans to the subareas can provide a scientific reference to rational distribution of economic development, elaborate management of water environment as well as regional sustainable development in the future [35].

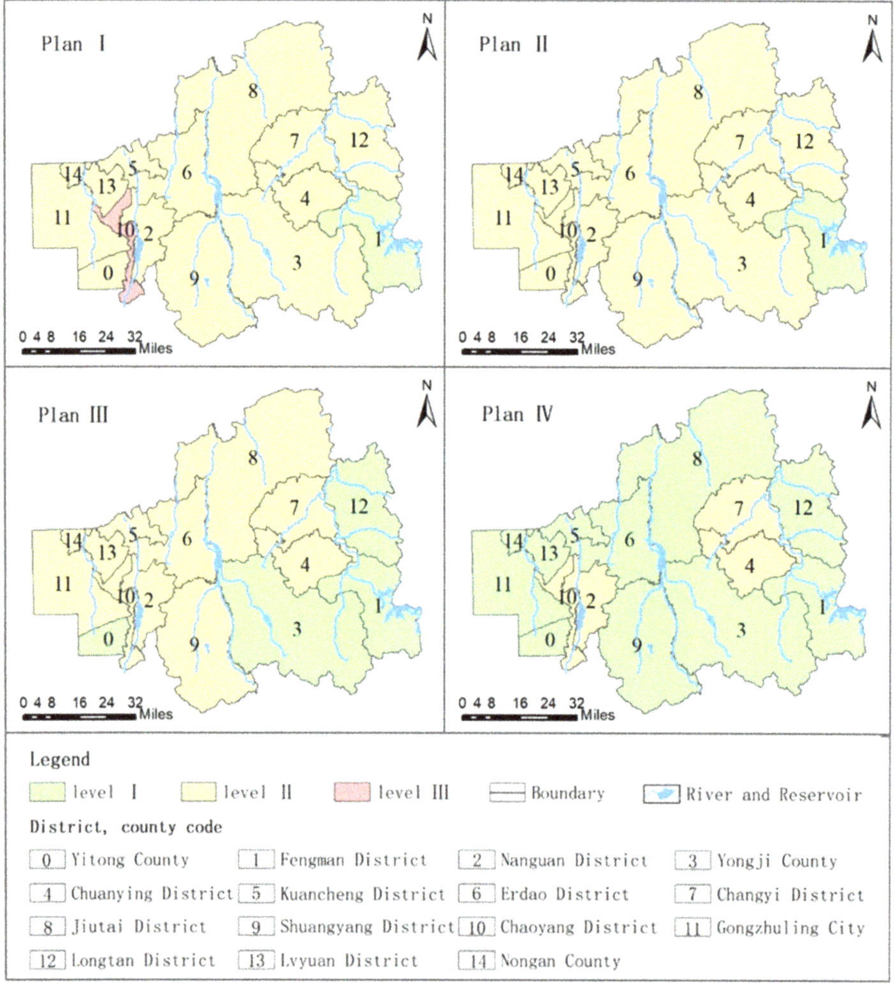

Figure 4. The level division of WRCC evaluation value in each plan (2020).

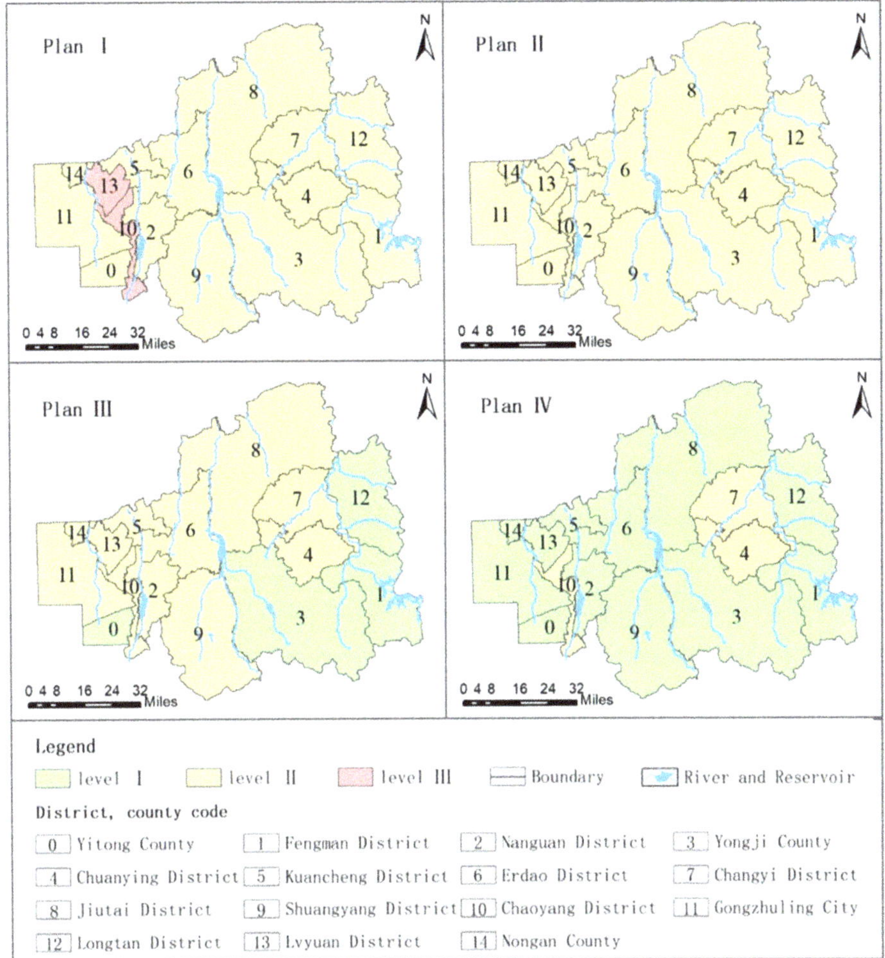

Figure 5. The level division of WRCC evaluation value in each plan (2030).

Throughout each plan in 2020 and 2030, plan III will greatly improve the WRCC of the whole region, with the eastern part of the study area having a significantly higher WRCC than the western part. Some areas in Jilin have strong WRCC, and each administrative region has a certain amount of development and utilization potential. Plan IV shows that the WRCC of the whole district is obviously improved, compared to the Plan I. By this time, the WRCC of the whole district will be strong, with most areas being at level I. Furthermore, the water resources would be able to provide certain guarantees for social and economic development.

The measures of water-saving, sewage treatment, water diversion projects, and transit water utilization mean that WRCC is constantly changing. Based on the above-mentioned various measures, the results of Plan IV show that over 84% of the regions have a relatively large development potential. As regional development progresses, most regions would develop slowly if no measures are taken (Plan I). These results further validate the intuition and visualization of the WRCC classification and provide a basis for the government to rationally allocate water resources [10,22].

5.3. Limitations and Future Research Directions

This study selects the GCA, MLR, and FCE combined model to evaluate the WRCC of the Chang-Ji Economic Circle. This combined model makes up for the shortcomings of traditional indicator evaluation methods and achieves qualitative and quantitative evaluation. Moreover, the coupling of the GCA and MLR model reduces the interference of human factors and reduces the error value of the predicted evaluation index. Nevertheless, this study still has certain limitations.

(1)　Although this study has considered relevant subsystems related to the WRCC as much as possible, such as social economy, water resources quantity, water resources quality, water ecological environment subsystems. The research results provide favorable information guidance for the future development of Chang-Ji Economic Circle. However, only the representative evaluation indicators in each subsystem are selected, and the number of indicators selected is relatively small [39]. Future research should be gradually improved.

(2)　The coupling evaluation model selected this time is based on the index evaluation. Although qualitative and quantitative analysis can be achieved, it is difficult to achieve negative feedback adjustment. However, the SD model can make up for this deficiency [34,40]. We believe that integrating the coupling assessment model established in this research into the dynamic system of the SD model can be the focus of future research [19].

6. Conclusions

The study established a hybrid model to analyze the WRCC of the Chang-Ji Economic Circle. First of all, in order to make up for the shortcomings of traditional trend analysis, the GCA and MLR coupling model can predict the changing trend of WRCC's influencing factors [28,41,42]. Accurate and quantitative evaluation index trend prediction can increase the credibility of the evaluation results of the WRCC [9]. Then, using the FCE model, the WRCC of each region is evaluated. Finally, based on the coupling results of the hybrid model, the future WRCC of the districts and counties in the Chang-Ji Economic Circle are compared in terms of time and space. It is worth noting that according to the actual situation of the Chang-Ji Economic Circle, the impact of water quality on regional WRCC is considered. The water environmental capacity is taken as a new evaluation index and the WRCC evaluation system is proposed based on water quantity and quality, social economy, and ecological environment. This makes up for the shortcomings of the existing evaluation indicators and can more realistically reflect the status quo of regional development. The presented research results allowed us to draw the following conclusions.

Four different water intake plans are considered to assess the WRCC of the study area in 2020 and 2030. The study aims to eliminate potential problems in the societal development of Chang-Ji Economic Circle through various plans and improve the affordability of economic development. Considering the spatial heterogeneity and spatial evolution, the spatial and temporal dynamic changes of WRCC are analyzed. From the perspective of time changes, the WRCC of Plan I and Plan II remains at a medium level. Affected by the constraints of supply and demand, the WRCC will continue to decline. The improvement of the WRCC in Plan III was better than the abovementioned scenarios, yet the potential development potential of the region is still hindered. Water saving measures and sewage treatment can relieve the pressure of WRCC. In order to achieve sustainable development of the region, Plan IV comprehensively considers the advantages of the above plans and increases the amount of transit water to make up for the shortage of resource-based water shortage. The WRCC of the whole region is generally good, and water resources can support the rapid development of the social economy in Plan IV. From the perspective of space changes, the WRCC of Plan I in Chaoyang and Lvyuan District will become a poor level and lack the potential of water resources exploitation in the future. The improvement thought Plan II will still not be sufficient compared with Plan III. It shows that the WRCC in the eastern area of Chang-Ji Economic Circle is significantly higher than others. The WRCC of the whole district Plan IV is significantly improved in comparison with the current conditions. Plan IV is proposed as the final recommendation through comprehensive analysis and research. Strengthening sewage treatment and proper use of transit water resources are more conducive to the rapid development of Chang-Ji Economic Circle.

Author Contributions: Conceptualization, G.W.; data curation, C.X. and X.L.; formal analysis, G.W., Z.Q., F.M., and Y.S.; funding acquisition, C.X. and X.L.; methodology, G.W.; project administration, C.X. and X.L.; supervision, X.L.; writing—original draft, G.W. All authors have read and agreed to the published version of the manuscript.

Funding: The study was financially supported by Natural Science Foundation of China (No. 41572216), the China Geological Survey Shenyang Geological Survey Center "Chang-Ji Economic Circle Geological Environment Survey" project (121201007000150012), the Provincial School Co-construction Project Special-Leading Technology Guide (SXGJQY2017-6), and the Jilin Province Key Geological Foundation Project (2014-13).

Institutional Review Board Statement: "Not applicable" for studies not involving humans or animals.

Informed Consent Statement: "Not applicable" for studies not involving humans.

Data Availability Statement: Data is contained within the article or supplementary material.

Acknowledgments: We would like to thank the anonymous reviewers and the editor.

Conflicts of Interest: The authors declare no conflict of interest.

Nomenclature

WRCC	Water resource carrying capacity
FCE	Fuzzy comprehensive evaluation
SD	System dynamics
GCA	Gray correlation analysis
MLR	Multiple linear regression models
WEC	Water environment capacity
COD	Chemical oxygen demand

References

1. Wang, X.K.; Wang, Y.T.; Wang, J.Q.; Cheng, P.F.; Li, L. A TODIM-PROMETHEE II Based Multi-Criteria Group Decision Making Method for Risk Evaluation of Water Resource Carrying Capacity under Probabilistic Linguistic Z-Number Circumstances. *Mathematics* **2020**, *8*, 1190. [CrossRef]
2. Harris, J.M.; Kennedy, S. Carrying capacity in agriculture: Global and regional issues. *Ecol. Econ.* **1999**, *29*, 443–461. [CrossRef]
3. Joardar, S.D. Carrying capacities and standards as bases towards urban infrastructure planning in India: A case of urban water supply and sanitation. *Urban Infrastruct. Plan. Indian.* **1998**, *22*, 327–337.
4. Rijisberman, M.A.; Van De Ven, F.H.M. Different approaches to assessment of design and management of sustainable urban water system. *Environ. Impact Assess. Rev.* **2000**, *29*, 333–345. [CrossRef]
5. Clarke, A.L. Assessing the carrying capacity of the Florida Keys. *Popul. Environ.* **2002**, *23*, 405–418. [CrossRef]
6. Liu, H.B.; Liu, Y.F.; Li, L.J. Study of an evaluation method for water resources carrying capacity based on the projection pursuit technique. *Water Sci. Technol. Water Supply* **2017**, *175*, 1306–1315. [CrossRef]
7. Zarghami, M. Urban water management using fuzzy-probabilistic multi-objective programming with dynamic efficiency. *Water Resour. Manag.* **2010**, *24*, 4491–4504. [CrossRef]
8. Cheng, K.; Fu, Q.; Meng, J.; Li, T.X.; Pei, W. Analysis of the Spatial Variation and Identification of Factors Affecting the Water Resources Carrying Capacity Based on the Cloud Model. *Water Resour. Manag.* **2018**, *32*, 2767–2781. [CrossRef]
9. Wang, G. *Research on the Assessment and the Carrying Capacity of Water Resources in Chang-Ji Economic Circle*; Jilin University: Changchun, China, 2018. (In Chinese)
10. Liu, T.; Yang, X.h; Geng, L.H. A Three-Stage Hybrid Model for Space-Time Analysis of Water Resources Carrying Capacity: A Case Study of Jilin Province, China. *Water* **2020**, *12*, 426. [CrossRef]
11. Zhou, X.Y.; Lei, K.; Meng, W.; Khu, S.-T.; Zhao, J.; Wang, M.N.; Yang, J.F. Space-time approach to water environment carrying capacity calculation. *J. Clean. Prod.* **2017**, *149*, 302–312. [CrossRef]
12. Song, X.M.; Kong, F.Z.; Zhan, C.S. Assessment of Water Resources Carrying Capacity in Tianjin City of China. *Water Resour. Manag.* **2011**, *25*, 857–873. [CrossRef]

13. Zhang, J.; Zhang, C.L.; Shi, W.L. Quantitative evaluation and optimized utilization of water resources-water environment carrying capacity based on nature-based solutions. *J. Hydrol.* **2019**, *568*, 96–107. [CrossRef]
14. Gong, L.; Jin, C.L. Fuzzy Comprehensive Evaluation for Carrying Capacity of Regional Water Resources. *Water Resour. Manag.* **2009**, *23*, 2505–2513. [CrossRef]
15. Wang, Y.Y.; Huang, G.H.; Wang, S.; Li, W.; Guan, P.B. A risk-based interactive multi-stage stochastic programming approach for water resources planning under dual uncertainties. *Water Resour.* **2016**, *94*, 217–230. [CrossRef]
16. Luo, M.; Huang, E.; Ding, R.; Lu, X. Research on water resources carrying capacity based on maximum supportable population. *Fresenius Environ. Bull.* **2019**, *28*, 100–110.
17. Wang, C.H.; Hou, Y.L.; Xue, Y.J. Water resources carrying capacity of wetlands in Beijing: Analysis of policy optimization for urban wetland water resources management. *J. Clean. Prod.* **2017**, *161*, 1180–1191. [CrossRef]
18. Zomorodian, M.; Lai, S.H.; Homayounfar, M.; Ibrahim, S.; Fatemi, S.E.; El-Shafie, A. The state-of-the-art system dynamics application in integrated water resources modeling. *J. Environ. Manag.* **2018**, *227*, 294–304. [CrossRef]
19. Mashaly, A.F.; Fernald, A.G. Identifying Capabilities and Potentials of System Dynamics in Hydrology and Water Resources as a Promising Modeling Approach for Water Management. *Water* **2020**, *12*, 1432. [CrossRef]
20. Dai, D.; Sun, M.D.; Lv, X.B.; Lei, K. Evaluating water resource sustainability from the perspective of water resource carrying capacity, a case study of the Yongding River watershed in Beijing-Tianjin-Hebei region, China. *Environ. Sci. Pollut. Res.* **2020**, *27*, 21590–21603. [CrossRef]
21. Chi, M.B.; Zhang, D.S.; Fan, G.W.; Zhang, W.; Liu., H.L. Prediction of water resource carrying capacity by the analytic hierarchy process-fuzzy discrimination method in a mining area. *Ecol. Indicat.* **2019**, *96*, 647–655. [CrossRef]
22. Wang, H.; Ji, F.Q.; Pang, Y. Fluctuation of River Network Water Environmental Carrying Capacity in a Complicated River-Lake Syatem. *Environ. Eng. Manag. J.* **2018**, *17*, 1511–1520. [CrossRef]
23. Zhang, X.Y.; Du, X.F.; Li, Y.B. Comprehensive evaluation of water resources carrying capacity in ecological irrigation districts based on fuzzy set pair analysis. *Desalin. Water Treat.* **2020**, *187*, 63–69. [CrossRef]
24. Lu, Y.; Xu, H.W.; Wang, Y.X.; Yang, Y. Evaluation of water environmental carrying capacity of city in Huaihe River Basin based on the AHP method: A case in Huai'an City. *Water Resour. Ind.* **2017**, *18*, 71–77. [CrossRef]
25. Meriem, N.A.; Ewa, B.A. Water resources carrying capacity assessment. The case of Algeria's capital city. *Habitat Int.* **2016**, *58*, 51–58.
26. Gao, Y.; Zhang, S.; Xu, G.W.; Su, H.M.; Zhang, Y. Study on Water Resources Carrying Capacity in Hefei City. *Adv. Mater. Res.* **2012**, *610*, 2701–2704. [CrossRef]
27. Sun, Q.F.; Guo, X.D.; Tian, H.; Yu, H.M.; Li, X.G.; Liang, X.J.; Xiao, C.L.; Zhang, Q.; Wang, G.; Qi, L.L. *Comprehensive Research on Water Resources and Geological Environment of Chang-Ji Economic Circle*; China University of Geosciences Press: Changchun, China, 2020. (In Chinese)
28. Zhang, Q. *Research on Rational Water Resources Allocation in Chang-Ji Economic Circle*; Jilin University: Changchun, China, 2017. (In Chinese)
29. Pang, Y.; Lu, G.H. *Theory and Application of Water Environment Capacity Calculation*; Science Press: Beijing, China, 2010. (In Chinese)
30. Han, L.X.; Yan, F.F.; Peng, H.; Gao, J.J.; Pan, M.M. Methods for Calculation of Water Environment Capacity of Small and Medium River Channels. *Adv. Mater. Res.* **2013**, *610*, 2745–2750. [CrossRef]
31. Lou, Y. *Jilin City Water Resources Carrying Capacity Evaluation Research*; Jilin University: Changchun, China, 2017. (In Chinese)
32. *Standards for Local Standard Water Use in Jilin Province*; DB22/T 389-2010; Quality and Technical Supervision Bureau of Jilin Province: Changchun, China, 2020. (In Chinese)
33. Huang, B.S.; Hong, C.H.; Du, H.H. Quantitative study of degradation coefficient of pollutant against the flow velocity. *J. Hydrodyn.* **2017**, *29*, 118–123. [CrossRef]
34. Zhang, Z., Lu, W.X., Zhao, Y., Gong, W.D. Development tendency analysis and evaluation of the water ecological carrying capacity in the Siping area of Jilin Province in China based on system dynamics and analytic hierarchy process. *Ecol. Model.* **2014**, *275*, 9–21. [CrossRef]
35. Jia, Z.M.; Cai, Y.P.; Chen, Y.; Zeng, W.H. Regionalization of water environmental carrying capacity for supporting the sustainable water resources management and development in China. *Resour. Conserv. Recycl.* **2018**, *134*, 282–293. [CrossRef]
36. Cui, Y.; Feng, P.; Jin, J.L.; Liu, L. Water Resources Carrying Capacity Evaluation and Diagnosis Based on Set Pair Analysis and Improved the Entropy Weight Method. *Entropy* **2018**, *20*, 359. [CrossRef]
37. Chen, Z.H.; Wei, S. Application of System Dynamics to Water Security Research. *Water Resour. Manag.* **2014**, *28*, 287–300. [CrossRef]
38. Ren, C.F.; Guo, P.; Li, M.; Li, R. An innovative method for water resources carrying capacity research e Metabolic theory of regional water resources. *J. Environ. Manag.* **2016**, *167*, 139–146. [CrossRef]
39. Cai, Y.P.; Huang, G.H.; Tan, Q.; Liu, L. An integrated approach for climate-change impact analysis and adaptation planning under multi-level uncertainties. Part II. Case study. *Renew. Sustain. Energy Rev.* **2011**, *15*, 3051–3073. [CrossRef]
40. Ali, A.B.; Hossein, A.; Jürgen, S. A system dynamics model of smart groundwater governance. *Agric. Water Manag.* **2019**, *221*, 502–518.

41. Du, Z.; Hu, Y.G.; Buttar, N.A. Analysis of mechanical properties for tea stem using grey relational analysis coupled with multiple linear regression. *Sci. Hortic.* **2020**, *260*, 108886. [CrossRef]
42. Wu, H.W.; Su, D.W.; Huo, X.S.; Hu, S.; Wang, Z.D.; Sun, K.Q. The Research of Mid-Long Forecasting Based on MGM (l, N) Model with Multiple Linear Regression Analysis in Nanjing Core Area. In Proceedings of the Asia-Pacific Power and Energy Engineering Conference, Xi'an, China, 25–28 October; 2016; pp. 38–42.

 *water*

Article

System Dynamics Modeling for Supporting Drought-Oriented Management of the Jucar River System, Spain

Adria Rubio-Martin *, Manuel Pulido-Velazquez *, Hector Macian-Sorribes and Alberto Garcia-Prats

Research Institute of Water and Environmental Engineering (IIAMA), Universitat Politècnica de València, 46022 Valencia, Spain; hecmasor@upv.es (H.M.-S.); agprats@upv.es (A.G.-P.)
* Correspondence: adrumar@upv.es (A.R.-M.); mapuve@hma.upv.es (M.P.-V.)

Received: 25 March 2020; Accepted: 10 May 2020; Published: 15 May 2020

Abstract: The management of water in systems where the balance between resources and demands is already precarious can pose a challenge and it can be easily disrupted by drought episodes. Anticipated drought management has proved to be one of the main strategies to reduce their impact. Drought economic, environmental, and social impacts affect different sectors that are often interconnected. There is a need for water management models able to acknowledge the complex interactions between multiple sectors, activities, and variables to study the response of water resource systems to drought management strategies. System dynamics (SD) is a modeling methodology that facilitates the analysis of interactions and feedbacks within and between sectors. Although SD has been applied for water resource management, there is a lack of SD models able to regulate complex water resource systems on a monthly time scale and considering multiple reservoir operating rules, demands, and policies. In this paper, we present an SD model for the strategic planning of drought management in the Jucar River system, incorporating dynamic reservoir operating rules, policies, and drought management strategies triggered by a system state index. The DSS combines features from early warning and information systems, allowing for the simulation of drought strategies, evaluating their economic impact, and exploring new management options in the same environment. The results for the historical period show that drought early management can be beneficial for the performance of the system, monitoring the current state of the system, and activating drought management measures results in a substantial reduction of the economic impact of droughts.

Keywords: water management; resources; system dynamics; drought management; drought impacts

1. Introduction

Drought is a natural hazard and, as such, has to be understood as a natural feature of climate. Whether or not a drought becomes a disaster depends on its social, economic, and environmental impacts [1]. Therefore, the key to understanding drought is to acknowledge its different dimensions. Drought affects both surface and groundwater resources and can lead to reduced water supply for in-home consumption and agricultural and industrial activities. Furthermore, it can deteriorate water quality by rising nitrate, ammonium, and phosphate concentrations, and disturb riparian habitats [2,3]. Agriculture is the most affected sector by droughts, but many other sectors may suffer relevant losses, including energy production, tourism and recreation, transportation, urban water supply, and the environment. Sustained drought can cause social, economic, and energy crises, even leading to migration from affected zones (often rural and agricultural-focused) to other regions or nearby countries [4]. Drought is not the only issue that water resource systems have to face regarding water

management policies and rules. To assess the truthfulness of the quantitative models they are validated by comparing their results to the available historical records.

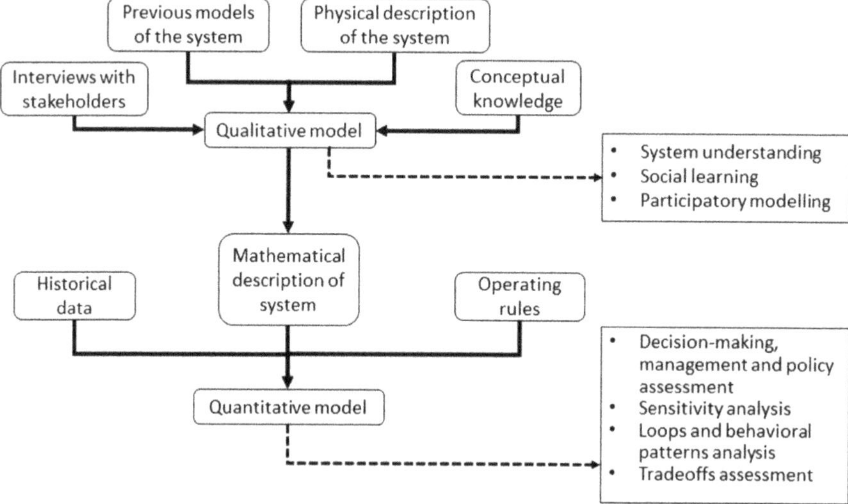

Figure 1. System dynamics modeling framework.

Traditionally, water resources management models were designed with a one-dimensional optimal engineering approach, performed with little regard for social, environmental, or cultural aspects [32]. However, the increased recognition of complexity and uncertainty has promoted the use of more flexible simulation-based tools such as the ones provided by system dynamics [28]. System dynamics provides tools for the graphical representation of systems, facilitates flexible and transparent modeling, eases the holistic understanding of the problem, captures long-run behavioral patterns and trends, facilitates clear communication of model structure and results, promotes sharing modeling, facilitates sensitivity analysis, and it is suitable for policy assessment and selection [25]. System dynamics modeling environments include Powersim (Powersim Corp., 1993), Simile (Simulistics, 2002), Stella (High Performance Systems, 1992), and Vensim (Ventana Systems, 1996). Nowadays, these environments are able to assist modelers and can handle many variables, delays, and interdependent subsystems, allowing the creation of modular object-oriented models, therefore increasing interchangeability and reusability.

The application of system dynamics in water resource management has grown since the 90s. Nowadays, we find applications of system dynamics modeling to study a large variety of water resource issues [29]. They range from region-scale models with multiple demands and frequent water scarcity events [33,34], to models coupling surface and groundwater dynamics for a basin [35], flood management or predicting models [36,37], reservoir operation and water supply for multiple water users [38], and the design of water pricing policies [39]. However, system dynamics application to simulate the management of highly regulated water resource systems integrating multiple reservoirs, operating rules, dynamic drought management, groundwater use, and conflicting water demands remains very limited. Yet all these features are required to analyze the issue of drought early warning and management in complex water resource systems.

Drought management is a multidimensional concept that includes meteorological, ecological, hydrological, environmental, and socioeconomic perspectives. The development of DSS for improving drought management requires the combination of several models [6]. Coupling and analyzing the interactions between these models is often a difficult issue. System dynamics is a methodology that provides a common playground for the interaction of different subsystems and submodels, facilitating the analysis of the existing relationships and providing a holistic view of the issue.

2.2. Case Study: Drought Management in the Jucar River System

The Jucar River system is located in Easter Spain. The system is subjected to a tight equilibrium between total water demand (1505 Mm^3/year, 2009–2015 period average) and water resource availability (1548 Mm^3/year) [40]. Agriculture is the largest water use by far (89%), followed by urban (9%) and industrial uses (2%). The Jucar is the main source of urban water supply to the city of Valencia and its metropolitan area (about 1,500,000 inhabitants, third largest municipality in Spain). Water from the Jucar is diverted to the Turia River through a 60 km canal (Canal Jucar-Turia), also used for irrigation of mainly citrus and vegetables (Figure 2). Furthermore, there is an intense water use for irrigation in the lower Jucar, downstream of Tous reservoir, with traditional irrigation districts holding senior water rights dating back to the Middle Ages. Non-consumptive water demands include minimum ecological instream flows and hydropower generation.

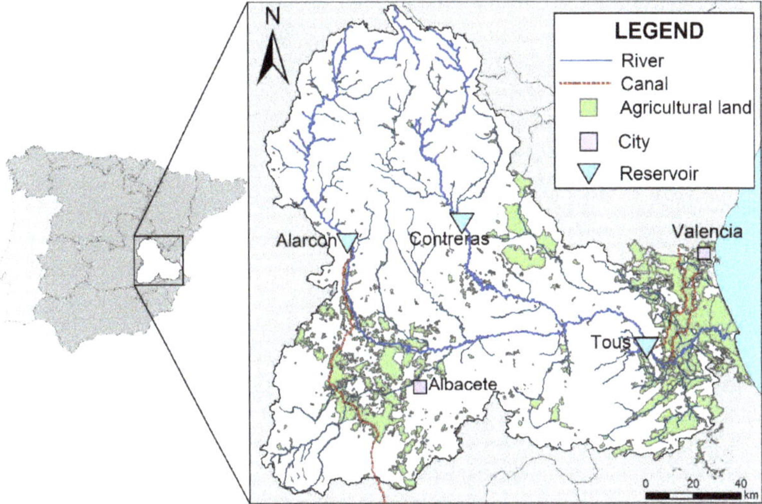

Figure 2. Main features of the Jucar River system included in the model.

The main surface reservoirs are Alarcon (1112 Mm^3 of capacity), Contreras (463 Mm^3 of useful capacity), and Tous (378 Mm^3). The regulation capacity of these reservoirs is mainly multi-annual: Alarcon and Contreras are devoted to consumptive uses, while Tous is mostly used for flood protection. The intense overexploitation of the main groundwater body, the Mancha Oriental aquifer (middle basin, near Albacete), for irrigation since the 1970s has shifted the stream-aquifer interaction between Alarcon and Tous from gaining to losing river, diminishing downstream surface water availability. The sustainable use of this aquifer is one of the challenges in the management of the system [41,42]. During droughts, the Plana de Valencia Sur aquifer, located in the lower basin (downstream of Tous), is used as an alternative water source.

Water scarcity, irregular hydrology, and groundwater overdraft result in droughts with significant economic, social, and environmental consequences. This situation is expected to be exacerbated by the impacts of climate and socioeconomic (global) changes and increasing institutional impediments from political disputes among the two main riparian regions, Castilla-La Mancha (upstream; mainly Albacete province) and Valencia (middle and downstream basin). A range of different innovative solutions are considered to face the main water management issues, such as pumping-water right acquisitions during droughts, increasing wastewater reuse, "in lieu" recharge (providing surplus surface water to groundwater users, keeping groundwater in storage for later use), water-saving in agriculture through drip irrigation, new water allocation mechanisms, water banks, water pricing, and irrigated crop drought insurances (among others), which makes this case a real lab for analyzing risk management strategies to cope with drought, extreme events, and climate change [43].

The operation of the system, managed by the Jucar River Basin Authority (Confederacion Hidrografica del Jucar, CHJ), is subject to physical, environmental, and legal constraints. The main physical constraints correspond to the reservoir, river, and canal capacities. The environmental constraints are the minimum flows prescribed in certain river reaches and the inflow requirements of the Albufera wetland. The main legal constraint in the Jucar River system is the Alarcon Agreement, signed between the Spanish Ministry of the Environment and the senior users of the Jucar River—mainly farmers—gathered together in the Unidad Sindical de Usuarios del Jucar (USUJ). The agreement divides Alarcon in two zones by a rule curve. If the water level in Alarcon is above the threshold, water can be freely allocated, but if the storage is below certain value, water in the system is reserved exclusively for the USUJ members. In this case, other water users who want to access water from the Jucar River would have to pay a financial compensation to USUJ. The operators also follow additional criteria to decide the releases during the irrigation period (May–September): not causing undesired spills from Tous (the downstream reservoir), not storing more than 450 Mm^3 in Contreras to avoid stability problems, and not storing more than 72 Mm^3 in Tous at the end of the summer to avoid flood damage during autumn due to intense rainfall events [42].

The Jucar River basin, as most Mediterranean and south-eastern basins of Spain, is very vulnerable to droughts [11]. The recurrence of these events is also an important factor when considering the management of the system, as a high-frequency appearance of droughts do not allow the system to properly recover water storage to face future water-scarcity events. The latest drought periods (1991–1995; 1997–2000 and 2004–2009) were classified as extreme droughts using the SPI index [14]. During the drought period 2005–2008, surface water available for agriculture decreased by up to 40% compared to the average. Because of this, drought emergency wells in the lower basin were activated. Despite of these efforts, the drought caused an important economic impact, especially to agriculture activities. The situation is expected to be exacerbated by the impact of climate change [8,44].

A key feature of drought management plans is the indices that define the different drought stages and trigger mitigation measures. Drought indices should capture the state of the water resource system as a whole, allowing the planner to active measures to reduce its impact. Some of the measures for drought management include conjunctive use of surface and groundwater, awareness campaigns to promote domestic water savings, economic tools, control of the supply to agricultural demands from reservoirs, and water reuse [16]. Traditionally, the management of droughts in the Jucar Basin was regulated as an emergency, and the application of Royal Decrees was necessary to mitigate their impacts [12]. From 2007, drought management in the Jucar River system is regulated by a drought management plan [14,40] that establishes a state index to monitor the system and a set of drought management measures triggered by the different drought stages. This index is calculated using different variables distributed in the area of the river basin, including reservoir storages, groundwater levels, streamflow, precipitation, and reservoir inflows. The state index takes values between 0 and 1, with four system states: normal, pre-alert, alert, and emergency. Then, different drought management measures are applied depending on the system's state index. These measures can be divided into 2 groups, (1) control of water supply for urban and agricultural uses and, (2) increase of water availability by drought emergency wells use and increasing water reuse.

2.3. System Dynamics for the Jucar River System

The system dynamics model developed for the Jucar River system represents its current management with a monthly time step, including the state index of the system and the management measures linked to this state. The software Vensim Pro [45] has been used for the creation of the model. The Jucar model was divided into 5 subsystems:

1. General view of the system: defines the system structure, its three main reservoirs, the connections, intakes, and outflows from the river (Figure 3).
2. Mancha Oriental aquifer: simulates the aquifer using a two-cell embedded multi-reservoir model, in line with the one used by the CHJ in its water resource management models [40,42].
3. Water demands: defines the different monthly water demands, the distribution of water, and the system deliveries and deficits.
4. Reservoir operation: defines the seasonal operating rules of the system.
5. State index: calculates the state index and defines the management measures to take based on it.

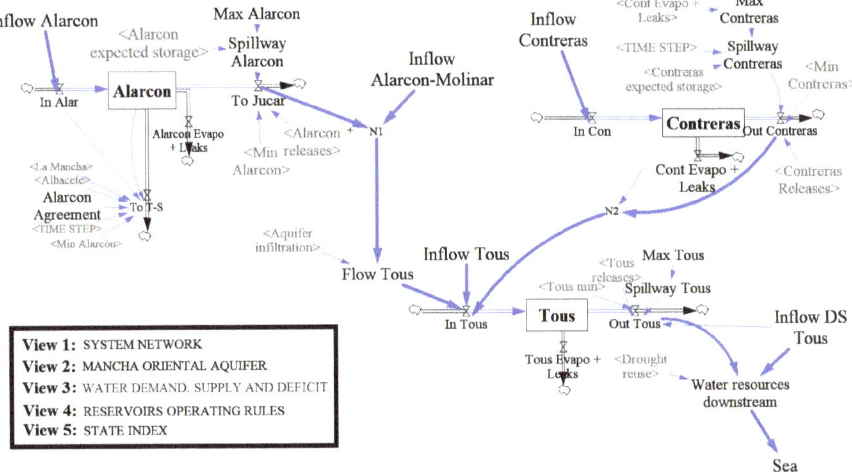

Figure 3. Subsystem for the general view of the system.

The model incorporates monthly water inflows in 5 sub-basins where data from CEDEX [46], the Spanish institution responsible for collecting and supplying data on civil engineering and water, has been obtained and processed. The main view of the model (Figure 3) captures the water flows through the Jucar system, including water infrastructures and stream-aquifer interaction with the Mancha Oriental aquifer. This structure is based on previous models for the area [42], and provides a general framework to visualize the system's network and to allow the integration of other sub-models. The model incorporates a submodel that simulates the current operation of the system (Figure 4), based on historical records and trends of the main variables.

The operating rules of the three reservoirs are defined at the monthly scale, mimicking the operation of the system for the 2003–2013 period, and introducing the constraints that bind the seasonal operation of the Jucar River system. The rules were obtained using fuzzy rule-based systems (FRB), co-developed with the experts from the Operation Office of the Jucar River Basin Authority [42]. A series of workshops and surveys were used to extract the decision-making processes followed in the seasonal operation of the Jucar River system. The implicit operation of the system was encoded into two FRB systems that were validated against historical records on reservoir storages and releases, streamflows, and deliveries to consumptive demands for the 2003–2013 period. The developed FRB were introduced into the SD model through piecewise linear regressions equations (Figure 5). Some flexibility is lost in the process of transforming the FRB rules into linear regressions, as it can be observed in the figure.

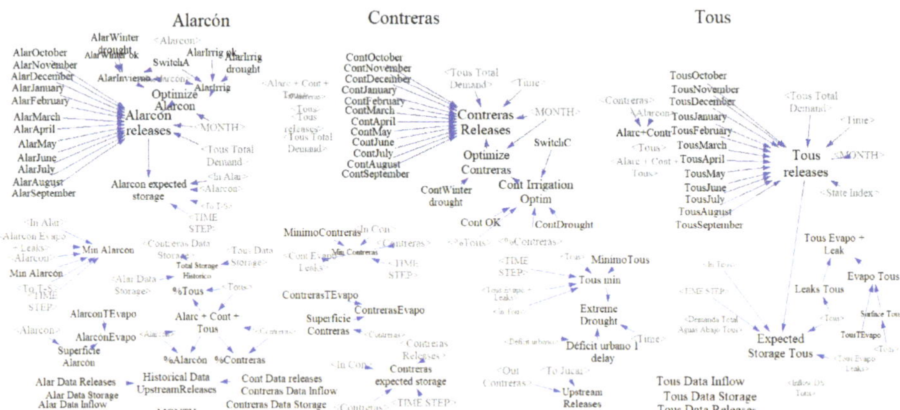

Figure 4. Reservoir operating rules subsystem that incorporates the monthly operating rules for each reservoir, variables, and seasonal parameters that determine final releases.

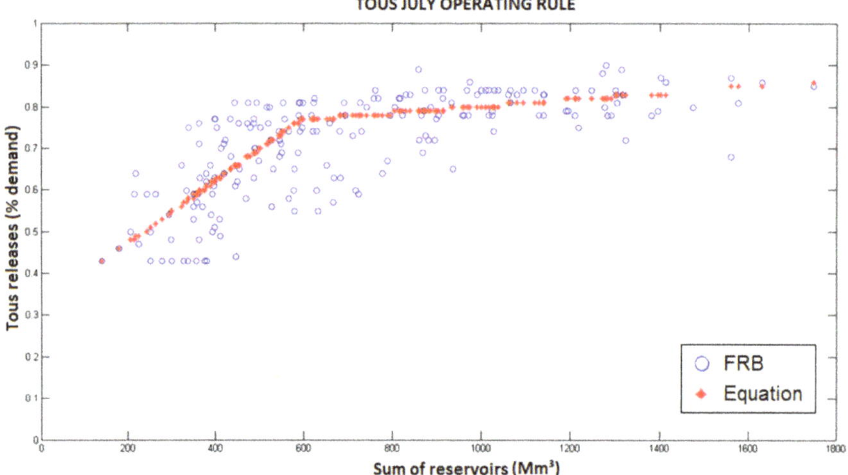

Figure 5. Graphical representation of the simulated operating rule of Tous reservoir in July. Blue dots represent the values of releases using fuzzy logic. Red crosses show values for the piecewise linear regression introduced into the SD model.

To compensate this loss of flexibility, the obtained rules for Alarcon and Contreras reservoirs were adjusted using seasonal factors (depending on whether it was or not irrigation season) and a scarcity factor different for both winter and summer seasons, to account for differences observed in the management of the system that were not correctly captured by the calculated piece-wise linear equations. Releases from Tous were computed as the minimum value between the downstream demand and the releases calculated by the piecewise linear equations. This implies that the system will not release more water from Tous than needed, minimizing unwanted releases to the sea while still capturing the seasonal behavior provided by the operating rules. The Alarcon Agreement was explicitly introduced into the model's formulation.

The water demands considered by the model are divided into urban and agricultural demands and were located and compiled from the public information provided by the CHJ [47]. Most of them are located downstream Tous, although the model also accounts for the demands located in the middle basin, one of them being a groundwater demand that affects the stream-aquifer interaction. The current operating rules of the system prioritizes water allocation to urban uses. Environmental requirements have been considered as a restriction and are captured by the operating rules of the reservoirs.

The model simulates stream-aquifer interaction between the Mancha Oriental aquifer and the Jucar River using a two-cell Embedded Multi-reservoir Model (Figure 6) [48]. Its formulation is based on the analytical solution of the stream-aquifer flow equation applied to linear systems, as well as it analogy with the state equation. Groundwater discharge can be expressed as the theoretical sum of an infinite number of linear reservoirs whose discharge is linearly proportional to the stored volume. In normal conditions, a limited number of linear reservoirs is enough to adequately reproduce groundwater discharge. Although the EMM does not calculate spatially-distributed heads and internal groundwater flows, it can provide an accurate representation of stream-aquifer interactions, even in karstic aquifers [49,50] and it is used in some general DSS services for water resource management [42,51]. Groundwater flow is calculated as the integration of the outflow of 2 linear reservoirs in which the discharge is linearly proportional to the volume stored. The EMM built for the Mancha Oriental aquifer represents exclusively the impacts of the anthropic stresses on stream-aquifer interaction, since the natural discharge was already included in the natural inflow time series of the model [42]. The anthropic-induced net recharge corresponds to the agricultural percolation minus groundwater abstractions. As shown by Macian-Sorribes et al., 2017, the calibrated EMM was able to capture well both the over-year trend and the seasonal variation of the historical values.

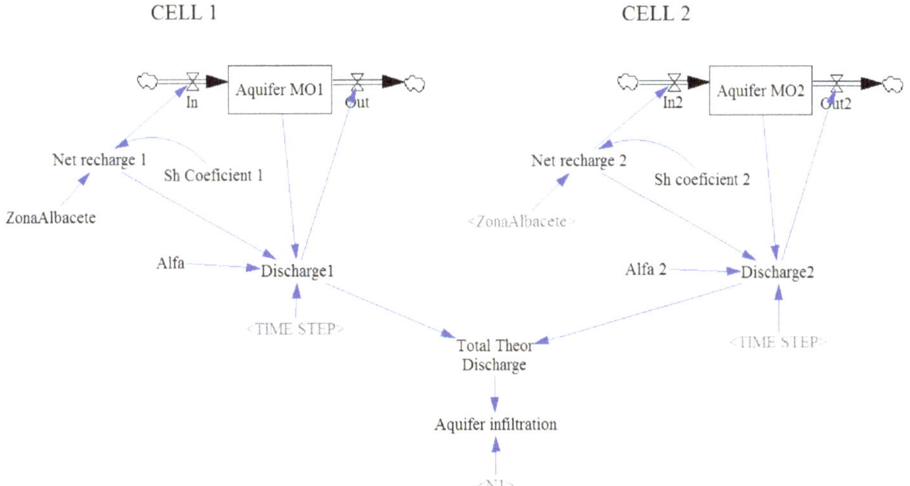

Figure 6. Subsystem for the stream aquifer interaction between the Jucar River and the Mancha Oriental aquifer.

The model also implements a state index subsystem. This subsystem checks the state of the system each time-step during the simulation, as does the state index used by the CHJ on a monthly basis. The equations defining the relationship between past and present system states are taken from the Jucar drought management plan [14,52].

The monthly system state index (S_i) has the following expression:

$$S_i = \frac{1}{2}\left[1 + \frac{V_i - V_{av}}{V_{max} - V_{av}}\right] \text{ if } V_i \geq V_{av}$$

$$S_i = \frac{V_i - V_{min}}{2(V_{av} - V_{min})} \text{ if } V_i < V_{av}$$

where V_i is the value of the variable at the beginning of the month i and V_{av}, V_{max} y V_{min} are the recorded average, maximum and minimum monthly values of the variable since 1982. In the case of the SD model, the subsystem uses historical data of the average, maximum, and minimum value of water storage for each one of the three reservoirs and compares the recorded values to the current state of the system. Although the evaluation of the system state index executed by the water authority for the Jucar River basin takes into account 9 additional variables other than the water storage (including piezometric levels and water inflows), in regulated systems the volume stored in the reservoirs is regarded as a good approximation of the actual status of the whole system [53].

The state index subsystem is able to trigger drought management measures depending on the current state of the system (Figure 7).

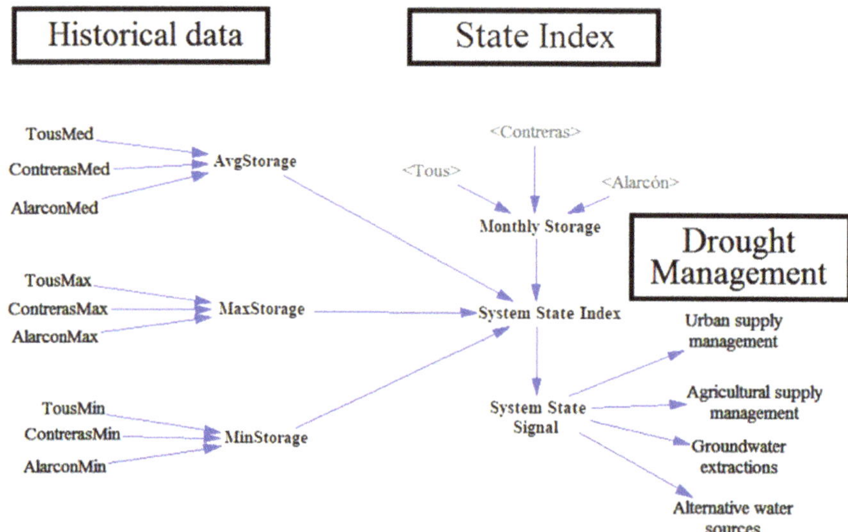

Figure 7. State index subsystem to calculate the system's state comparing with historical data and incorporating drought management strategies.

The system state index takes values that range from 0 to 1. Each month, the model transforms the system state index (a floating-point number) to the corresponding integer state (normal, pre-alert, alert, and emergency) applying the thresholds defined by the water authority. Drought management strategies defined in this subsystem are introduced as actions into their respective subsystems using shadow variables. The measures implemented consider both supply and demand side solutions. For instance, when triggered, the variable "Agricultural supply management" is linked to the agricultural supply on the "Water demand, supply and deficit" subsystem applying a restriction of 20% or 40% on the deliveries to the agricultural demands, depending on the state index. "Urban supply management" restricts the water delivered to the urban demand in alert or worse situations by 5%, reproducing the estimated effect of the water-saving awareness campaigns proposed by the water authority [14,52]. "Groundwater extractions" and "Alternative water sources" variables simulate the use of wells and the reuse of wastewater respectively for agricultural supply; the intensity of both actions depends on the monthly state of the system. All the values and management measures represented in the state index subsystem are based on the current drought management plan for the system.

3. Results and Discussion

3.1. Model Evaluation

The system dynamics model of the Jucar River system was evaluated using the 2003–2013 period. The comparison between the model's results and historical records showed that the model is able to capture the observed operation (Figure 8). Residual plots for the same variables can be found in Appendix A (Figure A1). Total storage was closely reproduced by the model, as can be observed in the plot and in the R-squared index. The Alarcon and Contreras releases were adequately reproduced on a broader view, due to the resemblance of the intra-annual patterns. However, the model results depart from the historical observations in some years. This is due to the fact that the middle basin is modeled in less detail than the lower one. For instance, hydroelectric production has not been included in the middle basin. In any case, storages in Alarcon and Contreras are adequately reproduced (Figure A2) and the overall in-year dynamics of the system was matched, so these deviations do not have a significant impact on the performance of the model. Tous releases results correctly fit the available data. These releases have a major importance for the model since the majority of the surface water demands are located downstream. As for water supply deficits, the simulated values matched the observed data adequately, including the main peaks associated with the 2005–2008 drought, especially during the years when the drought was more severe. Differences between observed and simulation results can also be explained by the fact that the model assumes a constant annual demand for the whole simulation period while, actually, demands changed due to population change, variation in irrigated areas, and shift from gravity to drip irrigation [47].

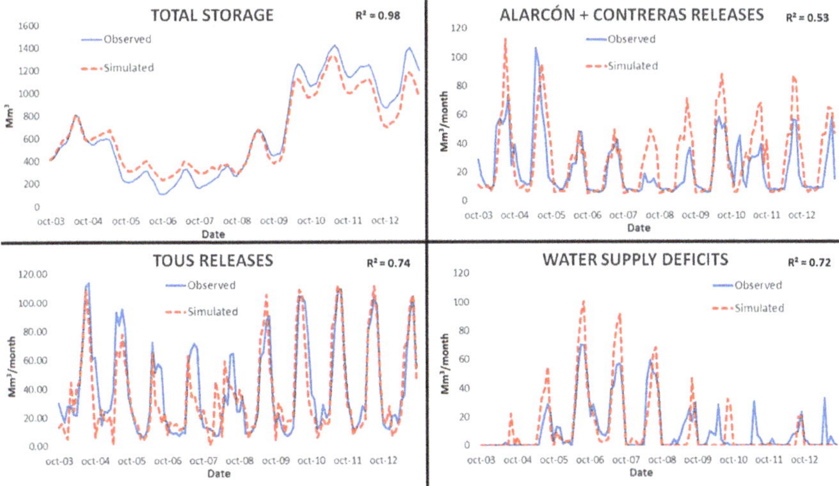

Figure 8. Comparison between observed and simulated values for key variables of the system.

Releases from Alarcon and Contreras are cumbersome to model because of the uncertainties of the middle basin, changes in downstream demands, and varied criteria of releases and management over the simulated period. Although it would be possible to introduce variable demands into the model, there is no available data to represent the variation of all the demands during the simulation period. Furthermore, although the model incorporates monthly operating rules for the reservoirs based on a fuzzy logic representation of the system operation reported by the managers [42], those rules cannot reproduce discretionary changes in the operation of the system during the simulation period.

Once verified that the developed model matches adequately the historical behavior of the Jucar River system, further simulations were launched to test different management assumptions and scenarios.

3.2. System State Index and Drought Management Strategies

The SD model has been applied to study the interaction between the previously indicated drought management strategies and other variables of the system. A comparison between simulations with and without the drought management strategies introduced into the management in 2007 was performed. Results obtained when applying the drought management measures show improvements for the state index of the system and for the system's total water storage (Figure 9).

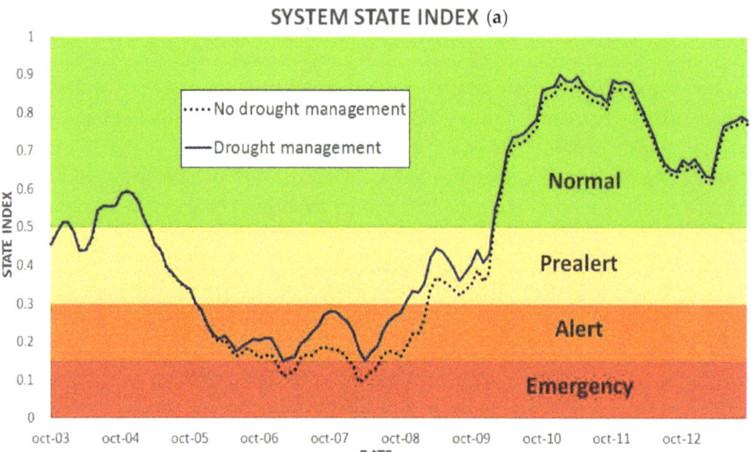

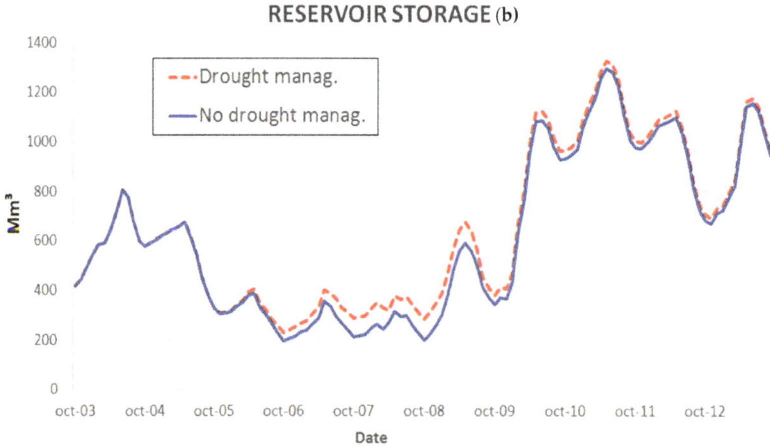

Figure 9. (**a**) State index and (**b**) total storage with and without drought management strategies.

The system state index benefits from applying the drought management strategies defined in the state index subsystem. Thanks to them, the system state does not drop into an emergency state during the 2005–2008 drought. It also recovers earlier from the alert state during that drought, and it enters the prealert stage months before than the scenario with no drought management measures.

After the system enters a normal state, it is worth pointing out that the state index is higher for the drought managed model, even when the drought is over (from 2010 onwards). According to the model, water storage in reservoirs is increased significantly when drought management measures are applied. The difference is up to almost 100 Mm^3 during October 2008. This is the result of the management strategies taken in anticipation thanks to the state index and the four threshold levels defined. The anticipated management also allows to reduce the system vulnerability by 62% in comparison with the scenario without drought management and considering vulnerability as the ratio between total water supply deficit and the number of failures to meet the demands during the whole period. A reduction in vulnerability means that the average water shortage is lower, although the frequency of these shortages may increase. These drought management measures entail the use of drought emergency wells for water abstraction within a maximum of 98 Mm^3/year (Figure 10) following the plan defined by the water authority [14,52].

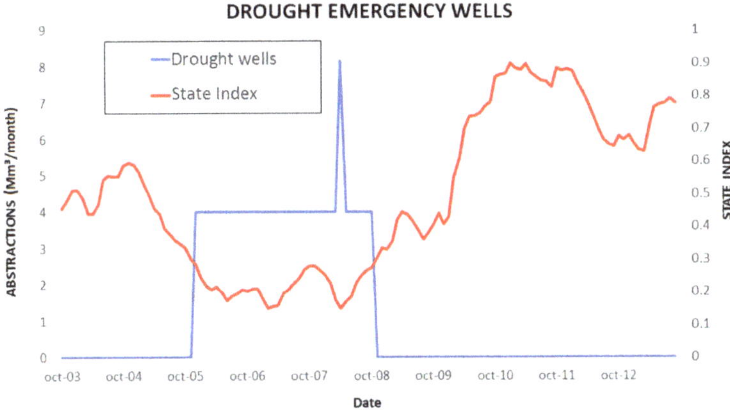

Figure 10. Water abstraction from emergency wells during drought compared to the system state index.

Water pumping from drought emergency wells located in the lower basin compensates the reduced surface water supply and alleviate the drought impact on agriculture. These groundwater abstractions are activated when the system falls into the alert state, and water abstraction scale up above 8 Mm^3/month if the emergency state is reached (Figure 10).

3.3. Economic Impact of Droughts

Results show that the total reservoir storage of the basin improves when drought management measures are applied. It is to expect that the gained storage will benefit the early recovery of the system allowing for more regular deliveries to agricultural demands. Indeed, it is possible to calculate the economic losses associated with the mismanagement of droughts for the 2003–2009 period (Figure 11).

Economic losses were calculated by economically characterizing the monthly demands of the system defined as targets [47] using demand curves or functions obtained by Positive Mathematical Programming (PMP) [54] for the different agricultural demands [55]. Benefits were obtained as the integration of the demand function between zero and the level of supply. It can be observed (Figure 11) that economic losses concentrate on the drought period (2005–2008), particularly when the system state index stays in alert for several months (2006–2008). During the irrigation season in drought periods is when economic losses rise due to water scarcity. The fact that, as defined by the drought management strategies subsystem, in alert and emergency states the water supply for agriculture is reduced by up to 40% its original demand could be thought of as detrimental for agricultural interests. However, according to the simulations, the water saved helps a faster recovery of the system, guarantees urban water supply, and reduces the long-term impact of droughts. In the model, the economic impact of the

2005–2008 drought was reduced from 89 M€ to 29 M€ thanks to the drought management strategies implemented. Due to conjunctive use of superficial and groundwater, agricultural activities suffer lower impact even considering the significant restrictions they suffer during the alert and emergency states. When the amount of available water is scarce, using groundwater to supply crops under deficit irrigation guarantees the survival of the plantations and minimizes economic losses.

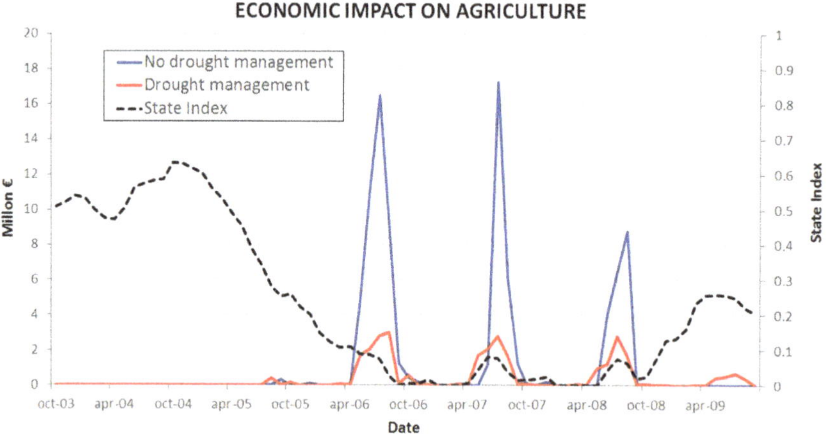

Figure 11. Estimation of economic losses for agriculture compared to the system state index during the 2000's drought.

4. Conclusions

This paper presents a system dynamics DSS for drought management of the Jucar River system, taking into account the combination of a state index and several drought management strategies. The resulting DSS showed the potential of system dynamics for simulating the management of multi-reservoir systems, integrating monthly-defined operating rules for the reservoirs, stream-aquifer interaction, conflicting water demands, and drought management strategies. The model adequately reproduces the operation of the system and is able to produce accurate quantitative results, as shown by the comparison with the historical records.

The DSS takes advantage of the holistic concept that drives the methodology and incorporates components from different disciplines (hydrology, economics, social sciences, laws, etc.) into its modular structure. The state index subsystem is an example of how it is possible to integrate policies and management strategies into a water resource model using a system dynamics approach. Likewise, water policy or legislation has been incorporated into the model—e.g., the Alarcon agreement.

The DSS opens up the possibility of analyzing different drought management strategies and assessing the interactions, feedbacks, and impacts within and between multiple sectors and variables.

Results showed that drought management strategies have a net positive effect in the Jucar River system from both the economic (agriculture) and the water management perspective. The defined measures lowered agricultural losses for the 2005–2008 drought period and increased the amount of stored water during drought allowing the faster recovery of the system. Although the model provides quantitative results similar to the historical data available, the main goal of a system dynamics model is neither to forecast nor to optimize, but studying patterns, trends, and interactions between different variables of the model [24]. Modeling and dynamically simulating the change in water resources over time provides a scientifically defensible basis for proactive management strategies, enhancing our prospects to maximize the adaptive capacity of the system as a whole [29].

Moreover, the same methodology used to study drought management strategies can be applied to study the impact of different realities and inputs into the system. The DSS model developed for the Jucar River system uses a quantitative approach for its simulation. Consequently, it requires numeric data and well-tuned equations to capture the behavior of the system in detail. Qualitative variables and inputs can also be implemented in this kind of model. Qualitative modeling often introduces "soft" variables to study the general patterns of behavior of the model, rather than precise numbers [56]. In this case, qualitative modeling can be restricted to new subsystems for the testing of different non-easily quantifiable hypothesis.

The model herein presented was successfully developed for the Jucar case study and it could be replicated in any basin or system where enough information and data are available. The development of quantitative system dynamics models requires the use of a large volume of data coming from different fields (from hydrological to economic and reservoir data) as well as a deep understanding of the system structure and behavior. Very often, the most complex issue of this type of model is the development of the monthly operating rules for the reservoirs. In this case, the final rules were inferred using fuzzy logic, but additional tests showed that it is possible to simulate the operation of the system using other approaches and calibrating the rules with the historical records for the releases and water storage of the reservoirs. Although the model is able to reproduce the stream aquifer interaction between the Jucar River and the Mancha Oriental aquifer, it simulates neither groundwater heads nor aquifer storage. Groundwater head specifically is a determinant factor for the Mancha Oriental aquifer, as it has suffered continuous drops in groundwater levels due to intense pumping since the early 1970s. To assess the effect of drought policies on groundwater levels, it would be necessary to apply a detailed groundwater model, such as finite-difference model, coupling it with the system dynamics model either through scripting, wrapping, or spreadsheet coupling [57].

The model developed using system dynamics for the Jucar River system has the potential to grow and increase its scope by integrating new dynamics that can modify the behavior of the whole system. Future lines of work include linking the agricultural demand subsystem and a land-use subsystem, which would allow for introducing changes in agricultural land use based on economic benefit from previous years and on changes in land-use policies. System dynamics provides an excellent framework to study trade-offs that land use changes can introduce in specific sectors and communities [58]. Furthermore, it is already possible to activate population growths or losses over time to study how changes in urban demand can affect the system. These functionalities are required to test the effect of different climate change narratives within the next decades, which is also a future line of research to explore.

Author Contributions: Conceptualization, A.R.-M. and M.P.-V.; methodology, H.M.-S. and M.P.-V.; software, A.R.-M. and H.M.-S.; validation, A.R.-M. and H.M.-S.; formal analysis, A.R.-M. and H.M.-S.; investigation, A.R.-M., H.M.-S. and M.P.-V.; resources, A.G.-P. and M.P.-V.; data curation, A.R.-M. and H.M.-S.; writing—original draft preparation, A.R.-M.; writing—review and editing, A.G.-P., H.M.-S. and M.P.-V.; visualization, A.G.-P., A.R.-M., H.M.-S. and M.P.-V.; supervision, A.G.-P. and M.P.-V.; project administration, A.G.-P. and M.P.-V.; funding acquisition, A.G.-P. and M.P.-V. All authors have read and agreed to the published version of the manuscript.

Funding: The data used in this study was obtained from the references included. We acknowledge the European Research Area for Climate Services consortium (ER4CS) and the Agencia Estatal de Investigación for their financial support to this research under the INNOVA project (Grant Agreement: 690462; PCIN-2017-066). This study has also been partially funded by the ADAPTAMED project (RTI2018-101483-B-I00) from the Ministerio de Ciencia, Innovación y Universidades (MICIU) of Spain.

Conflicts of Interest: The authors declare no conflict of interest.

Appendix A

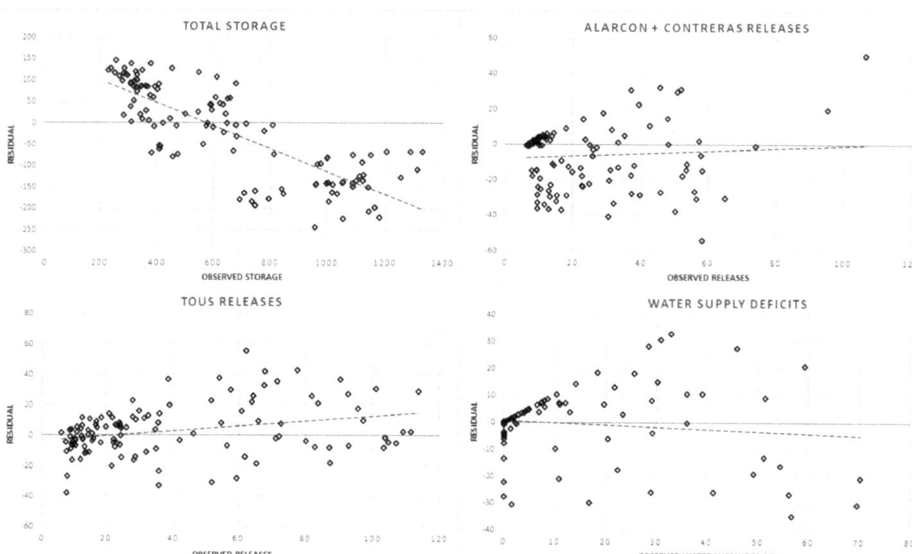

Figure A1. Residual plots of the variables presented in Figure 8.

The residuals for the variables of Alarcon & Contreras releases, Tous releases, and water supply deficit show a lack of general pattern and are distributed pretty symmetrically around the 0 line. Total storage, however, shows a pattern that was already observed in Figure 8: the model tends to store more water at the beginning of the decade and during the drought period, and it storages less water towards the late period. Several reasons have been given to explain this pattern. As most water resource management models, stationary conditions have been assumed for water demand and reservoir operation during the whole period. However, in reality, water demand and the operation of the reservoirs was changing during the 10-year period. There is not available data to correctly represent the variation of all the water demands, but we know that the demand at the beginning of the decade was greater than during the last years, due to changes in regulation and the improvement of control. We have assumed an average water demand based on the available data. This may explain in part why the model has more water than the observed at the beginning (in reality, the water demand was greater than the introduced) and less at the end (the water demand introduced is greater in the model). The same trend can be observed in Figure A2. Regarding the impact of the operating rules in the results, the rules are based on interviews and analysis performed in collaboration with the decision-makers [42] and are, in some regard, influenced by the knowledge gained during the decade simulated in our model. In reality the logic behind the operation of the reservoirs was evolving and changing during the whole period.

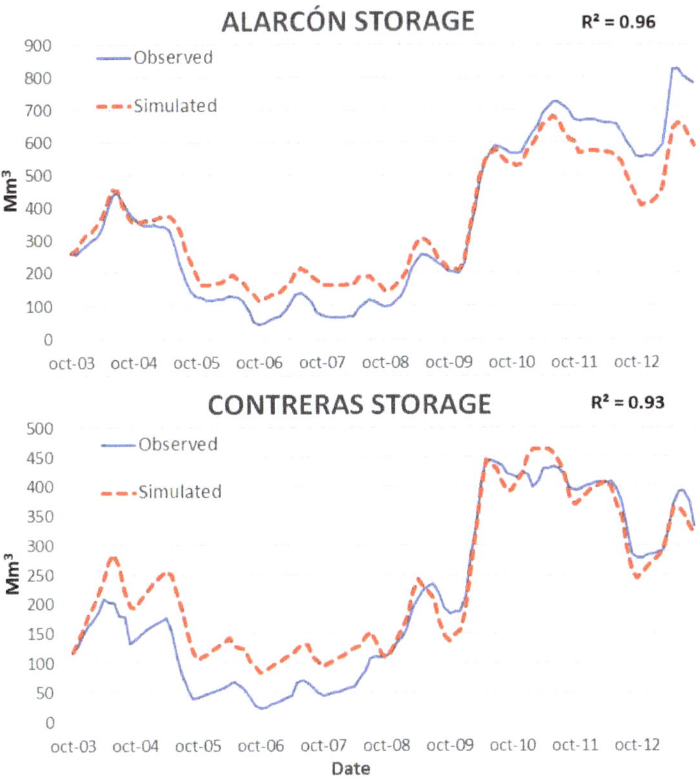

Figure A2. Observed vs simulated storage in Alarcon and Contreras reservoirs.

References

1. Wilhite, D.A.; Buchanan-Smith, M. Drought as Hazard: Understanding the Natural and Social Context. In *Drought and Water Crisis. Science Technology and Management Issues*; CRC Press Taylor and Francis: Boca Raton, FL, USA, 2005; pp. 3–27. ISBN 9780367393205.
2. Mishra, A.K.; Singh, V.P. A review of drought concepts. *J. Hydrol.* **2010**, *391*, 202–216. [CrossRef]
3. Momblanch, A.; Paredes-Arquiola, J.; Munné, A.; Manzano, A.; Arnau, J.; Andreu, J. Managing water quality under drought conditions in the Llobregat River Basin. *Sci. Total Environ.* **2015**, *503–504*, 300–318. [CrossRef] [PubMed]
4. Agrawala, S.; Barlow, M.; Heidi, C.; Lyon, B.; IRI International Research Institute for Climate and Society. *The Drought and Humanitarian Crisis in Central and Southwest Asia: A Climate Perspective*; IRI Spec. Rep. NO. 01-11; IRI: Palisades, NY, USA, 2001; Volume 20.
5. Van Loon, A.F.; Van Lanen, H.A.J. Making the distinction between water scarcity and drought using an observation-modeling framework. *Water Resour. Res.* **2013**, *49*, 1483–1502. [CrossRef]
6. Mishra, A.K.; Singh, V.P. Drought modeling - A review. *J. Hydrol.* **2011**, *403*, 157–175. [CrossRef]
7. Wilhite, D.A.; Sivakumar, M.V.K.; Pulwarty, R. Managing drought risk in a changing climate: The role of national drought policy. *Weather Clim. Extrem.* **2014**, *3*, 4–13. [CrossRef]
8. Marcos-Garcia, P.; Lopez-Nicolas, A.; Pulido-Velazquez, M. Combined use of relative drought indices to analyze climate change impact on meteorological and hydrological droughts in a Mediterranean basin. *J. Hydrol.* **2017**, *554*, 292–305. [CrossRef]
9. European Commission. Addressing the challenge of water scarcity and droughts in the European Union. *J. Chem. Inf. Model.* **2013**, *53*, 1689–1699.

10. Strosser, P.; Dworak, T.; Garzón Delvaux, P.A.; Berglund, M.; Schmidt, G.; Mysiak, J.; Kossida, M.; Iacovides, I.; Ashton, V. *Gap Analysis of the Water Scarcity and Droughts Policy in the EU. SWD 380 Final*; European Commission: Brussels, Belgium, 2012.

11. Estrela, T.; Vargas, E. Drought Management Plans in the European Union. The Case of Spain. *Water Resour. Manag.* **2012**, *26*, 1537–1553. [CrossRef]

12. Pedro-Monzonís, M.; Solera, A.; Ferrer, J.; Estrela, T.; Paredes-Arquiola, J. A review of water scarcity and drought indexes in water resources planning and management. *J. Hydrol.* **2015**, *527*, 482–493. [CrossRef]

13. Zaniolo, M.; Giuliani, M.; Castelletti, A.F.; Pulido-Velazquez, M. Automatic design of basin-specific drought indexes for highly regulated water systems. *Hydrol. Earth Syst. Sci.* **2018**, *22*, 2409–2424. [CrossRef]

14. CHJ (Confederación Hidrográfica del Júcar). *Plan Especial de Alerta y Eventual Sequía en la Confederación Hidrográfica del Júcar*; Confederación Hidrográfica del Júcar: Valencia, Spain, 2007; p. 185.

15. Carmona, M.; Máñez Costa, M.; Andreu, J.; Pulido-Velazquez, M.; Haro-Monteagudo, D.; Lopez-Nicolas, A.; Cremades, R. Assessing the effectiveness of Multi-Sector Partnerships to manage droughts: The case of the Jucar river basin. *Earth's Futur.* **2017**, *5*, 750–770. [CrossRef]

16. Iglesias, A.; Cancelliere, A.; Wilhite, D.A.; Garrote, L.; Cubillo, F. *Coping with Drought Risk in Agriculture and Water Supply Systems. Drought Management and Policy Development in the Mediterranean*; Iglesias, A., Garrote, L., Cancelliere, A., Cubillo, F., Wilhite, D.A., Eds.; Springer Netherlands: Dordrecht, The Netherland, 2009; ISBN 978-1-4020-9044-8.

17. Pallottino, S.; Sechi, G.M.; Zuddas, P. A DSS for water resources management under uncertainty by scenario analysis. *Environ. Model. Softw.* **2005**, *20*, 1031–1042. [CrossRef]

18. Sechi, G.M.; Sulis, A. Drought mitigation using operative indicators in complex water systems. *Phys. Chem. Earth* **2010**, *35*, 195–203. [CrossRef]

19. Svoboda, M.D.; Fuchs, B.A.; Poulsen, C.C.; Nothwehr, J.R. The drought risk atlas: Enhancing decision support for drought risk management in the United States. *J. Hydrol.* **2015**, *526*, 274–286. [CrossRef]

20. Buttafuoco, G.; Caloiero, T.; Ricca, N.; Guagliardi, I. Assessment of drought and its uncertainty in a southern Italy area (Calabria region). *Meas. J. Int. Meas. Confed.* **2018**, *113*, 205–210. [CrossRef]

21. Iglesias, A.; Garrote, L. Adaptation strategies for agricultural water management under climate change in Europe. *Agric. Water Manag.* **2015**, *155*, 113–124. [CrossRef]

22. Lewandowski, J.; Meinikmann, K.; Krause, S. Groundwater-surface water interactions: Recent advances and interdisciplinary challenges. *Water (Switzerland)* **2020**, *12*, 296. [CrossRef]

23. Forrester, J.W. Industrial Dynamics—After the First Decade. *Manag. Sci.* **1968**, *14*, 398–415. [CrossRef]

24. Sušnik, J.; Molina, J.L.; Vamvakeridou-Lyroudia, L.S.; Savić, D.A.; Kapelan, Z. Comparative Analysis of System Dynamics and Object-Oriented Bayesian Networks Modelling for Water Systems Management. *Water Resour. Manag.* **2013**, *27*, 819–841. [CrossRef]

25. Mirchi, A.; Madani, K.; Watkins, D.; Ahmad, S. Synthesis of System Dynamics Tools for Holistic Conceptualization of Water Resources Problems. *Water Resour. Manag.* **2012**, *26*, 2421–2442. [CrossRef]

26. Simonovic, S.P. World water dynamics: Global modeling of water resources. *J. Environ. Manag.* **2002**, *66*, 249–267. [CrossRef]

27. Saysel, A.K.; Barlas, Y.; Yenigün, O. Environmental sustainability in an agricultural development project: A system dynamics approach. *J. Environ. Manag.* **2002**, *64*, 247–260. [CrossRef]

28. Sterman, J.D. *Business Dynamics: Systems Thinking and Modeling for a Complex World*; McGraw-Hill Education: New York, NY, USA, 2000; ISBN 0072311355.

29. Winz, I.; Brierley, G.; Trowsdale, S. The use of system dynamics simulation in water resources management. *Water Resour. Manag.* **2009**, *23*, 1301–1323. [CrossRef]

30. Nikolic, V.V.; Simonovic, S.P. Multi-method Modeling Framework for Support of Integrated Water Resources Management. *Environ. Process.* **2015**, *2*, 461–483. [CrossRef]

31. Madani, K.; Mariño, M.A. System dynamics analysis for managing Iran's Zayandeh-rud river basin. *Water Resour. Manag.* **2009**, *23*, 2163–2187. [CrossRef]

32. Gleick, P.H. A Look at Twenty-first Century Water Resources Development. *Water Int.* **2000**, *25*, 127–138. [CrossRef]

33. Qaiser, K.; Ahmad, S.; Johnson, W.; Batista, J. Evaluating the impact of water conservation on fate of outdoor water use: A study in an arid region. *J. Environ. Manag.* **2011**, *92*, 2061–2068. [CrossRef]

34. Sušnik, J.; Vamvakeridou-Lyroudia, L.S.; Savić, D.A.; Kapelan, Z. Integrated System Dynamics Modelling for water scarcity assessment: Case study of the Kairouan region. *Sci. Total Environ.* **2012**, *440*, 290–306. [CrossRef]

35. Sehlke, G.; Jacobson, J. System dynamics modeling of transboundary systems: The river basin model. *Ground Water* **2005**, *43*, 722–730. [CrossRef]

36. Ahmad, S.; Simonovic, S.P. Modeling Dynamic Processes in Space and Time—A Spatial System Dynamics Approach. In *Proceedings of the Bridging the Gap—World Water and Environmental Resources Congress 2001, Orlando, FL, USA, 20–24 May, 2001*; American Society of Civil Engineers: Reston, VA, USA, 2001; Volume 48, pp. 1–20.

37. Li, L.; Simonovic, S.P. System dynamics model for predicting floods from snowmelt in north American prairie watersheds. *Hydrol. Process.* **2002**, *16*, 2645–2666. [CrossRef]

38. Ahmad, S.; Prashar, D. Evaluating Municipal Water Conservation Policies Using a Dynamic Simulation Model. *Water Resour. Manag.* **2010**, *24*, 3371–3395. [CrossRef]

39. De Araujo, W.C.; Oliveira Esquerre, K.P.; Sahin, O. Building a system dynamics model to support water management: A case study of the semiarid region in the Brazilian northeast. *Water (Switzerland)* **2019**, *11*, 2513.

40. CHJ (Confederación Hidrográfica del Júcar). *Plan Hidrológico de la Demarcación Hidrográfica del Júcar Ciclo 2015–2021*; Confederación Hidrográfica del Júcar: Valencia, Spain, 2015; pp. 286–392.

41. Apperl, B.; Pulido-Velazquez, M.; Andreu, J.; Karjalainen, T.P. Contribution of the multi-attribute value theory to conflict resolution in groundwater management - Application to the Mancha Oriental groundwater system, Spain. *Hydrol. Earth Syst. Sci.* **2015**, *19*, 1325–1337. [CrossRef]

42. Macian-Sorribes, H.; Pulido-Velazquez, M. Integrating historical operating decisions and expert criteria into a DSS for the management of a multireservoir system. *J. Water Resour. Plan. Manag.* **2017**, *143*, 1–12. [CrossRef]

43. CHJ. *Plan Hidrológico de la Demarcación Hidrográfica del Júcar. Memoria –Anejo 10. Programa de Medidas. Ciclo 2015–2021*; Confederación Hidrográfica del Júcar: Valencia, Spain, 2015; pp. 26–37.

44. Escriva-Bou, A.; Pulido-Velazquez, M.; Pulido-Velazquez, D. Economic Value of Climate Change Adaptation Strategies for Water Management in Spain's Jucar Basin. *J. Water Resour. Plan. Manag.* **2017**, *143*, 04017005. [CrossRef]

45. Ventana Systems. *Vensim User's Guide*; Ventana Systems, Inc., Ed.; Ventana Systems: Harvard, MA, USA, 2019.

46. CEDEX (Centro de Estudios y Experimentación de Obras Públicas). *Anuario de Aforos by the Minister of Agriculture Food and Environment*; Centro de Estudios y Experimentación de Obras Públicas: Madrid, Spain, 2016.

47. CHJ (Confederación Hidrográfica del Júcar). *Plan Hidrológico de la Demarcación Hidrográfica del Júcar. Memoria–Anejo 3. Usos y Demandas. Ciclo 2015–2021*; Confederación Hidrográfica del Júcar: Valencia, Spain, 2015; pp. 144–161.

48. Pulido-Velazquez, M.A.; Sahuquillo-Herraiz, A.; Camilo Ochoa-Rivera, J.; Pulido-Velazquez, D. Modeling of stream–aquifer interaction: The embedded multireservoir model. *J. Hydrol.* **2005**, *313*, 166–181. [CrossRef]

49. Sahuquillo, A. An eigenvalue numerical technique for solving unsteady linear groundwater models continuously in time. *Water Resour. Res.* **1983**, *19*, 87–93. [CrossRef]

50. Estrela, T.; Sahuquillo, A. Modeling the Response of a Karstic Spring at Arteta Aquifer in Spain. *Ground Water* **1997**, *35*, 18–24. [CrossRef]

51. Andreu, J.; Capilla, J.; Sanchis, E. AQUATOOL_Generalized_Decision-support_System_for_Water-resources. *J. Hydrol.* **1996**, *177*, 269–291. [CrossRef]

52. CHJ (Confederación Hidrográfica del Júcar, O.A.). *Plan Especial de Sequía Demarcación Hidrográfica del Júcar*; Confederación Hidrográfica del Júcar: Valencia, Spain, 2018; pp. 151–166.

53. Haro-Monteagudo, D.; Solera, A.; Andreu, J. Drought early warning based on optimal risk forecasts in regulated river systems: Application to the Jucar River Basin (Spain). *J. Hydrol.* **2017**, *544*, 36–45. [CrossRef]

54. Howitt, R.E. Positive Mathematical Programming. *Am. J. Agric. Econ.* **1995**, *77*, 329–342. [CrossRef]

55. Pulido-Velazquez, M.; Perez-Martin, M.A.; Solera, A.; Collazos, G.; Deidda, D.; Alvarez-Mendiola, E.; Benitez, A.; Andreu, J. Desarrollo y aplicacion de metodologias y herramientas en la cuenca piloto del rio Jucar para los analisis economicos requeridos en la Directiva Marco Europea del Agua (in Spanish). In *Final Report of EPTISA S.A. and UPV Research Project*; Research Institute of Water and Environmental Engineering (IIAMA), Universitat Politècnica de València: Valencia, Spain, 2006; pp. 10–21.

56. Coyle, G.; Road, C.; Sn, S. Qualitative Modelling in System Dynamics or What are the Wise Limits of Quantification? In Proceedings of the 17th International conference of the system dynamics society, Wellington, New Zealand, 20–23 July 1999; pp. 1–22.

57. Malard, J.J.; Inam, A.; Hassanzadeh, E.; Adamowski, J.; Tuy, H.A.; Melgar-Quiñonez, H. Development of a software tool for rapid, reproducible, and stakeholder-friendly dynamic coupling of system dynamics and physically-based models. *Environ. Model. Softw.* **2017**, *96*, 410–420. [CrossRef]

58. Vidal-Legaz, B.; Martínez-Fernández, J.; Picón, A.S.; Pugnaire, F.I. Trade-offs between maintenance of ecosystem services and socio-economic development in rural mountainous communities in southern Spain: A dynamic simulation approach. *J. Environ. Manag.* **2013**, *131*, 280–297. [CrossRef]

 *water*

Article

Flood Control Versus Water Conservation in Reservoirs: A New Policy to Allocate Available Storage

Ivan Gabriel-Martin, Alvaro Sordo-Ward *, David Santillán and Luis Garrote

Civil Engineering Department: Hydraulics, Energy and Environment, Universidad Politécnica de Madrid, ES 28040 Madrid, Spain; i.gabriel@alumnos.upm.es (I.G.-M.); david.santillan@upm.es (D.S.); l.garrote@upm.es (L.G.)
* Correspondence: alvaro.sordo.ward@upm.es

Received: 6 February 2020; Accepted: 26 March 2020; Published: 1 April 2020

Abstract: The aim of this study is to contribute to solving conflicts that arise in the operation of multipurpose reservoirs when determining maximum conservation levels (MCLs). The specification of MCLs in reservoirs that are operated for water supply and flood control may imply a reduction in the volume of water supplied with a pre-defined reliability in the system. The procedure presented in this study consists of the joint optimization of the reservoir yield with a specific reliability subject to constraints imposed by hydrological dam safety and downstream river safety. We analyzed two different scenarios by considering constant or variable initial reservoir level prior to extreme flood events. In order to achieve the global optimum configuration of MCLs for each season, we propose the joint optimization of three variables: minimize the maximum reservoir level (return period of 1000 years), minimize the maximum released outflow (return period of 500 years) and maximize the reservoir yield with 90% reliability. We applied the methodology to Riaño Dam, jointly operated for irrigation and flood control. Improvements in the maximum reservoir yield (with 90% reliability) increased up to 10.1% with respect to the currently supplied annual demand (545 hm^3) for the same level of dam and downstream hydrological safety. The improvement could increase up to 26.8% when compared to deterministic procedures. Moreover, dam stakeholders can select from a set of Pareto-optimal configurations depending on if their main emphasis is to maintain/increase the hydrological safety, or rather to maintain/increase the reservoir yield.

Keywords: hydrological dam safety; initial reservoir level; maximum conservation level; water conservation volume; flood control volume; yield reliability; regular operation; stochastic methodology

1. Introduction

Owing to increasingly risk-averse societies, stronger hydrological safety requirements are being imposed on existing dams in order to fulfil new regulations and prevent dam failures [1]. This problem can be addressed with two different kinds of technical solutions: hard solutions, as the alteration of dam spillways or elevation of dam crest and soft solutions, as the allocation of additional flood control volumes in the reservoir. Hard solutions increase the flood control capacity while maintaining the reservoir storage available for water supply. However, their main drawback is the need to allocate resources for infrastructure works. On the other hand, the implementation of soft solutions is easier and quicker, but can reduce the available volume to supply water with a specific reliability in a water system.

Allocation of flood control volume is addressed by defining a maximum conservation level (MCL), also known as flood-limited water level [2] or flood control level. MCL is the maximum operating level that the reservoir is allowed to reach under regular operation conditions. This reservoir level is below or equal to the maximum normal operating level (MNL) and can vary along the seasons of the year.

MCL is the most significant parameter in the trade-off established between flood control and water supply when increasing flood control volumes by soft solutions [3]. Traditionally, practitioners defining MCL only focused on hydrological dam and downstream safety. They frequently neglected other purposes of the reservoir, such as water supply and the economic consequences derived from loss of water yield reliability [4].

Some authors [3,5,6] have focused on accounting simultaneously for both regular (associated to water supply purposes of the reservoir) and flood control dam operations when defining MCLs. These studies analyzed hydrological dam safety by applying deterministic procedures, in which the return period associated to dam and downstream safety is assumed to be equal to the one associated to the flood event. Several authors [7,8] pointed out that hydrological dam safety and downstream safety should be assessed by analyzing the return periods of the maximum reservoir water levels and maximum outflows respectively.

Another relevant factor is that practitioners usually define MCLs assuming that the reservoir is at the maximum level under normal operating conditions prior to flood arrival [9–11]. This hypothesis results in conservative hydrological safety assessments. However, it can reduce the volume of water available to satisfy the demands of the water supply system because it influences the definition of MCLs. Accounting for the variability of the initial reservoir level in a hydrological dam and its downstream safety leads to more realistic results [8,12–14], which consequently can improve the definition of MCLs. Within this study, we propose a stochastic methodology to determine seasonal MCLs. The methodology combines three main innovative aspects:

- Stochastic assessment of hydrological dam and downstream river safety through return periods related to maximum reservoir levels and maximum outflows.
- Determination of MCLs that increase/maintain the water yield for a specific reliability while maintaining/improving hydrologic dam safety.

Determination of MCLs accounting for the variability of initial reservoir level prior to flood events. The methodology is illustrated through its application to a gated spillway dam located in Spain.

2. Materials and Methods

Maximum conservation levels represent the linking variable between flood control operation and water conservation operation of the reservoir (Figure 1).

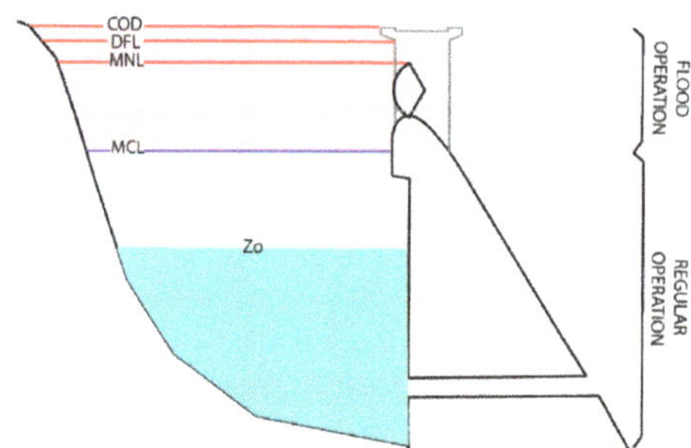

Figure 1. Conceptual scheme representing the role of maximum conservation levels (MCLs). COD represents the level of the crest of dam, DFL the design flood level, MNL the maximum normal operating level, and Zo the reservoir level prior to the flood event. Volume above MCL corresponds to the flood control operation volume and below MCL corresponds to the water conservation operation volume.

We propose a stochastic methodology to obtain the optimal set of seasonal MCLs accounting for both hydrological safety (dam and downstream) and water supply with a specific reliability. Figure 2 shows a scheme relating the main elements and procedures developed and applied in the methodology.

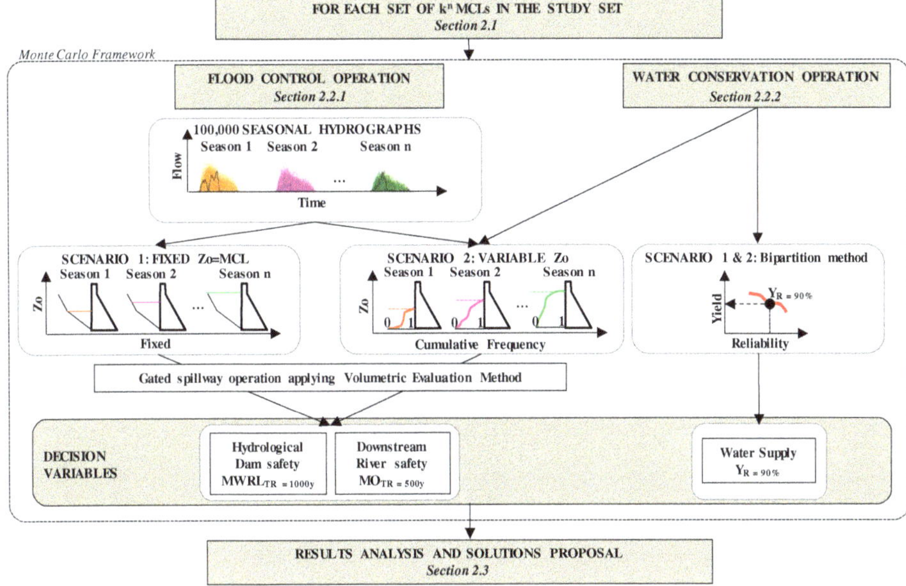

Figure 2. Scheme of the methodology proposed.

2.1. Study Set of Seasonal Maximum Conservation Levels

We defined a study set of seasonal MCLs representative of all possible configurations. The number of possible configurations of MCLs varies according to the number of seasons identified (n) and the number of possible maximum conservation levels selected (k) for a proper discretization. As the season of the MCLs matters (it is not the same to have the same MCL in one season or another), the number of possible configurations is k^n. The number of seasons is defined as described in Section 2.2.1.

2.2. Reservoir Operation Simulations

For each configuration of seasonal MCLs of the study set, we carried out two different reservoir operation simulations: simulation of flood control operation and simulation of water conservation operation.

2.2.1. Simulation of Flood Control Operation

Gabriel-Martin et al. [13] presented a stochastic methodology that enabled us to obtain stochastic inflow hydrographs representative of the observed daily annual floods (Figure 3). The main steps used to generate the inflow hydrographs were as follows:

- Generation of 100,000 pairs of flood duration (D) with their associated maximum annual flood volume (V). Pairs of 100,000 flood event durations were generated following the empirical probability distribution of historical floods. For each element of the 100,000 generated durations, the corresponding hydrograph volume was obtained following the probability distribution of the associated duration within a Monte Carlo framework.

- Generation of 100,000 values of cumulated precipitation depth. The value of the cumulated net precipitation was obtained by dividing the volume of each hydrograph by the area of the study basin. By applying the curve number method inversely [15] to the cumulated net precipitation, the value of the cumulated precipitation depth was obtained (Figure 3a)

- Temporal distribution of the 100,000 cumulated precipitation depth values. Each cumulated rainfall depth was distributed temporally by applying an autoregressive moving average (ARMA) (2,2) model [10]. Thus, 100,000 hourly hyetographs were obtained (Figure 3b).
- Generation of 100,000 hourly-distributed hydrographs. By applying the curve number method [15] and the soil conservation service dimensionless unit hydrograph procedure [15], 100,000 hydrographs were generated, which followed the empirical probability distributions of volume and duration (Figure 3b).

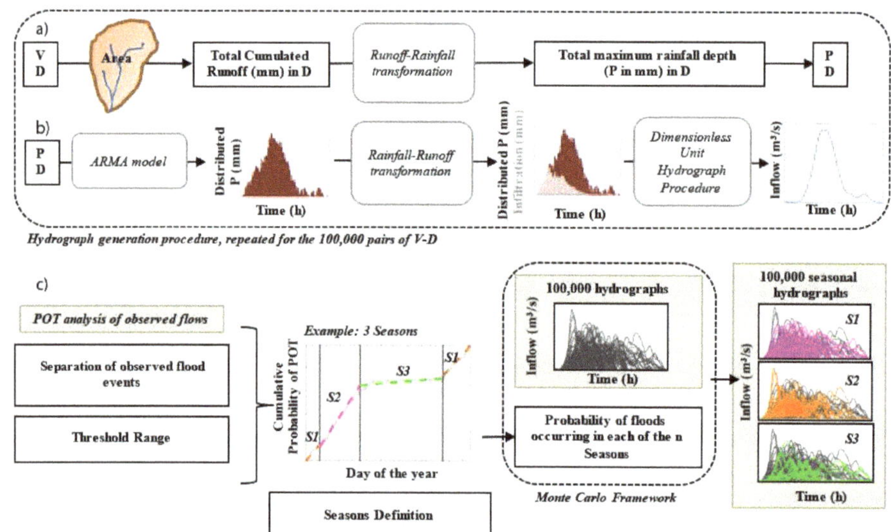

Figure 3. The procedure to generate the ensemble of inflow hydrographs (**a,b**) and its seasonal characterization (**c**). D represents the flood durations, whereas V represents the maximum annual flood. S1, S2, and S3 represent the different seasons.

A detailed description of the method to generate representative hydrographs can be found in Gabriel-Martin et al. [13]. Each hydrograph was associated to one of the seasons defined (Figure 3c). We identified the distinct seasons by applying a graphical test to the observed daily inflows proposed by Ouarda [16] and Ouarda et al. [17]. This test was based on a peak-over-threshold (POT) analysis. In order to identify the seasons, Ouarda et al. [17] tested different threshold values while assuring independence between the selected POT. We applied the criteria recommended by Lang et al. [18] to define the range of threshold values to be considered and the criteria proposed by the Water Resources Council [19] to assure the independence between two consecutive floods. We plotted the cumulative empirical probability of POT during the year against the time of the year for each threshold value tested. The slope changes within the plot indicated the significant seasons. A detailed description of the method to identify the seasons can be found in Gabriel-Martin et al. [20] and is summarized in Figure 3c.

For each configuration of MCLs, we obtained the maximum water level in the reservoir corresponding to a return period of 1000 years ($MWRL_{TR=1000y}$) and the maximum outflow corresponding to a return period of 500 years ($MO_{TR=500y}$). These values correspond to the set of 100,000 seasonal maximum annual inflow hydrographs generated by Gabriel-Martin et al. [13,20].

The analysis was carried out in two different scenarios of the initial reservoir level prior to the flood event (Zo) (Figure 2):

- Scenario 1 (Sc.1): Zo is constant and corresponds to the seasonal MCL. The reservoir is assumed to be at its maximum operating level when the maximum annual flood occurs.
- Scenario 2 (Sc.2): Zo is variable, and the reservoir can be at any level when the maximum annual flood occurs. Zo is randomly sampled from the cumulative probability distribution of Zo associated to the season of occurrence of the maximum annual flood event. This distribution is obtained from the simulation of the water conservation operation of the reservoir (Section 2.2.2).

The operation of the dam gates was simulated by applying the volumetric evaluation method (VEM), fully described in Giron [21] and Sordo-Ward et al. [11,22].

2.2.2. Simulation of Water Conservation Operation

We considered the yield reliability (Y_R) as the ratio of total volume supplied and the total volume demanded [23–26] during the period analyzed. For each configuration of MCLs (k), we obtained the reservoir yield with a reliability of 90% ($Y_{R\,=\,90\%}$) and the cumulative probability distribution of Zo, by developing a monthly water balance model. Reservoir storage is large compared to monthly inflows, and thus the monthly time scale is appropriate. The model manages the dam as an isolated element and applies rules of operation validated in Gabriel-Martin et al. [13]. The water balance considers time series of monthly inflows, environmental flow restrictions, evaporation rates, monthly demand distribution, storage–area–height reservoir curves, and dead storage volume (data extracted from "Duero National Water Master Plan" [27]). The main purpose of the reservoir is irrigation and therefore we adopted a required yield reliability of 90% ($Y_{R\,=\,90\%}$) which is adequate for irrigation demands in the region [28].

To identify the maximum amount of water that can be supplied to satisfy a regular demand with a specified reliability, a bipartition method was applied. Excessive values of demands were set (for example, similar to mean monthly runoff) and the simulation was carried out. The deficits were obtained and specified yield reliability requirements were checked. If the specified reliability requirements were not fulfilled, the demand was reduced by half and simulated again. If the specified reliability requirements were satisfied, half of the difference was added and simulated again and so on, until the deficit (or gain) was smaller than a pre-set tolerance (e.g., 0.1 hm^3/year). In addition, we simulated the operation of the reservoir with the associated mean annual current demand. We repeated the procedure for both Sc.1 and Sc.2 scenarios.

2.3. Results Analysis and Solutions Proposal

In order to propose the optimal configurations of MCLs within the case study, as exposed, we selected three main decision variables: $MWRL_{TR\,=\,1000y}$, $MO_{TR\,=\,500y}$, and $Y_{R\,=\,90\%}$. We assumed that the MCLs configuration would not fulfil the standards if $MWRL_{TR=1000y}$ was above the design flood level (DFL) and/or $MO_{TR\,=\,500y}$ was greater than the emergency flow (O_{EMER}). We compared $MWRL_{TR=1000y}$, $MO_{TR\,=\,500y}$, and $Y_{R\,=\,90\%}$ for all the configurations in the study set of MCLs by identifying configurations that are non-inferior solutions (Pareto framework) in terms of maximum volume of water supplied (with a specific yield reliability of 90%) and hydrological safety.

Determination of Possible MCLs by Applying a Pareto Analysis

The selected variables for conducting the Pareto analysis were $MO_{TR\,=\,500y}$ and $Y_{R\,=\,90\%}$ (Figure 4). In this study, we assumed that higher levels in the reservoir (and above MNL) imply greater outflows. This assumption is always fulfilled either for dams with fixed crest spillways or if the VEM is applied in gated spillways (as in this study). It should be noted that this assumption may not hold for all possible specific characteristics of dams, all rules of operation adopted, or all specific values adopted for the hydrological dam and downstream safety. Moreover, in this study, the outflow corresponding to the DFL condition is higher than that corresponding to O_{EMER}, that is, the $MO_{TR = 500y}$ is the most restrictive variable. Therefore, for each scenario (Sc.1 or Sc.2), we had a set of k^n MCLs configurations with k^n pairs of values $MO_{TR = 500y}$ and $Y_{R = 90\%}$. The purpose was to obtain the configurations of MCLs that minimize the value of $MO_{TR = 500y}$ while maximizing $Y_{R = 90\%}$, which implies a two-objective minimization problem (Equation (1)):

$$\text{Min}\{f_1(x_i), f_2(x_i)\}, \tag{1}$$

in which $x_i = [MCL_{S1}, MCL_{S2}, \ldots, MCL_{Sn}]$, $f_1(x_i) = [MO_{TR=500yi}]$ and $f_2(x_i) = [-Y_{R=90\%i}]$; being $i = 1,2 \ldots, k^n$. The solution of Equation (1) consisted of a set of non-dominated solutions. Therefore, following the procedure proposed in Chong and Zak [29] we conducted the mentioned analysis. Once the non-dominated solutions were identified for Sc.1 and Sc.2, we eliminated from both scenarios those that did not fulfil the hydrological safety regulation standards (both for $MWRL_{TR = 1000y}$ and/or $MO_{TR = 500y}$) and those providing a $Y_{R = 90\%}$ lower than the annual demand that is currently satisfied. Afterwards, we compared the proposed solutions, providing dam stakeholders with a set of possible configurations depending on whether their main objective was to increase hydrological dam and downstream river safety or increase the water supply with a specific reliability within the system.

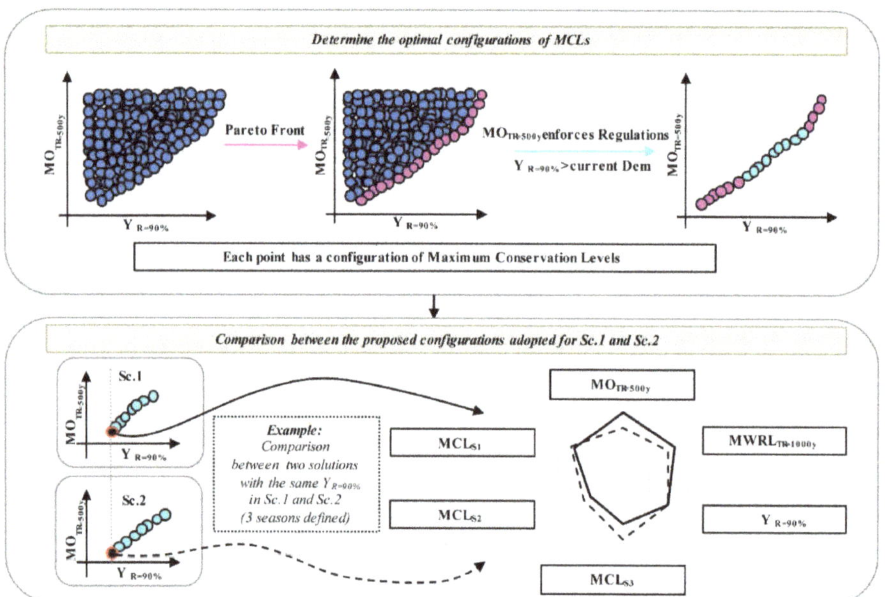

Figure 4. Scheme of the procedure for solutions proposal.

2.4. Case Study

We applied the methodology to Riaño Dam (Table 1). It belongs to the Esla basin water resources system, which is managed by the Duero River Basin Authority (west region of mainland Spain). The main purpose of the reservoir is irrigation. The mean monthly water demands for irrigation/urban water supply are as follows (in hm^3): April 10.9/0.08, May 54.5/0.08, June 81.7/0.08, July 158/0.16, August 147.1/0.16, and September 92.6/0.16. From October to March, the demands were 0/0.08 hm^3. The capacity of the reservoir is 651 hm^3 at MNL (the maximum reservoir level that water might reach under normal operating conditions) [30], with a bottom dead storage of 78 hm^3. The main characteristics of the Riaño Reservoir and its basin are summarized in Table 1.

Table 1. Main characteristics of the Riaño basin, dam, and reservoir.

Basin Features	Value	Dam Features	Value
Basin area	582 km^2	Maximum normal level (MNL)	1100.1 m
Concentration time	11 h	Design flood level (DFL)	1101.1 m
Mean annual Runoff	680 hm^3	Crest of dam (COD)	1102.5 m
Annual Current Demand (Da)	545 hm^3	Gated spillway capacity at MNL	621 m^3/s
Emergency downstream flow (O$_{EMER}$)	700 m^3/s	Auxiliary spillway capacity at DFL	98.7 m^3/s

Riaño Dam has two spillways. The main spillway is controlled by two tainter gates, each eight meters wide and seven meters high. There is a second spillway for emergency purposes. It is a fixed-crest spillway with the crest located at the MNL. Riaño Dam also has two bottom and two intermediate outlets, which we assumed were closed during the floods. Flood damage analyses summarized in the Dam Master Plan concluded that discharges above 700 m^3/s (O$_{EMER}$) could produce damage over urban settlements with more than five inhabitants and infrastructures in the downstream reach. The following data were used to perform the study:

- Simulation of flood control operation. Besides the flood control structures and dam configuration shown, we used 30 years of unaltered daily flow series data from a gauge located right downstream the Riaño reservoir (from the years 1954 to 1984 and prior to the existence of the dam). With this time series, the 100,000 seasonal synthetic flood hydrographs were generated.
- Simulation of water conservation operation. We used a monthly time series of naturalized inflows from 1940 to 2013, environmental flow restrictions, evaporation rates, monthly demand distribution, storage–area–height reservoir curves and dead storage volume (all data obtained from the Duero River Basin Management Plan [27]) and the reservoir characteristics previously stated.

2.5. Limitations of the Methodology

We applied this methodology to one basin and dam configuration. This might limit the generalization of the results obtained. Furthermore, the water resources management model focused on the regular operation of the dam as an isolated element. This methodology could be extended to take into consideration the interaction with other infrastructures within the system, using suitable water resources management models (e.g., AQUATOOL [31] and WEAP [32]).

3. Results and Discussion

3.1. Determination of the Study Set of MCLs to be Analyzed

We studied a set of MCLs that ranged from 651 hm^3 (volume at MNL) to 400 hm^3. For the sake of simplicity, we estimated the flood hydrograph volume of Tr = 5000 years (247 hm^3) and defined a maximum flood control volume of 251 hm^3. We discretized the ranges of reservoir volumes in intervals of 5 hm^3. Thus, we defined a set of k = 51 possible MCLs per season (associated to a volume in the reservoir of 400, 405, ... , 645, 651 hm^3).

According to Gabriel-Martin et al. [20], three characteristic seasons (regarding maximum annual floods) were identified for the location of Riaño: season 1 (S1) from the beginning of November to the end of January; season 2 (S2) from the beginning February to the end of April; and season 3 (S3) from the beginning of May to the end of October. Therefore, as n = 3 seasons, we had a set of 51^3 = 132,651 configurations of MCLs per scenario analyzed (Sc.1 and Sc.2.)

3.2. Simulation of the Water Conservation and Flood Operation of the Dam

Once the configurations of MCLs were defined, for each configuration we obtained the values of MWRL$_{TR = 1000y}$ and MO$_{TR = 500y}$ by simulating the flood operation of the reservoir for Sc.1 and Sc.2.

By simulating the water conservation operation of the reservoir, we obtained $Y_{R\,=\,90\%}$. Figure 5 shows the values of $MWRL_{TR\,=\,1000y}$, $MO_{TR\,=\,500y}$, and $Y_{R\,=\,90\%}$ with respect to the MCLs of each season.

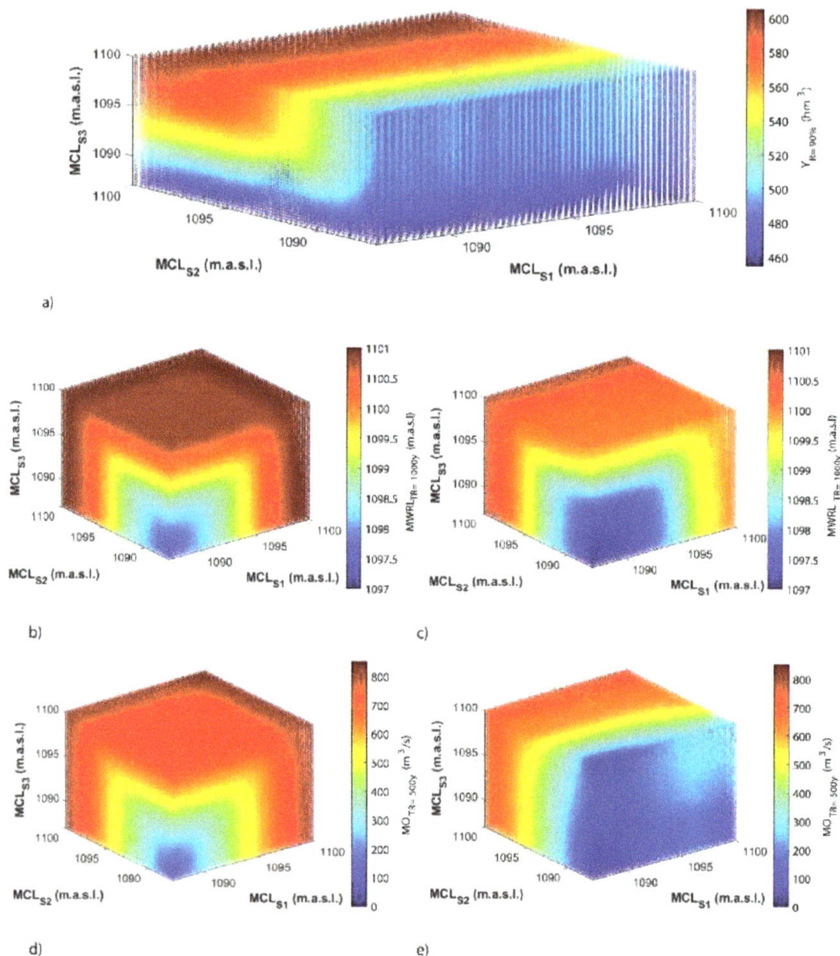

Figure 5. Representations of the 132,651 values of $MWRL_{TR\,=\,1000y}$, $MO_{TR\,=\,500y}$, and $Y_{R\,=\,90\%}$ for each of the seasons (S1, S2, and S3). MCL_{S1}, MCL_{S2}, and MCL_{S3} represent the maximum conservation levels in S1, S2, and S3 respectively. (**a**) Representation of the values $Y_{R=90\%}$, with a color bar, which are the same in Sc.1 (initial reservoir level equal to MCL) and Sc.2 (variable initial reservoir level). (**b,c**) Representation of the values $MWRL_{TR=1000y}$ with a color bar in Sc.1 (**b**) and Sc.2 (**c**). (**d,e**) Representation of the values $MO_{TR\,=\,500y}$ with a color bar in Sc.1 (**d**) and Sc.2 (**e**).

Figure 5a shows that variation of MCLs in S1 did not affect $Y_{R=90\%}$. This is because the water demands associated with the reservoir in S1 were less than 1% of the annual demand (Da), and, as the irrigation season in the Riaño system extends from May to September (98% of Da), the reservoir was able to recover the reduced volume in S1 within the months previous to irrigation (February, March, and April). Figure 5b,d shows that, for the case of Sc.1, the effects of MCLs were similar in the three seasons in terms of hydrological dam safety (Figure 5b) and downstream river safety (Figure 5d). However, in Sc.2 (Figure 5c,e), as regular operation was linked to the flood control operation by the initial reservoir level, MCLs in S1 and S3 did not affect either the hydrological dam safety or downstream river safety. Variations of MCL_{S2} are the main affection to values $MWRL_{TR = 1000y}$ and $MO_{TR = 500y}$ in Figure 5c,e, respectively.

3.3. Solutions Proposal

First, we identified the configurations which did not fulfil hydrological dam safety ($MWRL_{TR = 1000y} > DFL$) and/or downstream safety ($MO_{TR = 500y} > O_{EMER.}$) in both scenarios. A total of 826 configurations (0.6% of 132,651 configurations) had a value of $MWRL_{TR = 1000y}$ higher than DFL in Sc.1 (red dots in Figure 6a), while none of the configurations had $MWRL_{TR = 1000y}$ values higher than the DFL in Sc.2 (Figure 6c). It should be noted that the 826 configurations that did not fulfil hydrological dam safety in Sc.1 also did not fulfil the downstream safety condition. On the other hand, 59,013 configurations (44.5% of the total number of configurations) had a value of $MO_{TR = 500y}$ higher than $O_{EMER.}$ in Sc.1, whereas 4944 (3.7%) in Sc.2 (Figure 6a,c) shows the pair of values $Y_{R = 90\%}$ and $MO_{TR = 500y}$ (grey points) for each analyzed configuration in Sc.1 and Sc.2, respectively.

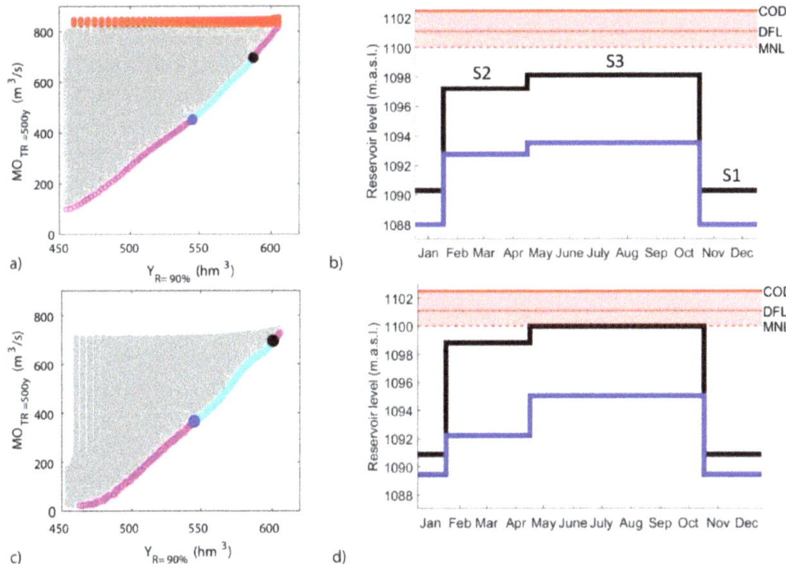

Figure 6. (**a**,**c**) Grey dots represent pair of values $Y_{R = 90\%}$ and $MO_{TR = 500y}$ for Sc. 1 and Sc.2, respectively. Red dots (**a**) represent the configurations in which $MWRL_{TR = 1000y} > DFL$. Magenta dots indicate the pareto front. Cyan dots represent the proposed solutions. The black dot represents the solution that maximizes the water volume supplied with a reliability of 90% and $MO_{TR = 500y} = O_{EMER}$. The blue dot represents the solution that minimizes the maximum outflow released (being $Y_{R = 90\%} = Da$). (**b**,**d**) MCLs for each month/season which corresponds to the proposed solutions previously selected (blue and black lines correspond to blue and black dots in (**a**) and (**c**)), for Sc.1 and Sc.2, respectively. The red dashed line represents the maximum normal level (MNL), the red dashed–dotted line represents the design flood level (DFL), and the red continuous line represents the crest of dam (COD).

In Sc.1, the Pareto-solutions consisted of 277 configurations (Figure 6a, magenta points). One those, 98 configurations satisfied the following: $Y_{R = 90\%} \geq Da$, $MO_{TR = 500y} \leq O_{EMER}$, and $MWRL_{TR = 1000y} \leq DFL$ (proposed solutions, cyan in Figure 6a). For the same analysis in Sc.2, we identified 247 configurations (magenta points in Figure 6c) and 135 (cyan points in Figure 6c), respectively. For both scenarios, the extreme proposed solutions were identified. On one hand, in the case of $MO_{TR = 500y} = O_{EMER} = 700$ m^3/s (Figure 6a,c, black point), $Y_{R = 90\%} = 587$ hm^3 (for Sc.1) and 600 hm^3

(for Sc.2) representing an improvement (compared to Da) of 7.7% and 10.1%, respectively. On the other hand, in the case of $Y_{R=90\%} = D_a = 545$ hm^3 (Figure 6a,c, dark blue point), $MO_{TR=500y} = 452$ m^3/s (for Sc.1) and 366 m^3/s (for Sc.2) representing a 64.2% and 52.3% of O_{EMER}, respectively. It is important to point out that, in the case of using a deterministic procedure focused on hydrological dam and downstream safety (conventional procedure), any studied configuration of MCLs could be a potential solution (grey dots within Figure 6a,c). Thus, if $MO_{TR=500y} = O_{EMER}$, the proposed stochastic procedure presented improvements of up to 146 hm^3 (26.8% of D_a) compared to the worst regular operation configuration ($Y_{R=90\%} = 454$ hm^3 in Sc.1, Figure 6a). The corresponding configurations of MCLs for the extreme proposed solutions are represented in Figure 6b (Sc.1) and Figure 6d (Sc.2) with the same color scheme as in Figure 6a,c, respectively. In both scenarios, the highest MCLs were associated with S3, while the lowest was associated with S1.

3.4. Comparison between the Proposed Configurations in the Two Scenarios

We compared the limit proposed solutions in both scenarios (Sc.1 and Sc.2; Figure 7). In the case of $MO_{TR=500y} = O_{EMER} = 700$ m^3/s, $MWRL_{TR=1000y}$ was 1100.6 m.a.s.l. The hydrological dam and downstream river safety were invariant for Sc.1 and Sc.2. However, higher MCLs were obtained for Sc.2, which increased the $Y_{R=90\%}$ of the system. In the case of $Y_{R=90\%} = Da = 545$ hm^3, the differences between Sc.2 and Sc.1 for $MWRL_{TR=1000y}$ and $MO_{TR=500y}$ were 0.2 m.a.s.l and 86 m^3/s, respectively. Moreover, the MCL of season two were lower when the variable initial reservoir level was considered. This is because of the increase of MCLs in the other seasons. Despite this, the demand supplied with a reliability of 90% was the same, accounting for lower values of $MWRL_{TR=1000y}$ and $MO_{TR=500y}$ if variable initial level was considered.

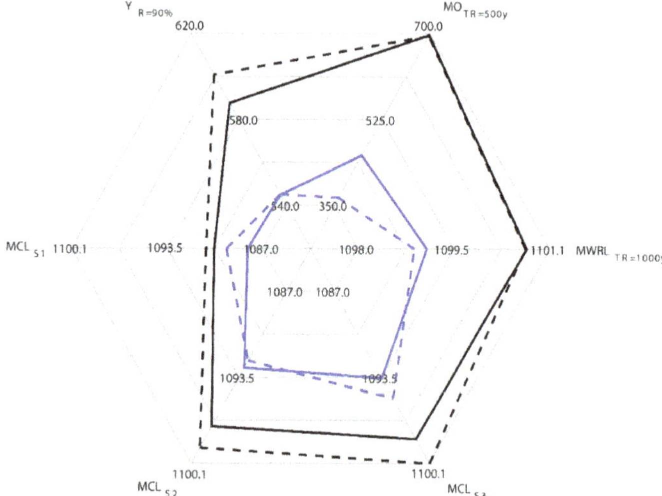

Figure 7. Spider plot comparing the value of the maximum conservation levels for different seasons (MCLsx), $MWRL_{TR=1000y}$, $MO_{TR=500y}$, and $Y_{R=90\%}$. In black, solutions that maximize the volume supplied with $MO_{TR=500y} = O_{EMER}$. In blue, the solutions that supply the current annual demand minimize the maximum outflow released. The continuous line corresponds to Sc.1 and the dashed line to Sc.2.

Within the framework of the present study, accounting for the variability of initial reservoir level implied a reduction of flood control volumes (increase of MCLs) maintaining the risk of overtopping and downstream river safety. Consequently, the volume for satisfying the demands increased, providing a more reliable system in terms of regular operation.

4. Conclusions

The main conclusions extracted from this study are as follow:

- The use of a stochastic methodology allowed us to assess hydrological dam safety and downstream safety by obtaining the frequency curves of outflow and maximum reservoir water levels, while accounting for the variability in hydrological loads with respect to deterministic procedures. As a drawback, it implied a more complex procedure and computational effort.

- We proposed a set of 98 non-inferior solutions while considering the initial reservoir level equal to the MCL for each season and 135 possible configurations while considering variable initial reservoir level. From the proposed configurations, dam stakeholders are able to decide which configuration to use depending on whether their preference is to increase dam and downstream hydrological safety or to increase water supply (with a specific reliability) in the water resources system. In the Riaño case study, the presented procedure showed improvements in the regular operation that satisfied an increase of up to 10.1% of the current annual demand of 545 hm^3 (with a reliability of 90%) while maintaining the same level of hydrological dam safety.

- Accounting for initial reservoir variability resulted in the possibility of supplying an extra demand of 13 hm^3 (2.4% of the current annual demand) compared to the optimal solution without accounting for initial reservoir level variability.

- The proposed stochastic procedure can improve the results obtained by deterministic procedures, increasing supply up to 26.8% of the current annual demand, from the worst regular operation configuration (not accounting for initial reservoir level variability) to the optimal configuration (accounting for initial reservoir level variability) of MCLs.

Author Contributions: Conceptualization, I.G.-M., A.S.-W., D.S. and L.G.; Data curation, I.G.-M. and A.S.-W.; Formal analysis, I.G.-M., A.S.-W., D.S. and L.G.; Funding acquisition, A.S.-W.; Investigation, I.G.-M., A.S.-W., D.S. and L.G.; Methodology, I.G.-M., A.S.-W., D.S. and L.G.; Resources, A.S.-W.; Software, I.G.-M.; Visualization, I.G.-M.; Writing – original draft, I.G.-M.; Writing – review & editing, A.S.-W., D.S. and L.G. All authors have read and agreed to the published version of the manuscript.

Funding: Universidad Politécnica de Madrid: VJIDOCUPM19AFSW.

Acknowledgments: The authors acknowledge the computer resources and technical assistance provided by the Centro de Supercomputación y Visualización de Madrid (CeSViMa) and the funds from Universidad Politécnica de Madrid in the framework of their Program "Ayudas para contratos predoctorales para la realización del doctorado en sus escuelas, facultad, centro e institutos de I+D+i" and "ayudas a proyectos de I+D de investigadores posdoctorales".

Conflicts of Interest: The authors declare no conflict of interest.

References

1. Bocchiola, D.; Rosso, R. Safety of Italian dams in the face of flood hazard. *Adv. Water Resour.* **2014**, *71*, 23–31. [CrossRef]
2. Li, X.; Guo, S.L.; Liu, P.; Chen, G.Y. Dynamic control of flood limited water level for reservoir operation by considering inflow uncertainty. *J. Hydrol.* **2010**, *391*, 124–132. [CrossRef]
3. Xie, A.; Liu, P.; Guo, S.; Zhang, X.; Jiang, H.; Yang, G. Optimal Design of Seasonal Flood Limited Water Levels by Jointing Operation of the Reservoir and Floodplains. *Water Resour. Manag.* **2018**, *32*, 179–193. [CrossRef]
4. Solera-Solera, A.; Morales-Torres, A.; Serrano-Lombillo, A. Cost estimation of freeboard requirements in water resources management. In *Risk Analysis, Dam Safety, Dam Security and Critical Infrastructure Management*; CRC Press: Boca Raton, FL, USA, 2012; pp. 9–14.
5. Ouyang, S.; Zhou, J.; Li, C.; Liao, X.; Wang, H. Optimal design for flood limit water level of cascade reservoirs. *Water Resour. Manag.* **2015**, *29*, 445–457. [CrossRef]
6. Moridi, A.; Yazdi, J. Optimal Allocation of Flood Control Capacity for Multi-Reservoir Systems Using Multi-Objective Optimization Approach. *Water Resour. Manag.* **2017**, *31*, 4521–4538. [CrossRef]
7. Michailidi, E.M.; Bacchi, B. Dealing with uncertainty in the probability of overtopping of a flood mitigation dam. *Hydrol. Earth System Sci.* **2017**, *21*, 2497. [CrossRef]

8. Micovic, Z.; Hartford, D.N.D.; Schaefer, M.G.; Barker, B.L. A non-traditional approach to the analysis of flood hazard for dams. *Stoch. Environ. Res. Risk Assess.* **2016**, *30*, 559–581. [CrossRef]

9. USACE. *Flood Control by Reservoirs. International Hydrological Decade*; USACE: Washington, DC, USA, 1976; Volume 7.

10. Sordo-Ward, A.; Garrote, L.; Martín-Carrasco, F.; Bejarano, M.D. Extreme flood abatement in large dams with fixed-crest spillways. *J. Hydrol.* **2012**, *466-467*, 60–72. [CrossRef]

11. Sordo-Ward, A.; Garrote, L.; Bejarano, M.D.; Castillo, L.G. Extreme flood abatement in large dams with gate-controlled spillways. *J. Hydrol.* **2013**, *498*, 113–123. [CrossRef]

12. Carvajal, C.; Peyras, L.; Arnaud, P.; Boissier, D.; Royet, P. Probabilistic Modeling of Floodwater Level for Dam Reservoirs. *J. Hydrol. Eng.* **2009**, *14*, 223–232. [CrossRef]

13. Gabriel-Martin, I.; Sordo-Ward, A.; Garrote, L.; Castillo, L.G. Influence of initial reservoir level and gate failure in dam safety analysis. Stochastic approach. *J. Hydrol.* **2017**, *550*, 669–684. [CrossRef]

14. Gabriel-Martin, I.; Sordo-Ward, A.; Garrote, L.; Granados, I. Hydrological Risk Analysis of Dams: The Influence of Initial Reservoir Level Conditions. *Water* **2019**, *11*, 461. [CrossRef]

15. USDA Soil Conservation Service. *National Engineering Handbook*; Section 4; Hydrology. U.S. Department of Agriculture: Washington, DC, USA, 1972.

16. Ouarda, T.B.M.J.; Ashkar, F.; Ejabi, N. Peaks over threshold model for seasonal flood variations. In *Engineering hydrology (Proc. of the Symposium of San Francisco, California, 25–30 July)*; Kuo, C.Y., Ed.; ASCE Publication: San Francisco, CA, USA, 1993; pp. 341–346.

17. Ouarda, T.B.M.J.; Cunderlik, J.M.; St-Hilaire, A.; Barbet, M.; Brunear, P.; Bobée, B. Data-based comparison of seasonality-based regional flood frequency methods. *J. Hydrol.* **2006**, *330*, 329–339. [CrossRef]

18. Lang, M.; Ouarda, T.B.M.J.; Bobee, B. Towards operational guidelines for over-threshold modeling. *J. Hydrol.* **1999**, *225*, 103–117. [CrossRef]

19. USWRC. *Guidelines for Determining Flood Flow Frequency*; United States Water Resources Council, Bulletin 17, of the Hydrology Subcommittee: Washington, DC, USA, 1976.

20. Gabriel-Martin, I.; Sordo-Ward, A.; Garrote, L. Influence of initial reservoir level on the allocation of seasonal maximum conservation levels. *Ingeniería del agua* **2018**, *22*, 225–238. [CrossRef]

21. Girón, F. The evacuation of Floods during the Operation of Reservoirs. In *Transactions Sixteenth International Congress on Large Dams*; Report 75; International Commission on Large Dams (ICOLD): San Francisco, CA, USA, 1988; Volume 4, pp. 1261–1283.

22. Sordo-Ward, A.; Gabriel-Martin, I.; Bianucci, P.; Garrote, L. A Parametric Flood Control Method for Dams with Gate-Controlled Spillways. *Water* **2017**, *9*, 237. [CrossRef]

23. Chavez-Jimenez, A.; Lama, B.; Garrote, L.; Martin-Carrasco, F.; Sordo-Ward, A.; Mediero, L. Characterisation of the sensitivity of water resources systems to climate change. *Water Resour. Manag.* **2013**, *27*, 4237–4258. [CrossRef]

24. Sordo-Ward, A.; Bianucci, P.; Garrote, L.; Granados, A. The influence of the annual number of storms on the derivation of the flood frequency curve through event-based simulation. *Water* **2016**, *8*, 335. [CrossRef]

25. Sordo-Ward, A.; Granados, A.; Iglesias, A.; Bejarano, M.D. Adaptation Effort and Performance of Water Management Strategies to Face Climate Change Impacts in Six Representative Basins of Southern Europe. *Water* **2019**, *11*, 1078. [CrossRef]

26. Sordo-Ward, A.; Granados, I.; Iglesias, A.; Garrote, L. Blue Water in Europe: Estimates of Current and Future Availability and Analysis of Uncertainty. *Water* **2019**, *11*, 420. [CrossRef]

27. CHD (Confederación Hidrográfica del Duero). *Plan hidrológico de la parte española de la D.H.Duero (2015–2021)*; Official Report of the Ministry of Agriculture, Food and Environment of Spain: Valladolid, Spain, 2015.

28. Garrote, L.; Iglesias, A.; Granados, A.; Mediero, L.; Martin-Carrasco, F. Quantitative assessment of climate change vulnerability of irrigation demands in Mediterranean Europe. *Water Resour. Manag.* **2015**, *29*, 325–338. [CrossRef]

29. Chong, E.; Zak, S.H. *An introduction to Optimization*; John Wiley & Sons: Hoboken, NJ, USA, 2004.

30. International Commission on Large Dams (ICOLD). *Technical Dictionary on Dams: A Glossary of Words and Phrases Related to Dams*; ICOLD: Paris, France, 1994; p. 365.

31.   Andreu, J.; Capilla, J.; Sanchís, E. AQUATOOL, a generalized decision support system for water-resources planning and management. *J. Hydrol.* **1996**, *177*, 269–291. [CrossRef]
32.   Yates, D.; Sieber, J.; Purkey, D.; Huber-Lee, A. WEAP21–A Demand-, Priority-, and Preference-Driven Water Planning Model. Part 1: Model characteristics. *Water Int.* **2005**, *30*, 487. [CrossRef]

 *water*

Article

A New Tool for Assessing Environmental Impacts of Altering Short-Term Flow and Water Level Regimes

María Dolores Bejarano [1,*], Jaime H. García-Palacios [2], Alvaro Sordo-Ward [2], Luis Garrote [2] and Christer Nilsson [3,4]

[1] Department of Natural Systems and Resources, Universidad Politécnica de Madrid, 28040 Madrid, Spain
[2] Department of Civil Engineering: Hydraulics, Energy and Environment, Universidad Politécnica de Madrid, 28040 Madrid, Spain; jaime.garcia.palacios@upm.es (J.H.G.-P.); alvaro.sordo.ward@upm.es (A.S.-W.); l.garrote@upm.es (L.G.)
[3] Landscape Ecology Group, Department of Ecology and Environmental Science, Umeå University, SE-901 87 Umeå, Sweden; christer.nilsson@umu.se
[4] Department of Wildlife, Fish and Environmental Studies, Swedish University of Agricultural Sciences, SE-901 83 Umeå, Sweden
* Correspondence: mariadolores.bejarano@upm.es; Tel.: +34-910671778

Received: 13 September 2020; Accepted: 10 October 2020; Published: 19 October 2020

Abstract: The computational tool InSTHAn (indicators of short-term hydrological alteration) was developed to summarize data on subdaily stream flows or water levels into manageable, comprehensive and ecologically meaningful metrics, and to qualify and quantify their deviation from unaltered states. The pronunciation of the acronym refers to the recording interval of input data (i.e., instant). We compared InSTHAn with the tool COSH-Tool in a characterization of the subdaily flow variability of the Colorado River downstream from the Glen Canyon dam, and in an evaluation of the effects of the dam on this variability. Both tools captured the hydropeaking caused by a dam operation, but only InSTHAn quantified the alteration of key flow attributes, highlighting significant increases in the range of within-day flow variations and in their rates of change. This information is vital to evaluate the potential ecological consequences of the hydrological alteration, and whether they may be irreversible, making InSTHAn a key tool for river flow management.

Keywords: fluvial ecosystems; hydropeaking; InSTHAn tool; short-term flow regimes; subdaily flows; sustainable river management

1. Introduction

Flow variables shape the dynamics of in-channel and floodplain conditions that determine fluvial ecosystem structure and functioning [1,2]. Whereas the ecological role of monthly and annual flow dynamics has been in focus for many years, less attention has been paid to flow variability within days [3].

Variation at such short time scales is altered by several human activities, such as land use and urbanization, and water management practices such as flood control, agricultural withdrawals and power generation [4,5]. Increasing instability of within-day flows and exacerbation of extreme flows may likely affect water quality [6], fluvial landforms [7] and aquatic and riparian organisms that are adapted to naturally less fluctuating conditions (review by Bejarano et al., 2018 [8]).

Subdaily flow regimes govern fish reproduction [9] by affecting egg viability and reproductive capacity. They also affect their behavior [10] and performance [11] by offering shelter and food, which affects their movements. Ultimately, subdaily flow regimes affect fish survival, by modulating fish energy balance with implications for growth rates and risk of illness, or due to stranding and drift [12]. Risk of desiccation [13] and catastrophic drift [14] of macroinvertebrates increases with more recurrent daily dry periods and peak flows. Highly fluctuating short-term flow regimes may also increase propagule dispersal of aquatic and riparian plants, and interfere with germination, growth and performance, thus likely hampering recruitment and increase mortality [15,16]. At the community level, alterations of short-term flows may ultimately result in removal of intolerant species and invasion by exotic species [17].

The rise of hydropower as a renewable energy source calls for a better understanding of the ecological consequences of altered flow regimes and associated hydraulic parameters at short time scales. Hydropeaking plants usually cause frequent and rapid fluctuations in flow and water level within the day [18], and this variation is superimposed upon the seasonal changes in flow regimes resulting from water storage in upstream reservoirs. The demand for hydropower is growing, especially in Southeast Asia, Africa and Latin America [19]. In Europe, hydropower is promoted by legislation such as the Renewable Energy Directive (RES; 82 2009/28/EC). Consequently, shifting flow regimes towards preindustrial conditions in rivers affected by hydropeaking without significantly affecting hydropower production is a challenge for river managers. To cope with this challenge, scientific studies focused on the short-term variation of flow regimes are needed.

The restoration of preindustrial flow regimes requires metrics comprising of the full range of flow components (i.e., magnitude, frequency, duration, timing and rise and fall rates; [1]) and temporal variability (i.e., long- and short-term variations) is essential. Whereas studies of seasonal and annual flow patterns have been common, analysis of short-term data have suffered from a lack of computational tools. To the best of our knowledge, the first metrics accounting for short-term variability of flow regimes appeared within the last two decades (e.g., [4]) and the most comprehensive approaches date from 2014 onwards (Table 1). Unlike the recent advance in the definition of subdaily metrics, computational tools supporting metric calculation have hardly been developed. The tools devised by Hass et al. [20] and Sauterleute and Charmasson [21] (Table 1) are the only ones we are aware of to date, and at the time of writing, the former tool was unavailable for use. This is unfortunate, because the management of series of flows or water levels recorded at such a fine resolution is challenging.

Our main goal is to develop a tool for computational time series analysis that assists in a comprehensive characterization of short-term stream flow and water level regimes and assesses the alterations of such regimes and, thus, their derived potential environmental impacts. We also want the tool to provide results through charts and graphs, which are easy to interpret by a wide range of users. Additionally, in this article we also aim to validate the devised tool by applying it to a case study. This manuscript will help to transmit the utility of the proposed tool to both the scientific and professional audience.

Table 1. Review of literature dealing with subdaily flows and water levels.

Reference	Time Interval between Records	Characteristics of the Subdaily Metrics	Characterization	Impact Assessment	Tool
Archer and Newson 2002 [2]	15 min	Metrics quantifying the frequency and duration of flow pulses per day	Yes	Yes	No
Topping et al., 2000 [22]	Several subdaily intervals	Metrics quantifying the subdaily discharge variability	Yes	Yes	No
White et al., 2005 [23]	1 h	Wavelet analysis	Yes	Yes	No
Meile et al. 2011 [24]	Any subdaily interval	Metrics quantifying the magnitude (maximum and minimum) and variability (ramping rate) of hourly flows per day	Yes	No	No
Zimmerman et al., 2010 [25]	1 h	Metrics quantifying magnitude (percentage of total flow), variation (coefficient of diel variation and flashiness) and frequency (reversals) of hourly flows per day	Yes	Yes	No
Bevelhimer et al., 2015 [26]	1 h	Metrics quantifying the magnitude (maximum, minimum and amplitude), variation (standard deviation, flashiness and maximum ramping rate) and frequency (reversals, rise and fall counts) of hourly flows per day	Yes	No	No
Haas et al., 2014 [20]	1 h	Statistics and metrics quantifying the variation (coefficient of variation, flashiness, rise and fall rates), magnitude (range), frequency and duration (path length) and timing (season) of hourly flows and flow pulses per day	Yes	No	Yes
Sauterleute and Charmasson 2014 [21]	Any subdaily interval	Metrics characterizing peaking events of subdaily flows or water levels through the magnitude (maximum and minimum), variation (rise and fall rate), timing (start time in the day), duration (duration between rapid increases or decreases) and frequency (counts of peaking events)	Yes	No	Yes
Carolli et al., 2015 [27]	1 h	Metrics related to the flow magnitude (maximum and minimum) and variation (a percentile of the discretized time derivative) of hourly flows per day	Yes	Yes	No
Chen et al., 2015 [28]	1 h	Metrics characterizing flow pulses per day by quantifying the magnitude (i.e., maximum and minimum), variation (i.e., maximum rise and fall rates), frequency (i.e., different or certain magnitude counts) and duration (i.e., duration of maximum and minimum)	Yes	Yes	No
Barbalić and Kuspilić 2015 [29]	1 h	Metrics quantifying the magnitude of hourly flows and associated water levels during a day (i.e., maximum and minimum)	Yes	Yes	No
Greimel et al., 2016 [30]	15 min	Metrics quantifying the duration, number and flow rates (i.e., maximum, mean and minimum) of flow events per day	Yes	No	No
Alonso et al., 2017 [31]	1 h	Graphical representation of commonly used metrics characterizing daily flow patterns based on hourly flow records related to the magnitude (i.e., amplitude), variation (i.e., fall rate) and frequency (i.e., reversals)	Yes	Yes	No
Bejarano et al., 2017 [32]	1 h	Metrics quantifying the magnitude (maximum, minimum and amplitude), variation (rise and fall rates), frequency (rise, fall and stability, minimum and maximum and reversals counts), duration (length of rise, fall and stability periods) and timing (day) of hourly flows per day	Yes	Yes	No
Ashraf et al. 2018 [33]	1 h	Two metrics that quantify the high-frequency variations at a given time and seasonal changes	Yes	No	No

2. Materials and Methods

2.1. InSTHAn's Development: Underlying Theory and Methods

We developed the new tool called InSTHAn: indicators of short-term hydrological alteration. InSTHAn allows the user to (i) summarize multiple, long series of subdaily flow or stage data into a manageable set of ecologically meaningful metrics (i.e., characterization), (ii) qualify and quantify the deviation of each series from the unaltered state to assess the hydrological alteration and its potential environmental impact and (iii) display both the short-term flow or stage pattern and its impact by using tables and graphs. The name informs on its ultimate purpose and time scale of the target regime. The pronunciation of the acronym refers to the required recording interval of the input data (i.e., instant flow or water level measured or modeled records).

2.1.1. Characterization of Short-Term Regimes

The first step when analyzing a subdaily flow or water level dataset is to describe its distinctive features. For this aim, the proposed tool computes a set of descriptors, here called short-term characterization indicators (STCI; Table 2). STCI meets two requirements: it (i) captures representative information on the magnitude, frequency, duration, timing and rates of change from the subdaily flow or water level dataset and (ii) is assumed relevant for the biotic composition of aquatic, wetland and riparian ecosystems [1,34].

Table 2. Short-term characterization indicators calculated in indicators of short-term hydrological alteration (InSTHAn). # means "number of".

STCI Name and Abbreviation	Units	Group	STCI 366 × *n* (366 Values per "*n*" Years)
Total Rise Records (TRR)	# records/day	Frequency	Within-day total records characterized by the rise in the variable
Total Fall Records (TFR)	# records/day	Frequency	Within-day total records characterized by the fall in the variable
Total Stability Records (TSR)	# records/day	Frequency	Within-day total records characterized by the stability in the variable
Total Change Records (TCR)	# records/day	Frequency	Within-day total records that are preceded and followed by different patterns in the variable
Total Reversals (TRev)	# reversals/day	Frequency	Within-day total times the hourly variable rises and falls
Total Minimum Records (TMinR)	# records/day	Frequency	Within-day total records when the variable equals that day's minimum
Total Maximum Records (TMaxR)	# records/day	Frequency	Within-day total records when the variable equals that day's maximum
Total Mean Records (TMeanR)	# records/day	Frequency	Within-day total records when the variable equals or exceeds that day's mean
Total Rise Periods (TRP)	# periods/day	Frequency	Within-day total periods characterized by a sustained over time rise in the variable
Total Fall Periods (TFP)	# periods/day	Frequency	Within-day total periods characterized by a sustained over time fall in the variable
Total Stability Periods (TSP)	# periods/day	Frequency	Within-day total periods characterized by a sustained over time stability in the variable
Total Stability Periods characterized by the Minimum (TMinSP)	# periods/day	Frequency	Within-day total periods characterized by a sustained over time that day's stability periods minimum
Total Stability Periods characterized by the Maximum (TMaxSP)	# periods/day	Frequency	Within-day total periods characterized by a sustained over time that day's stability periods maximum
Total Stability Periods characterized by the Mean (TMeanSP)	# periods/day	Frequency	Within-day total periods characterized by a sustained over time that day's stability periods mean

Table 2. *Cont.*

STCI Name and Abbreviation	Units	Group	STCI 366 × *n* (366 Values per "*n*" Years)
Duration Rise Periods (DurRP)	# records/day	Duration	Within-day average duration of the periods characterized by a sustained over time rise in the variable
Duration Fall Periods (DurFP)	# records/day	Duration	Within-day average duration of the periods characterized by a sustained over time fall in the variable
Duration Stability Periods (DurSP)	# records/day	Duration	Within-day average duration of the periods characterized by a sustained over time stability in the variable
Duration Stability Periods characterized by the Minimum (DurMinSP)	# records/day	Duration	Within-day average duration of the periods characterized by a sustained over time that day´s stability periods minimum
Duration Stability Periods characterized by the Maximum (DurMaxSP)	# records/day	Duration	Within-day average duration of the periods characterized by a sustained over time that day´s stability periods maximum
Duration Stability Periods characterized by the Mean (DurMeanSP)	# records/day	Duration	Within-day average duration of the periods characterized by a sustained over time that day´s stability periods mean
Mean (Mean)	unitless or variable units	Magnitude	Within-day average of the variable
Standard Deviation (SD)	unitless or variable units	Magnitude	Within-day standard deviation of the variable
Minimum (Min)	unitless or variable units	Magnitude	Within-day minimum of the variable
Maximum (Max)	unitless or variable units	Magnitude	Within-day maximum of the variable
Amplitude (A)	unitless or variable units	Magnitude	Difference between within-day maximum and minimum of the variable
Minimum Stability Period (MinSP)	unitless or variable units	Magnitude	Within-day minimum of the periods characterized by a sustained over time stability in the variable
Maximum Stability Period (MaxSP)	unitless or variable units	Magnitude	Within-day maximum of the periods characterized by a sustained over time stability in the variable
Mean Stability Period (MeanSP)	unitless or variable units	Magnitude	Within-day mean of the periods characterized by a sustained over time stability in the variable
Rise Rate (RR)	variable units/*T*	Rate	Within-day average rise rate of the variable
Fall Rate (FR)	variable units/*T*	Rate	Within-day average fall rate of the variable

STCI was calculated based on an *n*-year long series of flows (Q) or water levels (L) recorded or modeled at any subdaily time scale, e.g., every 15, 30, 60 or 120 min, T being the time interval between records. Optionally, series of longer T can be derived from the original dataset upon request. For the purpose of defining indicators, each daily hydrograph (or limnograph) is divided into two characterization units: records (R; e.g., Q or L records; Figure 1) and periods (P). The number of records (R) per day varies according to T, which can be the same as that of the series input at least. Each record (R) of a series can be assigned one of the following patterns: (1) rise (RR), when $Q_{(T)} - Q_{(T-1)} > 0$; (2) fall (FR), when $Q_{(T)} - Q_{(T-1)} < 0$; (3) stability (SR), when $Q_{(T)} - Q_{(T-1)} = 0$; (4) change (CR), when $Q_{(T-1)} \neq$ the pattern in $Q_{(T+1)}$; (5) reversal (RR), when the pattern changes from FR to RR or vice versa, without considering the stability; (6) minimum (MinR), when $Q_{(T)} = Q_{(min)}$; (7) maximum (MaxR), when $Q_{(T)} = Q_{(max)}$ and (8) mean (MeanR), when $Q_{(T)} = Q_{(mean)}$. The threshold from which two consecutive records are considered different (or equal) may be set by the user. It could be similarly applied to L. Where T is the user-defined subdaily time interval and min, max and mean are the daily minimum, maximum and mean flows or water levels, respectively. Periods (P) denote within-day portions of time of a similar pattern among records (cf. above). There may be one to several *P* per day, lasting up to 24 h, and which can be classified according to the characteristic short-term pattern into periods of rise (*RP*), fall (*FP*), stability (*SP*), minimum (*MinP*), maximum (*MaxP*) and mean (*MeanP*). STCI provides quantitative information on magnitudes, rates of change and frequencies of *R* and *P* and on durations of *P*, from each day of the year (i.e., *i*th day of the year from 1 to 366). That STCI has daily values also implies information on timing (i.e., intra-annual and inter-annual) of *R* and *P*. STCI referred to *R* patterns is called record-based STCI, whereas STCI referred to *P* patterns is named period-based STCI. For comparisons of several short-term regimes, the record-based STCI must be calculated based on the same time interval between records (*T* of their *R*) for all series (Table 2).

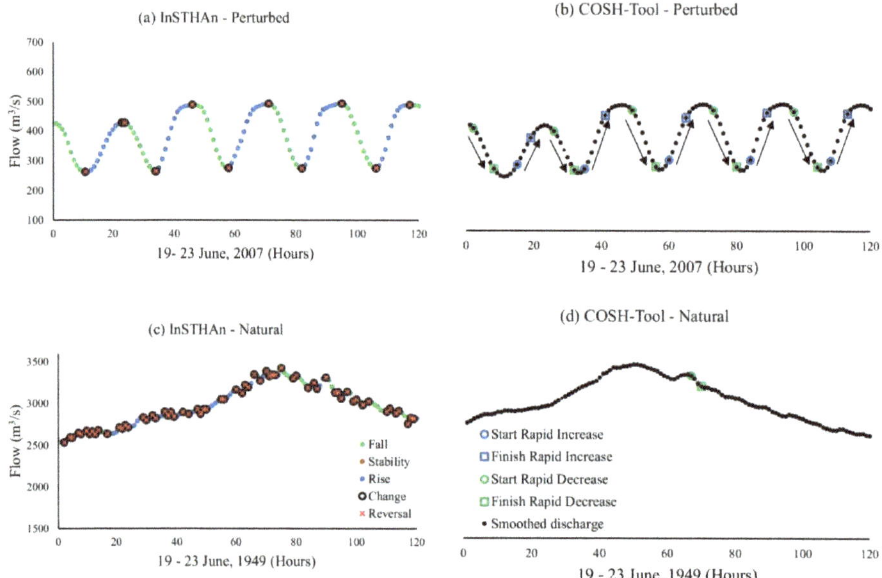

Figure 1. Patterns identified by InSTHAn (**a,c**; pre-dam) and COSH-Tool (**b,d**; post-dam) during five days in June, 2007 (**a,b**) and 1949 (**c,d**), in a hydrograph built on hourly flows recorded in the Colorado River reach downstream from the Glen Canyon dam. Dots represent the flow records, which are colored or marked according to their pattern for InSTHAn or to identify peaking events for COSH-Tool. The following figures were provided to COSH-Tool for peaking events identification: 4 and 96 as inferior and superior percentiles of the rate of change, 120 min as the minimum duration for a peak, 0.2 as the magnitude threshold to merge peaks and 180 min as the minimum duration between two consecutive peaks.

For several-year long series (*n* > 1; where *j*th denotes each year of the series from 1 to *n*), each indicator is ultimately computed as each day average for the whole *n* years dataset, getting 366 values per indicator (Equation (1); Table 2). The frequency and duration indicators report records a day of what it is being described by the indicator. Rate-related features report the rise or fall rates of the variable in its units per the time interval (*T*) between records (*R*). The units of the STCI magnitude-related indicators are the same of the selected variable (e.g., m³/s for flows or m for levels). Furthermore, for the calculation of STCI describing magnitude-related features, the series is also previously standardized by dividing between the mean flow or water level for the whole dataset. Consequently, InSTHAn also provides unitless magnitude-related indicators, which is useful when comparing series from different rivers. The tool calculates values for a total of 30 STCI, from which 14 are related to frequencies, 6 to durations and 10 to magnitudes and rates of change (Table 2).

$$STCI_{day(i)} = \frac{\sum\limits_{j=1}^{j=n} STCI_{day(i,j)}}{n} \tag{1}$$

Equation (1): $STCI_{day(i)}$: short-term characterization indicator for the ith day from 1 to 366 of the year; $\sum_{j=1}^{j=n} STCI_{day(i,j)}$: sum of the short-term characterization indicator for the ith day from 1 to 366 of the year jth of the several-year long dataset from 1 to n and n: total number of years of the dataset.

2.1.2. Assessment of Short-Term Hydrological Alteration and Environmental Impact

When assessing the impact of a perturbation we want to know whether the state of the perturbed system differs significantly from what it would have been in the absence of perturbation (natural onwards). Provided the difficulties in collecting direct ecological data both under perturbed and natural conditions, the here proposed tool is based on the widespread qualitative understanding of the ecological implications of the suite of hydrological indicators calculated by InSTHAn to derive the potential environmental impact of the alteration of the short-term flow or water level regimes. That is, the environmental impact is assumed in accordance with the degree and type of hydrological alteration, an assumption also applied by Bejarano et al. [35]. For the assessment of the hydrological alteration InSTHAn requires two datasets of subdaily flows or water levels to be compared, one representing the perturbed regime and the other the natural regime. The latter may come from the same location as the perturbed one as the preimpact period records or modeled records, or it may come from a comparable river reach.

The impact assessment involves a one-by-one comparison of the whole suite of STCI (record- and period-based STCI involving 366 values per indicator from each day of the averages for n years) from the perturbed and corresponding natural subdaily flow or water level datasets. InSTHAn's output is a suite of short-term impact indicators (STII, record- and period-based STII) obtained through Equation (2). Each impact indicator quantifies the deviation of the perturbed condition (*per*) from the natural condition (*nat*) of the corresponding characterization indicator (Equation (2)). Log$_{10}$ is applied to the quotient to avoid excessively high values when the averages of certain indicators in the natural conditions are very low (e.g., indicators related to flow rates of change). Impact indicators can take any positive and negative value and are unitless. Comparisons are not restricted to perturbed and natural series, but other comparisons between series may be made according to user needs.

$$STII_{day(i)} = sign\left(STCI^{nat}_{day(i)} - STCI^{per}_{day(i)}\right) \log_{10}\left[\frac{\left|STCI^{nat}_{day(i)} - STCI^{per}_{day(i)}\right|}{\left(\dfrac{\sum\limits_{i=1}^{i=366} STCI^{nat}_{day(i)}}{366}\right)} + 1\right] \tag{2}$$

Equation (2): $STII_{day(i)}$: short-term impact indicator for the ith day from 1 to 366 of the year; $sign\left(STCI^{nat}_{day(i)} - STCI^{per}_{day(i)}\right)$: sign function for the difference between the short-term characterization indicators for the ith day from 1 to 366 of the year from the natural (*nat*) and perturbed (*per*) series; $\left|STCI^{nat}_{day(i)} - STCI^{per}_{day(i)}\right|$: absolute value for the difference between the short-term characterization indicators for the ith day from 1 to 366 of the year from the natural and perturbed series and $\sum_{i=1}^{i=366} STCI^{nat}_{day(i)}$: sum of the short-term characterization indicator for the ith day from 1 to 366 of the year from the natural (*nat*) series.

2.2. InSTHAn's Application and Validation

We were interested in (i) characterizing the short-term flow variability of the Colorado River (USA) along the reach downstream from the Glen Canyon dam before and after its construction (i.e., 1966) and (ii) evaluating the impacts of the dam on this short-term flow regime and, thus, subsequent expected environmental impacts on the fluvial ecosystem. For this aim, and in order to verify InSTHAn's correct operation and demonstrate its advantages, we applied InSTHAn and the Computational Tool for the Characterization of Rapid Fluctuations in Flow and Stage (Sauterleute and Charmasson, 2014; COSH-Tool onwards), which was kindly provided by authors (v2016). We had two original flow (m^3/seg) data series (.xlsx files). The natural series corresponded to hourly flows measured between 1943 and 1951, whereas the perturbed series corresponded to every 15 min flow measured between 2003 and 2011, both at Lees Ferry (9,380,000 gauging station code; data from https://waterdata.usgs.gov/). The former file was characterized by one column (flow) without a heading and five decimal places measurements, and the latter was characterized by three columns (date, time, and flow) with their respective headings and two decimal place measurements.

3. Results

3.1. InSTHAn's Characteristics

InSTHAn has been developed in Matlab, and the code is created and executed based on a user's actions within the graphical user interface (GUI). This approach provides convenient access to the most relevant code functions via buttons in the GUI, but translates each user action into executable code that can be captured in a script. The distribution version of the tool is encapsulated into an executable file that does not require a Matlab license for the end user. Moreover, implementing scripting within the GUI enables immediate visualization of results via graph and table-based views of the data. InSTHAn supports the commonly used .xlsx and .txt data files containing flow and/or water level records in columns, measured at any subdaily time interval and provided in any consistent system of units defined by the user. The results are generated into excel files with open code macros to help the user to zoom into long series graphs. Finally, InSTHAn may be deployed on multiple platforms (Windows, Linux and Macintosh), the installation and calculations require little disk space and computing power, respectively, and graphics have satisfactory performance on commonly used processors. Specifically, the required disk space is 27 Mb for computers with Matlab v2018, but 1.56 additional Gb corresponding to the additional libraries distributed with the MCR_R2018a_win64_instaler.exe are necessary when Matlab is not installed. Concerning the computational power, it took four minutes to complete an impact analysis for the selected case study involving the management of records, in a i7, 20 Gb ram PC.

InSTHAn is organized into projects and analyses (Figure 2). A project consists of one to several analyses (e.g., Project 1 and Analyses 1, 2 and 3 in Figure 2). Any calculation of a set of indicators constitutes an analysis, being of two types: characterization analysis, aimed exclusively at characterizing a short-term flow or water level regime (calculation of STCI), and impact analysis, aimed at assessing the alteration of a short-term flow or water level regime (and thus inferring the derived environmental impact; calculation of STII). A folder is generated where specified in the computer to store the projects ("Project 1" directory; Figure 2) where data and all analyses run within the same project are stored, either in an automatically generated folder for the data files ("Excel" subdirectory), for the characterization analyses ("Characterization" subdirectory), or for the impact analyses ("Impact" subdirectory; Figure 2).

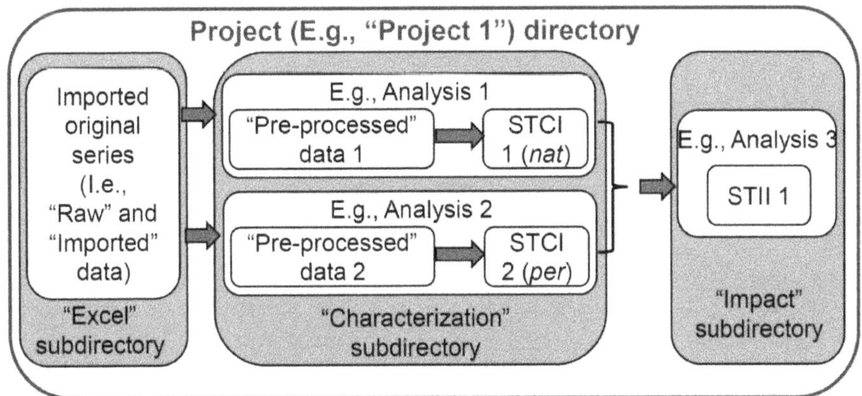

Figure 2. General organization of InSTHAn.

Analyses were linked to short-term data series (Figure 2) characterized by a set of flow or water level records measured (or modeled) at any subdaily time interval and from a specific time period, which was used for the calculation of indicators. Indicators may be calculated on the entire imported original data series (i.e., "Raw" data and "Imported" data), or on a preprocessed data series by changing the analysis period or the time interval between records with InSTHAn (i.e., "Pre-processed" data). Thus, each characterization analysis is linked to a single series, whereas each impact analysis is linked to two series, for example a perturbed (*per*) series and a comparable natural (*nat*) series (Figure 2). The impact assessment may be run (i) on a series of short-term flows or water levels, which can be split in InSTHAn into two independent (sub) series representing the preimpact (natural) and postimpact (perturbed) periods, or (ii) on two independent series representing the perturbed and the natural conditions. In any case, the previous characterization of each perturbed and natural series is necessary for the subsequent evaluation of the impact (Figure 2).

InSTHAn is organized into three modules corresponding to the steps that must be followed to set up and complete an impact assessment analysis, requiring the user to (i) create a project and import the data (Module I: Project management and data import; Supplementary Materials A: Figure S1), (ii) preprocess and analyze the data by calculating the STCI (Module II: Characterization; Supplementary Materials A: Figure S2) and (iii) calculate the STII (Module III: Impact assessment; Supplementary Materials A: Figure S3). Finally, outputs may be displayed in tables and graphs. Details on each module can be found in Supplementary Materials A.

3.2. InSTHAn's Functionality and Comparison with Other Tools

Both InSTHAn and COSH-Tool were launched from an executable file. Then, the main interface opened and allowed access to analysis of the time series. Both interfaces are simple and require no coding from the user (Table 3). With InSTHAn, two different projects named "ColoradoNat" and "ColoradoPer" were created (Supplementary Materials B: Figure S5). Two different characterization analyses were ran, one for the natural original series ("ColoradoNatCharacterization1") corresponding to the period before the construction of the dam, and the other for the perturbed original series ("ColoradoPerCharacterization1"), whose outputs were saved into their respective folders within "ColoradoNat" or "ColoradoPer" projects (Supplementary Materials B: Figures S4–S19). While importing the original data series we provided the required information on the series. Then, the two imported data series were preprocessed in order to set the entire available period of data as the characterization analysis period, and to round the flow measurements to two decimal places. The perturbed data series, originally characterized by every 15 min records, was also decimated in InSTHAn to get a measurement every hour.

Table 3. Comparison of the tools used in this article: InSTHAn (v2020) and COSH-Tool (v2016).

Characteristics		InSTHAn	COSH-Tool
General characteristics	Programming language	InSTHAn v2020 is programmed in Matlab, but it does not require a Matlab license and knowledge to deploy and customize output figures	COSH-Tool v2016 is programmed in Matlab and it requires a Matlab license and knowledge to deploy and customize output figures
	Graphical user interface (GUI)	Several windows, friendly user interface	Few windows, friendly user interface
	Languages	User selected between Spanish and English	Default English
Data loading, preparation and organization	File types supported	Excel and text files	Excel
	Number of variables per file	Up to four	One
	Data resolution	Intraday. It allows to change the time interval of records	Intraday. It does not allow to change the time interval of records
	Data units	User defined	User selected among options (stage (m), flow (m^3/s), unidentified)
	Navigation in the PC	Yes	No
	Organization of analyses	Hierarchical organization in projects and analyses, which may be open, consulted and modified anytime	No hierarchical organization. Analyses cannot be open, consulted and modified by the user
Data preprocessing	Preprocessing options	Selection of subperiods of analysis, data decimation (grouping records in larger time intervals), and data filtering (rounding the measurement figures)	Selection of subperiods of analysis, deletion of outliers, and data smoothing (moving average). No decimation (grouping records in larger time intervals) and data filtering (rounding the measurement figures)
Data analysis	Characterization	Based on patterns assigned to records and periods (within-day portions of time of similar pattern among records). They can be: rise, fall, stability change and reversals. No user requirements for patterns identification	Based on peaking events. They can be: rapid increase and rapid decrease. Peaking events identification is conditional on the provision of several figures by the user (the inferior and superior percentiles of the rate of change, a minimum duration for a peak, the magnitude threshold to merge peaks and the minimum duration between two consecutive peaks)
		Through metrics and statistics relating to the major flow components (i.e., magnitude, frequency, duration and rate of change). Deepening the duration of patterns. Information on stability and change patterns. See Table 2 for details (named STCI)	Through metrics and statistics relating to the major flow components (i.e., magnitude, frequency, duration and rate of change). No deepening the duration of peaking events. No information on stability and change patterns. See Table 1 in Sauterleute and Charmasson [21] for details
	Impact	Through comparisons of characterization metrics (STCI) from natural and perturbed series (named STII)	No
Outputs	Outputs format	Comprehensive tables and many figures in excel. Easy customization of figures through Excel	Simplified tables in excel. Many figures deployed in Matlab. Customization of figures and access to the data represented by the figures through Matlab
	Outputs scale	It captures each day´s subdaily patterns of the series, from which the user may derive longer-scale patterns	It captures daylight, monthly, seasonal and annual patterns

The natural and perturbed series were also loaded and prepared with COSH-Tool. Apart from small differences between the tools related to restrictions on the navigation in the PC, or on allowed variables, units and languages (Table 3), a notable difference of COSH-Tool is the non-organization of the outputs within projects or/and analyses where they may be easily found and consulted (Table 3). With a purpose similar to rounding in InSTHAn, smoothing was required by COSH-Tool at this stage. Smoothing, however, depends on a "smoothing factor" set by the user, which must be within a range of figures used during testing of the tool. Unlike InSTHAn, COSH-Tool is unable to modify the record interval of the input series, so the original every 15 min, perturbed series had to be turned into hourly time step series before loading to ensure that both natural and perturbed series had similar record intervals for later comparisons. Finally, for both natural and perturbed original series patterns were assigned to records (R) and periods (P) by InSTHAn (i.e., fall, rise, stability, change and reversal), but peaking events (i.e., rapid increases and decreases) were identified by COSH-Tool (Figure 1). Whereas the detection of such patterns in InSTHAn is based on differences between each previous and following rounded record and does not depend on predefined values, the detection of peaking events in COSH-Tool is conditional on the provision of several figures by the user, such as the inferior and superior percentiles of the rate of change, a minimum duration for a peak, the magnitude threshold to merge peaks, and the minimum duration between two consecutive peaks (Table 3). Since the subsequent characterization of the series is based on the patterns and peaking events previously identified by InSTHAn and COST-Tool, respectively, setting different figures in COSH-Tool may result in variations of the peaking events of a series, ultimately affecting its characterization (Figure 1). For the perturbed case, the whole flow series was split into many periods of rise and fall, and reversals and changes by InSTHAn (Figure 1). However, for the same series, the rapid increases and decreases were confined to the flow records that met the user-set (recommended by the users' manual) parameters (cf. above) by COSH-Tool (Figure 1). For the natural flow series, significantly more patterns through years were detected by InSTHAn compared to the almost non-existent peaking events found by COSH-Tool (Figure 1).

After data series loading and preparation, we required InSTHAn and COSH-Tool to characterize the natural and perturbed subdaily flow regimes. The records (R) and periods (P) previously assigned to different patterns were characterized by InSTHAn, whereas characterization of the identified peaking events was done by COSH-Tool. In both tools, characterization is done through metrics and statistics relating to the major flow components (i.e., magnitude, frequency, duration and rate of change; Table 3). However, a more thorough characterization representing all facets of the subdaily variation is achieved with InSTHAn, which goes into greater depth in duration metrics and provides information on periods of stability and reversals and changes (Table 3). Whereas InSTHAn's metrics (STCI) capture each day's subdaily patterns of the series, from which the user may derive longer-scale patterns through averaging the excel outputs, metrics from COSH-Tool characterize monthly, seasonal and annual patterns, which are displayed in figures (Table 3). Only a brief summary of the outputs for the whole analyzed period is provided in an excel template by COSH-Tool. Unlike InSTHAn, COSH-Tool also provides daylight patterns. Characterization metrics representative of each flow component (frequency, duration, magnitude and rate of change) have been chosen from each tool for Figure 3 (further outputs from InSTHAn can be consulted in Supplementary Materials B and in Alonso et al. [31] and Bejarano et al. [32]).

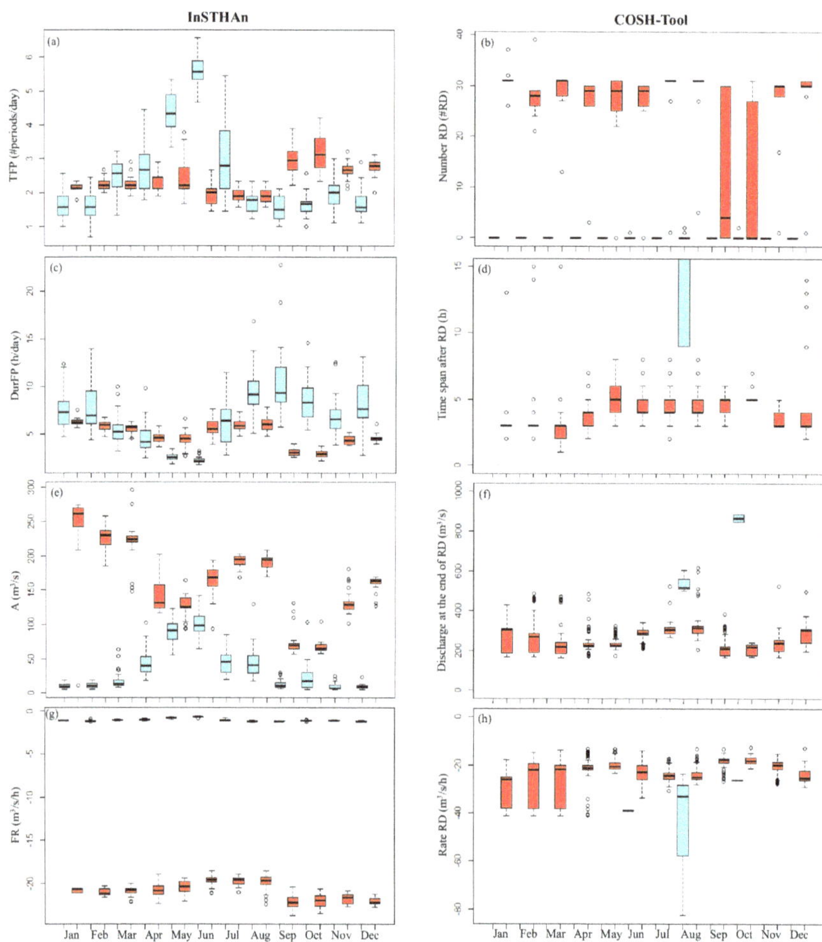

Figure 3. Box-and-whisker plots for selected outputs from the characterization analyses ran in InSTHAn and COSH-Tool for the pre- and post-dam (Glen Canyon dam) flow series (1943–1951 hourly flows, and 2003–2011 every-15 min flows, respectively) along the downstream reach of the Colorado River. *y*-axes represent the months in pre- (natural) and post-dam (perturbed) conditions, colored in blue and red, respectively. Black lines in the middle of the boxes are the median values for each group. The vertical size of the boxes is the interquartile range (IQR). The whiskers represent the minimum and maximum values that do not exceed 1.5 × IQR. The points are outliers. *x*-axes represent the characterization metrics related to frequency, duration, magnitude and rates of change provided by InSTHAn (i.e., short-term characterization indicators (STCI); **a,c,e,g**) and COSH-Tool (**b,d,f,h**). For InSTHAn, selected metrics are: (**a**) monthly average number of fall periods per day for the whole flow series, (**c**) monthly average duration of fall periods per day for the whole flow series, (**e**) monthly average amplitude per day for the whole series and (**g**) monthly average rate of flow decrease per day for the whole series. For COSH-Tool, the selected metrics are: (**b**) total number of rapid decreases per month for the whole series, (**d**) time span after rapid decreases per month for the whole series (not shown were three values in June, August and October for the natural period, which were higher than 15 h), (**f**) discharge after rapid decreases per month for the whole series (not shown was one value in June for the natural period, which was higher than 1000 m³/s) and (**h**) rate of flow decrease of rapid decreases per month for the whole series.

Both InSTHAn and COSH-Tool were able to capture the hydropeaking derived from the operation of the Glen Canyon dam in the perturbed flow series. In general, from both tools the user can derive that hydropeaking is associated to significantly frequent and short fall (and rise) periods (InSTHAn) or rapid decreases (and increases; COSH-Tool); fast hourly flow changes (highlighted by both tools) and high within-day flow amplitude (InSTHAn) and discharge (COSH-Tool; Figure 3). On average, InSTHAn identified three, 5 h fall periods per day during the whole year for regulated conditions (Figure 3). Other metrics (not shown) were consistent with these figures; the more frequent the fall (and rise) periods, the more frequent the flow changes and reversals, and the more frequent and shorter the stability periods. On average, COSH-Tool identified 25 rapid decreases per month for regulated conditions and described short time spans after rapid decreases (5 h on average) for regulated conditions (Figure 3). For the series subjected to hydropeaking, InSTHAn showed that the average daily amplitude was 162 m^3/s and the flow receded at a rate of (−) 21 m^3/s/h, whereas COSH-Tool showed an average discharge at the end of a decrease of 263 m^3/s and of rate of flow decrease per month of (−) 24 m^3/s/h (Figure 3). Conversely, the characterization of the natural series did vary significantly between the tools. Whereas the patterns of the flows used by InSTHAn for the characterization are also found in the series regardless of whether it is regulated or not, the peaking events used by COSH-Tool are restricted to artificial changes of the series, such as hydropeaking, and linked to exceptional natural peaking events (Figure 3). Consequently, hardly any peaking events were found by COSH-Tool throughout the natural flow series and, thus, most metrics were not applicable or equaled zero (Figure 3). The values for the metrics mentioned above obtained by applying InSTHAn to the natural series were in general (except for the spring values) significantly lower than the values from the perturbed series. Average values were as follows: four, 3 h fall periods per day and two, 8 h fall periods per day for the spring and the remaining seasons, respectively; a daily amplitude of 79 m^3/s during the flooding season and 21 m^3/s for the rest of the year and an hourly flow rate of 1 m^3/s/h (Figure 3).

In InSTHAn we ran an impact analysis named "ColoradoImpact1", whose outputs were saved into its corresponding folder within one of the existing projects (the project "ColoradoNat" in our case; Supplementary Materials B: Figures S20–S25). For the impact analysis we indicated the characterization files to compare natural and perturbed (i.e., "ColoradoNatCharacterization1" and "ColoradoPerCharacterization1") from the InSTHAn dropdown menu and the deviation from the naturalness of each metric for each day of an average year was calculated. Impact assessment is not available in COSH-Tool (Table 3). Described changes on each STCI are summarized by their respective STII, which evidence both the magnitude and the direction of the impact (a selection of STII is shown in Figure 4). On the one hand, the very positive STII values highlight the significant increase of the within-day flow amplitude and rates of change resulting from hydropeaking (Figure 4). On the other, the close-to-zero, positive and close-to-zero, negative STII values highlight the slight increase or decrease of the frequency and duration of the fall periods with regulation, respectively; the pattern is only unfulfilled during the flooding period (Figure 4).

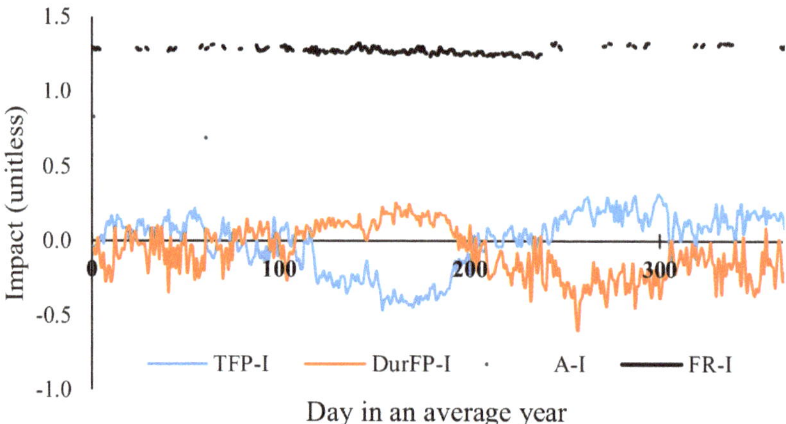

Figure 4. Outputs from the impact analyses ran in InSTHAn for the above mentioned characterization indicators (short-term impact indicators [STII]; -I denotes the impact on each indicator). Values around 0 mean a slight impact.

4. Discussion and Conclusions

4.1. Applicability

InSTHAn assists both scientists and river managers in describing and evaluating the naturalness of short-term flow/water level regimes, thus, eventually facilitating the understanding of the potential environmental impacts of the alterations of these regimes. Results from the application of InSTHAn to the analysis of the short-term flow variation in the Colorado River denote important modifications of certain key hydrological parameters at the subdaily scale due to the operation of the Glen Canyon dam. These would, otherwise, have gone unnoticed with other tools based on daily or larger time scale flow records. The derived consequences of these changes for the fluvial ecosystem may be severe. Particularly, significantly higher amplitudes of subdaily flows due to a regulation increase of the everyday wetted area, which may remove or move upwards on riparian areas plant species less tolerant to flooding while triggering the development of aquatic or amphibian species. Such consequences were described by Bejarano et al. [16] in rivers with hydropeaking from Northern Sweden, where *Betula pubescens* survival decreased significantly whereas *Salix* and *Carex* species were favored. Additionally, the significantly faster flow rates of change may result in fish/egg stranding, macroinvertebrate drift and obstruction of germination. For example, Casas-Mulet et al. [36] related the higher mortality of *Salmo salar* eggs in a river in central Norway to rapid dewatering, and Schülting et al. [37] observed macroinvertebrate drift proportions peaked during the up-ramping phase of water in an experimental flume. Although altered to a lesser extent, the more frequent and shorter inundations within a day may also cause scouring and burial, and soil surface clogging, damage or removal of sessile organisms or life stages and habitat deterioration and loss, which was already reported by Vanzo et al. [7].

Although based on different characterization units (patterns or peaking events), both InSTHAn and COSH-Tool were reliable for the characterization of short-term scale flow and water level series. The single characterization of the short-term natural and regulated flow regimes is valuable as it increases scientific knowledge on geographic patterns of hydrological variability [38,39], and helps to understand the influence of these patterns on biological communities and ecological processes [40]. InSTHAn's added contribution lies in its ability to quantitatively assess the short-term hydrological alteration by comparing identified patterns in natural and regulated conditions. Consequently, and unlike COSH-Tool, InSTHAn brings water managers and scientists closer to the potential ecological consequences of the hydrological alteration, and to whether consequences may be irreversible (when exceeding the ecosystem's thresholds), ultimately helping to determine the resistance and resilience of the river [41]. This knowledge is key for guiding any river management strategies [42], the assessment of its ecological status [34,43], prioritizing conservation efforts [44] and setting and measuring progress toward conservation or restoration goals [45]. Particularly, InSTHAn's results from the analyzed series would be useful when determining operational rules at the Glen Canyon plant and/or in-situ compensation measures aimed at harmonizing hydropower production and ecological integrity of the river [46]. Whatever the purpose, InSTHAn should be used in combination with other tools focused on longer time resolutions such as the IHA [34], in order to guarantee the comprehensiveness of the analyses by accounting for hydrological attributes at all time scales [47].

4.2. Merits and Limitations of InSTHAn in Relation to Other Tools

The appeal of InSTHAn is that it facilitates the analysis of long data series, which would otherwise be tedious. It offers several advantages and improvements over its peers. It allows different languages, reads widely used files of data from any source, records at any subdaily time scale and characterizes by a wide range of date styles and data units, and up to four variables in the same sheet can be imported; options that are more limited in existing tools. Additionally, InSTHAn provides a set of descriptive subdaily hydrological indicators comprehensive enough to account for the most ecologically determinant hydrological attributes [32], overcoming the limitations of other tools in duration metrics. Although it has been specially designed for flow and water level datasets, as included indicators make sense in the context of the field of stream hydrology, the user may consider it appropriate for other variable types recorded at similar short-term resolution, e.g., water temperature or water dissolved gasses in order to analyze the phenomena of thermopeaking [48] and saturopeaking [6], respectively. All these variables are usually affected by hydropower production, which has been the focus of this manuscript, but InSTHAn could be useful also in cases when flows are manipulated by dams with other purposes than electric power generation but also involving the alteration of the short-term flows.

An interesting novelty is that InSTHAn allows adaptive analyses by modifying the analysis periods (i.e., subperiods), the recording time intervals (i.e., to longer subdaily time steps) and the accuracy to detect subdaily patterns (i.e., thresholds from which a fluctuation is considered). The latter is crucial to avoid unreal fluctuations led by the influence of the accuracy of the measuring device or the model, or simply measurement or modeling errors [30], and which is lacking in existing tools. Finally, no tools to date enable the assessment of the alteration of short-term regimes (Table 1). Specifically, COSH-Tool founds the characterization of subdaily regimes on peaking events (to some extend similar to the so-called pulses by other authors) previously identified by the user based on subjectively defined thresholds (e.g., [4,21,28,30]; Table 1). As our results show, the use of peaking events as characterization units prevents the characterization of natural (or slightly affected) series usually lacking such events. This is not minor, as impact can only be assessed by comparing natural and perturbed series pairs. Characterization in InSTHAn, however, is based on patterns ultimately describing the records of the series. This, first, guarantees objectivity in the identification process of subdaily patterns, which, secondly, can be performed for any series regardless of the degree of alteration.

From a practical perspective, InSTHAn has been designed for a wide audience with different backgrounds and expertise. Although the decision-maker is often a water resources manager within a mandated organization, stakeholder participation, including water abstractors, wildlife campaigners and local community representatives, play a role in influencing decisions [49]. Unfortunately, reaching agreement is hindered by such a range of interested parties with usually conflicting goals, which can rely on InSTHAn outputs to set balanced thresholds. For this aim, InSTHAn is an easy installation tool, which requires little computer memory and optimizes the calculation time. The friendly windows within the GUI and clear results displayed through tables and graphs, which can be read and managed from Excel files, help to make the tool easy to use even for inexperienced users. Furthermore, it can be customized to change the language, units, and add/remove/zoom into graphs. Unfortunately, for the authors´ experience, the navigation through COSH-Tool and management of results was not as straightforward and intuitive.

With regards to the limitations of InSTHAn, we point out again that derived environmental impacts of short-term hydrological alterations are not directly provided by the tool but can be derived from the already understood ecological implications of the calculated hydrological indicators. Consequently, understanding of the ecological impacts from the outputs may require additional expertise and this may vary according to specific species, conservation objectives and site characteristics. Further research should address this issue. Another important limitation of InSTHAn derives from the requirements for the input data. Although InSTHAn may be run on daily (or longer intervals) data, results may not make sense at such time scales as indicators are focused exclusively on capturing subdaily patterns. Results should be analyzed with caution if subdaily records are few. In such cases, other tools could be more suitable (e.g., [34]). Further, for the case of hydrological datasets, measuring (especially in free-flowing rivers) and modeling at such fine resolution are still uncommon. This particularly affects the impact assessment module, which is dependent on free-flowing series. In the absence of data from free-flowing rivers, the solution would involve the restitution of the free-flowing regime at the study location. To accomplish this, at least one (representative) year of subdaily flows or water levels should be recorded at a comparable location (for example by using pressure-transducer loggers), which would provide the natural subdaily variability applied to model a longer period based on commonly available daily records (registered or modeled). In rivers with high interannual flow variability, more than one year of registered subdaily data would be desirable. A last restriction on the input data is that, with any subdaily registering interval allowed, this interval must remain constant throughout the whole study period. Finally, in the spirit of InSTHAn being a user-friendly tool that attracts a wide range of users, those who are more experienced may not like that actions are restricted to windows and cannot be ordered through commands.

4.3. Future Versions

We are working on completing existent modules and introducing new modules of InSTHAn. The modular structure and the tool architecture allow the inclusion of new modules that may extend the tool functions in future versions. Within the characterization and impact modules, new indicators will be added in future versions such as measures of central tendency and dispersion for the indicators. In addition, subdaily patterns will be summarized at other time spans apart from the daily basis (i.e., currently, indicators take an average value for each day). For example, subdaily flow fluctuations caused by hydropeaking along northern regions are higher during daytime, workdays or cold seasons following electricity demands [50]. Detecting these variations in subdaily flow patterns is key when planning strategies for sustainable hydropower management. In this regard, COSH-Tool already distinguishes between daytime and nighttime analysis. Limits on hydropower production could focus on situations when restrictions may result in great ecological gains but small economic losses. A module for the categorization of data series according to their subdaily patterns or impact will be built. We believe that it may facilitate management as similar management rules may be prescribed to all series pertaining to the same group [32]. Finally, extra ecological and economic modules, which provide the ecological and economic consequences of the already identified and quantified hydrological changes would round off the current version of the proposed tool. InSTHAn should be tested with other data series and improved accordingly. For this to be realized, our purpose is to make it generally accessible as soon as the patent is obtained by downloading it for free from a webpage with user registration as the only requirement. A user manual will be also available on the same webpage. The user may share his/her experience when using the tool, inform of the degree of satisfaction with it and ask doubts or suggest changes that could be included in future versions.

4.4. Conclusions

We introduced the new tool InSTHAn: indicators of short-term hydrological alteration. InSTHAn allows the user to (i) summarize multiple, long series of subdaily flow or stage data into a manageable set of ecologically meaningful metrics (i.e., characterization), (ii) qualify and quantify the deviation of each series from the unaltered state to assess the hydrological alteration and its potential environmental impact and (iii) display both the short-term flow or stage pattern and its impact by using tables and graphs. The name informs on its ultimate purpose and time scale of the target regime, whereas the pronunciation of the acronym refers to the required recording interval of the input data (i.e., instant records). InSTHAn represents an advance compared to existing tools. In the characterization stage, it guarantees objectivity in the identification of subdaily patterns from any (natural or altered) series, and provides a comprehensive set of ecologically meaningful hydrological indicators. In the impact stage, it enables the assessment of the alteration of short-term regimes. Finally, in terms of its functionality, it is characterized by the flexibility in the analyses (analysis periods, recording time intervals and accuracies to detect subdaily patterns) and in the supported languages, files and datasets properties (date styles, records time intervals and data units), and it is a friendly tool because its straightforward installation and use (windows within the GUI and clear display of results). InSTHAn responds to real-world needs in the fields of science and technology, and ultimately of society. By facilitating complex data management, it promotes the development of scientific studies on the short-term variability of river flows and levels—natural and altered by anthropogenic actions—underlying key ecological processes in rivers. By providing comprehensive and objective information on short-term stream flows and levels, this tool solves conflicting user perspectives and, hence, supports the sustainable integrated assessment and management of river systems. InSTHAn is particularly useful in the environmental management of rivers used for hydropower production, as it will assist in achieving the priority goal of maximizing hydroelectricity production while minimizing environmental losses.

Supplementary Materials: The following are available online at http://www.mdpi.com/2073-4441/12/10/2913/s1, Figure S1. Project management and data import module (InSTHAn's Module I), Figure S2. Characterization module (InSTHAn's Module II), Figure S3. Impact assessment module (InSTHAn's Module III), Figure S4. Start a new or load an existing project, Figure S5. Import the original data, Figure S6. Export "Raw" data and "Imported" data. Example from the post-dam flows, Figure S7. Export "Raw" data and "Imported" data, Figure S8. See "Imported" data. Example from the post-dam flows, Figure S9. Create a new or load an existing Characterization analysis, Figure S10. Select the Characterization analysis that we want to load from a list, Figure S11. Create and run a new Characterization analysis, Figure S12. Export Characterization analysis: main menu. Example from the post-dam flows, Figure S13. Export Characterization analysis: main results, Figure S14. Export Characterization analysis: extra results, Figure S15. See "Pre-processed" data. Example from the post-dam flows, Figure S16. See "RP Patterns" file: table sheet. Example from the post-dam flows, Figure S17. See "RP Patterns" file: graph sheet. Example from the post-dam flows (January, 2003 is represented), Figure S18. See "STCI 366" file: table sheet. Example from the post-dam flows, Figure S19. See "STCI 366" file: graph sheet. Example from the post-dam flows (The entire year values for two indicators are shown), Figure S20. Create a new or load an existing Impact analysis, Figure S21. Create and run a new Impact analysis, Figure S22. Export Impact analysis: main menu, Figure S23. Export Impact analysis, Figure S24. See "STII 366" file: table sheet, Figure S25. See "STII 366" file: graph sheet (The entire year values for two indicators are represented).

Author Contributions: Conceptualization, M.D.B.; methodology, J.H.G.-P. and A.S.-W.; investigation and formal analysis, M.D.B., J.H.G.-P. and A.S.-W.; resources and data curation, J.H.G.-P.; writing—original draft preparation, M.D.B.; writing—review and editing, C.N. and L.G.; visualization and supervision, C.N. and L.G.; funding acquisition, M.D.B. and A.S.-W. All authors have read and agreed to the published version of the manuscript.

Funding: This manuscript was supported by funding from: the Spanish Ministry of Science and Innovation (RIHEL; Ref. PID2019-111252RA-I00 CTA and SECA-SRH; Ref. PID2019-105852RA-I00); and Universidad Politécnica de Madrid (Programa Propio: Ayudas a Proyectos de I+D de Investigadores Posdoctorales) and Comunidad de Madrid (Convenio Plurianual con la Universidad Politécnica de Madrid) (Ref. APOYO-JOVENES-PHZKKU-148- SSPVMP).

References

1. Poff, N.L.; Allan, J.D.; Bain, M.B.; Karr, J.R.; Prestegaard, K.L.; Richter, B.D.; Sparks, R.E.; Stromberg, J.C. The natural flow regime. *BioScience* **1997**, *47*, 769–784. [CrossRef]
2. Poff, N.L.; Zimmerman, J.K. Ecological responses to altered flow regimes: A literature review to inform the science and management of environmental flows. *Freshw. Biol.* **2010**, *55*, 194–205. [CrossRef]
3. Biggs, B.J.; Nikora, V.I.; Snelder, T.H. Linking scales of flow variability to lotic ecosystem structure and function. *River Res. Appl.* **2005**, *21*, 283–298. [CrossRef]
4. Archer, D.; Newson, M. The use of indices of flow variability in assessing the hydrological and instream habitat impacts of upland afforestation and drainage. *J. Hydrol.* **2002**, *268*, 244–258. [CrossRef]
5. Eng, K.; Carlisle, D.M.; Wolock, D.M.; Falcone, J.A. Predicting the likelihood of altered streamflows at ungauged rivers across the conterminous United States. *River Res. Appl.* **2013**, *29*, 781–791. [CrossRef]
6. Pulg, U.; Vollset, K.W.; Velle, G.; Stranzl, S. First observations of saturopeaking: Characteristics and implications. *Sci. Total Environ.* **2016**, *573*, 1615–1621. [CrossRef] [PubMed]
7. Vanzo, D.; Zolezzi, G.; Siviglia, A. Eco-hydraulic modelling of the interactions between hydropeaking and river morphology. *Ecohydrology* **2015**, *9*, 421–437. [CrossRef]
8. Bejarano, M.D.; Jansson, R.; Nilsson, C. The effects of hydropeaking on riverine plants: A review. *Biol. Rev.* **2018**, *93*, 658–673. [CrossRef]
9. Casas-Mulet, R.; Alfredsen, K.; Boissy, T.; Sundt, H.; Rüther, N. Performance of a one-dimensional hydraulic model for the calculation of stranding areas in hydropeaking rivers. *River Res. Appl.* **2015**, *31*, 143–155. [CrossRef]
10. Boavida, I.; Harby, A.; Clarke, K.D.; Heggenes, J. Move or stay: Habitat use and movements by Atlantic salmon parr (*Salmo salar*) during induced rapid flow variations. *Hydrobiologia* **2017**, *785*, 261–275. [CrossRef]
11. Flodmark, L.E.W.; Vøllestad, L.A.; Forseth, T. Performance of juvenile brown trout exposed to fluctuating water level and temperature. *J. Fish Biol.* **2004**, *65*, 460–470. [CrossRef]
12. Nagrodski, A.; Raby, G.D.; Hasler, C.T.; Taylor, M.K.; Cooke, S.J. Fish stranding in freshwater systems: Sources, consequences, and mitigation. *J. Environ. Manag.* **2012**, *103*, 133–141. [CrossRef] [PubMed]
13. Holzapfel, P.; Leitner, P.; Habersack, H.; Graf, W.; Hauer, C. Evaluation of hydropeaking impacts on the food web in alpine streams based on modelling of fish-and macroinvertebrate habitats. *Sci. Total Environ.* **2017**, *575*, 1489–1502. [CrossRef]
14. Leitner, P.; Hauer, C.; Graf, W. Habitat use and tolerance levels of macroinvertebrates concerning hydraulic stress in hydropeaking rivers—A case study at the Ziller River in Austria. *Sci. Total Environ.* **2017**, *575*, 112–118. [CrossRef]
15. Gorla, L.; Signarbieux, C.; Turberg, P.; Buttler, A.; Perona, P. Effects of hydropeaking waves' offsets on growth performances of juvenile *Salix* species. *Ecol. Eng.* **2015**, *77*, 297–306. [CrossRef]
16. Bejarano, M.D.; Sordo-Ward, A.; Alonso, C.; Jansson, R.; Nilsson, C. Hydropeaking affects germination and establishment of riverbank vegetation. *Ecol. Appl.* **2020**, *30*, e02076. [CrossRef]
17. Schmutz, S.; Bakken, T.H.; Friedrich, T.; Greimel, F.; Harby, A.; Jungwirth, M.; Melcher, A.; Unfer, G.; Zeiringer, B. Response of fish communities to hydrological and morphological alterations in hydropeaking rivers of Austria. *River Res. Appl.* **2015**, *31*, 919–930. [CrossRef]
18. Jackson, D.C.; Brown, A.V.; Davies, W.D. Zooplankton transport and diel drift in the Jordan dam tailwater during a minimum flow regime. *Rivers* **1991**, *2*, 190–197.
19. Zarfl, C.; Lumsdon, A.E.; Berlekamp, J.; Tydecks, L.; Tockner, K. A global boom in hydropower dam construction. *Aquat. Sci.* **2015**, *77*, 161–170. [CrossRef]
20. Haas, N.A.; O'Connor, B.L.; Hayse, J.W.; Bevelhimer, M.S.; Endreny, T.A. Analysis of daily peaking and run-of-river operations with flow variability metrics, considering subdaily to seasonal time scales. *J. Am. Water Resour. Assoc.* **2014**, *50*, 1622–1640. [CrossRef]
21. Sauterleute, J.F.; Charmasson, J. A computational tool for the characterisation of rapid fluctuations in flow and stage in rivers caused by hydropeaking. *Environ. Model. Softw.* **2014**, *55*, 266–278. [CrossRef]
22. Topping, D.J.; Schmidt, J.C.; Vierra, L.E. *Computation and Analysis of the Instantaneous-Discharge Record for the Colorado River at Lees Ferry, Arizona, 8 May 1921, through 30 September 2000*; USGS: Reston, FL, USA, 2003.
23. White, M.A.; Schmidt, J.C.; Topping, D.J. Application of wavelet analysis for monitoring the hydrologic effects of dam operation: Glen Canyon Dam and the Colorado River at Lees Ferry, Arizona. *River Res. Appl.* **2005**, *21*, 551–565. [CrossRef]

24. Meile, T.; Boillat, J.L.; Schleiss, A.J. Hydropeaking indicators for characterization of the Upper-Rhone River in Switzerland. *Aquat. Sci.* **2011**, *73*, 171–182. [CrossRef]

25. Zimmerman, J.K.H.; Letcher, B.H.; Nislow, K.H.; Lutz, K.A.; Magilligan, F.J. Determining the effects of dams on subdaily variation in river flows at a wholebasin scale. *River Res. Appl.* **2010**, *26*, 1246–1260. [CrossRef]

26. Bevelhimer, M.S.; McManamay, R.A.; O'Connor, B. Characterizing sub-daily flow regimes: Implications of hydrologic resolution on ecohydrology studies. *River Res. Appl.* **2015**, *31*, 867–879. [CrossRef]

27. Carolli, M.; Vanzo, D.; Siviglia, A.; Zolezzi, G.; Bruno, M.C.; Alfredsen, K. A simple procedure for the assessment of hydropeaking flow alterations applied to several European streams. *Aquat. Sci.* **2015**, *77*, 639–653. [CrossRef]

28. Chen, Q.; Zhang, X.; Chen, Y.; Li, Q.; Qiu, L.; Liu, M. Downstream effects of a hydropeaking dam on ecohydrological conditions at subdaily to monthly time scales. *Ecol. Eng.* **2015**, *77*, 40–50. [CrossRef]

29. Barbalić, D.; Kuspilić, N. Indicators of sub-daily hydrological alterations. *Teh. Vjesn.* **2015**, *22*, 1345–1352.

30. Greimel, F.; Zeiringer, B.; Höller, N.; Grün, B.; Godina, R.; Schmutz, S. A method to detect and characterize sub-daily flow fluctuations. *Hydrol. Process.* **2016**, *30*, 2063–2078. [CrossRef]

31. Alonso, C.; Román, A.; Bejarano, M.D.; García de Jalón, D.G.; Carolli, M. A graphical approach to characterize sub-daily flow regimes and evaluate its alterations due to hydropeaking. *Sci. Total Environ.* **2017**, *574*, 532–543. [CrossRef]

32. Bejarano, M.D.; Sordo-Ward, A.; Alonso, C.; Nilsson, C. Characterizing effects of hydropower plants on sub-daily flow regimes. *J. Hydrol.* **2017**, *550*, 186–200. [CrossRef]

33. Ashraf, F.B.; Haghighi, A.T.; Riml, J.; Alfredsen, K.; Koskela, J.J.; Kløve, B.; Marttila, H. Changes in short term river flow regulation and hydropeaking in Nordic rivers. *Sci. Rep.* **2018**, *8*, 17232. [CrossRef]

34. Richter, B.D.; Baumgartner, J.V.; Powell, J.; Braun, D.P. A method for assessing hydrologic alteration within ecosystems. *Conserv. Biol.* **1996**, *10*, 1163–1174. [CrossRef]

35. Bejarano, M.D.; Sordo-Ward, A.; Gabriel-Martin, I.; Garrote, L. Tradeoff between economic and environmental costs and benefits of hydropower production at run-of-river-diversion schemes under different environmental flows scenarios. *J. Hydrol.* **2019**, *572*, 790–804. [CrossRef]

36. Casas-Mulet, R.; Saltveit, S.J.; Alfredsen, K. The survival of Atlantic salmon (*Salmo salar*) eggs during dewatering in a river subjected to hydropeaking. *River Res. Appl.* **2015**, *31*, 433–446. [CrossRef]

37. Schülting, L.; Feld, C.K.; Zeiringer, B.; Hudek, H.; Graf, W. Macroinvertebrate drift response to hydropeaking: An experimental approach to assess the effect of varying ramping velocities. *Ecohydrology* **2019**, *12*, e2032. [CrossRef]

38. Poff, N.L.; Olden, J.D.; Pepin, D.M.; Bledsoe, B.P. Placing global stream flow variability in geographic and geomorphic contexts. *River Res. Appl.* **2006**, *22*, 149–166. [CrossRef]

39. McManamay, R.A.; Orth, D.J.; Dolloff, C.A.; Frimpong, E.A. A regional classification of unregulated stream flows: Spatial resolution and hierarchical frameworks. *River Res. Appl.* **2012**, *28*, 1019–1033. [CrossRef]

40. Puckridge, J.T.; Sheldon, F.; Walker, K.F.; Boulton, A.J. Flow variability and the ecology of large rivers. *Mar. Freshw. Res.* **1998**, *49*, 55–72. [CrossRef]

41. McCluney, K.E.; Poff, N.L.; Palmer, M.A.; Thorp, J.H.; Poole, G.C.; Williams, B.S.; Williams, M.R.; Baron, J.S. Riverine macrosystems ecology: Sensitivity, resistance, and resilience of whole river basins with human alterations. *Front. Ecol. Environ.* **2014**, *12*, 48–58. [CrossRef]

42. Arthington, A.H.; Kennen, J.G.; Stein, E.D.; Webb, J.A. Recent Advances in Environmental Flows Science and water management—Innovation in the Anthropocene. *Freshw. Biol.* **2018**, *63*, 1022–1034. [CrossRef]

43. Dudgeon, D.; Arthington, A.H.; Gessner, M.O.; Kawabata, Z.I.; Knowler, D.J.; Lévêque, C.; Naiman, R.J.; Prieur-Richard, A.H.; Soto, D.; Stiassny, M.L.J.; et al. Freshwater biodiversity: Importance, threats, status and conservation challenges. *Biol. Rev.* **2006**, *81*, 163–182. [CrossRef] [PubMed]

44. Snelder, T.H.; Dey, K.L.; Leathwick, J.R. A procedure for making optimal selection of input variables for multivariate environmental classifications. *Conserv. Biol.* **2007**, *21*, 365–375. [CrossRef] [PubMed]

45. Palmer, M.A.; Bernhardt, E.S.; Allan, J.D.; Lake, P.S.; Alexander, G.; Brooks, S.; Carr, J.; Clayton, S.; Dahm, C.N.; Follstad Shah, J.; et al. Standards for ecologically successful river restoration. *J. Appl. Ecol.* **2005**, *42*, 208–217. [CrossRef]

46. Bruder, A.; Tonolla, D.; Schweizer, S.P.; Vollenweider, S.; Langhans, S.D.; Wüest, A. A conceptual framework for hydropeaking mitigation. *Sci. Total Environ.* **2016**, *568*, 1204–1212. [CrossRef]

47. Ward, J.V. The four-dimensional nature of lotic ecosystems. *J. N. Am. Benthol. Soc.* **1989**, *8*, 2–8. [CrossRef]

48. Zolezzi, G.; Siviglia, A.; Toffolon, M.; Maiolini, B. Thermopeaking in Alpine streams: Event characterization and time scales. *Ecohydrology* **2011**, *4*, 564–576. [CrossRef]
49. Petts, J.; Leach, B. *Evaluating Methods for Public Participation: Literature Review*; Environment Agency: Bristol, UK, 2000.
50. Jansson, R.; Nilsson, C.; Dynesius, M.; Andersson, E. Effects of river regulation on river-margin vegetation: A comparison of eight boreal rivers. *Ecol. Appl.* **2000**, *10*, 203–224. [CrossRef]

Publisher's Note: MDPI stays neutral with regard to jurisdictional claims in published maps and institutional affiliations.

 *water*

Article

Pump Efficiency Analysis for Proper Energy Assessment in Optimization of Water Supply Systems

Araceli Martin-Candilejo *, David Santillán and Luis Garrote

Departamento de Ingeniería Civil, Hidráulica, Energía y Medio Ambiente, Universidad Politécnica de Madrid, 28040 Madrid, Spain; david.santillan@upm.es (D.S.); l.garrote@upm.es (L.G.)

* Correspondence: araceli.martin@upm.es

Received: 29 November 2019; Accepted: 24 December 2019; Published: 31 December 2019

Abstract: Water supply systems need to be designed accounting for both construction and operational costs. When the installation requires water pumping, it is key for the operational costs to know how well the pump can perform. So far, pump efficiency has been considered using conservative values, in the absence of a better estimation. The aim of this paper was to improve determining the energy costs by clarifying what the value of the pump performance should be. For this, 226 commercial pumps were studied, registering the efficiency at the optimum operating point, as well as other variables such as the flow rate, height, and pump type. As a result, a strong relationship between the pump performance and the discharge flow was spotted. That allowed the generation of an empirical curve, which can be used by designers to anticipate what pump efficiency can be expected. The results are used in a simple case study using the Granados Optimization System. These achievements can be implemented in design policies for a better energy assessment in the optimization of water supply systems.

Keywords: pump efficiency; water distribution systems; water supply systems; optimization; design policies; design

---

1. Introduction

Under the climate change threat, becoming energetically efficient is now a crucial necessity for our society. A proper energy assessment is, therefore, a key factor to account for in the developing of new policies that look after sustainable designs of water supply systems [1]. The designing stage of a water distribution system requires considering not only the construction phase, but also the operation of the facility over its entire lifespan. With this approach, energy expenditure takes a huge fraction of the final cost, and it should not be dismissed from the calculations. There are many designing approaches that do incorporate this aspect in their analysis. Mala-Jetmarova et al. [2] summarized what other authors considered in their proposals. After their review of the state of art, it seems that the design of water distribution systems is increasingly emphasizing the importance of including operation assessments.

The operational costs are often included in optimization algorithms in a single economic function that would also include construction costs of the network. This approach is used in some classical studies, like [3], but it is still the most common way to assess energy costs in the design. That is the case of [4–9] or [10]. Nevertheless, a different perspective is to treat operational and construction costs as separated objectives, as [11] does. However, this assessment confirmed that such a perspective throws back very different design solutions: the optimum design for construction costs means the highest operational investment and vice versa. This approach makes the search of the optimum for the overall installation more difficult, and therefore it is more advisable to adopt a single cost function to optimize.

In order to achieve a more precise assessment of all costs involved, multiobjective algorithms have started to incorporate in the optimization function costs of maintenance [12,13], replacement [14–16], and greenhouse gas emissions [17–20], among others. Multiobjective methods offer a very complete revision of all costs, but many times they require complex programming, resulting in being computationally expensive and hard to use. The work of [21] explained the state of the art of the multiobjective techniques.

Because operation costs are distributed along the lifespan of the installation, they need to be computed at the design stage. The equivalence is calculated using an accumulative factor that depends on the duration of the operation of the water drive considered, the duration of the construction period, and the discount rate employed. Regarding the lifespan of the facility, some methodologies consider an operative life of 20 years [3,5], but other authors [15,16,20] prefer to carry out the analysis for 100 years.

Another key variable for the operation costs is the pump efficiency. One of the most widely used methods for designing water supply systems in Spain is the Granados System [22,23], which is a gradient-based procedure. Among all variables affecting the Granados System, the pump efficiency is the key factor for calculating the diameter of the hydraulic conduction. The aim of this paper is to define the relationship between the pump's efficiency and the other parameters involved in the procedure, such as the flow rate, the pumping head and the required power. This pump efficiency analysis does not only serve for the purposes of the Granados System, but it can also help in many other methods that require the value of the pump's efficiency. This is the case of [24], where they use an estimated value of 75% in their calculations; for Wu et al. [15], the values range from 81% to 84%; Gessler and Walski [25] estimate the efficiency as 75% and so do Alperovits and Shamir [26] and Featherstone and El-Jumaily [27].

Other studies have also decided to optimize the design using as key variables the pump location [28,29], pump capacity [4], type [19], power [3], pumping head [30–32], pumping schedule [24,33], or pressure [34,35]. Water supply system design is a complex task where many variables are involved [36] and inter-connected; the decision-making needs a full comprehension of how each factor affects the installation. Some of the relationships between variables are still unknown and only intuited by experience. For this purpose, sensitivity analyses are necessary. The work [37] carries out an exhaustive sensitivity analysis using Sobol's method (a variance-based approach) to determine the variables that most affect the installation. These variables could vary from pipe diameters to tank sizes, and the degree of influence depends on each case study. In their work, they prove the computational savings that can be obtained and, therefore, how beneficial the analysis is. This paper also intends to serve as a sensitivity analysis of how much pump efficiency is influenced by other factors and how the pump's efficiency can affect the final cost of a water supply system.

Since the pump efficiency can substantially affect the design of water supply systems due to the wide range of values they can adopt, with this study we carry out an extensive study of the pump performance in order to characterize its values. For this to be made, we analyze the features of commercial pumps available in the market. The conclusions are integrated in a gradient-based procedure, using as a base the Granados System, giving rise to an optimized design method that accounts for the construction and operational costs. This procedure is also compatible with adding other variables such as the carbon footprint [38] of the design or the GHG emissions. Our new methodology is applied to a case study to illustrate its computationally straightforward conception.

2. Materials and Methods

2.1. Brief Summary of the Granados System

Granados System is a pipe-sizing method indicated for the design of branched water distribution systems. It is a gradient based methodology. It is, in fact, based on the 'change gradient' concept:

The change gradient is defined as the cost of reducing one meter of head loss by increasing the pipe diameter from $\varnothing_i$ to the next bigger one $\varnothing_j = \varnothing_{i+1}$:

$$GC^q_{\varnothing_i \to \varnothing_j} = \frac{P_j - P_i}{\Delta h^q_i - \Delta h^q_j},\tag{1}$$

where P_i and P_j are prices of pipes of length L and diameters $\varnothing_i$ and $\varnothing_j$, respectively; Δh_i^q and Δh_j^q are the head losses of a pipe of length L and diameters $\varnothing_i$ and $\varnothing_j$, respectively, when a flow rate q is circulating. When the head losses are calculated using Manning's formulation, the change gradient's expression is:

$$GC^q_{\varnothing_i \to \varnothing_j} = \frac{P_j\,L - P_i\,L}{\dfrac{L\,n_i^2\,2^{20/3}\,q^2}{\pi^2\,\varnothing_i^{16/3}} - \dfrac{L\,n_j^2\,2^{20/3}\,q^2}{\pi^2\,\varnothing_j^{16/3}}} = \frac{\pi^2}{n^2\,2^{20/3}}\,\frac{(P_j - P_i)}{\left(\dfrac{n_i}{\varnothing_i^{16/3}} - \dfrac{n_j}{\varnothing_j^{16/3}}\right)}\frac{1}{q^2} = K_{GC}\frac{1}{q^2},\tag{2}$$

with n being Manning's friction coefficient.

The definition of the change gradient also needs to include some other associated costs such as excavation expenditure or reinforcements against chemically aggressive environment or hydraulic transients. During the optimization stage, hydraulic transients are usually taken into account in a simplified way, for instance, extra thickness of the pipes to withstand water hammer transient pressure. Once the optimum design is achieved, proper assessment of water hammer should be carried out to verify the initial hypothesis. These associated cost factors increase the pipe price, and therefore, they have to be accounted in the design through P_i and P_j.

Increasing the pipe's diameter means a reduction in the head loss along the pipeline, but it has the extra cost of the wider and more expensive tube. Since the change gradient is the cost of reducing the head loss by one meter, it needs to be compared to the cost of the energy required for pumping the water at one meter height throughout the entire life of the facility, C_{E1}, which reads:

$$C_{E1} = Ca_{E1} \cdot f_A = f_A\,9.81\,\frac{V}{3600}\,\frac{1}{\mu_B}\,\frac{1}{\mu_M}\,p_{E},\tag{3}$$

$$f_A = \frac{(1+i)^{n_U} - 1}{(1+i)^{n_U}\,i} \cdot \frac{1}{(1+i)^{n_C}},\tag{4}$$

where Ca_{E1} is the annual energy cost per meter, f_A is the discount factor, n_c is the duration of the construction period, n_u is the useful life of the installation, i is the discount rate, V is the annual volume of water to pump, μ_B and μ_M are the pump and engine efficiency, and lastly p_E is the unit price of energy.

As it is justified in Granados' work, with the exception of the pump performance μ_B, the other variables in the previous equation are relatively known data: V is the total demanded volume to pump, p_E is the unit price of the energy that has been hired, and f_A is calculated from the discount rate i, the service life n_U of the water pipeline which depends on the chosen material, and the construction period n_C. Regarding the engine efficiency μ_M, although it varies theoretically depending on the model chosen and the operating point of the pump, the variations in engine performance are so small that it can be considered constant across different models and manufacturers. Therefore, the above equation can be simplified to:

$$C_{E1} = KC_{E1}\,\frac{1}{\mu_B}.\tag{5}$$

As previously said, Granados System consists of comparing the cost of building a wider pipe (change gradient) and reducing the head loss by a meter, to the cost of pumping one meter of head loss (C_{E1}). The reasoning is the following:

- If $GC^q_{\varnothing_i \to \varnothing_{i+1}} < C_{E1} \Rightarrow$ Diameter $\varnothing_{i+1}$ is preferable to $\varnothing_i$.
- If $GC^q_{\varnothing_i \to \varnothing_{i+1}} > C_{E1} \Rightarrow$ Diameter $\varnothing_i$ is preferable to $\varnothing_{i+1}$.

But since C_{E1} depends on the pump efficiency μ_B, the procedure changes to:

- If $GC^q_{\varnothing_i \to \varnothing_{i+1}} < C_{E1} \Rightarrow GC^q_{\varnothing_i \to \varnothing_{i+1}} < KC_{E1} \frac{1}{\mu_B} \Rightarrow \mu_B < \dfrac{KC_{E1}}{GC^q_{\varnothing_i \to \varnothing_{i+1}}} \Rightarrow$ Move to $\varnothing_{i+1}$.
- If $GC^q_{\varnothing_i \to \varnothing_{i+1}} > C_{E1} \Rightarrow GC^q_{\varnothing_i \to \varnothing_{i+1}} > KC_{E1} \frac{1}{\mu_B} \Rightarrow \mu_B > \dfrac{KC_{E1}}{GC^q_{\varnothing_i \to \varnothing_{i+1}}} \Rightarrow$ Keep $\varnothing_i$.

This means that, whenever there is a pump on the market whose efficiency can be greater than the calculated μ_B, the optimum diameter will be $\varnothing_i$. In the event that no commercial pump can reach that performance because it is very high, it will be necessary to move to the next diameter $\varnothing_{i+1}$. Therefore, to apply this method it is necessary to know the maximum efficiency that pumps can reach. This is an uncertainty of the Granados System, and up to now, typically, pump efficiency values around 80% are already indicative.

2.2. Methodology

Among all variables affecting Granados System, the pump performance is the key factor for calculating the diameter of the hydraulic conduction. Nevertheless the results may vary significantly depending on the pump efficiency; indeed, μ_B presents a wide range of possible values that typically goes from 70% up to 90%. This means a variability of almost 20% in the estimation of the cost, which is a substantial difference. Therefore the aim of this paper is to define the relationship between the pump's performance and the other parameters involved in the procedure, such as the flow rate, the pumping head or the required power. For this to be done, we select 400 commercial pumps from the catalogues of several manufactures. We discard custom-made pumps since the cost of these pumps is much higher than the ordinary ones listed and the offering in commercial catalogues is already very wide. The selected pumps vary from each other in their type, impeller diameter, number of stages, rotation speed (electrical current frequency, number of poles), brand, etc. For all these models, we study the pump efficiency; in particular, we register the optimum value together with the correspondent flow rate, head and power consumed. Nevertheless, some pumps are ruled out of the sample and presentation of the results because they either were similar to those of other manufacturers, or because they were very specific for some industrial or sanitary engineering uses. In the end, the sample consists of 226 hydraulic pumps.

For the detailed study of the pump performance, this research has focused on the most common type of pump for the applications in civil engineering (supply, irrigation, sanitation, etc.)—centrifugal pumps. Within centrifugal pumps, both horizontal and vertical axes are selected, mostly with a radial flow configuration, with the exception of submersible pumps, for which the axial arrangement is more common. The sample includes regular horizontal and vertical pumps, split case pumps, multistage and submersible pumps. The manufacturers used for this analysis were IDEAL, WILO, ESPA and HASA. More specifically, the commercial models were:

- Split case pumps: CP/CPI/CPR series.
- Horizontal pumps (normalized in the European Union): RNI/RN series.
- Multistage horizontal pumps: APM series.
- Vertical pumps: VS/VG series.
- Submersible vertical pumps: SVA/SVH series.

Multistage pumps perform with the same efficiency for a specific flow rate and different heights (which is the number of stages multiplied for the unitary head). To avoid this dispersion that could make it difficult to draw conclusions, it was decided to only use the optimum operating point correspondent to a single stage.

Using the data collected for the optimum operating points for all the 226 hydraulic pumps previously mentioned (pump's optimum efficiency and associated flow rate, head, speed, power, frequency, diameter, etc.) we carried out an analysis to establish relationships among the design variables of a water drive.

3. Results and Discussion

Figure 1a shows the values of the optimum efficiency of the pump and the flow rate correspondent to that point. From it, it can be seen that there is a relationship between both variables: the performance of the pump improves as the discharge increases, although it seems to reach a horizontal asymptote for μ_B around 90%. In this same figure, it can be spotted that the relationship between the two variables is more precise for flow rate values that are greater than 500 L/s. Below this value, as it can be seen in Figure 1b, there is more dispersion. Also, when the discharge is at least 100 L/s, the pump efficiency can be expected to be better than 80%, but under it, the dispersion is stronger. On the other hand, it has been studied whether the type of pump has any relation with the efficiency. For that, Figure 1 also shows the distribution of the operating points for the split case pumps, horizontal and multistage (both horizontal and vertical).

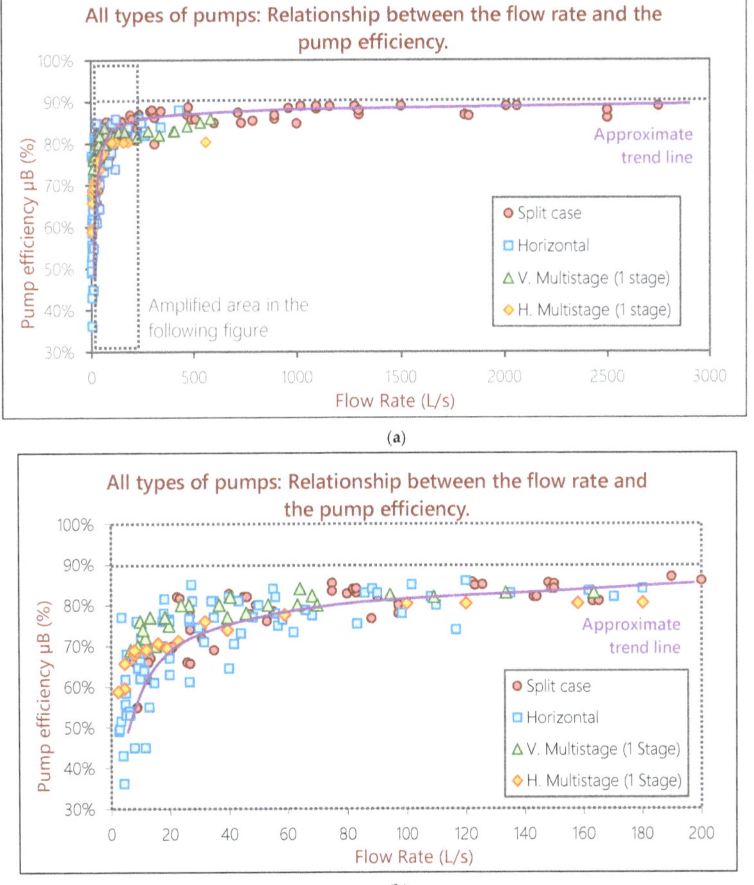

Figure 1. (**a**). Optimum pump efficiency and the correspondent flow rate at that operating point, classified by the pump type. (**b**) Detail of the previous figure for smaller discharges.

Regarding the pump type, we conclude that:

- Split case pumps are, in general, the ones that provide the best performance for flow rates over 500 L/s.
- Between 100 L/s and 500 L/s, both split case and horizontal offer the best results.
- Under 100 L/s, the distribution is very heterogeneous and disperse, but as a general rule, there is always a horizontal pump that can provide the best efficiency (but also the lowest, due to the dispersion). Vertical pumps also show great performances, but they can generally be exceeded by a horizontal model.
- Vertical multistage pumps always give a better efficiency than horizontal multistage pumps.
- For small flow rates, the differences between the pump efficiency among models are very high. Different models can almost double other pump's efficiency, meaning that the energy cost can be almost twice as much if the pump is not well selected. This is why the pump analysis is more complicated for small discharge values.

When the pumping height and the efficiency are plotted together, as shown in Figure 2, there is no clear sign of a relationship between the two variables since the dispersion is too strong for any head value. There is a very light tendency of high pump performances (around 80%) for under 40 m head, but over 60 m the distribution of μ_B is too scattered.

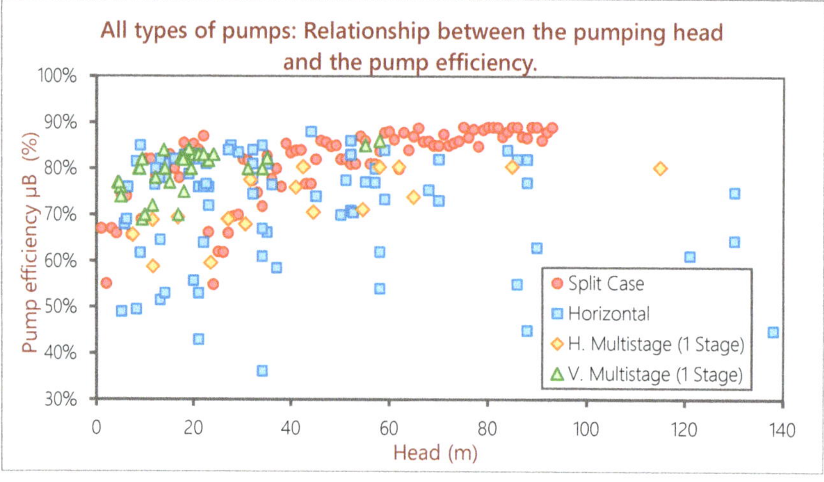

Figure 2. Optimum pump efficiency and the correspondent pumping height at that operating point.

To test out what the relationship between the optimum μ_B and the required power for that operating point is, Figure 3 was elaborated.

The main conclusions drawn from Figure 3 are:

- There is a relationship between the pump efficiency and the required power for the optimum operating point.
- The relation seems to be clearer from 400 kW and on, and values greater than 80% efficiency can be expected.
- This relationship seems to be weaker than the existing one between the flow rate and the pump efficiency. The interpreted reason for this fact is that, since the power is calculated from the flow rate and the pumping height, the good relationship that the flow rate transfers to the power is weakened by the poor bond that the pumping head and efficiency have. The authors of [3] had already presented similar conclusions. In their work, they empirically proved that the cost of pumping depended on power; which indeed agrees with these results.

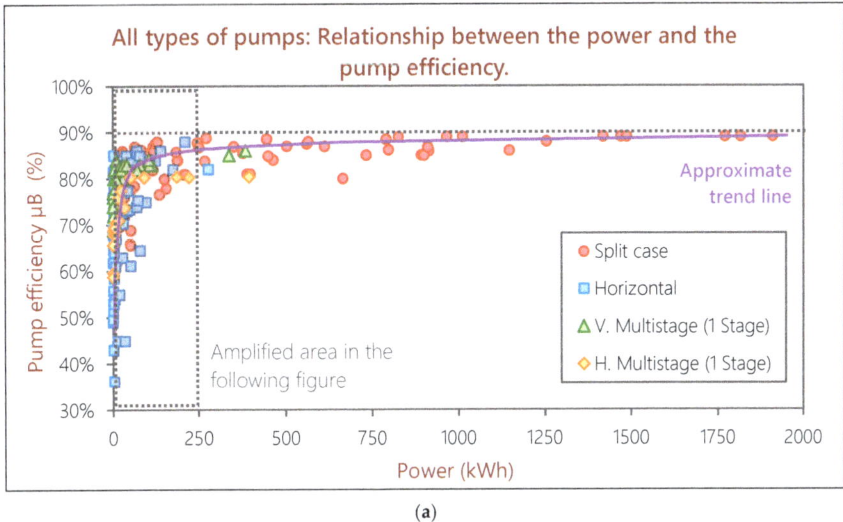

(**a**)

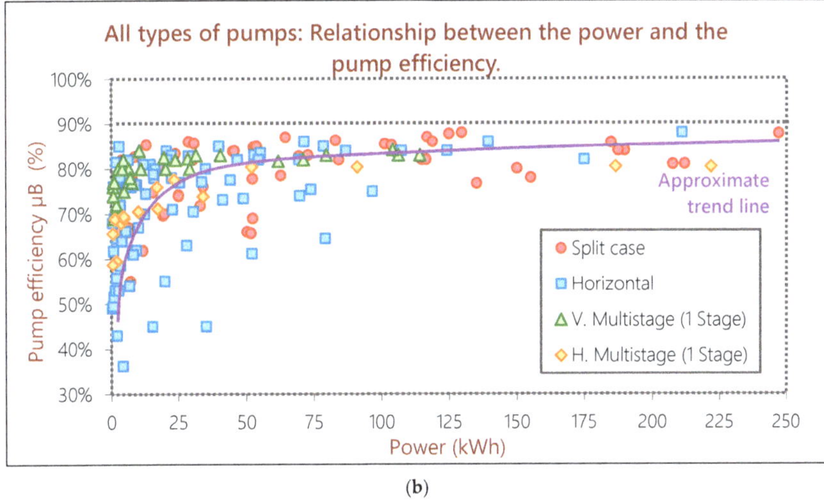

(**b**)

Figure 3. (**a**) Optimum pump efficiency and the correspondent pump power at that operating point. (**b**) Detail of the previous figure.

To summarize, this first assessment concludes that the strongest of the relationships with the pump efficiency is that of the flow rate, especially when the flow is greater than 100 L/s and it is almost linear.

On the other hand, the relationship between the specific speed of the pump and the efficiency is shown in Figure 4. The specific speed of the pump is defined as:

$$n_s = \frac{n_r \cdot q^{\frac{1}{2}}}{H^{\frac{3}{4}}},\tag{6}$$

where n_s is the specific speed; q, H and n_r are the flow rate, pumping head and the rotation speed, respectively, at the optimum operating point [39]. It is interpreted as the rotation speed that a geometrically similar pump should have in order to elevate a discharge of 1 m^3/s at 1 m height.

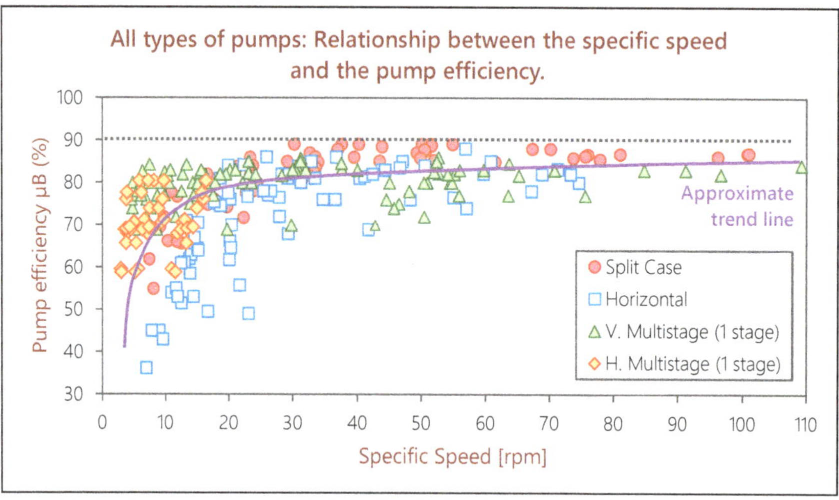

Figure 4. Optimum pump efficiency and the correspondent specific speed of the pump, classified by the pump type.

Low specific speed indicates that the pump is suitable for small flow rates and great heights. This means that, since small flow rates have a poorer relationship with the pump efficiency, pumps with a low specific speed will have the worst performance values. This is shown in Figure 4. On the contrary, high specific speed is an indicator of a pump suitable for great flow rates and low heights. According to the previous analysis made from Figure 1, greater discharges correspond to better performances, and therefore, a high specific speed can be associated with a good µB, as Figure 4 shows. The scattering of the figure is caused by the poor relationship that the pumping height and the rotation speed have with the pump efficiency. However there is a diffuse tendency of increasing µB as the specific speed grows. As far as pump type is concerned, the types of pumps that operate at higher flow rates will be the ones with a greater specific speed and therefore, better performance: These are split case pumps, as shown in Figure 4. Vertical multistage pumps also give good results in the smaller flow rate ranges.

3.1. Relationship between the Flow Rate and the Pump Efficiency

From the previous figures, it seems that there are three flow rate ranges where the relationship with the pump efficiency is different. The first one would be from 0 to 100 L/s, the second from 100 to 500 L/s and the last one over 500 L/s. Nevertheless, in order to be more precise, instead of three zones, the curve has been fitted using up to fourteen flow rate subdivisions. Table 1 shows the average optimum pump efficiency for each interval, as well as the maximum and minimum found among the studied pumps.

Table 1. Elaboration of the adjusted curves of the pump efficiency versus the flow rate.

Interval Number	Pump Efficiency for Each Flow Rate Interval					
	Flow Rate Range			Pump Efficiency		
	Minimum	Maximum	Average	Average	Maximum	Minimum
	q (L/s)	q (L/s)	q (L/s)	μ_B (%)	μ_B (%)	μ_B (%)
1	0	5	2.5	55.3%	77.0%	36.2%
2	5	10	7.5	63.5%	76.0%	45.0%
3	10	15	12.5	65.7%	77.0%	45.0%
4	15	25	20	74.2%	82.0%	63.0%
5	25	40	32.5	75.0%	85.0%	61.2%
6	40	65	52.5	78.6%	84.0%	70.5%
7	65	100	82.5	81.0%	85.4%	75.4%
8	100	150	125	83.1%	86.0%	74.0%
9	150	200	175	82.8%	87.0%	80.4%
10	200	300	250	83.9%	88.0%	81.0%
11	300	500	400	84.9%	88.8%	79.9%
12	500	1000	750	85.5%	88.5%	80.4%
13	1000	2000	1500	88.3%	89.0%	86.7%
14	2000	3000	2500	88.2%	89.0%	86.2%

When these values are adjusted through a doubly logarithmic curve, the relationships for both the average and maximum values fit satisfactorily ($r^2 > 98\%$ and $r^2 > 90\%$, respectively), as can be seen in Figure 5. The empirical equations that relate the optimum pump efficiency and the flow rate are:

$$\mu_B^{\text{Average}} = 0.1286 \ln\left(2.047 \ln q - 1.7951\right) + 0.5471 \quad r^2 > 98\%, \tag{7}$$

$$\mu_B^{\text{Maximum}} = 0.0576 \ln\left(2.047 \ln q - 1.7951\right) + 0.741 \quad r^2 > 90\%, \tag{8}$$

where q is the flow rate in liters per second (L/s).

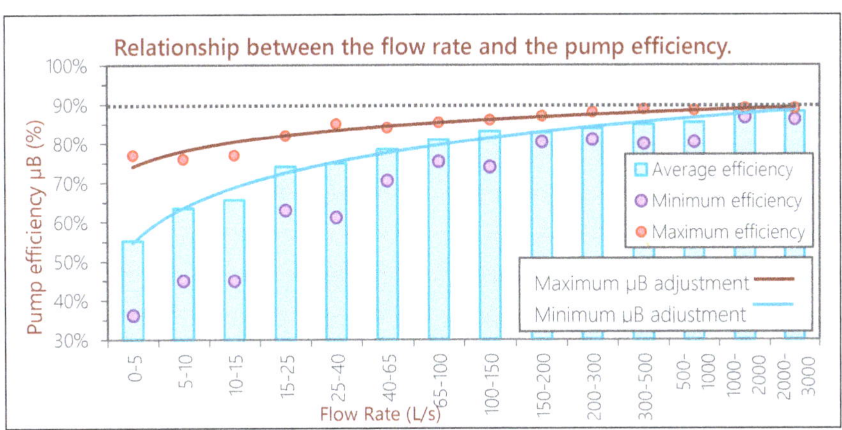

Figure 5. Adjusted curves. The curves are elaborated using 14 intervals of the collected data.

Finally, Figure 6 shows the adjusted curves along with the collected data. Both curves fit satisfactorily for most flow rate ranges, especially for the bigger ones; nevertheless the dispersion is higher for the small discharge values (under 50 L/s), but still, it gives a valid reference. To facilitate the visualization, Figure 6b shows the results in logarithmic scale.

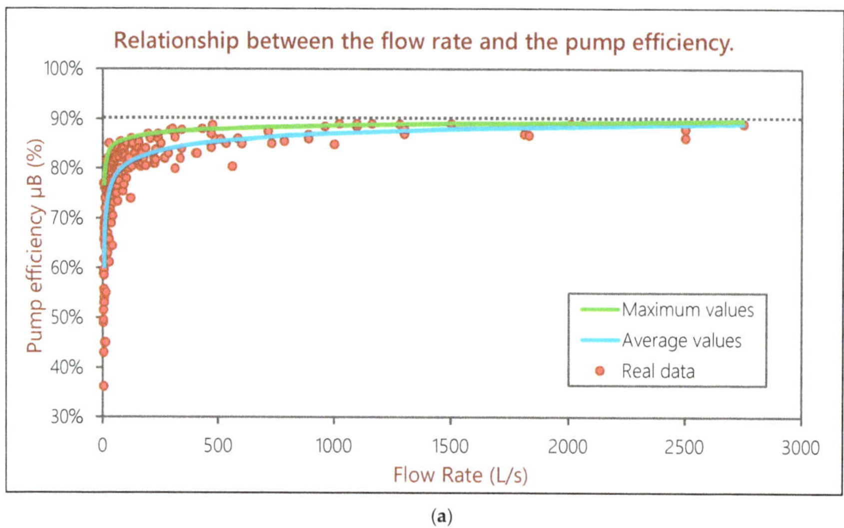

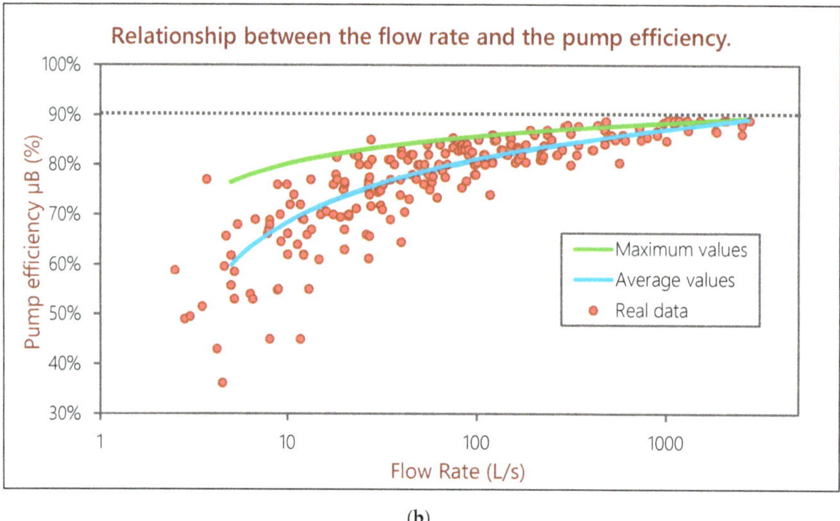

Figure 6. (**a**) Optimum pump efficiency curve: database and mathematical adjustment; (**b**) mathematical adjustment in logarithmic scale.

3.2. Application of the Pump Efficiency Curves to the Granados System

For a better comprehension, the full procedure is shown in Figure 7. On the figure, all variables affecting the change gradient and the energy cost are graphically represented: flow rate, pump efficiency, energy price, annual volume of water, engine efficiency, useful life of the installation, construction period, discount rate, commercial diameters, pipe prices and pipe roughness.

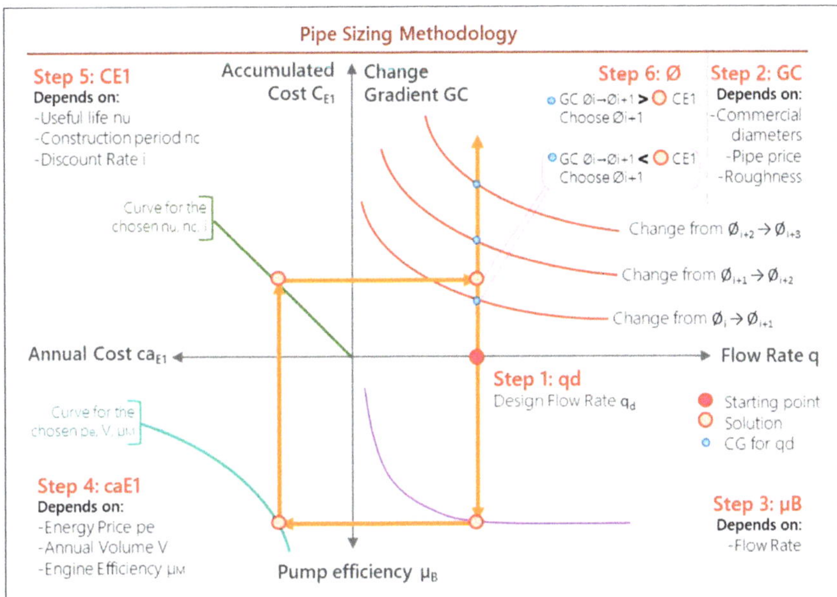

Figure 7. Pipe sizing methodology representing the steps for the design and the shapes that curves might present depending on the variables.

The design process represented in Figure 7 consists of the following steps:

1. Calculate the design demand flow rate of the hydraulic drive.
2. Calculate the change gradient curves for a commercial series of diameters with Equation (2).

   Since the aim is to compare the construction costs to those of the energy, excavation costs also must be considered and added to these figures.

3. Calculate the expected pump efficiency using the average pump efficiency equation (Equation (7)).

   It is always more conservative to use the average pump efficiency equation rather than the maximum one, nevertheless this one can also be used but it would require a much more exhaustive pump search.

4. Calculate the annual energy cost for one meter height Ca
5. $_{E1}$ using Equation (3).

   At this point, different alternatives might be considered for various energy prices, etc.

6. Calculate the total accumulated energy cost for one meter height C_{E1} with Equations (3) and (4).
7. Compare the change gradient and the energy cost C_{E1} and select the pipe diameter: when $GC^q_{\varnothing_i \to \varnothing_{i+1}} < C_{E1}$ select the wider diameter $\varnothing_{i+1}$; but when $GC^q_{\varnothing_i \to \varnothing_{i+1}} > C_{E1}$ select diameter $\varnothing_i$. When $GC^q_{\varnothing_i \to \varnothing_{i+1}} = C_{E1}$, it is always preferable to build a bigger diameter $\varnothing_{i+1}$, in case energy price, discount rate, etc. change.

3.3. Case Study: Navas del Marqués

To exemplify the application of the previous design procedure, the method will be used in a case study. It is based on the dam project in the Navas del Marqués locality, Ávila, Spain, which was carried out by the Tagus Hydrographic Confederation [40]. In this section we intend to compare the design results obtained in the project to the ones that would be obtained with the proposed procedure. The project included the following works:

- New dam.
- New pumping station.
- New pipe replacing the first part of the water drive. There was a tunnel section from the previous water supply system that will remain as it was.
- New raw water deposit, previous to the water treatment station.
- New regulation deposit, after the water treatment station.
- New distribution network.

Figure 8 shows the schematic works that the project included, along with some relevant altitudes, and input data. This paper is concerned about the sizing of first part of the water drive, i.e., Pipe 1 (remember that the tunnel section, Pipe 2, was to remain in the original state).

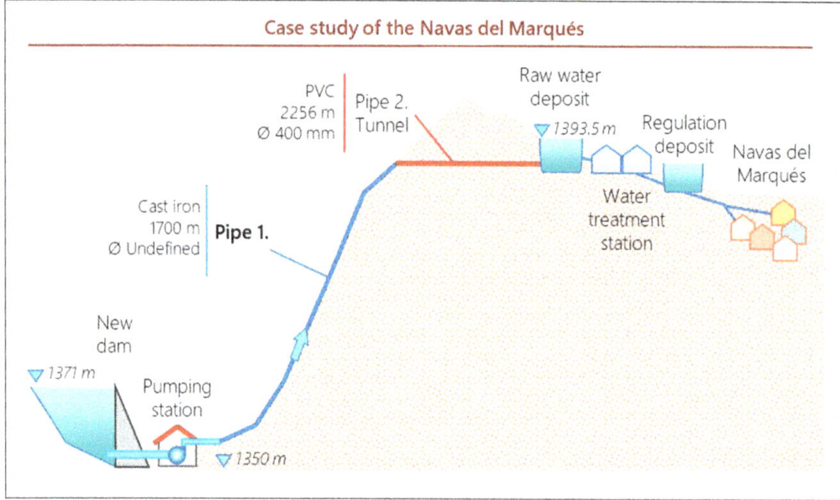

Figure 8. Case study at Navas del Marqués.: Pipe 1 is the only one to replace.

The project data that are used can be summarized in the following list:

- Design flow rate: 0.164 m^3/s.
- Annual volume of water: 1.5 hm^3/year.
- Pipe diameters and the correspondent prices. These are compiled in Table 2.
- Energy price: 0.072 €/kWh.
- Duration of the construction period: 30 months.
- Although it is not specified in the project, pump engine efficiency is fixed at 94%.

Table 2. Pipe prices series and change gradients calculations.

Pipe Diameter	Change Gradients			
	Canal de Isabel II		Project Diameters	
	Construction Cost	Change Gradient	Construction Cost	Change Gradient
mm	€/m	€/m	€/m	€/m
60	55	0.11		
80	64	0.26		
100	69	2.11		
125	79	4.84		
150	86	30	103 [1]	36
200	105	233	127	256
250	134	796	159	1044
300	161	3070	194	2373
350	197	6670	221	6593
400	227	15,972	252	11,226
450	260	33,781	295	64,309
500	295	98,565	341	76,920
600	378		405	

[1] The project includes a limited series of ductile cast iron pipes in the estimated budget documents. These start from 150 mm.

This design process can be also followed throughout all series of Figure 9. These figures follow the same color code and symbology as Figure 7. The calculation followed the steps listed above:

1. The design discharge flow is the one used by the original project which is $q = 0.164$ m^3/s.

2. For the calculation of the change gradients, the material for the pipe is ductile cast iron. Two different studies have been elaborated for two different pipe prices. Firstly, the pipe prices used in the project have been considered. Secondly, we used the pipe prices proposed by the Canal de Isabel II. Canal de Isabel II is the public authority in charge of the integral water supply of Madrid, Spain, and they published the average pipe prices they use in their infrastructure. Since the Tagus Hydrographic Confederation is also a public company, these prices have also been considered for a sensitivity analysis. On the other hand, it must be kept in mind that the prices used for the change gradient include that of the full construction cost (including excavation, joining, etc.). Table 2 shows the pipe prices and the change gradients.

3. Using Equations (7) and (8), the average pump efficiency obtained is 82.44% and the maximum value is 86.52%. This same result can be obtained from Figure 6a with the design flow rate $q = 0.164$ m^3/s.

4. The calculation of the annual energy cost per meter has been made for different values of the energy price. Engine efficiency was fixed at 94% and the demanded annual volume of the project is 1.5 hm^3. Energy prices used for the calculations are 0.072 €/kWh, which, using Equation (3), gives an annual energy cost of 380 €/m/year; and also 0.125 €/kWh, for which Ca$_{E1}$ is 659 €/m/year.

   This will allow a sensibility analysis to see how robust the design is against energy price changes. The first price, 0.072 €/kWh, corresponds to the energy price used in the project, and the second alternative, 0.125 €/kWh, corresponds to a higher energy price. These results can be seen in Figure 9a.

5. Once the annual unit pumping cost is obtained, the total accumulated cost is calculated using Equation (3). We assume a life span for the cast iron pipe of 31 years. Likewise, the duration of the expected construction period according to the project is 30 months, which corresponds to 2.5 years. We analyzed the sensibility of the design for the discount rate. In this line we adopt three values of the flow rate, where i equals 3%, 4% and 5%. With all this, the results obtained in each situation of energy prices and discount rate are shown in Table 3 below and in Figure 9b.

6. Once the full energy cost has been obtained, as well as the change gradients, the comparison is made, following the reasoning previously described. The selected pipes are compiled in Table 3 and shown in Figure 9c,d.

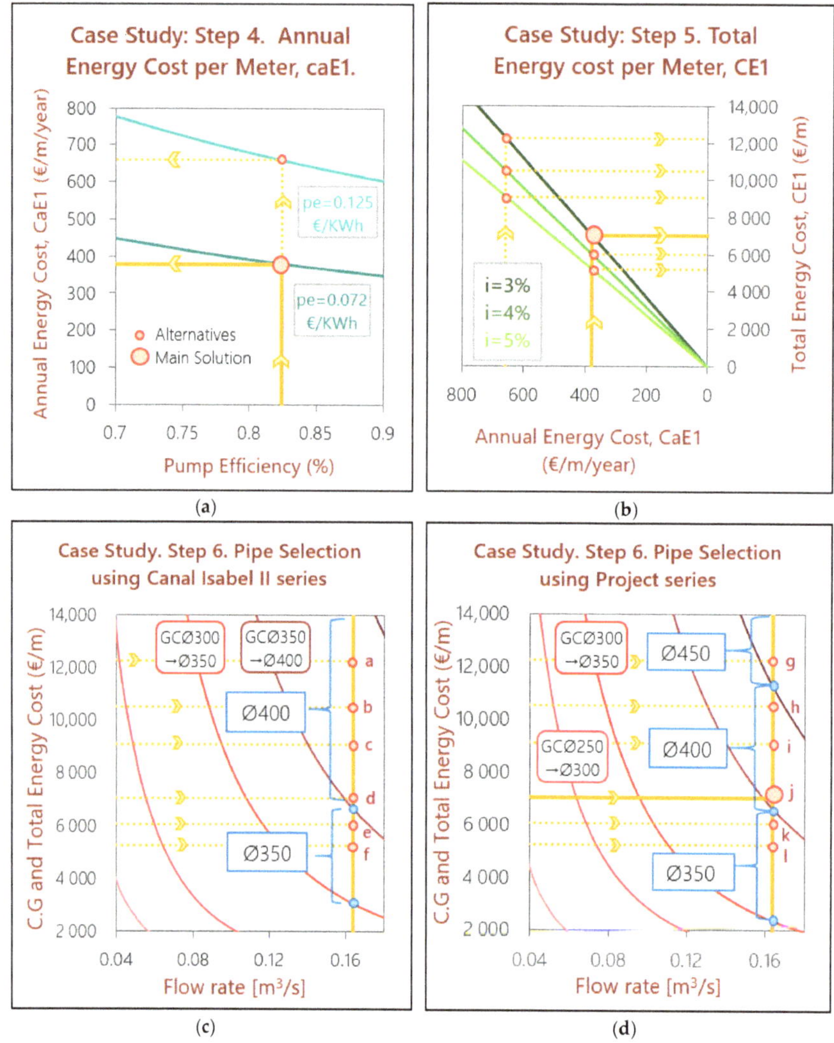

Figure 9. Case study of the Navas del Marqués. (**a**) Step 4: Calculation of the annual energy cost; (**b**) Step 5: Calculation of the capitalized energy cost throughout the entire life of the installation. (**c,d**) Both panels show steps 1 and 5 of the procedure. This means that both the change gradient calculation (red curves) and the pipe selection is shown in the figures. In (**c**), the change gradients are calculated using the Canal de Isabel II commercial diameters, and in (**d**), they used the project data. The big orange circle in (**d**) represents the main solution, whilst the small orange circles represent the solutions for the other alternatives evaluated in the case study. For the main solution, as the energy cost per meter C_{E1} is greater than CGØ350→Ø400, it is more convenient to choose Ø400 mm. But because C_{E1} is smaller than CGØ400→Ø450, it should not be passed onto Ø450 mm and remain with Ø400 mm. The same reasoning applies to the other solution alternatives.

Table 3. Selection of the pipe diameter, depending on the energy prices, diameter series and discount rate considered.

					Pipe Selection		
Annual Unit Energy Cost		Total Unit Energy Cost				Selected Diameter	
Energy Price	Ca_{E1}	i	fa	C_{E1}		Canal Isabel II Pipes	Project Pipes
	€/m/year	%		€/m		mm	mm
0.125 €/kWh	659	3	18.6	12,248		(a) 400 *	(g) 450
		4	15.9	10,513		(b) 400	(h) 400
		5	13.8	9100		(c) 400	(i) 400
0.072 €/kWh	380	3	18.6	7055		(d) 400	(j) 400
		4	15.9	6056		(e) 350	(k) 350
		5	13.8	5242		(f) 350	(l) 350

* The letters in brackets indicate the graphical solutions in Figure 9c,d.

As it is shown, most of the alternatives throw back a pipe diameter solution of 400 mm; this is not too far from the solution taken in the project, which is 500 mm. The diameter selected in the project implies a facility with higher costs than the design obtained with the optimization of this research. However the project solution is energetically less consuming, and this is a positive fact because energy prices tend to increase with time, leading to significant increases in the energy cost. Nevertheless, when the sizing is made for project energy prices, the solutions oscillate between 350 and 400 mm. Although the project and presented method give close results, the new methodology can help to optimize the full cost of the facility.

An economic evaluation has been conducted using the project energy prices and pipe prices. This can be seen in Table 4 and Figure 10.

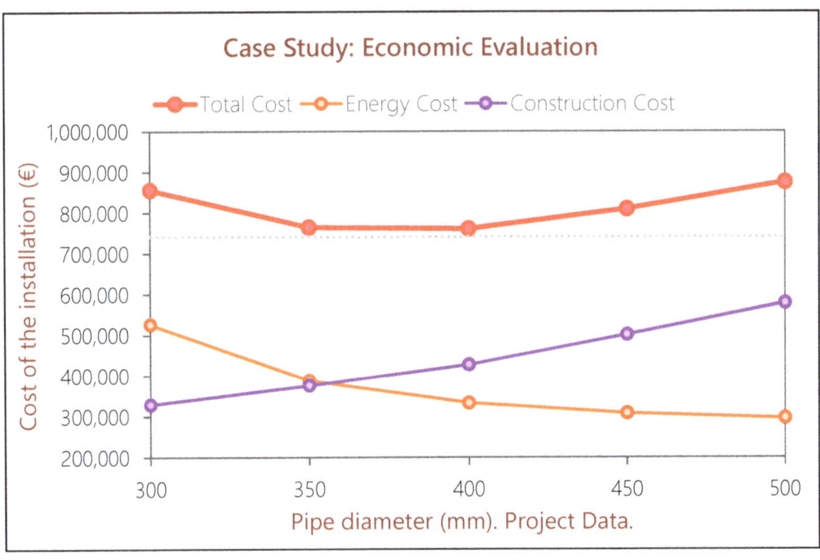

Figure 10. Economic evaluation of the case study at the Navas del Marqués.

Table 4. Economic evaluation for the case study using the project prices for the pipes, once a diameter of 400 mm has been selected, to prove the least cost result.

Economic Evaluation					
Diameter	Construction	Head Loss	Annual Cost	$i = 3\%$	Total Cost
Ø	Costs	Δh	caE	CE	
(mm)	€	(m)	(€/year)	(€)	(€)
300	329,886	35.01	28,374	527,076	856,962
350	376,444	15.39	20,922	388,646	765,091
400	428,121	7.55	17,946	333,353	761,474
450	501,646	4.03	16,609	308,516	810,162
500	578,985	2.30	15,951	296,302	875,288

Table 4 and Figure 10 show how diameters Ø400 mm and Ø350 mm give very close results for the total cost, since they only differ in approximately 4000 €. Nevertheless, the cost variation increases compared to diameter Ø500 mm and it is significantly higher, as it can be seen in Figure 10 even though the results only differ in 100 mm wide. When the sizing is made for higher energy prices, the most common result is 400 mm. As a conclusion, although the experience of the designer can never be replaced, this method is a very convenient tool to find the optimum diameter of a pipe.

4. Conclusions

The present pump analysis is a useful tool to help with the uncertainties regarding decision-making in water supply system design. Because the relationship of the pump efficiency with the other variables involved in the design process has been elucidated, the understanding of such a complex procedure as the design of a water supply system is improved.

The definition of the pump-efficiency–flow-rate curve reduces the conflict resolution involved in gradient-based methods, avoiding iterations regarding the pump efficiency. This automatically leads to less computational effort. In this sense, the Granados method is a straightforward procedure.

Some current methods are somewhat black-boxes, losing part of their utility by not being easy to comprehend for design engineers. The visual representation of our design methodology presented in this paper facilitates the understanding of the influence of each variable and allows to have a clear picture of the process. This simplicity is one of the main necessities to aim for when it comes to water supply system design.

As the case study shows, energy costs are proved to be a great fraction of the full cost of the installation, and they need to be considered from the first phase of the design. In contrast with other methodologies that do not consider these energy cost, this analysis aims for an integrated water assessment where all costs of the water–energy nexus are integrated.

Here, we obtain empirical equations that define the average and maximum optimal pump efficiency to expect depending on the flow rate of the installation. These results can be incorporated to elaborate a proper energy assessment of the operational costs. As the case study has shown, the design of a water supply system is also very much influenced by the construction costs (the design differs when Canal de Isabel II or project pipe prices are used) and by the energy rates (centesimal order variations will throw back different results). This reinforces the importance of carrying out a sensitivity analysis when it comes to designing a water supply system. These conclusions can be used for policy assessment in integrated water management as well as supply system design.

Author Contributions: Investigation, A.M.-C., D.S. and L.G. All authors have read and agreed to the published version of the manuscript.

Funding: This research was funded by CARLOS GONZÁLEZ CRUZ Grant.

Conflicts of Interest: The authors declare no conflict of interest.

References

1. Sanchis, R.; Díaz-Madroñero, M.; López-Jiménez, P.A.; Pérez-Sánchez, M. Solution Approaches for the Management of the Water Resources in Irrigation Water Systems with Fuzzy Costs. *Water* **2019**, *11*, 2432. [CrossRef]
2. Mala-Jetmarova, H.; Sultanova, N.; Savic, D. Lost in optimisation of water distribution systems? A literature review of system operation. *Environ. Model. Softw.* **2017**, *93*, 209–254. [CrossRef]
3. Kim, J.H.; Mays, L.W. Optimal rehabilitation model for water-distribution systems. *J. Water Resour. Plan. Manag.* **1994**, *120*, 674–692. [CrossRef]
4. Kang, D.; Lansey, K. Scenario-based robust optimization of regional water and wastewater infrastructure. *J. Water Resour. Plan. Manag.* **2012**, *139*, 325–338. [CrossRef]
5. Dziedzic, R.; Karney, B.W. Cost gradient–based assessment and design improvement technique for water distribution networks with varying loads. *J. Water Resour. Plan. Manag.* **2015**, *142*, 04015043. [CrossRef]
6. Samani, H.M.; Mottaghi, A. Optimization of water distribution networks using integer linear programming. *J. Hydraul. Eng.* **2006**, *132*, 501–509. [CrossRef]
7. Costa, A.; De Medeiros, J.; Pessoa, F. Optimization of pipe networks including pumps by simulated annealing. *Braz. J. Chem. Eng.* **2000**, *17*, 887–896. [CrossRef]
8. Vamvakeridou-Lyroudia, L.; Walters, G.; Savic, D. Fuzzy multiobjective optimization of water distribution networks. *J. Water Resour. Plan. Manag.* **2005**, *131*, 467–476. [CrossRef]
9. Spiliotis, M.; Tsakiris, G. Minimum cost irrigation network design using interactive fuzzy integer programming. *J. Irrig. Drain. Eng.* **2007**, *133*, 242–248. [CrossRef]
10. Jin, X.; Zhang, J.; Gao, J.-L.; Wu, W.-Y. Multi-objective optimization of water supply network rehabilitation with non-dominated sorting genetic algorithm-II. *J. Zhejiang Univ. Sci. A* **2008**, *9*, 391–400. [CrossRef]
11. Roshani, E.; Filion, Y. Event-based approach to optimize the timing of water main rehabilitation with asset management strategies. *J. Water Resour. Plan. Manag.* **2013**, *140*, 04014004. [CrossRef]
12. Perelman, L.; Ostfeld, A. An adaptive heuristic cross-entropy algorithm for optimal design of water distribution systems. *Eng. Optim.* **2007**, *39*, 413–428. [CrossRef]
13. Perelman, L.; Ostfeld, A.; Salomons, E. Cross entropy multiobjective optimization for water distribution systems design. *Water Resour. Res.* **2008**, *44*. [CrossRef]
14. Wu, W.; Simpson, A.R.; Maier, H.R. Multi-objective genetic algorithm optimisation of water distribution systems accounting for sustainability. *Proc. Water Down Under* **2008**, *2008*, 1750.
15. Wu, W.; Simpson, A.R.; Maier, H.R. Accounting for greenhouse gas emissions in multiobjective genetic algorithm optimization of water distribution systems. *J. Water Resour. Plan. Manag.* **2009**, *136*, 146–155. [CrossRef]
16. Wu, W.; Maier, H.R.; Simpson, A.R. Single-objective versus multiobjective optimization of water distribution systems accounting for greenhouse gas emissions by carbon pricing. *J. Water Resour. Plan. Manag.* **2009**, *136*, 555–565. [CrossRef]
17. Wu, W.; Simpson, A.R.; Maier, H.R. Sensitivity of optimal tradeoffs between cost and greenhouse gas emissions for water distribution systems to electricity tariff and generation. *J. Water Resour. Plan. Manag.* **2011**, *138*, 182–186. [CrossRef]
18. Wu, W.; Simpson, A.R.; Maier, H.R.; Marchi, A. Incorporation of variable-speed pumping in multiobjective genetic algorithm optimization of the design of water transmission systems. *J. Water Resour. Plan. Manag.* **2011**, *138*, 543–552. [CrossRef]
19. Stokes, C.S.; Simpson, A.R.; Maier, H.R. A computational software tool for the minimization of costs and greenhouse gas emissions associated with water distribution systems. *Environ. Model. Softw.* **2015**, *69*, 452–467. [CrossRef]
20. Stokes, C.S.; Maier, H.R.; Simpson, A.R. Effect of storage tank size on the minimization of water distribution system cost and greenhouse gas emissions while considering time-dependent emissions factors. *J. Water Resour. Plan. Manag.* **2015**, *142*, 04015052. [CrossRef]
21. Reed, P.M.; Hadka, D.; Herman, J.D.; Kasprzyk, J.R.; Kollat, J.B. Evolutionary multiobjective optimization in water resources: The past, present, and future. *Adv. Water Resour.* **2013**, *51*, 438–456. [CrossRef]
22. Carrasco, F.J.M.; de Marcos, L.G. *Dimensionamiento y Optimización de Obras Hidráulicas*, 4th ed.; Garceta Grupo Editorial: Malaga, Spain, 2013. (In Spanish)

23. Granados, A. *Problemas de Obras Hidráulicas*; Escuela de Ingenieros de Caminos, Canales y Puertos. Servicio de Publicaciones: Madrid, Spain, 1995. (In Spanish)

24. Ostfeld, A. Optimal design and operation of multiquality networks under unsteady conditions. *J. Water Resour. Plan. Manag.* **2005**, *131*, 116–124. [CrossRef]

25. Gessler, J.; Walski, T.M. *Water Distribution System Optimization*; Army Engineer Waterways Experiment Station Vicksburg Ms Environmental Lab: Washington, DC, USA, 1985.

26. Alperovits, E.; Shamir, U. Design of optimal water distribution systems. *Water Resour. Res.* **1977**, *13*, 885–900. [CrossRef]

27. Featherstone, R.E.; El-Jumaily, K.K. Optimal diameter selection for pipe networks. *J. Hydraul. Eng.* **1983**, *109*, 221–234. [CrossRef]

28. Dandy, G.; Hewitson, C. Optimizing hydraulics and water quality in water distribution networks using genetic algorithms. In *Building Partnerships, Joint Conference on Water Resource Engineering and Water Resources Planning and Management 2000, Minneapolis, MN, USA, 30 July–2 August 2000*; Hotchkiss, R.H., Glade, M., Eds.; ASCE: Reston, VA, USA, 2000; pp. 1–10.

29. Tsakiris, G.; Spiliotis, M. Fuzzy linear programming for problems of water allocation under uncertainty. *Eur. Water* **2004**, *7*, 25–37.

30. Gonçalves, G.M.; Gouveia, L.; Pato, M.V. An improved decomposition-based heuristic to design a water distribution network for an irrigation system. *Ann. Oper. Res.* **2014**, *219*, 141–167. [CrossRef]

31. Marques, J.; Cunha, M.; Savić, D. Using real options in the optimal design of water distribution networks. *J. Water Resour. Plan. Manag.* **2014**, *141*, 04014052. [CrossRef]

32. Schwartz, R.; Housh, M.; Ostfeld, A. Least-cost robust design optimization of water distribution systems under multiple loading. *J. Water Resour. Plan. Manag.* **2016**, *142*, 04016031. [CrossRef]

33. Ostfeld, A.; Tubaltzev, A. Ant colony optimization for least-cost design and operation of pumping water distribution systems. *J. Water Resour. Plan. Manag.* **2008**, *134*, 107–118. [CrossRef]

34. Vicente, D.; Garrote, L.; Sánchez, R.; Santillán, D. Pressure management in water distribution systems: Current status, proposals, and future trends. *J. Water Resour. Plan. Manag.* **2015**, *142*, 04015061. [CrossRef]

35. Capelo, B.; Pérez-Sánchez, M.; Fernandes, J.F.; Ramos, H.M.; López-Jiménez, P.A.; Branco, P.C. Electrical behaviour of the pump working as turbine in off grid operation. *Appl. Energy* **2017**, *208*, 302–311. [CrossRef]

36. Tsakiris, G.; Spiliotis, M. Uncertainty in the analysis of urban water supply and distribution systems. *J. Hydroinform.* **2017**, *19*, 823–837. [CrossRef]

37. Fu, G.; Kapelan, Z.; Reed, P. Reducing the complexity of multiobjective water distribution system optimization through global sensitivity analysis. *J. Water Resour. Plan. Manag.* **2011**, *138*, 196–207. [CrossRef]

38. Pérez-Sánchez, M.; Sánchez-Romero, F.-J.; López-Jiménez, P.A. Huella energética del agua en función de los patrones de consumo en redes de distribución. *Ingeniería Agua* **2017**, *21*, 197–212. [CrossRef]

39. Pérez-Sánchez, M.; López-Jiménez, P.A.; Ramos, H.M. Modified affinity laws in hydraulic machines towards the best efficiency line. *Water Resour. Manag.* **2018**, *32*, 829–844. [CrossRef]

40. Cabreros, J.M. *Proyecto de Presa Navas del Marqués*; Tajo, C.H.d., Ed.; Ministerio de Medio Ambiente: Ávila, Spain, 1999. (In Spanish)

Article

A Nation-Wide Framework for Evaluating Freshwater Health in China: Background, Administration, and Indicators

Chen Xie [1], Yifan Yang [2,*], Yang Liu [1], Guoqing Liu [1], Ziwu Fan [1] and Yun Li [1]

[1] Hydraulic Engineering Department, Nanjing Hydraulic Research Institute, Nanjing 210029, China; cxie@nhri.cn (C.X.); liuyang@nhri.cn (Y.L.); gqliu@nhri.cn (G.L.); zwfan@nhri.cn (Z.F.); yli@nhri.cn (Y.L.)

[2] Department of Civil and Environmental Engineering, The University of Auckland, Auckland 1010, New Zealand

* Correspondence: yyan749@aucklanduni.ac.nz

Received: 31 August 2020; Accepted: 15 September 2020; Published: 17 September 2020

Abstract: This study reviewed the existing experience of implementing the nation-wide freshwater health evaluation in China and around the world and proposes a new framework that works in collaboration with the River Chief System (RCS). The institutional context of China with intertwined political and scientific considerations makes it essential to establish a concise and quantitative approach to assess the effectiveness of the RCS as well as local freshwater health conditions that can be easily understood by non-experts for decision-making. To fulfil this objective, we reconstructed the indicator categories based on the best practices in major western countries and the existing regional standards in China. The new indicator framework includes two main aspects: Ecosystem integrity (physical habitat, water quantity, water quality, and aquatic life) and non-ecological performance (social services and water governance). Specifically, the non-ecological attributes of freshwaters are in accordance with the purposes of the RCS and are usually ignored in many countries. The final health grade for a specific water body is determined by a weighted averaging method; this grade is the core element of an evaluation protocol designed to produce reliable data for adaptable water resources governance in China. The research findings in this study will also be integrated into the new national standard to be issued by the Ministry of Water Resources of China in late 2020.

Keywords: freshwater health; river chief system; ecological integrity; social services; water governance; national standard

1. Introduction

There has been a longstanding recognition of the need for maintaining and improving the status of freshwater ecosystems across the world. Freshwater resources, mainly consisting of rivers and lakes, are closely related to not only the natural environment but also the human society as well as how human beings are engaged with the environment. The United Nations' Sustainable Development Goal 6 (SDG6), 'Clean Water and Sanitation' [1] sets goals of ensuring global water access and safety by 2030 by investing in adequate infrastructure and protecting and restoring water-related ecosystems. SDG 6 recognizes that water has multiple values and that they are all related, i.e., the success in one may depend on another. For example, a healthy freshwater ecosystem entails not only good water quality (less pollution) and low water stress (reasonable consumption), but also the improved water governance. This worldwide consensus was reached after decades of exploration and efforts made by many countries to find a well-balanced combination of 'led-by-science' and 'led-by-administration', while this exploration still continues.

The well-known Clean Water Act of 1977 was enacted in response to the dramatic pollution in the 1970s in the USA and is the world's first comprehensive and systematic environmental law for managing freshwater health [2]. Since the 1990s, researchers have contributed to developing advanced freshwater health frameworks beyond the original non-modern working modes; and the governments of many countries are working with scientists and the public to explore innovative approaches. For example, the European Water Framework Directive 2000 (WFD) was a valuable attempt to establishing a large transboundary cooperation framework, although the outcomes are not satisfactory in some aspects due to the unrealistic timeframe and the lack of functional indicators [3,4]. The flawed performance of the WFD also highlights the importance of a system with consistent administrative supervisions and enforcements; for example, the United Kingdom, even Scotland itself, have their own detailed assessment approaches [5–7]. In New Zealand and Australia, officials and scientists have also been aware of the urgency and made remedial efforts to upgrade the existing protocols [8,9].

Understanding the meaning of river (lake) health is another obstacle for countries with goals of sustainable development. Many researchers have been describing and debating the concept of river (lake) health since the 1990s [10–14]. As usefully defined by Meyer [15], a healthy water body can keep its ecological functions while maintaining the needs of the society. The models proposed by Boulton and Lackey [12,16] provided an important point of view that the human use of rivers does not automatically mean that the ecosystem health is being degraded. A healthy river (or lake) should contain both ecological values (ecological integrity and resilience to stress) and human values (social services and benefits). Therefore, ecological integrity is one part of a healthy ecosystem but not the only part; the lack of human values might also be detrimental. Usually, ecological integrity includes the physical, chemical, and biological components [17] and has long been treated as the major (sometimes the only) aspect of freshwater assessments. Obviously, incorporating social indicators is of significance, as detailed by Hanna et al. [18].

In the past decades, the health status of rivers and lakes in China has undergone significant deteriorations due to urban development and the increasing water use. Severe outcomes, e.g., water pollution, hydrologic mutation, the damaged physical structures, and the degenerated ecological diversity and functions, have attracted wide attention from governments and researchers. For example, in the 2000s, the water pollution in Guangdong Province, one of the most prosperous region in China, significantly affected the sustainability of the regional social economy while was rarely contained by local officials [19]. The outbreak of blue-green algae in Wuxi, Jiangsu Province, in 2007 eventually pushed the central government of China to establish a nation-wide system to contain pollution with a consistent protocol [20]. This is the origin of the River Chief System (RCS) known by the public nowadays. Since 2016, China has been fully implementing the RCS to build an efficient and productive framework that involves not only administrative innovations but also refined scientific guidelines to ensure the adequacy of specific evaluations. The principle of the RCS is using a hierarchical administrative structure with well-defined responsibilities and accountabilities to avoid the inaction of officials at different levels. Therefore, evaluating and improving freshwater health in China has become an issue with intertwined political and scientific drivers that make the original working models much less effective than expected. At the moment, China is still lacking a comprehensive evaluation framework due to the vast regional differences and the insufficient administrative experience of managing the massive RCS system. This study is conducted based on the request of the Department of River and Lake Management of the Ministry of Water Resources of China in 2019. As stated, it is essential to clarify the meaning and criteria of river/lake health and propose a systematic approach that fits for the characteristics of the RCS to improve the freshwater health in China.

The objectives of this paper include: (1) Reviewing the relevant research on the RCS comprehensively and providing a clear picture of its working mechanism; (2) comparing the best practices in China and the rest of the world to highlight the key aspects that are ignored and the lessons learnt; (3) analyzing the existing freshwater health indicators adopted by researchers and practitioners to help understand their strengths and weaknesses; and (4) proposing a new evaluation framework for the future implementation within the framework of the RCS in China. The achievement made in this study will be integrated into the new national standard to be issued by the Ministry of Water Resources of China. The specific methods for field sampling and data collection, aggregation, harmonization, and integration are beyond the scope of this paper.

2. Methodology

The literature review in this study is mainly focused on two aspects: The research on the RCS in China, and the research and application of freshwater health indicators both in China and around the world. Regarding the RCS, only published materials in English are reviewed in this paper to reach a wider audience. Nevertheless, a large number of scholars in China have made significant contributions by publishing articles in Chinese, which are recommended to be read extensively, if possible, for a better understanding of the development of the RCS. Regarding freshwater health indicators, the authors have collected most of the systematic and well-structured regional standards/guidelines in China, which are either formal documents or unpublished drafts, by extensive data search and enquiry. Data sources without a clearly defined indicator framework are not included. International frameworks reviewed in this paper are restricted to those that have significant influence across the world and have been well studied and improved. Therefore, the selected frameworks include the European WFD and those applied in major English speaking countries, including the USA, the UK, Australia, New Zealand, and Canada. The practices of other countries or regions are also reviewed but not included. The indicators used by the reviewed materials are further summarized and analyzed later in this paper to produce a new evaluation framework. In regard to data integration, similar categories, sub-categories, or individual indicators that have different names are compared carefully in terms of their definitions, and only the most widely adopted names are kept to avoid confusing readers. For example, the term 'physical habitats' is used to replace all the similar terms like 'forms' or 'physical environments'; nitrogen-related indicators, e.g., total nitrogen and dissolved nitrogen, are all replaced by the generic term 'nitrogen'.

The methodology of this paper is also subjected to limitations. The detailed approaches to determine the characteristic value of each indicator are not addressed so that this paper intends to provide an overview of freshwater health evaluation instead of technical instructions. Furthermore, the practical experience gained in the developed world, as well as the working model of the RCS, might not be suitable for less developed countries or regions, as the tradition and awareness of preserving water resources differ, and their needs for boosting economy are much more urgent. We argue that protecting water resources and ecosystems does not automatically entail the sacrifice of economic development, and it is also inequal to enforce the same level of restrictions to countries of different levels of prosperity. More efforts should be made to facilitate the coordination of global sustainable development under the framework of the UN's SDG6.

3. River Chief System (RCS) in China

3.1. Recent Literature in English

The implementation of River Chief System has facilitated the effective integrated freshwater management and water protection campaigns in China in the last decade and also attracted extensive research interests both domestically and internationally. Dai [21] introduced the history of the River Chief System and provided a perspective to the water governance in China as well as what role formal laws have played during the transition to the RCS. Huang and Xu [22] argued that the public participation of RCS mainly depends on local government, which, although ensures effective integration of resources, might impede the real public participation and supervision due to the political complexity of water governance in China. Wang et al. [23] stated that, although RCS can improve the work efficiency of water governance in the short-term, more efforts are still needed to the problems of organizational logic and the responsibility dilemma that remain due to the persistence of vertical coordination of the hierarchical system. Liu and Richards [20] summarized the structure of the RCS and the progress of implementation and identified future challenges. Similar to Wang et al. [23], Liu and Richards [20] also pointed out the flaw of the existing RCS regarding trans-regional collaboration and accountability. Liu et al. [24] presented a case study of Foshan in Fujian Province and showed that the RCS could establish a considerably sophisticated and effective management structure. More recently, Wang and Chen [25] argued that the institutional context and motivations are the external conditions influencing the collaborative governance regime and thus the outcomes of the RCS. In general, numerous studies in recent years have noticed and identified the internal flaws of the existing RCS that may influence its sustainability and efficacy in the long-term. Compared with the water governance policies in many other western countries (e.g., USA, UK, Australia, and New Zealand), the political and administrative complexity of China makes it necessary to come up with a more structured and practical approach to management the River/Lake Chiefs at different levels and assess their performances.

3.2. Policy

As summarized by Wang and Chen [25] the institutional context of China contains three major elements: The centralized political authority [26], the party-state hierarchy [27,28], and the cadre responsibility system [29]. This context enables the central government of China to play the role of 'policy issuer' while maintaining sufficient authority of enforcement at different administrative levels through a top-down performance assessment structure. Since the full implementation of the RCS in 2016, the central government (mainly the Ministry of Water Resources) has issued multiple instructive documents in regard of the responsibility, accountability, major problems to be addressed, and the principle of assessing the performance of Chiefs [30–32]. Meanwhile, the local governments (mainly in provincial level) are responsible for making their own regulations under the prescribed framework due to the absence of national legislation for the RCS. In contrast, many western countries tend to leave water resources management relying on a relatively flat structure by establishing managerial agencies at the level of individual river basins, which is partly due to the lack of central authority. For example, the European WFD provides motivations to initiate waterways monitoring programs as well as directions for its EU member states on data processing and reporting so that their results will be comparable. However, the duties of managing specific river basins, e.g., Rhine River Basin, still belongs to relatively independent organizations like The International Commission for the Protection of the Rhine against Pollution (ICPR), which is co-chaired by ministers from stakeholder countries. This working model is more about cross-border collaboration and coordination and is obviously not suitable for China that needs a well-defined procedure for top-down supervision.

3.3. Assessing the Performance of Chiefs

By the end of 2018, RCS had been established and fully implemented in all the 31 provinces and regions in China with over 300,000 River (Lake) Chiefs being appointed at provincial, municipal, county, and township levels. Thus, as mentioned above, it is necessary to build a comprehensive and effective approach to assess the performance of those River (Lake) Chiefs, and this approach is much different to the assessment of freshwater health itself that is mainly determined by scientific criteria. More considerations are raised due to the political and administrative complexity of the RCS. The transition from 'rule by men' to 'rule by law' [25] entails not only assessing to what extent the RCS has been established and supervised by the heads in different administrative levels, but also how intertwined scientific and political requirements are satisfied during this implementation process. Therefore, a practical and straightforward assessment approach was urgently needed and have been proposed accordingly based on the existing outlines [33], as summarized in Table 1. This approach was officially released in 2019, and, together with the freshwater health scoring approach to be discussed later in this paper, comply well with the principle of using quantifiable measures for assessments. We also believe this approach can be well applied in other countries after adaption.

Table 1. The approach to assessing the implementation of the River Chief System (RCS) with detailed scoring criteria.

No.	Tier-1 Categories		Tier-2 Categories	Scoring Criteria
I	Building RCS hierarchy and organizations (25 points)	1.	Establishment of RCS heads in provincial, municipal, county and township levels. (4/25)	• Establish provincial Chiefs and announcement (newspaper, television, internet, etc.) (1/4) • Establish municipal, county and township Chiefs and announcement (newspaper, television, internet, etc.) (3/4)
		2.	Establishment of River Chiefs and announcement. (9/25)	• Duty assignment for rivers/lakes in provincial level and announcement (newspaper, television, internet, etc.) (3/9) • Duty assignment for rivers/lakes in municipal, county and township levels and announcement (newspaper, television, internet, etc.) (6/9) • Deduct all the Establish provincial Chiefs and announcement (newspaper, televisionscores if any one of the River (Lake) Chiefs is not established properly.
		3.	Establishment of RCS offices and administrative system. (9/25)	• Employment of full-time and part-time RCS staff with a valid contract; staff in position. (6/9) • Sufficient work fund. (1/9) • Appropriate workplace and signage (2/9)
		4.	Set-up of RCS information notice boards. (3/25)	• Notice boards showing the duties and information of River (Lake) Chiefs, basic conditions of rivers/lakes, and contact information. (3/3) • Spot check shall be performed for at least ten sites; information should be updated up to one month after the appointment of new chiefs.
II	Building RCS regulations and mechanism (15 points)	5.	Establishment of RCS regulations in provincial, municipal and county levels. (4/25)	• Establish, announce and implement RCS rules in provincial, municipal and county levels, including chief meeting, information sharing, information reporting, supervision, accountability/incentives, and acceptance inspection.
		6.	Establishment of organizational mechanism (8/25)	• Coordination mechanism between different departments. (4/8) • Public participation (supervision, volunteering, photographing, education, scientific popularization, etc.). (2/8) • Capital investment mechanism. (2/8)
		7.	Determination of responsibility and accountability (3/15)	• In provincial level. (1/3) • In municipal and county levels. (2/3)
III	Duties performed by River Chiefs (12 points)	8.	Main issues (8/12)	• RCS meeting for work deployment and actions (4/8) • RCS meeting for coordinating main issues and supervision. (4/8)
		9.	Daily issues (4/12)	• Sufficient patrol and inspection; problems are spotted and addressed. (2/4) • Appraisal of subordinate chiefs and RCS departments. (2/4)

Table 1. *Cont.*

No.	Tier-1 Categories		Tier-2 Categories	Scoring Criteria
IV	Organizational works (16 points)	10.	Supervision and appraisal (6/16)	• Supervising subordinate chiefs and RCS department; works are improved; inaction and default are punished. (3/6) • Appraisal scheme is available; outcome is considered when assessing the performance of local government. (3/6)
		11.	Basic works (6/16)	• Compiling, printing and issuing RCS policy documents in provincial level. (2/6) • Establishment of river/lake archive. (1/6) • Establishment of the RCS information system. (3/6)
		12.	Publicity and training (4/16)	• Publicity in provincial, municipal and county levels (television, internet, newspaper, social media, etc). (2/4) • Training of RCS staff at different levels. (2/4)
V	The outcome of water protection and management (32 points)	13.	Water quality of rivers, lakes and centralized drinking water sources (9/32)	• The ratio of surface water areas of good quality meets the national criterion. (3/9) • The ratio of surface water areas of bad quality meets the national criterion. (3/9) • The ratio of drinking water sources (municipal level or beyond) of good quality meets the national criterion. (3/9)
		14.	Remediation of odorous urban water (4/32)	• Over 90% of odorous water bodies in urban areas are eliminated. (4/4)
		15.	Protection of shoreline and riparian zone (9/32)	• Over 80% of illegal occupation, mining, piling and construction are eliminated; works accomplished are archived; appropriate shoreline and riparian management plan are available. (6/9) • Administrative zones are determined. (3/9)
		16.	Integrated management of ecosystem (5/32)	• Integrated ecosystem remediation works are implemented at the provincial level. (2/5) • Contamination management of aquaculture, rural waterway and waste, and livestock breeding. (3/5)
		17.	Public satisfaction (5/32)	• Public satisfaction >90% (5/5); >80% (4/5); >70% (3/5); >60% (2/5); >50% (1/5); <50% (0/5).

4. Integrated Indicator Framework

Evaluating the health status of freshwater also heavily relies on the establishment of a practical and comprehensive scientific framework based on the best practices around the world. In this study, we aim to investigate how ecological and non-ecological indicators are adopted and incorporated in different regions of China and other major western countries with substantial experience. A comprehensive data and literature research are therefore conducted and will be presented in detail in this section.

4.1. Data Sources

Although the River Chief System has already been established across the country, the lack of a nation-wide operational standard forced the local government to resort to their existing experience or routines for implementing assessments; those routines usually show enormous inconsistency and even contradiction sometimes. Currently, only a few standards have been issued officially in provincial or municipal levels, while some other regions still rely on "drafts", "trial documents", or informal manuals that are subjected to many limitations. Table 2 summarized the existing regional standards, guidelines, and specifications being used in China [34–48]. Furthermore, we also reviewed the frameworks implemented in some major western countries for comparative analysis, such frameworks including the United States (National Aquatic Resources Surveys (NARS)) [49–51], New Zealand (Cawthron's Freshwater biophysical ecosystem Health Framework) [9], Australia (Integrated Ecosystem Condition Assessment) [8], and the United Kingdom (Common Standards Monitoring Guidance for Rivers/Lakes) [5,6]. Some other national or regional frameworks known by the public (e.g., Canadian Aquatic Biomonitoring Network (CABIN), the Murray-Darling Basin Sustainable Rivers Audit of Australia, and the National Environmental Monitoring and Reporting of New Zealand) are also reviewed but are not included in this section as they only focus on limited aspects of freshwater health or are relatively outdated and have been superseded in their own countries. More details can be found in [52–55] regarding those frameworks. It should also be noted that although the European Water Framework Directive (WFD) streamlines the legislation across Europe and provides guidance to member states on what component indicators are necessary (i.e., biological quality, hydro-morphology, and physicochemical attributes), the implementation is still inconsistent and varies widely from country to country. Therefore, in this study, we only selected the UK guidelines as representatives as they have been widely accepted and well examined. The overseas frameworks included in Table 2 are the best practices in the corresponding countries and have been well revised and modified for better effectiveness.

Table 2. The national, regional, and industrial standards, guidelines, and specifications used for summarizing the existing freshwater health indicators.

Source	Abbr.
Local Standard of Beijing City: DB/11T 1722/2020 Technical regulations for ecological health on aquatic ecosystem assessment [34]	DB11
Local Standard of Liaoning Province: DB21/T 2724/2017 Liaoning provincial evaluation guidelines for river and lake (reservoir) health [35]	DB21
Local Standard of Jiangsu Province: DB32/T 3674-2019 Specification for ecological river and lake status assessment [36]	DB32
Local Standard of Suzhou City: DB3205T 2019 Indicator system of river and lake health assessment (Unpublished draft) [37]	DB3205
Local Standard of Shandong Province: DB37/D 3018-2017 Shandong provincial evaluation standard for ecological river [38]	DB37
Report of Zhejiang Institute of Hydraulics and Estuary Evaluation of main rivers and lakes in Zhejiang Province [39]	ZJ

Table 2. *Cont.*

Source	Abbr.
Report of Fujian Normal University: Indicators and methods for assessing river health in Fujian Province [40]	FJ
Report of Guizhou Normal University: Guideline for assessing river (lake) health in Guiyang City [41]	GY
Standard of China Association for Engineering Construction Standardization: Technical Guidelines for Evaluating Water Quality of Urban Rivers and Lakes (draft) [42]	CECS
Draft of Ministry of Water Resources (for consultation purpose): Guideline for river and lake health assessment [43,44]	SL
Report of Ministry of Water Resources: Indicators, standard and method for assessing river and lake health (for pilot work) [45]	2010
"Happy River" indicators of Ministry of Water Resources (internal documents)	XFH
The United States Environmental Protection Agency: National Aquatic Resource Surveys [49–51]	USA
Cawthron Institute report prepared for Ministry for the Environment, New Zealand: Freshwater Biophysical Ecosystem Health Framework [9]	NZ
Australian Department of the Environment and Energy: Aquatic Ecosystems Toolkit. Module 5: Integrated Ecosystem Condition Assessment [8]	AU
The United Kingdom Joint Nature Conservation Committee (JNCC): Common Standards Monitoring Guidance for Rivers/Lakes [5,6]	UK

4.2. Key Components of an Evaluation Framework

After reviewing the documents listed in Table 2, we summarized the core components of a freshwater health framework, as shown in Table 3. The definition of a healthy freshwater ecosystem in New Zealand is stated by the National Policy Statement for Freshwater Management (NPS-FM) [56] as that reflects the importance of physical and chemical as well as biological elements of ecosystems. In the United States, NARS divides the basic indicators into four categories: Biological, physical, chemical, and recreational; the last category is a measurement of the pathogen and pollutant substances that can threaten people's wellbeing during recreational activities. The UK and Australian guidelines [5–8] further introduced more indicators in regard to social services; the former concerns more about the direct disturbance to human beings while the latter about the social functions and values provided. In general, biophysical ecosystem health can be represented by a measure of 'ecological integrity' [9], but it is only one part of an assessment of the ability of a freshwater ecosystem to support multiple freshwaters values. Even for ecological integrity itself, a well-balanced assessment cannot be achieved by a single biophysical measure because an ecosystem is a complex network of interacting biological communities and their physical environment. Therefore, the core components of a freshwater health framework should include the following sectors: Aquatic life (fauna, flora and microbes), water quality (physio-chemical features and pathogens), water quantity (hydrology), physical habitat (forms), ecological processes, and, as mentioned above, social services and values. Specifically, "ecological processes" are the interactions among aquatic lives and their environment and can be represented by a series of individual indicators in other categories to suggest the status of dynamic biochemical processes [57,58]. The assessment of ecological processes entails advanced knowledge and experience and may be relatively more difficult than other categories.

It should also be noted that, as discussed previously, the unique administrative hierarchy in China make freshwater health a political mission that is directly related to the performance of local officials. Therefore, water governance indicators should also be included in this framework with quantifiable items and practical calculation methods.

Table 3. Indicator categories of a freshwater evaluation framework and definitions.

Category		Criteria for Being "Healthy"
Ecological integrity	Physical habitat	The physical form and extent of the water body and the surrounding riparian areas are capable of supporting diverse flora and fauna throughout their life cycle.
	Water quality	The physical and chemical properties as well as other components of the water body are in natural status and may support diverse flora and fauna. For example, contaminants are scarce or absent.
	Water quantity	The water level, extent and flow regime are sufficient to support diverse flora and fauna during their full life cycle. As defined by Clapcott et al. [9], 'flow regime' includes the floods and droughts that ensure the surface water connectivity between the fresh waters and surrounding terrestrial habitat and other freshwaters (e.g., rivers and their floodplains, and wetlands), the regulation of biotic production and diversity, and that shape the morphology of physical habitat.
	Aquatic life	A diverse range of native species of flora and fauna persist; invasive alien species are scarce or absent; rare native species can be seen.
	Ecological processes	The normal interactions among aquatic lives and their environment persist (e.g., metabolism) with an optimized level of organic matter cycling, e.g., the retention, transformation and uptake of carbon and other nutrients.
Social services / values		A healthy water body can provide a wide range of non-ecological functions and benefits to human beings and the society, including cultural services (e.g., recreation and aesthetic beauty), provisioning services (e.g., water supply and aquaculture), regulating services (e.g., flood protection), etc.
Water governance		A healthy water body entails a set of managerial protocols that can effectively facilitate the implementation of specific water protection activities and more efficient water governance in different administrative levels.

4.3. Comparison

Table 4 shows the detailed comparison of indicators used by the standards and guidelines summarized in Table 2. The following issues should be noted:

- For the four overseas frameworks included, only the Australian guideline introduced social service indicators. The recreation indicators used by NARS are actually still biological and physio-chemical indicators or pathogens and could be more reasonably integrated into those corresponding categories.
- Many standards in China (e.g., Jiangsu, Shandong, Fujian, and Suzhou City) suggest referring to other national standards (GB3838-2002: Environmental quality standards for surface water; SL395-2007: Technological regulations for surface water resources quality assessment) for instructing water quality assessment. We also recommend this approach as those two water quality standards have been well examined during practices.
- "Ecological processes" is a relatively abstract and unintuitive concept compared with other categories and is only recommended by New Zealand's framework. Each ecological process can actually be represented by the values of specific physio-chemical or biological indicators, e.g., nutrient loads and BOD_5.
- Zoning indicators include: (1) The classification of water function zones that will be discussed later in this paper; (2) the clarification responsibilities; (2) and (3) the implementation of water quality management in different water function zones.

By comparison, we find that 11 indicators are mentioned over 10 times, including riparian vegetation coverage, dissolved oxygen, eco-water (environmental flow), water mobility and connectivity, riparian naturalness, macroinvertebrate, phosphorous, ammonia or nitrate, heavy metals, flood protection, and zoning indicators. The most significant distinction between Chinese standards and overseas frameworks will be further discussed in the following section.

Table 4. Comparison of indicators of different frameworks, guidelines, and regional specifications. The dots show in which framework the inidicators are adopted.

Category	Sub-Category	Indicator	DB11	DB21	DB32	DB3205	DB37	ZJ	FJ	GY	CECS	SL	2010	XFH	NZ	AU	UK	USA	Total
Physical habitat	Shape	Bank/channel form & stability		●		●		●		●		●	●		●	●	●		9
	Substrate	Substrate stability													●				1
		Substrate composition													●	●	●	●	4
		Substrate contamination				●	●					●				●			4
	Connectivity	Water mobility & connectivity	●	●	●	●		●	●			●	●		●	●	●		11
		Instream structures											●				●		2
	Riparian	Riparian naturalness	●	●	●	●			●	●		●	●		●	●	●		11
		Riparian vegetation coverage	●	●	●	●	●	●	●	●		●	●	●	●	●	●	●	15
	Other	Naturalness of water area & form			●	●	●									●			4
		Erosion & sedimentation		●				●				●		●		●	●		6
		Debris														●	●		2
		Wetland status						●				●		●					3
		Habitat extent & structure													●			●	2
Water quality	Physical	Temperature	●										●		●				3
		Dissolved oxygen	●	●	●	●	●		●	●	●	●		?	●	●	●	●	13
		Conductivity																●	1
		Clarity/turbidity	●						●	●				?	●	●		●	6
		Sediment load													●	●	●		3
		Salinity	●											?		●		●	3
		Oxidation Reduction Potential									●								1
	Chemical	pH & acidity	●		●	●	●	●						?	●	●	●	●	9
		Nitrogen	●		●	●	●	●	●					?	●		●	●	9
		Phosphorous	●		●	●	●	●	●					?	●	●	●	●	10
		Ammonia or nitrate	●		●	●	●	●	●	●			●	?	●	●			10
		Permanganate	●		●	●	●	●	●				●	?					7
		Oxygen needs (COD, BOD_5, etc)		●	●	●	●	●	●			●	●						8
		Other organic composites															●	●	2
		Nutrient loads		●				●				●		?	●	●			5
		Trophic diatom															●	●	2
		Heavy metals		●	●	●	●	●	●				●	?	●		●	●	10
		Chlorophyll a	●						●					?		●	●		4
		Sulfide			●	●	●	●						?					4
		Other (cyanide, fluoride, etc)			●	●	●	●								●			5

Table 4. Cont.

Category	Sub-Category	Indicator	DB11	DB21	DB32	DB3205	DB37	ZJ	FJ	GY	CECS	SL	2010	XFH	NZ	AU	UK	USA	Total
	Pathogen	E. coli			●	●	●		●					?					4
		Enterococci																●	1
Water quantity	Magnitude	Water depth & area	●									●			●		●		4
		Water volume														●			1
		Discharge & velocity												●	●		●		3
	Flow	Eco-water (environmental flow)		●	●	●	●	●	●	●		●	●	●		●	●		12
		Mean (annual) flow													●				1
		Long-term and seasonal variability		●				●		●		●	●		●	●	●		8
	Flood and drought	Flood occurrence													●				1
		Drought occurrence														●			2
Aquatic life		Fish	●	●			●		●			●	●		●			●	8
		Benthonic (macro)invertebrate	●	●	●	●				●		●	●		●	●	●	●	11
		Macrophytes	●				●					●			●	●	●	●	7
		Periphyton			●	●									●			●	4
		Plankton	●		●	●				●		●				●		●	7
		Microbes													●			●	2
		Waterbirds		●			●	●	●						●	●			6
		Terrestrial animals		●												●			2
		Fauna diversity						●										●	2
		Invasive species								●			●				●		3
		Rare species											●			●	●		3
Social service		Flood protection		●	●	●	●	●	●			●	●	●		●			10
		Shipping & navigation						●											1
		Water supply & consumption			●	●			●			●		●		●			6
		Scenic & aesthetic				●	●	●											3
		Culture & history		●			●	●								●			4
		Recreation & tourism														●		●	2
		Sensory & human comfort				●	●	●											3
		Public satisfaction		●	●	●	●	●	●			●	●						8
		Agriculture & food supply											●	●		●			3
		Industrial benefit												●					1

Table 4. *Cont.*

Category	Sub-Category	Indicator	DB11	DB21	DB32	DB3205	DB37	ZJ	FJ	GY	CECS	SL	2010	XFH	NZ	AU	UK	USA	Total
Water governance		Zoning & zonal status		●	●	●	●	●	●	●		●	●	●					10
		Professional team & plan						●											1
		Disturbance to society			●	●	●	●				●					●	●	7
		Utilization of water resources		●			●	●	●	●		●	●						7
		Facility, instruments & signboards						●											1
		Sewage regulation				●	●	●											3
		Illegal activities				●	●	●											3
Ecological processes		Biota interaction & food web													●				1
		Ecosystem metabolism													●			●	2
		Organic matter processing													●				1

4.4. Indicators Highlighted

Based on the comparison of detailed indicators shown in Table 4, the major difference between those widely used in China and in major western countries are mainly non-ecological indicators, e.g., social services and water governance indicators. This difference is attributed to the different social cultures and administrative structures between China and western countries. The adoptions of ecosystem-related indicators are generally similar and consistent, as mentioned in the last section. This section discusses the details of those discrepancies as well as how those indicators are adapted in China.

4.4.1. Indicators of Social Services and Values

Current frameworks used by major countries across the world have paid much attention to ecological values when evaluating the health status of rivers, lakes, and wetlands. In contrast, very few countries have made attempts to include human values. The Australian Integrated Ecosystem Condition Assessment (IECA) [8] recommends to include cultural, regulating, and provisioning services in the evaluation process but does not provide detailed indicator sets. The National Aquatic Resource Survey (NARS) of the United States [49–51] uses four indicators (Algal toxins, Cyanobacteria, Enterococci, and fish tissue contaminants) to assess the status of recreational functions, but does not pay much attention to the broader concepts of human values. In general, although a few studies have emphasized the importance of social services, more works are still needed to convert the existing research accomplishments of scientists to guide the actions of governments and authorities.

The first systematic attempt to define human services was made by the Millennium Ecosystem Assessment [59]. In this report, ecosystem services are divided into four categories, including cultural ecosystem services, provisioning ecosystem services, regulating ecosystem services, and supporting ecosystem services; the first three categories resemble those prescribed by the Australian IECA guideline. Specifically, cultural ecosystem services include non-material benefits such as recreational activities (e.g., swimming), the aesthetic beauty of rivers, and their spiritual significance among many communities. Provisioning ecosystem services include the products obtained, e.g., drinking water supply, fish, etc. Regulating ecosystem services include the benefits obtained from the regulation such as erosion prevention, pollution reduction, and flood protection. Furthermore, supporting ecosystem services include processes that help the production of other services, such as nutrient cycling and habitat provision, which are closely related to specific water quality indicators. As a recent and representative study, Hanna et al. [18] systematically reviewed 89 relevant studies and summarized the human services provided by rivers. It is found that the most studied and discussed social services are recreation and tourism and water supply that can provide visible monetary benefit. Regarding the categories, provisioning and regulating indicators have been much more studied than the other two categories (cultural and supporting). However, in China, more attention has been paid to flood protection engineering and public satisfaction as those measures can be more easily quantified and related to the performance of officials. Therefore, it is essential to combine the most emphasized indicators both in China and overseas to build a refined social services indicator set.

4.4.2. Indicators of Water Governance

As mentioned previously, one distinct feature of evaluating freshwater health in China is the close connection between science and politics. The establishment of the River Chief System facilitates the clarification of the responsibility and accountability of local government officials, who, however, lack the necessary scientific background and expertise to actively engage with and manage the specialist teams and track their performances. The political structure and hierarchy in China could not grant sufficient authority to any stand-alone managerial committee that supervises the condition of specific water bodies. Therefore, introducing a set of water governance indicators is obviously necessary and may assist the local RCS heads (i.e., Chiefs) to involve with the detailed scientific programs in a practical and concise way.

One of the most typical water governance indicators is the water condition in different water function zones, which enable the RCS heads to gain a general understanding of the spatial-temporal distribution of health status without going through too many details. The definition of water function zones in China is prescribed by the Ministry of Water Resources [60] according to the purpose of water usage, as shown in Table 5. Many existing standards in China (see Table 4) only implicitly prescribed that "water quality status should meet the criteria in different water function zones". The criteria in Table 5 should be widely used for the classification of main water areas in China and clearly suggest the suitable working objectives of Chiefs in different regions.

Other water governance indicators in Table 4 are also closely related to the operation of the RCS and are rarely used by common guidelines oversea.

4.4.3. Indicators of Ecosystem Resilience

The definition of resilience shows significant ambiguity regarding the restoring capability of an ecosystem and usually to some extent overlaps with some other individual indicators that can be more explicitly measured or quantified. Resilience is a measure of persistence; ecological resilience measures the magnitude of disturbance that a system can absorb before it undergoes changes in structure and function [61]. As stated by Davies et al. [54], ecosystems incorporate the properties of the living and non-living components with "emergent properties", e.g., diversity and resilience. Those properties are more like attributes of the system rather than its components and cannot be directly observed or measured intuitively. A typical ecosystem health monitoring may examine components and processes that are sensitive to disturbances over a range of spatial-temporal scales [13]. Those disturbances might include losses of native flora and fauna, and the invasion of alien species may further complicate the conditions. For example, Scotland's Environment Protection Agency (EPA) uses the condition of peatland and the number of a few alien species (e.g., grey squirrel and American mink) to determine the recovery capacity of an ecosystem. Another approach is to monitor the products of ecological processes operating over a range of scales [62], e.g., dissolved oxygen and chemical oxygen demand (COD), which are also key components of water quality monitoring. Although being of scientific significance, using resilience indicators needs balanced scientific judgement and may unnecessarily complicate the evaluating process. The tradeoff between science and administration is one of the core components of the RCS, and thus the formal application of resilience indicators might not be suitable but can provide supplementary information for experts.

4.4.4. Eco-Water (Environmental Flow)

Eco-water, or environmental flow, is another key indicator to measure the level of satisfaction of water quantity needed for maintaining the basic ecosystem functions and processes. A large number of studies has been published in recent two decades regarding this topic. Tharme [63] provided a comprehensive review of the concept of environmental flow as well as the existing environmental flow methodologies around the world. The methodologies can usually be divided into four categories: Hydrological, hydraulic rating, habitat simulation, and holistic methodologies. Specifically, hydrological methodologies that rely on arbitrary low flow indices are the most widely applied and easy-to-use category. Kuriqi et al. [64,65] provided new insights into the interaction between hydropower energy yields and environmental flow, as well as the needs for maintaining the balance of the ecosystem. Although the European WFD does not use the term "environmental flow" explicitly, the required biological status can only be achieved when the necessary hydrological regimes are maintained [66]. The UK also has set environmental standards for water abstraction limits and appropriate water release from reservoirs. Poff and Zimerman [67] reviewed the relevant publications in the last decades and emphasized that the risk of ecological deterioration may increase with increasing magnitude of flow alteration. Many studies [68–70] have also paid attention to the significance of environmental flow for water governance.

Table 5. The definition and classification of water function zones under the framework of the RCS in China.

Tier-1 Category	Tier-2 Category	Description
Protection zone		A protection zone is a water area of significance for the protection of water sources, drinking water, nature reserves, scenic locations, and the protection of rare and endangered species. It is prohibited to build, rebuild, expand, or engage in water-related activities that are not related to protection within the core areas of nature reserves and tier-1 water source areas.
Reservation zone		A reservation zone is a water area reserved and protected for future development and water resources utilization. Activities that may have a significant impact on water quantity, water quality, and water ecology should be strictly limited and managed.
Buffer zone		A buffer zone is a water area designated for the following purposes: (1) coordinating water-use relations among provinces and areas with prominent conflicts of benefit; (2) connecting inland/marine zones and protection/development zones that are designated for different purposes. In a buffer zone, all types of water-related activities should be strictly managed to prevent adverse effects on adjacent water function zones. All water-related activities that may be detrimental to the protection of water functional areas in the buffer zone at the provincial boundary shall be notified to the basin management agency in advance.
Development zone	Drinking water zone	A drinking water zone is a water area delimited or reserved for providing drinking water for urban and rural areas. In the areas that have been supplying water, further protection areas should be delimited for subsided water sources to preserved the water quality and volume. It is prohibited to build, rebuild or expand any types of sewage outfalls. In the areas reserved for future water use, the discharge of pollutants should be strictly controlled, and no new discharge into the river is allowed.
	Industrial water zone	An industrial water zone is an area designated to meet industrial water demand. Priority should be given to the specified water usage, and any type of water intake should be strictly managed. Any installation of sewage outfall should not influence the water quality required for the specified zonal functions.
	Agricultural water zone	An agriculture water zone is an area designated to meet the water demand for irrigation. Priority should be given to the specified water usage, and any type of water intake should be strictly managed. Any installation of sewage outfall should not influence the water quality required for the specified zonal functions.
	Fishery zone	A fishery zone is an area designated for protecting aquatic life (e.g., fish). The basic water demand for fishery should be maintained, and important physical environments (e.g., habitats for natural species, spawning beds, wintering beds, feeding grounds and migration passages) should be protected. Water pollution should be strictly controlled by the units and individuals engaged in aquaculture.
	Scenic and recreation zone	A scenic and recreation zone is the area designated to meet the needs of landscape, entertainment, and various leisure activities. Any activity shall not influence the water quality status in the area.
	Transitional zone	A transition zone is an area designated to connect adjacent function zones with different water quality requirements. Transition zone should be managed to guarantee the water quality in the downstream function zone. Any water-related activity that may damage the self-purification capacity of the water body should be strictly controlled.
	Sewage control zone	A sewage control zone is the area designated to receive intensive domestic and industrial wastewater while restraining the adverse impacts on the functions of downstream zones.

In China, most of the regional standards and guidelines reviewed in this study adopt eco-water/environmental flow methods to evaluate the health status of local rivers and lakes (see Table 4). Considering the difficulty of introducing complex methodologies within the framework of the RCS, a certain level of simplification is required. Currently, the standard issued by the Ministry of Water Resources of China [71] provides a simplified scoring method, based on the ratio of the minimum daily averaged discharge to the multi-year average discharge in the corresponding period. The final score is determined by calculating the lowest score in the periods of October to March and April to September, respectively. This approach, with the further involvement of aquatic life indicators, may reflect the local ecosystem-related hydrological conditions and is thus recommended to be carried on to the new framework.

5. Discussion

5.1. Lessons Learnt

5.1.1. Well-Defined Objectives and Motivations

A strong policy driver from authorities may provide the purpose for, and clear direction on, setting well-defined and descriptive objectives for regional water governance and specific monitoring and reporting activities. This driver is essential for maintaining the long-term sustainability and consistency of the existing framework. For example, the European Union's WFD has stated that the objectives are protecting and enhancing the health of aquatic ecosystems while successfully maintaining social and economic systems [72]. As requested, all European Union member states should achieve at least 'good ecological status' for all 'natural' water bodies by 2015 and at the latest by 2027. Although being too strict regarding the timeframe, the WFD's objectives have facilitated the comprehensive implementation of freshwater surveys and thus successfully obtained a large dataset for scientific judgement and further policy adjustment. In the USA, the objective of the Clean Water Act is restoring and maintaining the chemical, physical, and biological integrity of the waters, and thus the aims of the National Aquatic Resource Survey (NARS) is to provide robust data to help the corresponding assessments. In China, as defined by the central government, the main objectives of water governance at the current phase include mitigating human disturbance to water bodies (e.g., illegal mining, occupation, disposing, and constriction), enhancing local and regional regulations, and improving the public satisfaction and people's living quality. Therefore, the RCS was established accordingly to build a hierarchical administrative structure with clear responsibility and accountability. In general, the detailed approach for implementation should fit for the high-level goals set by authorities.

5.1.2. Comprehensive and Practical Indicator Metrics

Using a wide range of indicator metrics ensures that all essential aspects of ecosystem health are addressed. For those well-accepted frameworks outside China, biological (aquatic life) indicators are the core components that also reflect the need to measure the combined response to multiple pressures in a catchment. Those indicators are usually assessed in addition to water quality and habitat indicators and sometimes water quantity and ecosystem processes. As a result, the status of ecological integrity can be accurately assessed, but social values are rarely mentioned. In contrast, the previous practices in China overemphasized the importance of administration and management while not treating different aspects of an ecosystem as a dynamic entity.

Furthermore, during the establishment and implementation of the RCS in China in recent years, we found that the most complained part of the initial RCS framework is that the poorly trained local RCS heads had to be held accountable for what they did not fully understand. Overstretching the scientific aspect does not automatically lead to a better outcome.

5.1.3. A Standard Data Acquisition and Analysis Protocol

A standardized data approach ensures the consistency of evaluation results, which is crucial to the nation-wide implementation of the RCS. Some overseas frameworks (e.g., NARS and CABIN) are based on standardized approaches including field survey design, data collection and processing, and reporting. Some other frameworks (e.g., European WFD and the UN's SDG 6) integrate data from different methods and make comparisons by data harmonization. In New Zealand and Australia, scientists have also been making efforts to establish standardized measurement protocols, but more works are still required on determining how to incorporate different components of river health into assessment programs in terms of network design and reporting [9]. Therefore, we intend to propose a functional and standardized protocol that is suitable for the RCS in China, which will be discussed in detail in the next section.

5.1.4. Adaptability

An adaptable framework is capable of adjusting or "upgrading" itself through a structured and iterative process that supports robust decision-making in the face of uncertainty. This feature is especially crucial for a country with typical hierarchical administrative structures that might cause unintended delay during decision-making. As a national standard, the indicator framework proposed in this study (see following sections) that is about to be officially issued this year will be reviewed and revised as a routine of two years.

5.2. A New Evaluation Protocol

As mentioned previously, a new protocol is proposed in this study to fit for the nation-wide implementation of RCS in China, which is displayed in Figure 1. This protocol is now going through the consultation process and will be officially issued by the end of 2020. Figure 1 shows the three core components of the protocol, including technical preparation, survey and monitoring, and report preparation. Specifically, "technical preparation" provides detailed instruction for preparing for field operation and data acquisition, and this sector is beyond the scope of this study. The following section will present and discuss the newly proposed indicator framework used as the core component of "survey and monitoring". In addition, the results in the complied final report will be concise and understandable to non-experts and can be used to support the assessment of Chiefs.

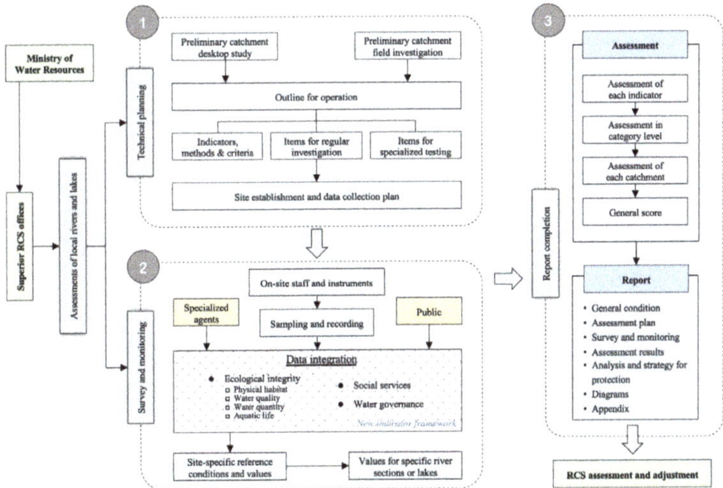

Figure 1. The new evaluation protocol proposed for nation-wide implementation in China.

5.3. Modified Indicator Framework

Table 6 summarizes the refined indicator framework to be adopted in China as well as the details of data sources and collection method. More specific sampling and testing methods are not discussed in this paper as they are beyond the scope of this study. It can be found that the indicators are classified into three main categories: ecological integrity, social service and water governance. Specifically, ecological integrity can further be classified into four sub-categories (physical habitat, water quantity, water quality and aquatic life) based on the existing experience and best practices around the world.

We recommend that the assessment of water quality follows the existing national standard implemented in China (GB3838-2002 and SL395-2007) [46,47] instead of building another new protocol, as the existing practices have been proved efficient with the reliable outcome. Specifically, GB3838-2002 [46] provides an extensive list of the substances (and threshold values) to be examined for general surface waters and, more importantly, centralized drinking water sources. The included items are more comprehensive than most of the overseas guidelines, and the testing can be conducted by the existing specialized agencies to produce concise reference reports for decision-making officials or Chiefs. SL395-2007 [47] provides an extra composite measure for lakes to determine the degree of eutrophication based on s series of individual water quality indicators (e.g., total nitrogen, total phosphorous, chlorophyll a, permanganate, clarity, etc). Those two standards have also been adopted by some pioneering provinces in China, e.g., Jiangsu, Shandong, and Fujian Provinces (see Table 4). Therefore, a seamless transition to the new nation-wide framework can be accomplished and save administrative resources significantly. Furthermore, the calculation of eco-water (environmental flow) demands can be made based on the existing standard [71]. The daily averaged discharge for rivers or daily averaged water level for lakes are compared with the calculated minimum demands to evaluate to what extent the necessary environmental flows are satisfied.

As shown in Table 6, weights (in parentheses) are assigned to each category and sub-category. A general health score will firstly be given to each selected river (lake) section. The determination of the score is based on weighted averaging approach. The equation can be expressed as:

$$RHI_i = \sum_{m=1}^{m} \left\{ WC_m \times \sum_{n=1}^{n} \left[WS_n \times \sum_{k=1}^{k} (WI_k \times G_k) \right] \right\} \tag{1}$$

where RHI_i = the general health score of the ith river (lake) section, W_k = the weight of the kth indicator, G_k = the score of the kth indicator, WS_n = the weight of the nth sub-category, and WC_m = the score of the mth category.

For the entire river or lake, the final score is averaged based on the score at each section, whose weights are determined by the corresponding ratio of length (for rivers) or area (for lakes). The equation for the final score can be expressed as:

$$RHI = \frac{\sum_{i=1}^{R_s} (RHI_i \times A_i)}{\sum_{i=1}^{R_s} A_i} \tag{2}$$

where RHI = the final score of a river (lake), A_i = the length (area) of the ith river (lake) section, and R_s = the amount of the total sections. Based on the final score, a certain river or lake can be categorized as "very healthy" ($RHI'' 90$), "healthy" ($75 \leq RHI < 90$), "subhealthy" ($60 \leq RHI < 75$), "unhealthy" ($40 \leq RHI < 60$), or "hazardous" ($RHI < 40$).

5.4. Limitation and Future Development

It should be mentioned that the main limitation of the new framework is the over-simplification of some indicators (e.g., eco-water/environmental flow) for the convenience of implementing the RCS by non-experts. Furthermore, those more sophisticated indicators (e.g., ecological processes and ecosystem resilience) excluded from the framework should not be ignored by researchers, as they can better reflect the freshwater health as a dynamic status with multiple interrelated constituents.

Adopting a simplified framework does not imply that the scientific principles are compromised; on the contrary, ongoing research should be carried out to monitor the efficacy of the framework and make adjustments accordingly.

Although the RCS has been proved as an effective system to manage freshwater resources and improve the health status, challenges still exist due to the fast urbanization in China and the increasing pollutant emission and water consumption, which may significantly influence people's lives. The future development of the RCS should be based on the essential goal that freshwater health status meets people's expectation for an enjoyable living environment, which is obviously more fundamental than any "index" or "grade" that describe the condition of water bodies in an unintuitive way. Some pilot works [73] have been done to explore a method to quantify the sensory aspect of rivers and lakes. In 2019, the Ministry of Water Resources in China had also set the objective of the RCS as "unblocked rivers, clear waters, green banks, beautiful scenery, people in harmony". Priority will be given to a few cities, regions, and catchments of significance to build healthy freshwater ecosystems that meet ecological, sensory, and safety demands under the framework of the RCS; those examples can further be the role models for the rest of China.

Table 6. The proposed new indicator framework. The weights for categories and sub-categories are shown in parentheses.

Category	Sub-Category	Indicator	Selection	More Description	Data Source
Ecosystem integrity (0.7)	Physical habitat (0.2)	Water connectivity	Optional	Instream obstructions (rivers) and incoming discharge (lakes)	Engineering record; field investigation; remote sensing
		Riparian naturalness	Mandatory	Bank stability and vegetation coverage	Field investigation; remote sensing
		Riparian wideness[R]	Optional	Width during the dry season	
		Lake shrinking ratio[L]	Mandatory	Compared with the area in the 1950s	
	Water quantity (0.15)	Eco-water satisfaction	Mandatory	Follow SL/Z 712-2014	Hydrologic monitoring live data; on-site monitoring; the historical records
		Water quantity mutation	Optional	Use monthly surface discharge deviation for rivers / Use monthly incoming discharge deviation for lakes	
	Water quality (0.15)	Water quality	Mandatory	Follow GB3838-2002 and SL395-2007 [46,47]	Local water quality reports; live data or on-site monitoring
		Substrate contamination	Optional	Follow GB15168-2018 [48]	
		Eutrophication[L]	Mandatory	Follow SL395-2007 [47]	
	Aquatic life (0.2)	Macroinvertebrate *	Optional	Use the Benthic Index of Biological Integrity (B-IBI) [74,75]	Field investigation by specialists
		Fish	Mandatory	Compared with the status before 1980	Data from aquaculture departments; field investigation by specialists
		Waterbirds	Optional	Population, diversity and the existence of rare birds	Data from forestry departments or environment protection agencies; field investigation by specialists
		Plankton density[L]	Mandatory	Compared with the status before 1980	FIeld investigation by specialists
		Macrophytes	Optional	Compared with the status of specific periods in history	Field investigation; remote sensing
Non-ecological performance (0.3)	Social services (0.2)	Flood protection	Optional	Length of the embankment and other engineering interventions	Data from water conservancy departments
		Water supply	Optional	How often the daily averaged discharge (or water level) is higher than the threshold for abstraction	Reports of local water authorities
		Land development	Optional	The ratio of developed riparian land to the reference value	Data from local planning consents
		Navigation[R]	Optional	The ratio of navigable days in a year	Open hydrologic data or official reports
		Public satisfaction	Mandatory	Public opinion about sensory/recreational aspects, the general condition of physical habitats and the surrounding environment	On-site or online questionnaires
	Water governance (0.1)	Drinking water zones **	Optional	The ratio of drinking water zones with the required water quality.	Local water quality reports; live data or on-site monitoring
		Illegal activities ("four chaos")	Mandatory	Illegal mining, occupation, disposing, and construction are managed.	Reports of local administration
		Sewage regulation	Mandatory	The outfalls should be registered and well managed with reasonable layouts that do not influence drinking water sources	

[R] Applicable to rivers; [L] Applicable to lakes;. * Macroinvertebrate status is assessed using the B-IBI method proposed by [74,75]. ** The concept of water functions zones is simplified to focus on drinking water zones and avoid ambiguity.

6. Summary and Conclusions

In this paper, we analyze the administrative and scientific aspects of evaluating freshwater health in China under the framework of the River Chief System (RCS) and compare those features with the best practices around the world. The most significant distinctions between the principles in China and those overseas are discussed particularly. A new indicator framework is proposed together with a well-structured protocol for implementation. The research findings in the present study will be integrated into China's new national freshwater evaluation standard to be issued this year.

Regarding the administrative aspect, the unique institutional context in China (centralized authority, party-state hierarchy, and cadre responsibility system) make it is essential to adopt the top-down mechanism of accountability to avoid the inaction of local officials in different levels. Thus, the performance of the RCS should be graded in a concise and quantitative way to simplify the workload of local Chiefs who are usually non-experts. Regarding the scientific aspect, social services and water governance indicators should be appropriately integrated into the new framework, as those indicators emphasize some attributes that are particularly valued in China and also sometimes ignored in western counties. Those attributes, for example, include the effectiveness of water resources management by using a centralized model and the harmony between individuals, the society and the environment.

From this study, we argue that a series of issues should be paid attention to when establishing a widely implemented freshwater health evaluation framework, including well-defined objectives and motivations, comprehensive and practical indicators metrics, standardized data acquisition and analysis protocols, and the adaptability for future development. Regarding the last issue, the new framework proposed in this study is expected to be revised for every two years to ensure its appropriateness.

Author Contributions: Conceptualization, C.X., Y.Y., and Y.L. (Yun Li); formal analysis, C.X. and Y.Y.; funding acquisition, Y.L. (Yun Li); methodology, C.X. and Y.Y.; project administration, C.X. and Y.L. (Yun Li); resources, Y.L. (Yang Liu), G.L., and Z.F.; writing—original draft, C.X. and Y.Y.; writing—review and editing, Y.Y., Y.L. (Yang Liu), G.L., and Z.F.. All authors have read and agreed to the published version of the manuscript.

Funding: This research was funded by National Key R&D Program of China (2018YFC0407205); Shanghai Chengtou Science and Technology Innovation Program (CTKY-ZDXM-2020-003) and the Special Research Fund of the Nanjing Hydraulic Research Institute (Y119016).

Acknowledgments: The authors would like to thank the research group from The University of Auckland, including Bruce Melville, Asaad Shamseldin, Naresh Singhal, and Graham Macky, for their contributions to the completion of this project.

Conflicts of Interest: The authors declare no conflict of interest. The funders had no role in the design of the study; in the collection, analyses, or interpretation of data; in the writing of the manuscript, or in the decision to publish the results.

References

1. UN-Water. Integrated Monitoring Guide for Sustainable Development Goal 6 on Water and Sanitation Targets and Global Indicators. Geneva, Switzerland, 2017. Available online: http://www.unwater.org/publications/sdg-6-targetsindicators/ (accessed on 30 August 2020).
2. Adler, R.W.; Landman, J.C.; Cameron, D.M. *The Clean Water Act 20 YEARS Later*; Island Press: Washington, DC, USA, 1993.
3. Moss, B. The Water Framework Directive: Total environment or political compromise? *Sci. Total Environ.* **2008**, *400*, 32–41. [CrossRef] [PubMed]
4. Hering, D.; Borja, A.; Carstensen, J.; Carvalho, L.; Elliott, M.; Feld, C.K.; Heiskanen, A.-S.; Johnson, R.K.; Moe, J.; Pont, D.; et al. The European Water Framework Directive at the age of 10: A critical review of the achievements with recommendations for the future. *Sci. Total Environ.* **2010**, *408*, 4007–4019. [CrossRef] [PubMed]
5. Joint Nature Conservation Committee. *Common Standards Monitoring Guidance for Rivers*; Joint Nature Conservation Committee: Peterborough, UK, 2015.

6.   Joint Nature Conservation Committee. *Common Standards Monitoring Guidance for Lakes*; Joint Nature Conservation Committee: Peterborough, UK, 2016.

7.   Scotland's environment. Ecosystem Health Indicators. 2020. Available online: https://www.environment.gov.scot/our-environment/state-of-the-environment/ecosystem-health-indicators (accessed on 14 August 2020).

8.   Department of the Environment and Energy. *Aquatic Ecosystems Toolkit. Module 5: Integrated Ecosystem Condition Assessment*; Department of the Environment and Energy: Canberra, Australia, 2017.

9.   Clapcott, J.; Young, R.; Sinner, J.; Wilcox, M.; Storey, R.; Quinn, J.; Daughney, C.; Canning, A. *Freshwater Biophysical Ecosystem Health Framework*; Technical Report prepared for Ministry for the Environment of New Zealand No. 3194; Cawthron Institute: Nelson, New Zealand, 2018; p. 89.

10. Rapport, D.J.; Costanza, R.; McMichael, A.J. Assessing ecosystem health. *Trends Ecol. Evol.* **1998**, *13*, 397–402. [CrossRef]

11. Norris, R.H.; Thoms, M.C. What is river health? *Freshw. Biol.* **1999**, *41*, 197–209. [CrossRef]

12. Lackey, R.T. Values, policy, and ecosystem health. *Bioscience* **2001**, *51*, 437–443. [CrossRef]

13. Vugteveen, P.; Leuven, R.S.E.W.; Huijbregts, M.A.J.; Lenders, H.J.R. Redefinition and elaboration of river ecosystem health: Perspective for river management. *Hydrobiologia* **2006**, *565*, 289–308. [CrossRef]

14. Elosegi, A.; Gessner, M.O.; Young, R.G. River doctors: Learning from medicine to improve ecosystem management. *Sci. Total Environ.* **2017**, *595*, 294–302. [CrossRef]

15. Meyer, J.L. Stream health: Incorporating the human dimension to advance stream ecology. *J. N. Am. Benthol. Soc.* **1997**, *16*, 439–447. [CrossRef]

16. Boulton, A.J. An overview of river health assessment: Philosophies, practice, problems and prognosis. *Freshw. Biol.* **1999**, *41*, 469–479. [CrossRef]

17. Schallenberg, M.; Kelly, D.; Clapcott, J.; Death, R.; MacNeil, C.; Young, R.G.; Sorrell, B.; Scarsbrook, M. Approaches to assessing ecological integrity of New Zealand freshwaters. *Sci. Conserv.* **2011**, *307*, 84.

18. Hanna, D.E.L.; Tomscha, S.A.; Dallaire, C.O.; Bennett, E.M. A review of riverine ecosystem service quantification: Research gaps and recommendations. *J. Appl. Ecol.* **2018**, *55*, 1299–1311. [CrossRef]

19. Ediger, L.; Hwang, L. Water quality and environmental health in southern China. In Proceedings of the BSR Forum, Guangzhou, China, 15 May 2009; Volume 15, p. 2009.

20. Liu, D.; Richards, K. The He-Zhang (River chief/keeper) system: An innovation in China's water governance and management. *Int. J. River Basin Manag.* **2019**, *17*, 263–270. [CrossRef]

21. Dai, L. A new perspective on water governance in China: Captain of the River. *Water Int.* **2015**, *40*, 87–99. [CrossRef]

22. Huang, Q.; Xu, J. Rethinking environmental bureaucracies in River Chiefs System (RCS) in China: A critical literature study. *Sustainability* **2019**, *11*, 1608. [CrossRef]

23. Wang, L.; Tong, J.; Li, Y. River Chief System (RCS): An experiment on cross-sectoral coordination of watershed governance. *Front. Environ. Sci. Eng.* **2019**, *13*, 64. [CrossRef]

24. Liu, H.; Chen, Y.D.; Liu, T.; Lin, L. The river chief system and river pollution control in China: A case study of Foshan. *Water* **2019**, *11*, 1606. [CrossRef]

25. Wang, Y.; Chen, X. River chief system as a collaborative water governance approach in China. *Int. J. Water Resour. Dev.* **2020**, *36*, 610–630. [CrossRef]

26. Zhu, X. Mandate versus championship: Vertical government intervention and diffusion of innovation in public services in authoritarian China. *Public Manag. Rev.* **2014**, *16*, 117–139. [CrossRef]

27. Cartier, C. Territorial urbanization and the party-state in China. *Territ. Politics Gov.* **2015**, *3*, 294–320. [CrossRef]

28. Edin, M. Remaking the communist party-state: The cadre responsibility system at the local level in China. *China Int. J.* **2003**, *1*, 1–15. [CrossRef]

29. Burns, J.P. Local cadre accommodation to the "responsibility system" in rural China. *Pac. Aff.* **1985**, *58*, 607–625. [CrossRef]

30. The State Council of China. Opinions on Comprehensively Promoting the River Chief System. 2014. Available online: http://www.gov.cn/gongbao/content/2014/content_2707859.htm (accessed on 30 August 2020). (In Chinese)

31. The State Council of China. Opinions on Strengthening the Management of Rivers and Lakes. 2017. Available online: http://www.gov.cn/gongbao/content/2017/content_5156731.htm (accessed on 30 August 2020). (In Chinese)

32. The State Council of China. Opinions on Promoting the Implementation of the River Chief System: From Nominal to Actual. 2017. Available online: http://www.gov.cn/gongbao/content/2019/content_5363080.htm (accessed on 30 August 2020). (In Chinese)

33. Development Research Center. Technical outline for comprehensive implementation of the assessment of the River (Lake) chief system. In *Report Completed in Collaboration with Hohai University and North China University of Water Resources and Electric Power*; Ministry of Water Resources: Beijing, China, 2019. (In Chinese)

34. Local Standard of Beijing City. *DB/11T 1722/2020: Technical Regulations for Ecological Health on Aquatic Ecosystem Assessment*; Beijing Municipal Bureau of Market Supervision and Administration: Beijing, China, 2020. (In Chinese)

35. Local Standard of Liaoning Province. *DB21/T 2724/2017: Liaoning Provincial Evaluation Guidelines for River and Lake (Reservoir) Health*; Liaoning Provincial Bureau of Quality and Technical Supervision: Shenyang, China, 2017.

36. Local Standard of Jiangsu Province. *DB32/T 3674-2019: Specification for Ecological River and Lake Status Assessment*; Jiangsu Provincial Bureau of Market Supervision and Administration: Nanjing, China, 2019. (In Chinese)

37. Local Standard of Suzhou City. *Indicator System of River and Lake Health Assessment*; Suzhou Municipal Bureau of Market Supervision and Administration: Suzhou, China, ubpublished. (In Chinese)

38. Local Standard of Shandong Province. *DB37/D 3018-2017: Shandong Provincial Evaluation Standard for Ecological River*; Shandong Provincial Bureau of Quality and Technical Supervision: Jinan, China, 2017. (In Chinese)

39. Zhejiang Institute of Hydraulics and Estuary. *Evaluation of Main Rivers and Lakes in Zhejiang Province*; Zhejiang Institute of Hydraulics and Estuary: Hangzhou, China, September 2018. (In Chinese)

40. Fujian Normal University. *Indicators and Methods for Assessing River Health in Fujian Province*; Fujian Normal University: Fuzhou, China, 2019. (In Chinese)

41. Guizhou Normal University. *Guideline for Assessing River (Lake) Health in Guiyang City*; Guizhou Normal University: Guiyang, China, February 2019. (In Chinese)

42. Standard of China Association for Engineering Construction Standardization. *Technical Guidelines for Evaluating Water Quality of Urban Rivers and Lakes*; Tsinghua University: Beijing, China, ubpublished. (In Chinese)

43. Standard of Water Conservancy Industry of China. *Guideline for River and Lake Health Assessment (Draft)*; Ministry of Water Resources: Beijing, China, unpublished. (In Chinese)

44. Peng, W. Research on river and lake health assessment indicators, standards and methods. *J. China Inst. Water Resour. Hydropower Res.* **2018**, *16*, 394–416. (In Chinese)

45. Ministry of Water Resources. *Indicators, Standard and Method for Assessing River and LAKE Health (for Pilot Work) (Version 1.0)*; National Technical Document for Evaluating River and Lake Health; Ministry of Water Resources: Beijing, China, October 2010. (In Chinese)

46. National Standard of China. *GB3838-2002: Environmental Quality Standards for Surface Water*; State Environmental Protection Administration: Beijing, China, 2002. (In Chinese)

47. Standard of Water Conservancy Industry of China. *SL395-2007: Technical Regulations for Surface Water Resources Quality Assessment*; Ministry of Water Resources: Beijing, China, 2007. (In Chinese)

48. National Standard of China. *GB15168-2018: Soil Environmental Quality Risk Control Standard for Soil Contamination of Agricultural Land*; Ministry of Ecology and Environment: Beijing, China, 2018. (in Chinese)

49. USEPA. *National Lakes Assessment 2018/19: Field Operations Manual*; Office of Water and Office of Environmental Information: Washington, DC, USA, 2017.

50. USEPA. *National Rivers and Streams Assessment 2018/19: Field Operations Manual Non-Wadable*; Office of Water and Office of Environmental Information: Washington, DC, USA, 2019.

51. USEPA. *National Rivers and Streams Assessment 2018/19: Field Operations Manual Wadable*; Office of Water and Office of Environmental Information: Washington, DC, USA, 2019.

52. Environment Canada. *Canadian Aquatic Biomonitoring Network Laboratory Methods: Processing, Taxonomy, and Quality Control of Benthic Macroinvertebrate Samples*; Environment Canada: Ottawa, ON, Canada, 2012.

53. Environment Canada. *Canadian Aquatic Biomonitoring Network Field Manual: Wadeable Streams*; Environment Canada: Ottawa, ON, Canada, 2012.

54. Davies, P.E.; Harris, J.H.; Hillman, T.J.; Walker, K.F. The Sustainable Rivers Audit: Assessing river ecosystem health in the Murray-Darling Basin, Australia. *Mar. Freshw. Res.* **2010**, *61*, 764–777. [CrossRef]

55. Hudson, N.; Ballantine, D.; Storey, R.; Schmidt, J.; Davies-Colley, R. *National Environmental Monitoring and Reporting (NEMaR): Indicators for National Freshwater Reporting*; NIWA Client Report HAM2012-025; Ministry for the Environment: Wellington, New Zealand, 2012.

56. Ministry for the Environment. *National Policy Statement for Freshwater Management 2014 (Amended 2017)*; Ministry for the Environment: Wellington, New Zealand, 2017.

57. Young, R.G.; Matthaei, C.D.; Townsend, C.R. Organic matter breakdown and ecosystem metabolism: Functional indicators for assessing river ecosystem health. *J. North Am. Benthol. Soc.* **2008**, *27*, 605–625. [CrossRef]

58. Clapcott, J.E.; Collier, K.J.; Death, R.G.; Goodwin, E.O.; Harding, J.S.; Kelly, D.; Leathwick, J.R.; Young, R.G. Quantifying relationships between land-use gradients and structural and functional indicators of stream ecological integrity. *Freshw. Biol.* **2012**, *57*, 74–90. [CrossRef]

59. Millennium Ecosystem Assessment. *Ecosystems and Human Well-Being: Synthesis*; Island Press: Washington, DC, USA, 2005.

60. National Standard of China. *GB 50594-2010: Standard for Water Function Zoning*; Ministry of Housing and Urban-Rural Development: Beijing, China, November 2010. (In Chinese)

61. Holling, C.S. Engineering resilience versus ecological resilience. In *Engineering within Ecological Constraints*; Schulze, P., Ed.; National Academy Press: Washington, DC, USA, 1996; pp. 31–44.

62. Parsons, M.; Thoms, M.; Capon, T.; Capon, S.; Reid, M. *Resilience and Thresholds in River Ecosystems*; Waterlines Report; National Water Commission: Canberra, Australia, 2009.

63. Tharme, R.E. A global perspective on environmental flow assessment: Emerging trends in the development and application of environmental flow methodologies for rivers. *River Res. Appl.* **2003**, *19*, 397–441.

64. Kuriqi, A.; Pinheiro, A.N.; Sordo-Ward, A.; Garrote, L. Flow regime aspects in determining environmental flows and maximizing energy production at run-of-river hydropower plants. *Appl. Energy* **2019**, *256*, 113980.

65. Kuriqi, A.; Pinheiro, A.N.; Sordo-Ward, A.; Garrote, L. Water-energy-ecosystem nexus: Balancing competing interests at a run-of-river hydropower plant coupling a hydrologic–ecohydraulic approach. *Energy Convers. Manag.* **2020**, *223*, 113267. [CrossRef]

66. Acreman, M.C.; Ferguson, A.J.D. Environmental flows and the European water framework directive. *Freshw. Biol.* **2010**, *55*, 32–48.

67. Poff, N.L.; Zimmerman, J.K. Ecological responses to altered flow regimes: A literature review to inform the science and management of environmental flows. *Freshw. Biol.* **2010**, *55*, 194–205.

68. King, J.; Brown, C. Environmental flows: Striking the balance between development and resource protection. *Ecol. Soc.* **2006**, *11*, 1–26.

69. Hirji, R.; Davis, R. *Environmental Flows in Water Resources Policies, Plans, and Projects: Findings and Recommendations*; The World Bank: Washington, DC, USA, 2009.

70. Pahl-Wostl, C.; Arthington, A.; Bogardi, J.; Bunn, S.E.; Hoff, H.; Lebel, L.; Nikitina, E.; Palmer, M.; Poff, L.N.; Richards, K.; et al. Environmental flows and water governance: Managing sustainable water uses. *Curr. Opin. Environ. Sustain.* **2013**, *5*, 341–351. [CrossRef]

71. Standard of Water Conservancy Industry of China. *SL/Z 712/2014: Specification for Calculation of Environmental Flow in Rivers and Lakes*; Ministry of Water Resources: Beijing, China, 2014. (In Chinese)

72. Antunes, B.; Carvalho, L.; Geamana, N.; Giuca, R.; Grizzetti, B.; Leone, M.; Liquete, C.; McConnell, S.; Preda, E.; Santos, R.; et al. Ecosystem services for water policy: Insights across Europe. *Environ. Sci. Policy* **2016**, *66*, 179–190.

73. Zhu, Q.; Pan, Y.; Xia, J.; Jia, H.; Xi, J. Sensory Evaluation and Analysis of River Water Quality in Suzhou City. Unpublished manuscript. (In Chinese).

74. Karr, J.R. Rivers as sentinels: Using the biology of rivers to guide landscape management. In *The Ecology and Management of Streams and Rivers in the Pacific Northwest Coastal Ecoregion*; Naiman, P.P., Bilby, R.E., Eds.; Springer: New York, NY, USA, 1996.

75. Karr, J.R.; Chu, E.W. *Restoring Life in Running Waters: Better Biological Monitoring*; Island Press: Washington, DC, USA, 1998.

 *water* 

Article

An Investigation of Stormwater Quality Variation within an Industry Sector Using the Self-Reported Data Collected under the Stormwater Monitoring Program

Maryam Salehi *, Khashayar Aghilinasrollahabadi and **Mitra Salehi Esfandarani**

Department of Civil Engineering, University of Memphis, Memphis, TN 38152, USA;
kghlnsrl@memphis.edu (K.A.); mslhsfnd@memphis.edu (M.S.E.)
* Correspondence: mssfndrn@memphis.edu

Received: 22 October 2020; Accepted: 10 November 2020; Published: 14 November 2020

Abstract: Storm runoff pollutants are among the major sources of surface water impairments, globally. Despite several monitoring programs and guidance on stormwater management practices, there are many streams still impaired by urban runoff. This study evaluates an industry sector's pollutant discharge characteristics using the self-reported data collected under Tennessee Multi Sector Permit program. The stormwater pollutant discharge characteristics were analyzed from 2014 to 2018 for an industry sector involving twelve facilities in West Tennessee, USA. The data analysis revealed the presence of both organic and inorganic contaminants in stormwater samples collected at all twelve industrial facilities, with the most common metals being magnesium, copper, and aluminum. The principal component analysis (PCA) was applied to better understand the correlation between water quality parameters, their origins, and seasonal variations. Furthermore, the water quality indexes (WQIs) were calculated to evaluate the stormwater quality variations among studied facilities and seasons. The results demonstrated slight variations in stormwater WQIs among the studied facilities ranging from "Bad" to "Medium" quality. The lowest seasonal average WQI was found for spring compared to the other seasons. Certain limitations associated with the self-reported nature of data were identified to inform the decision makers regarding the required future changes.

Keywords: stormwater; industrial facilities; BMPs; water quality; run off; self-reported data

1. Introduction

Storm runoff pollutants are among the major sources of surface water impairments, globally [1–3]. However, there are several types of guidance on stormwater best management practices (BMPs) and pollution prevention plans (SWPPPs), but there are many streams still impaired by runoff contaminants [4–6]. Storm runoff from urban areas, roads, agricultural and constructional sites, atmospheric depositions, and acid drainage from abandoned mines are considered as the major sources of pollution threatening the surface water quality [6]. Deteriorated stormwater quality is linked to the human health due to the acute and chronic illnesses from exposure through drinking water, seafood, and contact recreation [7–9]. To protect and enhance the quality of ground and surface water resources as a top priority for both public and governments, it is essential to understand the pollutant discharge by different sources. Stormwater pollutants originate from a variety of sources in the urban environment including residential and commercial landscapes, construction sites, roads and highways, parking lots, and industrial sites [10,11].

The proportion of pollutants in urban stormwater that originated from industrial activities has been reported to be significant compared to the other sources [12]. Manufacturing, shipping, and storage operations that are exposed to the rain release several pollutants such as heavy metals, nitrate, and organic materials into the stormwater. Despite the extensive research in stormwater quality at urban areas like roads, residential or construction sites [8–10], industrial facilities' runoff constituents, pollutants loading, and their impacts on receiving waters are less understood. Few studies conducted in this area reported the large numbers of organic and inorganic constituents' presence in the industrial facilities' stormwater [13,14]. Concentration of certain pollutants like heavy metals in industrial facilities' storm runoff has been found to be significantly greater than that from other land uses. Additionally, human health risks associated with stormwater from the industrial area is considerably greater than a residential or commercial area [15].

Several federal and state stormwater monitoring programs are implemented to encourage or enforce many types of industrial facilities taking steps toward protecting their stormwater quality [16,17]. These monitoring programs mainly aimed to identify the high risk dischargers and eventually reduce the stormwater pollution. However, due to the requirement and design of these monitoring programs, the precise data were not generated generally to satisfy the decision-making needs [18]. As demonstrated by Clean Water Act 303(d) section, numerous streams and surface waters in Tennessee have been listed as impaired by runoff contaminants [19]. As a part of the Tennessee Multi Sector Permit (TMSP), qualified industrial facilities located in Tennessee are required to develop the site-specific Stormwater Pollution Prevention Plan (SWPPP) to maintain the stormwater quality. The expected stormwater pollution sources are identified in SWPPP and the planned strategies to minimize the pollutant release to stormwater are described. As a part of the National Pollutant Discharge Elimination System (NPDES), TN Multi-Sector Stormwater Permit (TMSP), each facility is required to collect the stormwater run-off samples, conduct the laboratory analysis to quantify the certain constituents in stormwater, and report the analytical results to the Tennessee Department of Environment and Conservation (TDEC), annually. Table 1 demonstrates the benchmark's stormwater quality levels listed in TMSP, in addition to the associated USEPA drinking water and acute aquatic life limits for both fresh and saltwater. The purpose of this annual monitoring is both to assist the regulatory agency in understanding the effectiveness of management practices and evaluate aggregate pollutant loading by an individual facility to the watershed, and to help the facility operators understand their BMPs' effectiveness. However, limited collected data along with a lack of understanding of the long-term stormwater quality variations obscure the understanding of the BMPs' performance. Evaluation of regulations that impact thousands of industries would not be possible without understanding industrial facilities' stormwater pollutants' loadings to the watersheds. This study aimed to better understand an industry sector's pollutant discharge characteristics using the self-reported data collected under TMSP program. Specific objectives were to (1) investigate the stormwater quality characteristics in twelve industrial facilities within an industry sector, (2) examine the seasonal and temporal variations of stormwater quality within this industry sector, and (3) identify the limitations associated with the self-reported nature of stormwater quality to inform the decision makers and monitoring agencies regarding the required future changes.

Table 1. Tennessee Department of Environment and Conservation (TDEC) Tennessee Multi Sector Permit (TMSP) stormwater quality benchmark levels in 2018 and USEPA limits for drinking water and aquatic life ([1] SMCL: Secondary Maximum Contaminant Level; [2] NL: No Limit; [3] USEPA Action Limit, [4] NO_2^- as N; [5] NO_3^- as N).

Parameter	TDEC	Drinking Water	Aquatic (Acute Levels)	
			Fresh	Salt
Al (µg/L)	750	50 to 500 [1]	NL [2]	NL
Cd (µg/L)	16	5 [3]	2	33
Cyanide (µg/L)	64	200 [3]	22	1
Cu (µg/L)	18	1000 [1]	NL	5
Pb (µg/L)	150	15 [3]	82	140
Ag (µg/L)	32	100 [1]	3	2

Table 1. *Cont.*

Parameter	TDEC	Drinking Water	Aquatic (Acute Levels)	
			Fresh	Salt
Zn (µg/L)	395	5000 [1]	120	90
As (µg/L)	169	100 [3]	340	69
Se (µg/L)	239	50 [3]	NL	290
Mg (µg/L)	64	NL	NL	NL
Hg (µg/L)	2	2 [3]	1	2
Ni (µg/L)	875	NL	470	74
Fe (mg/L)	5	0.3 [1]	NL	NL
COD (mg/L)	120	NL	NL	NL
BOD_5 (mg/L)	30	NL	NL	NL
Oil and Grease (mg/L)	15	NL	NL	NL
NH_4^+ (mg/L)	4	NL	NL	NL
$NO_2^- + NO_3^-$ (mg/L)	0.7	1 [4], 1 [5]	NL	NL
Phosphorous (mg/L)	2	NL	NL	NL
pH	5–9	6.5–8.5	6.5–9.0	6.5–8.5
TSS (mg/L)	150	NL	NL	NL
Fluoride (mg/L)	1.8	4 [3]	NL	NL

2. Methods

2.1. Data Sources

The industrial facilities under TMSP permit are required to collect stormwater samples and conduct laboratory analysis to quantify certain constituents in stormwater, and report to the monitoring agency, annually. In this study, the self-reported stormwater quality for 12 industrial facilities within an industry sector with active coverage under TMSP during the period of 2014–2018 was analyzed. This industry sector's activities mainly consist of storage of utility related equipment as summarized in Table 2. The stormwater quality for this industrial sector was selected to be monitored quarterly instead of the usual annual monitoring requirement. The stormwater management team at the targeted industrial facilities collected the grab stormwater samples at designated outfalls at different drainage areas through the facilities. Further information regarding the location of outfall and potential pollution sources in each drainage area is described in the facilities' SWPPPs. The TMSP requires collection of the grab samples within the first 30 min of discharge at the outfall. Additionally, if more than one sample is collected and analyzed, the average concentration of that parameter should be reported. The water samples were mostly examined for pH, aluminum (Al), copper (Cu), iron (Fe), lead (Pb), magnesium (Mg) and zinc (Zn), chemical oxygen demand (COD), and biochemical oxygen demand (BOD_5) concentrations, ammonia (NH_4^+), total suspended solids (TSS), nitrate and nitrite nitrogen ($NO_2^- + NO_3^-$), and oil and grease concentrations. Additional information about these industrial sites was sought through other compliance documents such as Notices of Intent (NOI), Inspection Reports, SWPPPs, and through TDEC-Division of Water Resources, Memphis Environmental Field Office. Occasionally some stormwater quality data were not reported to the regulatory agency due to the dry seasons and lack of enough rain to collect the stormwater samples. The list of available self-reported data for the studied industrial facilities is shown in the supplementary materials (Table S1). Accounting for all 12 facilities, in total, 9363 data were analyzed in this study, but in total, no data were reported for 34% of quarters from 2014 to 2018. The greatest number of missing data belonged to 2015. Calculating the number of outfalls per industrial area revealed the lowest number of outfalls in facilities A9 (1 outfall/100,000 m^2), and the greatest number of outfalls was present in facility A8 and A10 (51 outfall/100,000 m^2). It should be noted that the reported area is only the area of property associated with the industrial activities and does not include the recreation area, office buildings, landscaping, employee parking, etc. This industry sector has been regulated under sector AD of the

TMSP, that requires monitoring for pH, BOD_5, COD, TSS, oil and grease, and ammonia concentrations in stormwater.

Table 2. The list of studied industrial facilities, their industrial area, and activities exposed to stormwater.

Site Label	Area (m^2)	Activity
A1	8903	Storage/Distribution Vehicle Storage Outside Waste Disposal
A2	2833	Storage/Distribution Outside Waste Disposal
A3	2428	
A4	70,415	Storage/Distribution
A5	118,978	
A6	6475	
A7	101,576	Vehicle Storage
A8	9712	
A9	441,512	Outside Waste Disposal
A10	9712	
A11	214,888	Vehicle Maintenance
A12	47,348	

2.2. Statistical Analysis

The statistical analysis was conducted using IBM SPSS version 26 software. The Normal distribution of stormwater quality data has been evaluated using Kolmogorov–Smirnov test [20]. Due to non-normal distribution, the non-parametric test of Mood's Median was applied to examine the significant variations of stormwater quality data [20,21]. The non-parametric Kruskal Wallis test with multiple comparisons was followed if only a significant difference between the median of tested variables was assessed using the Mood's Median test [22]. The difference was considered significant if p-value < 0.05.

The principal component analysis (PCA) was implemented to identify the stormwater pollution sources and correlation between stormwater quality parameters. Furthermore, PCA was used to assess seasonal correlations of stormwater quality parameters. For this purpose, the data were divided into four different temporal databases of winter (quarter 1), spring (quarter 2), summer (quarter 3), and fall (quarter 4). Due to a large number of missing water quality parameters at facilities A1, A2, A3, A6, and A9, these facilities were not considered for the PCA analysis. The PCA was only conducted for facilities A4, A5, A7, A8, A10, A11, and A12. The water quality parameters including Al, NH_4^+, Cu, Fe, Pb, Zn, COD, BOD_5, TSS, pH, $NO_2^- + NO_3^-$, oil and grease concentrations were utilized for the PCA analysis. As data were not normally distributed (p-value < 0.05), the lognormal transformation was applied to normalize the data prior to the PCA analysis. This transformation advantageously scaled the data to the range of 0 to 1. In PCA, the number of extracted components is equal to the number of input variables. There were 13 water quality characteristics, so 13 components were extracted, however, only the components with large total variance were considered as the most significant. Kaiser's criteria was used to identify the number of significant principal components [23].

2.3. Water Quality Index (WQI) Calculation

The water quality index (WQI) was utilized to account for a large number of water quality parameters into the simplest form in terms of water quality classification [24,25]. The *WQI* that was defined by Pesce and Wunderlin (2000), Reference [26] and calculated using Equation (1) was used in this study. The C_i is the value assigned to the parameter i after normalization, n is the number of water quality parameters, and P_i is the relative weight associated with ith water quality parameter. The relative weight (P_i) varies from 1 to 4, where the most impactful water quality parameter has the relative weight of 4. The relative weight of water quality parameters was adapted from the literature [26]. The water quality index was graded as excellent (91–100), good (71–90), medium (51–70), bad (26–50), and very bad (0–25) [24]. The average *WQIs* were calculated across different facilities and during different seasons using Equation (1).

$$WQI_{obj} = \frac{\sum_{i=1}^{n} C_i P_i}{\sum_{i=1}^{n} P_i} \tag{1}$$

3. Results and Discussion

3.1. The Industry Sector's Stormwater Quality Evaluation

Analyzing the self-reported data revealed that both organic and inorganic contaminants were released to the stormwater at studied industrial facilities. The percentage of collected data exceeded the state regulatory agency benchmark levels is shown in Table 3. The variation of the median concentrations of BOD_5, COD, NH_4^+, $NO_2^- + NO_3^-$, oil and grease and TSS across 12 different facilities within the studied industry sector is shown in Figure 1. These data revealed that oil and grease concentrations were mostly below the benchmark level (15 mg/L), despite the fact that some of these facilities were utilized for the vehicle storage and maintenance activities (Table 2). The median pH value at studied industrial facilities varied between 7.8 and 8.2. The individual stormwater samples' pH levels were generally within the USEPA recommended pH values of 6.5 to 9.0 for fresh water, except for some elevated pHs (>9.0) at facilities A4, A5, and A8. The median TSS concentration showed a very wide range of variations from 40 to 334 mg/L. The unvegetated and unpaved areas are known as the main sources contributing to the TSS in stormwater [27]. Furthermore, reviewing the recent inspection reports at facility A4 revealed the contribution of fill material storage sites to silt/sand release to the stormwater. The median NH_4^+ concentration varied between 0.2 and 0.6 mg/L, which might be released to the stormwater by decaying the nitrogen-containing organic matters. The median $NO_3^- + NO_2^-$ concentration varied from 0.4 to 1.3 mg/L as N. The excessive $NO_3^- + NO_2^-$ presence in freshwater could negatively impact the aquatic organisms [28]. The organic contaminant loadings in stormwater at studied industrial facilities were evaluated through COD and BOD_5 concentrations. The median BOD_5 concentrations varied between 8.0 and 20.0 mg/L, and median COD concentration varied between 55.0 and 199.0 mg/L. As demonstrated in Table 3, a greater number of stormwater samples exceeded the COD concentration benchmark level (120 mg/L) compared to the BOD_5 concentrations benchmark level (30 mg/L). The strong significant correlation ($r = 0.83$) between median COD and BOD_5 concentrations indicates that the major part of organic components was biodegradable. The correlation matrix for stormwater quality parameters is shown in Supplementary materials (Table S2).

Table 3. Percentage of data that exceeded the TDEC benchmark stormwater quality levels (NR: not reported).

	Water Quality		Facilities											
		A1	A2	A3	A4	A5	A6	A7	A8	A9	A10	A11	A12
Metals	Al	NR	NR	NR	81	87	64	64	55	NR	86	86	93
	Cu	NR	46	NR	70	52	NR	40	61	NR	7	77	90
	Fe	NR	NR	NR	27	50	NR	26	0	NR	45	44	NR
	Pb	NR	NR	NR	5	6	NR	NR	NR	NR	9	NR	NR
	Zn	NR	NR	25	25	23	NR	NR	49	NR	50	NR	33
	Mg	100	100	100	100	100	100	100	100	100	100	100	100
Organics	COD	31	46	33	65	48	45	41	24	26	36	63	57
	BOD_5	0	38	8	32	28	9	18	7	6	10	34	31
	Oil and Grease	0	8	0	4	5	0	2	4	2	2	3	6
Others	TSS	31	31	33	64	59	45	40	28	36	60	72	62
	pH	8	8	8	15	19	9	2	14	7	10	11	7
	NH_4^+	0	0	0	0	0	0	2	0	0	0	0	0

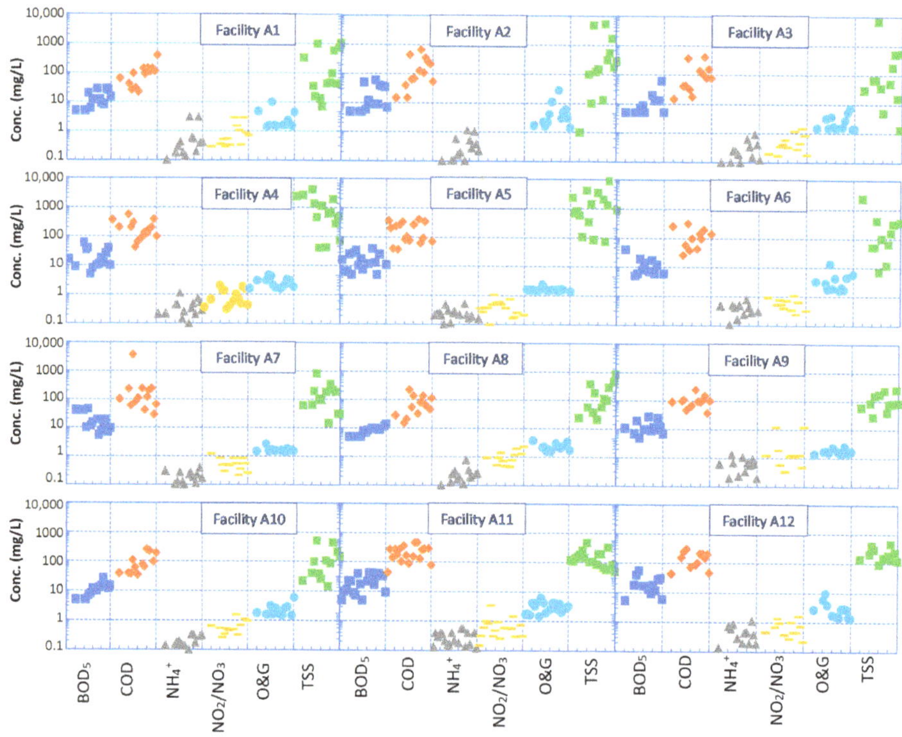

Figure 1. Scatter diagram of the quarterly median values for biochemical oxygen demand (BOD$_5$) (mg/L), chemical oxygen demand (COD) (mg/L), NH$_4^+$ (mg/L), NO$_2$ + NO$_3$ (mg/L as N), Oil and Grease (mg/L), and total suspended solids (TSS) (mg/L) concentrations at 12 industrial facilities from 2014 to 2018.

The variation of metals concentration within the studied industrial facilities is shown in Table 4. The total Mg, Al, and Cu mostly exceeded the benchmark levels (64 µg/L, 750 µg/L, 18 µg/L). The greater levels of Mg and Al in stormwater samples might be associated with their higher abundance in the silty soils which is common in this region [29,30]. Figure 2C shows the erosion around the facility (A3) fence, which may have contributed to the TSS presence in stormwater samples. A low percentage of stormwater samples collected at facilities A4, A5, A10 exceeded the benchmark Pb level (150 µg/L) although, Pb was not monitored at the other facilities. The equipment related to metallic utility like transformers (Figure 2A), pipes (Figure 2B), and rusted metal scrape (Figure 2D) may also contributed to the metal release to stormwater. This finding is in agreement with the literature that reported metals as the major contaminant released by industrial sources to the storm runoff [18]. The self-reported nature of data may result in insufficient quality controls which makes identification of the meaningful trends challenging [12]. Our analysis suggested that TSS should be considered as the priority in developing future stormwater pollution prevention plans for this industry sector, as it mostly exceeded the recommended benchmark levels. Furthermore, the metals such as Al and Mg which are associated with silt/sand presence in stormwater frequently exceeded the benchmark levels.

Table 4. The statistical description of metals concentration in stormwater samples collected at studied industrial facilities (STD: standard deviation, NR: not reported).

Facility Conc. (mg/L)		A1	A2	A3	A4	A5	A6	A7	A8	A9	A10	A11	A12
Mg	Median	1.66	1.35	1.09	4.80	4.87	1.44	2.31	0.80	2.45	6.93	3.44	1.67
	STD	2.24	4.10	9.53	18.9	29.77	3.05	7.34	1.17	40.78	19.75	10.59	2.28
	95tile	5.06	11.88	22.00	37.5	35.50	7.94	13.06	3.89	51.20	49.64	25.98	4.05
Al	Median	NR	NR	NR	2.25	3.82	1.04	2.00	0.98	NR	2.69	3.04	3.40
	STD	NR	NR	NR	3.32	11.11	2.19	4.30	1.51	NR	7.25	4.52	13.50
	95tile	NR	NR	NR	10.10	32.60	5.51	14.16	5.00	NR	20.42	14.36	22.78
Cu	Median	NR	0.02	NR	0.3	0.02	NR	0.01	0.04	NR	0.02	0.03	0.05
	STD	NR	0.02	NR	0.05	0.06	NR	0.02	0.05	NR	0.06	0.69	0.09
	95tile	NR	0.07	NR	0.11	0.17	NR	0.08	0.11	NR	0.12	0.23	0.31
Pb	Median	NR	NR	NR	0.02	0.02	NR	NR	NR	NR	0.04	NR	NR
	STD	NR	NR	NR	0.34	0.05	NR	NR	NR	NR	0.1	NR	NR
	95tile	NR	NR	NR	0.15	0.16	NR	NR	NR	NR	0.2	NR	NR
Zn	Median	NR	NR	0.21	0.23	0.18	NR	NR	0.39	NR	0.38	NR	0.28
	STD	NR	NR	0.83	0.44	0.47	NR	NR	0.53	NR	0.98	NR	0.40
	95tile	NR	NR	1.95	1.25	1.25	NR	NR	1.70	NR	1.57	NR	1.36
Fe	Median	NR	NR	NR	2.71	4.96	NR	2.18	0.10	NR	4.53	3.98	NR
	STD	NR	NR	NR	4.82	18.36	NR	6.38	0.04	NR	49.46	7.79	NR
	95tile	NR	NR	NR	13.86	49.57	NR	20.30	0.17	NR	38.15	20.20	NR

Figure 2. Various sources contributing to storm runoff contamination at (**A**) Facility A1, (**B,C**) Facility A3, and (**D**) Facility A6 (Photos obtained from State Regulatory Agency's inspection reports in 2015).

The statistical description of self-reported stormwater quality data for the studied industry sector is shown in Table 5. A comparison has been made between the stormwater quality at the studied industry sector and the data reported in the literature for several other industrial facilities, commercial and residential sites (supplementary materials Table S3). No record was found describing the stormwater quality for residential or commercial areas in the studied region thus, we compared our data with the other published literature. As demonstrated in Table S3, the Al, Pb, TSS, and $NO_2^- + NO_3^-$ concentrations were greater in the studied industry sector than values reported by Lau et al. (2001) for residential and commercial sites [31]. The stormwater quality investigation for the industrial facilities is limited in the literature and the type of industrial activities, and water sampling procedures were different which made the comparison challenging [19,32–34]. The research investigated the stormwater quality at an industrial log yard during eight run-off events that resulted in greater levels of heavy metals, TSS, and COD concentrations than our investigation [33]. However, sampling after eight events might be too small to capture the variation of pollutant release to the stormwater by sources present in the industry sector. The study conducted by Line et al. (1997), examined the first flush runoff quality at 20 industrial sites and found Zn and Cu as the most common metals found at these sites [13]. Although, our analysis revealed Mg, Cu, and Al as the most common metals present in stormwater at the 12 studied industrial facilities. It should be noted that rather than different industrial activities, their data acquisition was more controlled than self-reported data which are utilized in our study. These findings underscore the critical impacts of land uses and data acquisition practices on stormwater quality evaluations, which should be considered in the future urban planning and development of stormwater monitoring programs.

Table 5. The statistical description of stormwater quality parameters within the studied industry sector.

Parameter (mg/L)	n	Mean	STDV	CV%	Median	95th Percentile
Al	646	4.7	8.0	168	2.4	15.7
Cu	585	0.1	0.3	463	0.03	0.2
Fe	427	6.8	10.8	158	3.2	23.6
Pb	249	0.1	0.3	428	0.0	0.2
Zn	446	0.4	0.6	131	0.3	1.5
Mg	784	8.8	20.3	231	3.1	33.8
COD	783	208.8	332.2	159	118	600.0
BOD_5	787	23.2	34.1	147	12.0	60.0
Oil and Grease	779	5.0	16.9	338	1.8	11.6
TSS	779	428.2	969.2	226	191	1415
pH	761	7.8	1.1	14	8.0	9.2
NH_4^+	788	0.4	0.5	116	0.3	1.1
$NO_2^- + NO_3^-$	770	1.7	10.0	576	0.6	2.5

The metals released to the storm runoff at the studied facilities may not only be originated from the stored scrape metals or erosion materials (Figure 2C,D), but they also could be released from building materials, as previously suggested by Davis et al. (2001) [31]. In facility A11, the drainage area included the parking lots, storage area for drums, old pipes, wooden poles, and metal spools, as well as a paved driveway and warehouse. The main contaminants at this facility were Mg and Al which possibly were released due to the silt and sand release to the stormwater. The median concentration of Mg at this facility was the greatest among all facilities and it was 111 times greater than the benchmark level (64 µg/L). The median concentration of Al at this facility (3.0 mg/L) was four and 15 times greater than the TDEC benchmark level and USEPA drinking limit, respectively. The drainage area at facility A4 included the storage area for the equipment, treated wood, drums, metal scrap, and pipes. In addition, the recent inspection report of facility A4 indicated the silt and sand release from the fill material storage area. In this facility, Mg and Al median concentrations were 75 and three times greater than TDEC benchmark levels, respectively. The statistical analysis revealed significant differences among all organic and inorganic contaminants found in stormwater at these 12 industrial sites (p-value < 0.05) but not the Fe (Supplementary materials Tables S4 and S5). The greatest median

Pb and Zn levels (0.04 mg/L and 0.39 mg/L) were found in facility A10, however, the greatest median Fe and Al levels (5.0 mg/L and 3.8 mg/L) were identified in facility A5. Though all water quality data were not available for all facilities, the greatest median Mg and Cu levels (7.1 mg/L and 0.05 mg/L) belonged to facility A11 and A12. The high variation of metals concentration among studied facilities prevented identifying a specific trend in stormwater quality deterioration. It was noted by the State Regulatory Agency that the main activities at each facility were stagnant, however, ancillary activities, along with their associated pollutant sources, may have been changed daily. No record was found describing the types of ancillary activities, thus, it was challenging to identify the potential sources for the identified stormwater pollutants.

The variation of stormwater quality within the industrial facilities was investigated in terms of Water Quality Index (WQI). As demonstrated in Figure 3, the water quality within this industry sector varied from "Bad" to "Medium". The greatest water quality within this industry sector was evaluated as "Medium" and belonged to facilities A1, A3, A6, A7, A8, and A9 ($50 \leq WQI \leq 58$). The stormwater quality at the rest of the studied facilities was evaluated as "Bad" ($42 \leq WQI \leq 49$). The WQI at facility A3 was significantly greater than WQI at facilities A5, A10, and A11 (p-value < 0.05). The lowest average WQI was found for facility A11 as 42. It should be noted that as the authors followed the literature, some of the water quality parameters like metals concentration (Pb, Cu, Fe, Zn, Al) were not considered in WQI calculation, however those may impact the variation among facilities [26]. This information could be utilized by the decision makers in the industry sector developing their priority list to modify and develop their future SWPPPs. To have a complete assessment of reasons behind the variation of stormwater quality, the detailed description of the type of industrial activities conducted at each site is required. However, the SWPPPs that contain more descriptions regarding the facility characteristics and activities resulting in stormwater contamination were only accessible by authors for facilities A3, A4, and A11.

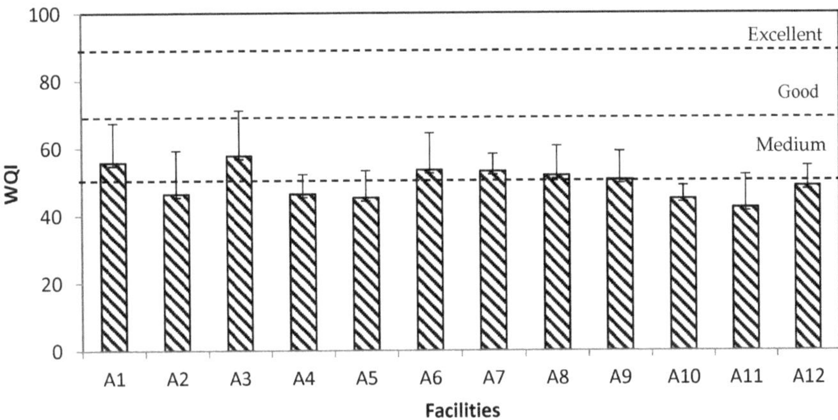

Figure 3. The variation of stormwater water quality index (WQI) at studied industrial facilities.

3.2. Implementation the Principle Component Analysis (PCA) to Identify the Stormwater Pollution Sources

As demonstrated in the scree plot of components (Supplementary materials Figure S1), only three components were retained which had the Eigenvalues greater than one. The PC loadings were categorized as "strong", "moderate", and "weak" according to their absolute loading values of >0.75, 0.75–0.50, and <0.5, respectively [23]. The loadings of water quality variables on principal components are shown in Table 6. With the Kaiser normalization, these components explained 74% of the total variance. The first component (PC1), accounting for 50% of the total variance and was strongly loaded by several inorganic contaminants (Al, Cu, Fe, Pb, Mg, TSS, and Zn). This factor could be attributed to the contaminant release by storage of utility-related equipment and fill materials at the industrial facilities. The second component (PC2), accounting for 17% of the total variance and showed a strong positive loading of nitrogenous contaminants (NH_4^+, $NO_2^- + NO_3^-$) and moderate loading with BOD_5 substances. This factor might be associated with biogenic pollution sources [23]. However, the third component (PC3), accounting only for 8% of the total variance and showed a strong positive loading by water pH, and negative weak loading on metals present in stormwater.

Table 6. The loading of water quality variables on principal components.

Variables	Components												
	1	2	3	4	5	6	7	8	9	10	11	12	13
Al	0.87	−0.28	−0.06	0.16	−0.16	−0.08	−0.13	0.06	0.23	−0.08	−0.02	0.12	−0.12
NH_4^+	0.23	0.79	0.04	0.33	0.04	−0.28	0.34	0.06	0.09	0.00	−0.06	−0.02	0.00
BOD_5	0.59	0.64	0.10	−0.21	−0.14	−0.20	−0.20	−0.07	0.01	0.23	0.18	0.02	0.00
COD	0.80	0.36	0.08	−0.13	−0.04	−0.16	−0.21	−0.12	−0.17	−0.24	−0.16	−0.08	0.00
Cu	0.82	0.02	−0.15	−0.04	0.38	−0.03	−0.08	0.35	−0.16	0.08	−0.03	0.07	0.00
Fe	0.90	−0.28	−0.14	0.13	−0.15	−0.05	−0.05	0.02	0.16	−0.03	−0.01	0.06	0.16
Pb	0.86	−0.26	−0.12	0.02	0.23	0.06	−0.01	−0.05	0.16	0.11	−0.01	−0.28	−0.02
Mg	0.88	−0.11	−0.08	0.11	−0.19	0.08	0.18	0.07	−0.17	−0.18	0.24	−0.07	−0.01
$NO_2^- + NO_3^-$	0.14	0.77	−0.01	0.41	0.06	0.44	−0.16	−0.03	0.03	−0.02	0.00	0.02	0.01
Oil and Grease	0.54	0.22	0.54	−0.52	−0.01	0.22	0.14	0.11	0.15	−0.06	−0.02	0.02	0.01
pH	0.08	−0.41	0.78	0.41	0.16	−0.10	−0.08	−0.02	−0.06	0.02	0.05	0.00	0.01
TSS	0.82	−0.19	0.08	0.09	−0.35	0.14	0.14	−0.04	−0.18	0.24	−0.16	0.02	−0.02
Zn	0.83	−0.09	−0.10	−0.09	0.37	0.03	0.16	−0.32	−0.03	−0.01	0.03	0.15	−0.01
Eigen value	6.44	2.25	1.00	0.85	0.59	0.43	0.35	0.27	0.26	0.22	0.15	0.14	0.04
% of Variance	49.57	17.3	7.67	6.55	4.56	3.33	2.71	2.05	1.98	1.73	1.19	1.04	0.32
Cumulative %	49.57	66.87	74.54	81.09	85.65	88.98	91.69	93.74	95.72	97.46	98.64	99.68	100

3.3. Seasonal Variation of Stormwater Quality at the Industrial Facilities

Considering the first two principal components, a general view of seasonal variation of stormwater quality parameters at studied industrial facilities was provided, as shown in Figure 4. The winter's first principal component described 52% of the total variance, in which all water quality parameters were positively influential. Winter's second principal component explained 19% of the total variance, in which NH_4^+, $NO_2^- + NO_3^-$, and pH were the most influential. Other inorganics like Fe, Zn, and Al were negatively impacted. In spring, water pH was slightly and negatively influential on the first component which explained 45% of total variance. In summer, the first component described 50% of the total variance, and was largely impacted by both organic and inorganic contaminants present in stormwater. Summer's second principal component which described 14% of the total variance was largely influenced by NH_4^+, $NO_2^- + NO_3^-$, and pH, however, other water quality parameters were not that influential. Fall's first component explains 54% of the total variance and revealed a similar pattern to winter as the majority of water quality parameters were influential. Its second component accounts for 12% of the total variance, and similar to the other seasons, was largely impacted by NH_4^+, $NO_2^- + NO_3^-$, and pH. This slight seasonal variability of stormwater quality could be due to environmental conditions like the number, duration, and intensity of precipitation events, the time between rainfalls, and the type of ancillary activities conducted at studied facilities [24,35].

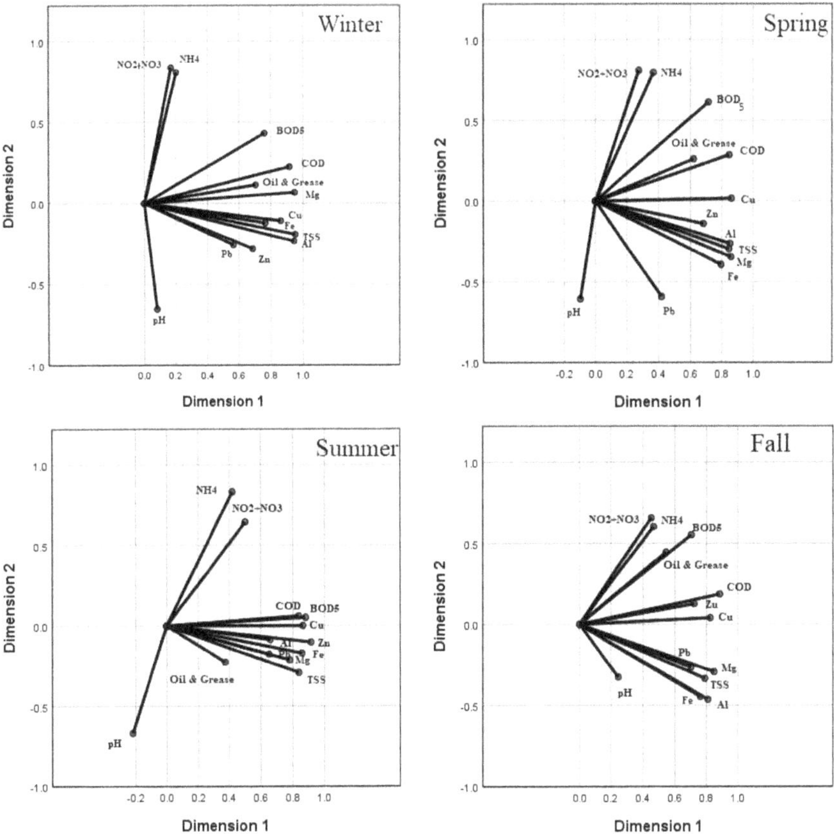

Figure 4. The seasonal coordinates of principal component axis based on the monitored stormwater quality data.

The seasonal variation of WQI was significant within this industry sector (p-value < 0.05). As demonstrated in Supplementary materials (Figure S2), most of the facilities experienced a slightly lower stormwater quality in spring (average WQI: 45) compared to the other seasons (average WQI: 50–51). However, it should be noted that heavy metals concentrations were not considered in calculation of WQI due to the large number of missing data, while they impacted the water quality. Analyzing the seasonal data revealed the lowest median concentrations of oil and grease, TSS, Al, Fe, Pb, Cu and Mg in stormwater samples collected during the summer (p-value < 0.05). The highest concentrations of NH_4^+, BOD_5 and COD concentrations were found in samples collected during spring with the median values of 0.4 mg/L, 18.0 mg/L, and 199.0 mg/L, respectively, which resulted in the lowest WQI for this season. This results are in a good agreement with the literature that reported a lower stormwater quality in spring due to the largest range of total phosphorous, dissolved phosphorus, soluble reactive phosphorous, COD, total Kjeldahl nitrogen (TKN), and total $NO_2^- + NO_3^-$ [36]. Few studies conducted in the Europe and U.S. reported the seasonality of stormwater pollutant loadings. The research investigated road runoff quality in Germany and highlighted a significant seasonal increase in Cu, Zn, TSS, pH, and TOC concentration during the cold seasons compared to the warm seasons [37]. The higher levels of heavy metals were also reported in urban storm runoff during spring and winter compared to summer [38]. The number of dry days before the storm event is a critical factor impacting the contaminant accumulation on the surface [39]. Investigating the temporal changes of stormwater quality during the study period (2014 to 2018) revealed the lowest median concentrations of oil and grease, Mg and TSS in 2015 compared to the other years. However, other stormwater quality parameters varied during the years, and no specific trend was identified. Calculation of WQIs using the data reported by this industry sector over the years revealed a better water quality in 2016 (average WQI:48), although it was not statistically different from other years (average WQI: 44 to 45). Rather than variation of type and extent of industrial activities, the alteration of rainfall pattern during the years may influence the stormwater quality characteristics [40]. The intensity and duration of rainfall influences dislodging and transporting the pollutant to stormwater [35].

4. Limitations and Policy Implications

Our review of published literature revealed that industrial facilities' stormwater quality characteristics are either investigated through a limited number of water samplings over a short duration of time [13], or through the large numbers of self-reported stormwater quality data collected over a longer period [12]. Conducting the systematic stormwater sampling is beneficial as it generates more reliable data source. However, collecting the limited number of water samples during the short duration of time limits the understanding of seasonal and temporal changes in contaminant loadings to the surface waters. Furthermore, limited or lack of access to the industrial facilities and inadequate financial resources mostly obscure the detailed and long-term investigation of pollutant discharge by industry sectors into the stormwater. Although a large volume of stormwater quality data are collected by industrial sectors to satisfy the regulatory agencies' monitoring requirements, only a few investigations have been conducted to examine the contaminant loadings and temporal variations in stormwater using these self-reported data. These self-reported data have been utilized to identify high polluting facilities, assess the pollutant loads to receiving water resources, and examine the facility improvement in reducing the pollutant discharges [41]. In our study due to the lack of access to the industrial sites, only the self-reported stormwater quality data and state regulatory agency compliance documents such as recent SWPPPs, inspection reports, and notices of violations were utilized. Thus, certain uncertainties associated with the self-reported nature of data like poor sampling or analysis practices should be considered. Although 9363 data points were analyzed in this study, there was a significant number of missing data (33.75%) that made the identification of specific trends difficult. Limited information regarding the type of industrial activities was reported in the Notice of Intent documents which created significant challenges in identifying the pollution sources. Despite the presence of heavy metals in stormwater at industrial facilities, this was not considered in the calculation of water quality index (WQI) as we applied the method reported in the literature [26]. However, future research is needed to specifically develop a WQI for industrial facilities' storm runoff, accounting for heavy metals and their appropriate relative weights in the calculations.

Decision makers' understanding of existing stormwater monitoring programs' effectiveness is limited due to the high variability of stormwater quality data. This variability even could rise due to the stormwater's natural variability, variation of industrial activities, poor sampling practices, and analysis. In our study, the state regulatory agency required the collection of grab samples within the first 30 min of discharge at outfall and not later than one hour, but sampling either late or early during storm event could have a significant impact on the result. Sampling at the beginning of storm event results in a collection of the first flush stormwater sample, which generally contains a greater concentration of contaminant [18]. This is an appropriate practice if the peak concentration of contaminant is required, otherwise the samples that are collected in the middle of storm event are more representative [18]. The duration of the dry period before the storm event could influence the level of stormwater contaminant. After a long dry period, the higher concentration of contaminant could be found in stormwater [42]. The results of this study suggested that the efficiency of current industrial stormwater monitoring programs could be improved by providing more detailed stormwater sampling guidelines and more strict requirements for the water quality reports to generate a more accurate database for future decision making and policy development. This research also highlights the importance of understanding stormwater quality data, to encourage industrial stakeholders to take serious steps toward protection of their stormwater quality. Industrial facilities' operators are encouraged to understand their stormwater quality data, detect the pollutants of concern, potential problems in contaminating the storm runoff and implement the efficient BMPs. In addition, our findings will assist the state and federal regulatory decision makers recognize the impacts of industrial activities on local creek and streams. It will allow the regulators to assess their progress toward managing the stormwater quality at industrial sites and protecting the watersheds to lower the pollutant discharge. The data could serve as the representative of industrial stormwater quality resulting from different types of activities to assist the regional, state, and local decision makers in protecting receiving waters effectively. We expect this work to catalyze the community and regulatory agencies into understanding this increasingly important, although not yet completely understood fact. Lack of information about the pollutants' types and their concentrations released to the stormwater by the specific type of industry, makes it difficult for the regulatory agencies to identify the specific pollutants that should be monitored. So, this research will provide background data about the type and concentrations of the pollutants we expect from a specific type of industry. In addition, the research outcomes could be used in issuing future industrial stormwater permits. Our study is aligned with the previous research that found high variability within the collected self-reported stormwater quality. The study conducted by Lee et al., (2007) evaluated several stormwater monitoring programs by investigating several data sets collected between 1991 and 2003 [18]. They found high variability in data collected by the industrial monitoring program and suggested reducing this variability from experimental error and artifacts in data collection to provide better guidance to the decision makers [18].

5. Conclusions

The pollutions originate from a variety of sources in industrial facilities are substantial contributors to surface water quality impairment. This study was conducted to better understand stormwater quality at 12 industrial facilities to evaluate the industry sector's pollutant discharge characteristics. In this study, the quarterly self-reported stormwater quality data reported by an industry sector during 2014–2018 were analyzed to identify the variation of stormwater quality within an industrial sector and examine the seasonal changes of stormwater quality. Implementation of principle component analysis (PCA) revealed three major components that were significantly loaded—inorganic contaminant (PC1), nitrogenous contaminant (PC2), and water pH (PC3)—and demonstrated their association with the industrial activities and biogenic pollution sources. A significant variation of stormwater quality parameters was found within the studied industrial facilities. The WQI calculation showed the variation of stormwater quality ranging from "Bad" to "Medium" quality among the industrial facilities. The seasonal variation of stormwater quality was analyzed using PCA and WQI calculations. The result demonstrated a lower water quality in spring compared to the other seasons. Several limitations were identified with respect to the self-reported nature of data and information on industrial activities. This research underscores the importance of understanding stormwater quality data to encourage the industry stakeholders to improve their best management practices to maintain the storm runoff quality. It also assists the policymakers in better understanding the impact of their regulatory strategies on protecting the water resources.

Supplementary Materials: The following are available online at http://www.mdpi.com/2073-4441/12/11/3185/s1. Table S1: The number of outfalls with available self-reported stormwater quality data. Table S2: The stormwater quality parameters correlation factors. Table S3: A comparison of median values of stormwater quality in studied industry sector with the levels reported in the literature. Table S4: The results of Mood's Median test. Table S5: The results of Kruskal–Wallis test and Normality test. Figure S1: The scree plot for the components. Figure S2: The seasonal variation of stormwater WQI at studied industrial facilities.

Author Contributions: Conceptualization, M.S.; methodology, M.S., M.S.E.; data analysis, K.A., M.S.; visualizations, K.A., M.S., M.S.E.; writing, M.S., K.A.; supervision, editing, funding acquisition, M.S. All authors have read and agreed to the published version of the manuscript.

Funding: This work is supported by U.S. Geological Survey (USGS) under grant No. G16AP00084.

Acknowledgments: The authors would like to thank Ronne Adkins, Joellyn Brazile, and Crystal Warren at the Tennessee Department of Environment and Conservation, Memphis Environmental Field Office for their feedback and assistance with the data resources. The authors also thank Raymonde Bell for her assistance with the data collection and Tanvir Ahamed for assistance with the data demonstrations. This work is supported by U.S. Geological Survey (USGS) under grant No. G16AP00084. The views and conclusions contained in this document are those of the authors and should not be interpreted as representing the opinions or policies of the U.S. Geological Survey. Mention of trade names or commercial products does not constitute their endorsement by the U.S. Geological Survey.

Conflicts of Interest: The authors declare no conflict of interest.

References

1. Rădulescu, D.; Racovițeanu, G.; Swamikannu, X. Comparison of urban residential storm water runoff quality in Bucharest, Romania with international data. *E3S Web Conf.* **2019**, *85*, 07019. [CrossRef]
2. Bichai, F.; Ashbolt, N. Public health and water quality management in low-exposure stormwater schemes: A critical review of regulatory frameworks and path forward. *Sustain. Cities Soc.* **2017**, *28*, 453–465. [CrossRef]
3. Jeong, H.; Choi, J.Y.; Lee, J.; Lim, J.; Ra, K. Heavy metal pollution by road-deposited sediments and its contribution to total suspended solids in rainfall runoff from intensive industrial areas. *Environ. Pollut.* **2020**, *265*, 115028. [CrossRef] [PubMed]
4. Toronto Region Conservation Authority (TRCA). *Low Impact Development Stormwater Management Practice Inspection and Maintenance Guide Version 1.0; Toronto.* 2016. Available online: https://sustainabletechnologies. ca/app/uploads/2016/08/LID-IM-Guide-2016-1.pdf (accessed on 11 November 2020).
5. DayWater. *Report 5.1. Review of the Use of Stormwater BMPs in Europe*; Middlesex University: London, UK, 2003; Available online: http://daywater.enpc.fr/www.daywater.org/REPORT/D5-1.pdf (accessed on 11 November 2020).
6. He, J.; Valeo, C.; Chu, A.; Neumann, N.F. Characterizing Physicochemical Quality of Storm-Water Runoff from an Urban Area in Calgary, Alberta. *J. Environ. Eng.* **2010**, *136*, 1206–1217. [CrossRef]
7. Gan, H.; Zhuo, M.; Li, D.; Zhou, Y. Quality characterization and impact assessment of highway runoff in urban and rural area of Guangzhou, China. *Environ. Monit. Assess.* **2007**, *140*, 147–159. [CrossRef] [PubMed]
8. Gaffield, S.J.; Goo, R.L.; Richards, L.A.; Jackson, R.J. Public Health Effects of Inadequately Managed Stormwater Runoff. *Am. J. Public Health* **2003**, *93*, 1527–1533. [CrossRef] [PubMed]
9. Mann, A.G.; Tam, C.C.; Higgins, C.D.; Rodrigues, L.C. The association between drinking water turbidity and gastrointestinal illness: A systematic review. *BMC Public Health* **2007**, *7*, 1–7. [CrossRef]
10. Xu, C.; Jia, M.; Xu, M.; Long, Y.; Jia, H. Progress on environmental and economic evaluation of low-impact development type of best management practices through a life cycle perspective. *J. Clean. Prod.* **2019**, *213*, 1103–1114. [CrossRef]
11. Aghilinasrollahabadi, K.; Salehi, M.; Fujiwara, T. Investigate the Influence of Microplastics Weathering on Their Heavy Metals Uptake in Stormwater. *J. Hazard. Mater.* **2020**, 124439. [CrossRef]
12. Duke, L.D.; Buffleben, M.; Bauersachs, L.A. Pollutants in storm water runoff from metal plating facilities, Los Angeles, California. *Waste Manag.* **1998**, *18*, 25–38. [CrossRef]
13. Line, D.E.; Wu, J.; Arnold, J.A.; Jennings, G.D.; Rubin, A.R. Water quality of first flush runoff from 20 industrial sites. *Water Environ. Res.* **1997**, *69*, 305–310. [CrossRef]
14. Lee, H.; Stenstrom, M.K. Utility of Stormwater Monitoring. *Water Environ. Res.* **2005**, *77*, 219–228. [CrossRef] [PubMed]
15. Ma, Y.; Egodawatta, P.; McGree, J.; Liu, A.; Goonetilleke, A. Human health risk assessment of heavy metals in urban stormwater. *Sci. Total Environ.* **2016**, *26*, 764–772. [CrossRef] [PubMed]
16. USEPA. National Pollutant Discharge Elimination System (NPDES). Available online: https://www.epa.gov/ npdes/stormwater-discharges-industrial-activities%0A%0A (accessed on 7 November 2020).
17. USEPA. Developing Your Stormwater Pollution Prevention Plan A Guide for Industrial Operators June 2015. Available online: https://www.epa.gov/sites/production/files/2015-11/documents/swppp_guide_industrial_ 2015.pdf (accessed on 7 November 2020).
18. Lee, H.; Swamikannu, X.; Radulescu, D.; Kim, S.-J.; Stenstrom, M.K. Design of stormwater monitoring programs. *Water Res.* **2007**, *41*, 4186–4196. [CrossRef]

19. USEPA. Final YEAR 2016 303 (d) LIST. 2017. Available online: https://www.adeq.state.ar.us/water/planning/integrated/303d/pdfs/2016/final-2016-303d-list.pdf (accessed on 7 November 2020).

20. Ullah, A.; Arshad, M.; Kächele, H.; Khan, A.; Mahmood, N.; Müller, K. Information asymmetry, input markets, adoption of innovations and agricultural land use in Khyber Pakhtunkhwa, Pakistan. *Land Use Policy* **2020**, *90*, 104261. [CrossRef]

21. Lien, L.T.Q.; Chuc, N.T.K.; Hoa, N.Q.; Lan, P.T.; Thoa, N.T.M.; Riggi, E.; Tamhankar, A.J.; Lundborg, C.S. Knowledge and self-reported practices of infection control among various occupational groups in a rural and an urban hospital in Vietnam. *Sci. Rep.* **2018**, *8*, 1–6. [CrossRef]

22. Więcław, H.; Kurnicki, B. Morphological variation of Platanthera chlorantha (Orchidaceae) in forest sites of NW Poland. *Acta Biol.* **2016**, *23*, 139–149. [CrossRef]

23. Barakat, A.; El Baghdadi, M.; Rais, J.; Aghezzaf, B.; Slassi, M. Assessment of spatial and seasonal water quality variation of Oum Er Rbia River (Morocco) using multivariate statistical techniques. *Int. Soil Water Conserv. Res.* **2016**, *4*, 284–292. [CrossRef]

24. Zeinalzadeh, K.; Rezaei, E. Determining spatial and temporal changes of surface water quality using principal component analysis. *J. Hydrol. Reg. Stud.* **2017**, *13*, 1–10. [CrossRef]

25. Mohd Hafiyyan, M.; Lee, K.E.; Mazlin, M.; Marfiah, A.W.; Goh, T.L.; Norbert, S.; Maria, M.H.; Azhar, A.H. Spatial distribution of water quality index in stormwater channel: A case study of Alur Ilmu, UKM Bangi campus. *Asia Pacific Environ. Occup. Health J.* **2017**, *3*, 33–38.

26. Pesce, S. Use of water quality indices to verify the impact of Córdoba City (Argentina) on Suquía River. *Water Res.* **2000**, *34*, 2915–2926. [CrossRef]

27. Han, Y.; Lau, S.-L.; Kayhanian, M.; Stenstrom, M.K. Characteristics of highway stormwater runoff. *Water Environ. Res.* **2006**, *78*, 2377–2388. [CrossRef] [PubMed]

28. Camargo, J.A.; Alonso, A.; Salamanca, A. Nitrate toxicity to aquatic animals: A review with new data for freshwater invertebrates. *Chemosphere* **2005**, *58*, 1255–1267. [CrossRef] [PubMed]

29. Gransee, A.; Fuhrs, H. Magnesium mobility in soils as a challenge for soil and plant analysis, magnesium fertilization and root uptake under adverse growth conditions. *Plant. Soil* **2012**, *368*, 5–21. [CrossRef]

30. Hugh, C.R.W.; Bennett, H.; Risden, T.; Allen, L.; Davis, V. *Soil Survey of Shelby County, Tennessee*; U.S. Department of Agriculture: Nashville, TN, USA, 1916. Available online: https://www.nrcs.usda.gov/Internet/FSE_MANUSCRIPTS/tennessee/shelbyTN1970/Shelby.pdf (accessed on 11 November 2020).

31. Lau, S.L.; Khan, E.; Stenstrom, M.K. Catch basin inserts to reduce pollution from stormwater. *Water Sci. Technol.* **2001**, *44*, 23–34. [CrossRef]

32. Campbell, C.G.; Mathews, S. An Approach to Industrial Stormwater Benchmarks: Establishing and Using Site-Specific Threshold Criteria at Lawrence Livermore National Laboratory. In Proceedings of the CASQA Stormwater 2006 Conference, Sacramento, CA, USA, 25–27 September 2006. UCRL-CONF-224278.

33. Kaczala, F.; Marques, M.; Vinrot, E.; Hogland, W. Stormwater run-off from an industrial log yard: Characterization, contaminant correlation and first-flush phenomenon. *Environ. Technol.* **2012**, *33*, 1615–1628. [CrossRef]

34. Davis, A.P.; Shokouhian, M.; Ni, S. Loading estimates of lead, copper, cadmium, and zinc in urban runoff from specific sources. *Chemosphere* **2001**, *44*, 997–1009. [CrossRef]

35. Greenway, M.; Le Muth, N.; Jenkins, G. Monitoring Spatial and Temporal Changes in Stormwater Quality through a Series of Treatment Trains. A Case Study—Golden Pond, Brisbane, Australia. *Glob. Solut. Urban. Drain.* **2002**, *44*, 1–16. [CrossRef]

36. Brezonik, P.L.; Stadelmann, T.H. Analysis and predictive models of stormwater runoff volumes, loads, and pollutant concentrations from watersheds in the Twin Cities metropolitan area, Minnesota, USA. *Water Res.* **2002**, *36*, 1743–1757. [CrossRef]

37. Helmreich, B.; Hilliges, R.; Schriewer, A.; Horn, H. Runoff pollutants of a highly trafficked urban road—Correlation analysis and seasonal influences. *Chemosphere* **2010**, *80*, 991–997. [CrossRef]

38. Zhang, L.; Zhao, B.; Xu, G.; Guan, Y. Characterizing fluvial heavy metal pollutions under different rainfall conditions: Implication for aquatic environment protection. *Sci. Total Environ.* **2018**, *635*, 1495–1506. [CrossRef] [PubMed]

39. Tsihrintzis, V.A.; Hamid, R. Modeling and Management of Urban Stormwater Runoff Quality: A Review. *Water Resour. Manag.* **2001**, *11*, 137–164. [CrossRef]

40. Wijesiri, B.; Liu, A.; Goonetilleke, A. Impact of global warming on urban stormwater quality: From the perspective of an alternative water resource. *J. Clean. Prod.* **2020**, *262*, 121330. [CrossRef]
41. Gleaton, K.L. *Effectiveness of Environmental Regulations: Monitoring by the Regulated Community under Clean Water Act. Industrial Stormwater Runoff Requirements*; University of South Florida: Tampa, FL, USA, 2006.
42. Barbosa, A.; Fernandes, J.; David, L. Key issues for sustainable urban stormwater management. *Water Res.* **2012**, *46*, 6787–6798. [CrossRef] [PubMed]

Publisher's Note: MDPI stays neutral with regard to jurisdictional claims in published maps and institutional affiliations.

Article

Necessity of Acknowledging Background Pollutants in Management and Assessment of Unique Basins

Maoqing Duan [1,2]**, Xia Du** [1,3]**, Wenqi Peng** [1,3,*]**, Shijie Zhang** [1,3] **and Liuqing Yan** [1,3]

[1]   Department of Water Environment, China Institute of Water Resources and Hydropower Research, Beijing 100038, China; 17694854017@163.com (M.D.); duxia@iwhr.com (X.D.); zsj@iwhr.com (S.Z.); ylqiwhr@163.com (L.Y.)
[2]   College of Hydrology and Water Resources, Hohai University, Nanjing 210098, China
[3]   State Key Laboratory of Simulation and Regulation of Water Cycle in River Basin, China Institute of Water Resources and Hydropower Research, Beijing 100038, China
*   Correspondence: pwq@iwhr.com

Received: 22 April 2019; Accepted: 23 May 2019; Published: 27 May 2019

Abstract: The limitations of water quality management and assessment methods in China can be ascertained by comparison with other countries. However, it is unreasonable to use a uniform standard to evaluate water quality throughout China because one standard cannot fully account for the regional differences in background water quality. This study aimed to provide a basis for environmental water management decision-making. Areas seriously affected by background pollutants were identified by comparing several factors across 31 provinces in China. By coupling an improved export coefficient model (ECM) with a mechanistic model, a suitable pollutant yield coefficient was determined and its rationality was analysed. The export coefficient model was applied to estimate the pollutant (chemical oxygen demand and ammonia nitrogen) output of the basin in 2015. The spatial distribution characteristics of the pollutants were determined by simulating the pollutant outputs of 22 sub-basins and nine water function zones. For the year 2020, the simulation results of pollutant outputs far exceed the sewage discharge limit in water function zones and the pollutant concentration was much higher than the standard. Considering background pollutant outputs, more reasonable sewage discharge limit and water quality evaluation method are proposed.

Keywords: water quality; background pollutants; export coefficient model; chemical oxygen demand; ammonia nitrogen

---

1. Introduction

The water environment is closely related to human life and production, and its protection, control, and management, as well as prevention of pollution, are the focus of scholars in the field of environmental research [1–5]. In order to understand the characteristics of regional water environments and control water pollution, a large number of models have been developed thus far to simulate non-point source pollution loads and provide a basis for regional water environment control and planning [6–9].

With the development of the economy in China, the government is paying increasingly more attention to environmental problems, further illustrating the importance of addressing environmental issues by expanding the functions of the Ministry of Ecology and Environment in the institutional reform program of the state council, 2018. In 2011, the strictest water resources management system (three red lines) was proposed by the Ministry of Water Resources of the People's Republic of China to control the total water consumption, improve water use efficiency, and limit sewage discharge. The sewage discharge limit implies the control of the regional water environment, that is, developing the economy while ensuring the safety of the water ecological environment.

A scientific and reasonable water quality assessment method is an important basis for water quality management and assessment [10–12]. The approaches and methods of water quality management in other countries have been considered; for example, the United States does not have a unified national water environmental quality standard [13], but the environmental protection agency (EPA) has developed a technical guide to determine water quality benchmarks. Each state can formulate its water quality standard through published technical guidelines for water quality benchmarks in combination with the actual situation of the state [14]. The Canadian Council of Ministers of the Environment (CCME) stipulates that the water quality of water sources should be evaluated using the Canadian Water Quality Index [15,16]. The scores of different water bodies (0–100) can be calculated by the index equation. Water bodies can be divided into five levels: clean (95–100), good (80–94), medium (60–79), passing (45–59), and poor (0–44) [15]. However, in China, the water quality evaluation standard and method lack objectivity and scientificity because of regional differences, and comprehensive indicators have not been taken into account. The current evaluation standard of surface water quality in China is based on the environmental quality standard for surface water (GB 3838-2002), which classifies water quality pollution into six categories based on pollutant concentration. The evaluation method is a single factor evaluation method in which water quality is considered to be below the standard and requiring pollution control if one of the water quality detection indicators exceeds the standard. However, it is unfair to use the same water quality standard to restrict water quality in different regions, which are affected by background pollutants with varying degrees.

Background values, as the basis for distinguishing the impact of the natural environment and human activities on the environment, were proposed early last century and widely used in the environmental field [17,18], particularly in soil background elements and groundwater [19–22]. In contrast, the background value of surface water has been less studied because of its drastic spatial and temporal variability and complex influencing factors. However, it is particularly important to understand background pollutants in the formulation of water quality standards and water management [23]. In some areas of China, this will lead to misinformation in developing a water pollution control scheme due to the degree that water pollution will be overestimated due to the background pollutant problem [24–27].

In this study, the Tangwang River Basin, which is prominently affected by the background values, is taken as the research object. According to standard for water function zoning (GB/T50594-2010), the study area was divided into 9 water function zones with different water quality standards and service objects to guarantee the sustainable utilization of water resources (Table 1). Perennial low temperature and short frost-free period decreased soil microbial respiration activity, causing the accumulation of humic substances (HSs). This underlying surface condition produces organic erosion in the process of runoff [28,29] and brings a large amount of organic matter into the river, which leads to the chemical oxygen demand (COD) and ammonia nitrogen (NH_3-N) exceeding the standard perennially (standard values are shown in Table 1). This also presents an obstruction for effective regional water quality control. Therefore, we have systematically studied the background value of the study area. The main contents are as follows: (1) Clear the influence of background values on the water quality in Heilongjiang by comparing the environmental characteristics and water quality standards of 31 provinces in China; (2) by coupling an improved export coefficient model (ECM) with a mechanistic model, the land use yield coefficients suitable for this study were determined and their rationalities were verified; (3) to quantitatively estimate the background pollutant outputs in 2015 using the improved ECM; (4) a reasonable and objective method for water quality management in the study area was proposed after considering background pollutant values.

Table 1. Types of water function zone and water quality standards for each type.

Number	Type	Standard	Permissible Maximum Value (mg/L)		Remarks
			COD	NH_3-N	
a	River source water reserve	II	15	0.5	*a*: Delimited waters of great significance for the protection of water resources, natural ecosystems and rare and endangered species; *b,e*: Water demarcated to meet industrial water use and agricultural irrigation needs; *c,f,h,i*: Water delimited for the purpose of accepting production and discharge from sewage outlets of domestic wastewater; the wastewater accepted does not have a significant adverse impact on the water environment; *d,g*: Water area delimited to meet the transition water quality standards in lines connecting adjacent water functional zones with great differences in water quality.
b	Agricultural and industrial water use zoning	IV	30	1.5	
c	Discharge control zoning	/			
d	Transition zoning	IV	30	1.5	
e	Industrial water use zoning	IV	30	1.5	
f	Discharge control zoning	/			
g	Transition zoning	V	40	2	
h	Discharge control zoning	/			
i	Discharge control zoning	/			

2. Materials and Methods

2.1. Study Area

The Tangwang River is one of the ten rivers in Heilongjiang Province, China. The catchment area of its basin is about 21,000 km^2 (Figure 1). The northern part of the basin is bounded by the Xiao Hinggan Mountains and connected with the left bank tributary of Heilongjiang River. The annual average temperature is 0.6 °C, and the annual average temperature is below 0 °C, which is observed for five months. The annual precipitation varies from 530 to 700 mm, with precipitation from May to September accounting for 75% of the total annual precipitation. The main types of soil are dark brown loam and swamp soil rich in organic matter, and the forest coverage in the basin is as high as 87.9%. Perennial freeze-thaw alternation prevents complete decomposition of litter and HSs accumulates on the surface. Moreover, litter decompose slowly and accumulate massively on the surface owing to the special climatic and hydrological conditions [30–32]. Consequently, a large amount of organic matter flows into the river with rainfall-runoff, resulting in degraded regional water quality.

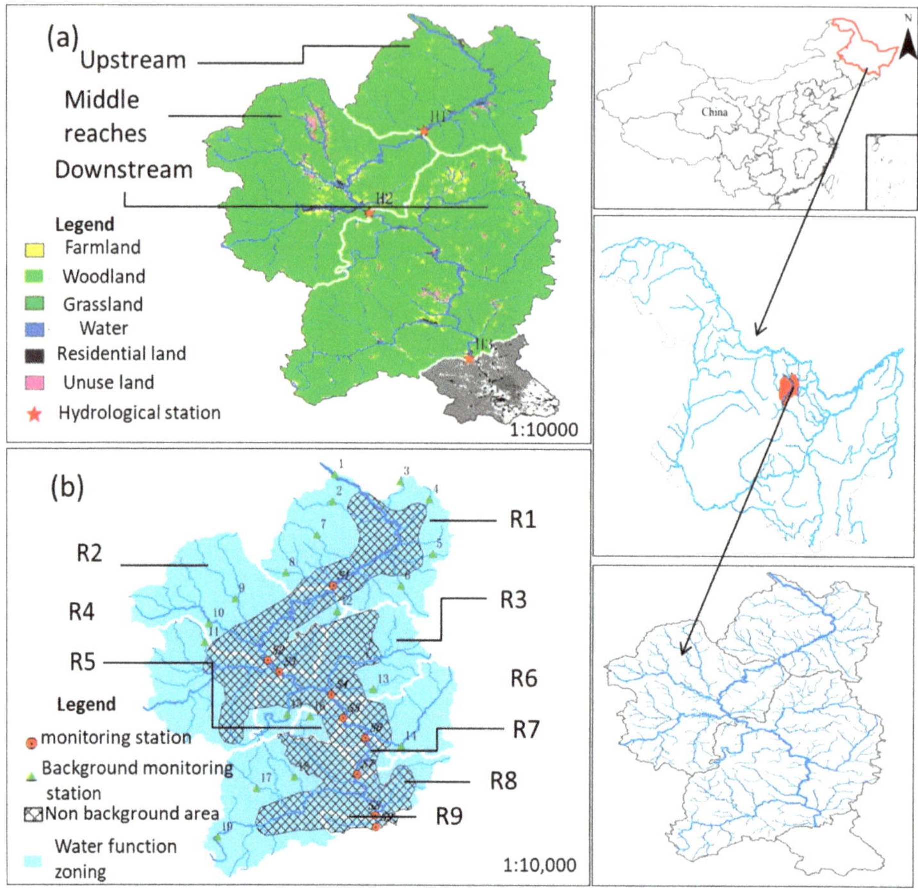

Figure 1. Partition of sampling sites in the Tangwang River Basin according to land-use types. (**a**) The river is divided into upper, middle, and lower reaches according to the control area of hydrological stations (H1–H3); (**b**) monitoring stations for water quality in water function zones (R1–R9); background monitoring stations (1–19) were used to monitor water quality in areas where human activity is scarce; Non-background areas include areas with high intensity of human activity; reference standard (GB/T50594-2010) for the division of water function zones.

2.2. Data Source

The main data used in this study included data on land-use types, precipitation, synchronous monitoring data of water quality and quantity and socio-economic statistics (Table 2). Spatial data were processed in ArcGIS 10.2.

Data from the hydrological stations (H1–H3) were for January 2011 to December 2013, and the synchronous monitoring of water quality and quantity at the hydrological stations was done twice a month in the dry season (January–March and October–December) and three times a month in the wet season (April–September). Data from the monitoring stations (S1–S9) were for January 2005 to December 2014 and done once a month. Data from the background monitoring stations (1–18) were for May 2015 and done only once.

Table 2. Main data sources and purposes.

Data Types	Purpose	Sources
Land-use types	Model structure	Landsat 8 OLI interpretation of satellite remote sensing images
Precipitation	Analysis of rainfall impact	National Meteorological Information Centre (http://data.cma.cn/)
Topographic data	Terrain division and watershed division	National Geographic Centre; Geospatial data cloud (http://www.gscloud.cn/sources/)
Water quality and quantity data	Estimation of pollutant output and evaluation of water quality	Yichun Water Environment Monitoring Centre (H1–H3); Heilongjiang Hydrological Bureau (S1–S9); Project Research Group (1–18)
Socio-economic data	Statistics of sewage discharge data	Statistical yearbook

2.3. Model Structure

The ECM has certain applicability to simulate the pollution load for areas with insufficient data. The improved ECM can increase the accuracy and applicability of estimating the pollution output in river basins [6,7,33,34]. The study area is located in a mountainous area with 90% of the slope above 85°. The improved ECM used in this study varies from that used in previous studies, in which terrain factors were not considered. The rainfall influence factor and runoff-migration influence factor were taken into account in the improved model. The basic equations are as follows:

$$L = \sum_{i=1}^{n} \alpha \gamma E_i A_i, \tag{1}$$

$$\alpha = \alpha_t \alpha_s, \tag{2}$$

$$\alpha_t = \frac{AverPre_{year}}{AverPre}, \tag{3}$$

$$\alpha_s = \frac{Pre_{a,b}}{AverPre_{a,b}}, \tag{4}$$

where E_i is the pollutant export coefficient for land use of type i, A_i is the total land use area for type i, α_t is the interannual variation factor for rainfall, α_s is the spatial variation factor for rainfall, $AverPre_{year}$ is the total annual precipitation for a year (mm), $AverPre$ is the average annual rainfall (11 years of rainfall data were selected), $Pre_{a,b}$ is the precipitation of grid a, b (mm), and $AverPre_{a,b}$ is the average annual rainfall of grid a, b (mm). γ is the runoff-migration influence factor because most of the study area is forestland and the effect of the underlying surface on runoff interception is obvious. This paper considers runoff coefficient rather than topographic factors. The formulas are as follows:

$$\gamma = \frac{Q \times 365.5 \times 24 \times 3600}{P} \tag{5}$$

where Q is the average annual discharge of the sub-basin (m³/s), and P is the total precipitation of the sub-basin (m³). The total annual pollutant output of the river is calculated according to the pollutant concentration and discharge at the outlet of the river basin.

$$L = \sum_{j=1}^{12} c_j Q_j \tag{6}$$

where c_j is the pollutant monthly average concentration for month j in the river basin, and Q_j is the pollutant monthly average discharge for j months in the river basin.

$$\alpha \gamma \sum_{i=1}^{n} E_i A_i = \sum_{j=1}^{12} c_j Q_j. \tag{7}$$

In the study area, land use is mainly forest and grassland, followed by farmland, residential land, and another land account for a relatively low proportion. Therefore, only the pollution sources from forest and grassland and farmland were considered in the ECM. The output of pollutants from residential land and other land was calculated according to local sewage discharge information.

$$\alpha\gamma \sum_{i=1}^{n} E_i A_i = \sum_{j=1}^{12} c_j Q_j - L_n \tag{8}$$

L_n is the discharge of human pollution sources in the basin for areas with a single land use type. The formulas can be translated as follows:

$$E = \frac{\sum_{j=1}^{12} c_j Q_j - L_n}{\alpha\gamma A}. \tag{9}$$

2.4. Management and Assessment Method

In 2011, the Central Committee released Document No. 1, which clearly stated that the most stringent water resource management system should be implemented, and the "three red lines" of total water use control, water use efficiency control, and restriction of pollution acceptance in water function zones should be established. The third one is to strengthen the management of water function zones for restricting pollution acceptance, such that the total amount of main pollutants entering rivers and lakes can be controlled within the scope of the pollution acceptance capacity of rivers, and the rate at which water function zones meet the water quality standard should be increased to more than 95%.

Water pollution acceptance capacity is an important basis for regional water pollution control and management. It consists of two parts: the first part is called target capacity, which is determined by the difference between water flow rate, environmental quality target, and baseline value; the second part is called degradation capacity [35,36]. There are many methods for calculating water pollution acceptance capacity. The following one-dimensional calculation formula is suitable for rivers with uniform pollutant mixing:

$$S = 86.4Q(C - C_0) + kCV$$

where S is the water pollution acceptance capacity, Q is the outlet discharge of basin, C is the standard value of pollutant concentration, C_0 is the outlet concentration, K is the comprehensive degradation coefficient of pollutants, and V is the regional environmental volume. S is closely related to water quality standards and management objectives and plays a restrictive role in regional socio-economic development.

According to GB/T50594-2010, the study area was divided into nine water function zones. The types and water quality standards are listed in Table 1. When the water quality monitoring value is higher than the water quality standard value (permissible maximum value), it will exceed the standard, and regional pollution discharge should be controlled.

3. Results and Discussion

3.1. Impact of Background Pollutants

There are 194 water function zones on the main rivers in Heilongjiang Province. Among them, the Tangwang River Basin is divided into nine water function zones. Figure 2 compares the population density, annual discharge of COD and NH_3-N per unit area, forest land coverage, and rate of water function zones reaching the standard in 31 provinces of China. The population density is 87 persons/km^2, the annual discharge of COD and NH_3-N per unit area are 0.86 and 0.12 t/km^{2}a, respectively, in Heilongjiang Province. In addition, the forest coverage is relatively high, indicating the good maintenance of the ecological environment in Heilongjiang Province. However, the rate of water function zones meeting the standard is insufficient (Figure 2e). According to the response of the local water quality management department, the main reason for the low rate is that COD and NH₃-N exceed the permissible maximum value.

The results shown in Figure 2 confirm that the water quality in Heilongjiang Province is affected by background pollutants. Because Heilongjiang Province is located in northeast China, low annual average temperatures and short frost-free periods inhibit biological activity, resulting in the accumulation of a large amount of partially decomposed litter to form a thicker humus layer. In addition, perennial freeze-thaw alternation causes the breakdown of plant residual cells and the easy dissolution of organic solutes [30]. Moreover, the influx of runoff into rivers leads to excessive pollutants in rivers, which further affects the management and assessment of regional water quality. Therefore, research on background pollutants is very important for the study area.

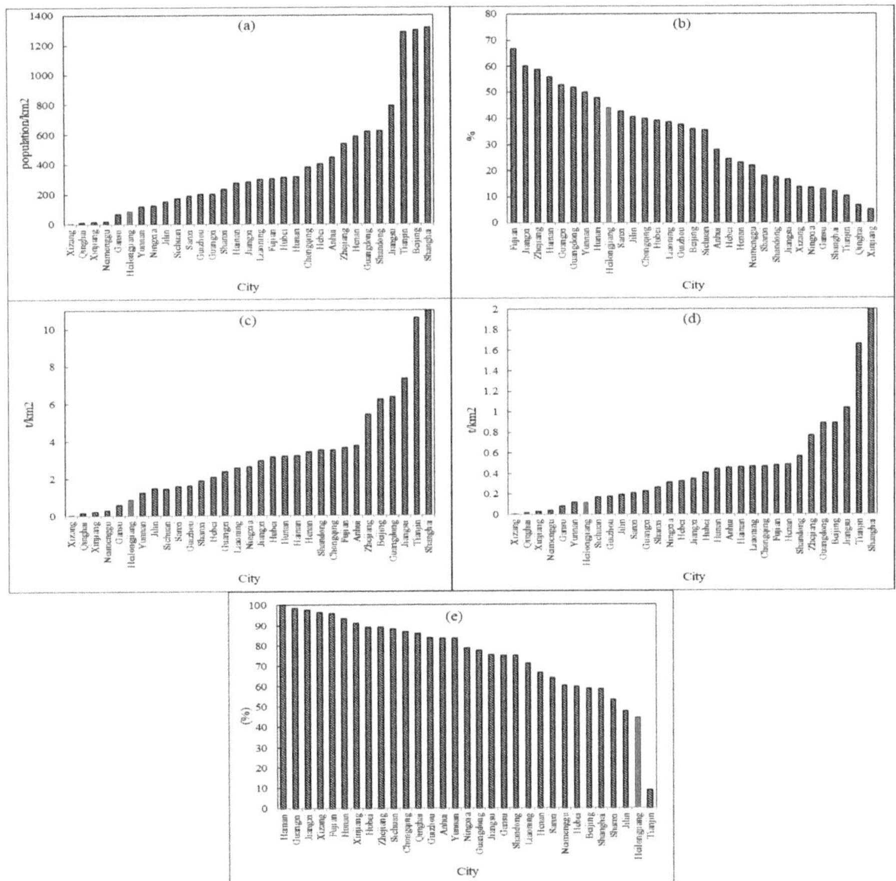

Figure 2. Comparison of factors in 31 provinces of China. (**a**) Population density; (**b**) forest land coverage; (**c**) annual discharge of COD; (**d**) annual discharge of NH₃-N; (**e**) the rate of water function zones meeting the standard. COD: chemical oxygen demand; NH₃-N: ammonia nitrogen.

3.2. Pollutant Yield Coefficient

Common methods for determining the export coefficient are literature consultation, field monitoring, and statistical data consultation; many scholars have based their export coefficients on previous studies [6,7]. The study area features a single type of land use, mainly forest land and farmland, which account for 93.9% of the total area. The special geographical location and hydro-climatic conditions have led to the formation of the underlying surface with high humus content, which results in the high pollutant yield under the action of runoff erosion. Therefore, if the export coefficient of this region is derived from land use in other regions, large errors will occur in the simulation results. According to the three water quality and quantity monitoring stations shown in Figure 1, the whole basin is divided into the upper, middle and lower parts. The land use in the upstream source area is forest and grassland, accounting for 95.8%. The pollutant yield coefficient for forest and grassland can be determined using Formula 9 rather than the export coefficient, considering the runoff-migration influence factor. Then, the pollutant yield coefficient for farmland can be calculated according to the middle and lower reaches.

The distribution of the rainfall influence factor in the study area for three years from 2011 to 2013 is shown in Figure 3. It ranges from 0.45 to 1.43, with greater values in 2012 and 2013 than in 2011.

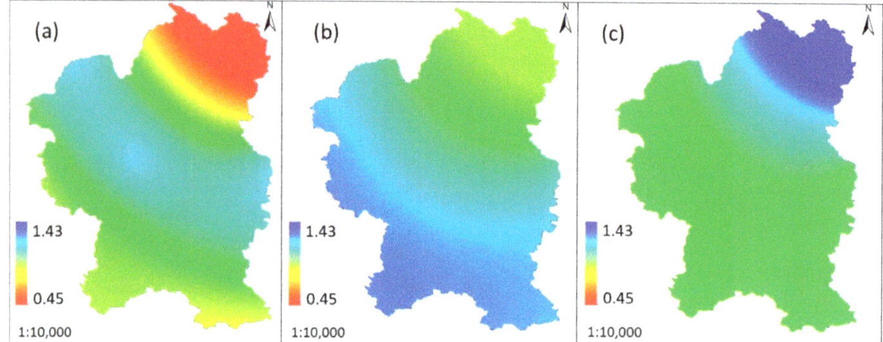

Figure 3. Distribution of rainfall influence factor in the study area ((**a**) 2011; (**b**) 2012; (**c**) 2013).

According to Formula (5), the distribution of the runoff-migration influence factor is calculated by using a grid calculator. The values range from 0.28 to 0.66 (Figure 4).

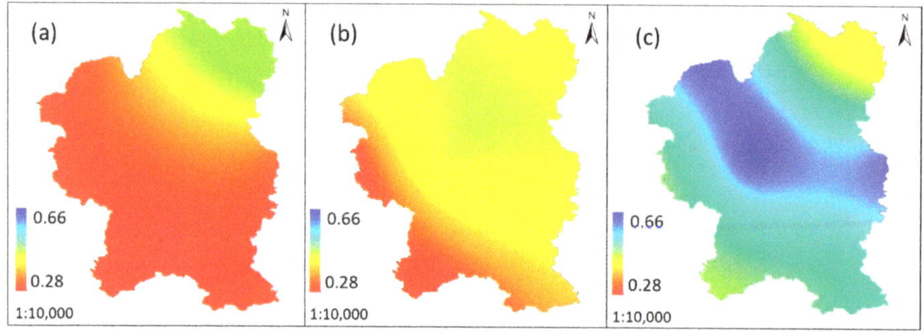

Figure 4. Distribution of runoff-migration influence factor in the study area ((**a**) 2011; (**b**) 2012; (**c**) 2013).

Based on the monitoring data from 2011 to 2013, the COD and NH_3-N yield coefficients for forest and grassland in the study area were calculated. The average values calculated over 2011 to 2013 were brought into the lower and middle reaches and the yield coefficient for farmland for 2011–2013 was determined using Formula 9 (Table 3). Compared with the export coefficient of land use in previous studies, the yield coefficient calculated was much larger than that in previous studies. This is due to the essential difference between the pollutant export coefficient and yield coefficient in terms of loss through runoff. The COD yield coefficients for forest and grassland, and farmland were 28.1 t/km^{2}a and 15.8 t/km^{2}a, respectively, and their NH_3-N yield coefficients were 0.47 t/km^{2}a and 1.77 t/km^{2}a, respectively.

Table 3. Pollutant yield coefficient according to land use.

Year	Area	Forest and Grassland (t/km^{2}a)		Farmland (t/km^{2}a)	
		COD	NH$_3$-N	COD	NH$_3$-N
2011		31.7	0.52	/	/
2012	Upstream reaches	22.6	0.51	/	/
2013		29.9	0.37	/	/
2011		28.1	0.47	15.5	1.93
2012	Middle reaches	28.1	0.47	14.4	3.07
2013		28.1	0.47	23.7	1.02
2011		28.1	0.47	7.2	0.54
2012	Lower reaches	28.1	0.47	21.3	2.96
2013		28.1	0.47	12.5	1.08
	Average	28.1	0.47	15.8	1.77

The land use yield coefficient is much higher than that listed in the literature [37,38], which is why the region is heavily affected by background pollutants. In order to demonstrate the applicability of the high yield coefficient obtained in this study, the whole basin was divided into the background and non-background areas according to land use, and water quality monitoring stations were set up in the background area (May 2015, sampling site: 1–19). The distributions of values exceeding the standard (Table 1) of water quality are shown in Figure 5. The upstream standard value is lower than that in the middle and lower reaches. Therefore, values exceeding the standard value at each monitoring point decrease gradually from upstream to downstream.

The concentration of pollutants also appears to be high in regions with almost no human activities, which shows that the high yield coefficient obtained is consistent to a certain extent. Distribution of values exceeding the standard further proves that water quality is affected by the background values of COD and NH$_3$-N and the degree of impact is greater. Therefore, it is unreasonable that higher concentrations of pollutants caused by natural factors are considered to be substandard.

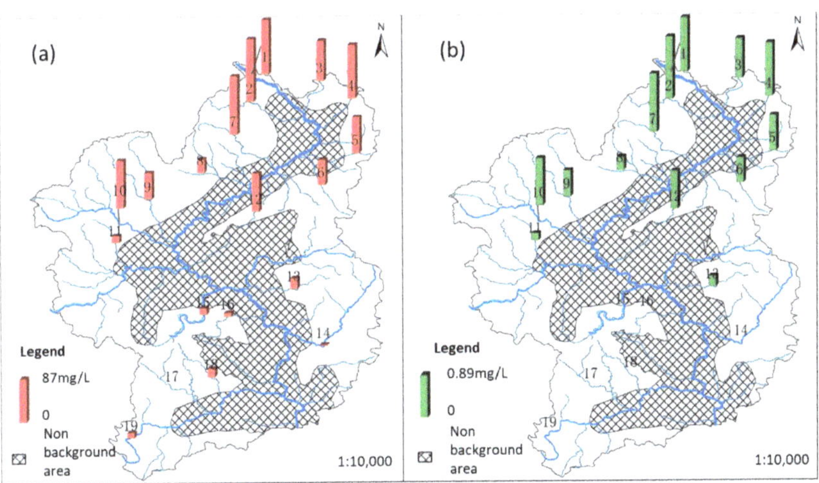

Figure 5. Distribution of values exceeding the standard (standard value subtracted from monitoring value). (**a**) Chemical oxygen demand; (**b**) ammonia nitrogen.

The correlation between perennial monthly mean pollutant concentration and flow was determined on the basis of the data from synchronous monitoring of water quality and quantity at three monitoring stations (S1, S3, S9) from 2001 to 2014 (Figure 6). The results show that the correlations between COD,

NH$_3$-N, and flow in the upstream region are high (0.69 and 0.60, respectively). In the middle and downstream regions, the correlation between COD and flow was lower than that in the upstream region (0.60). This may be attributed to sewage discharge in the middle and downstream regions. However, a negative correlation was observed between NH$_3$-N and flow, which may indicate that the impact of human sewage on NH$_3$-N output in the river basin is relatively large compared with that of COD.

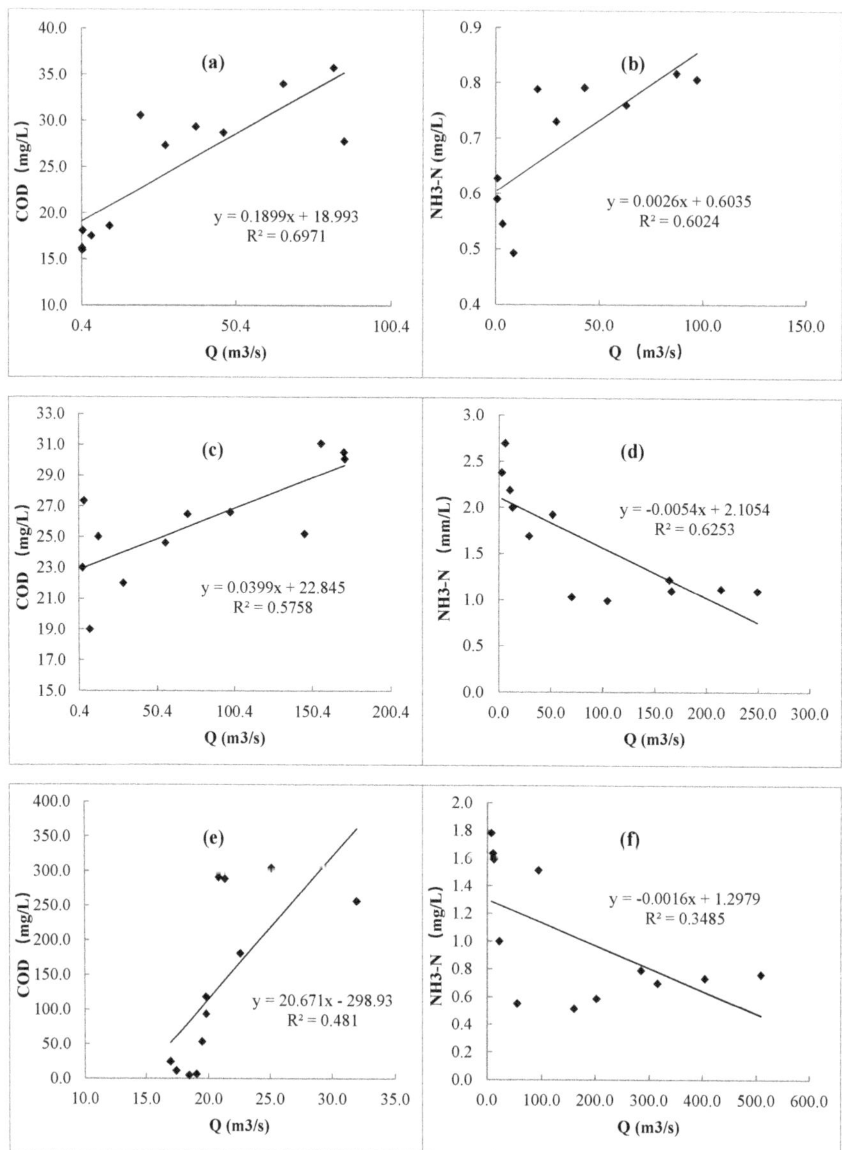

Figure 6. Correlation between perennial monthly mean concentration and flow. (**a**) Upstream reaches, COD and flow; (**b**) upstream reaches, NH$_3$-N, and flow; (**c**) middle reaches, COD, and flow; (**d**) middle reaches, NH$_3$-N, and flow; (**e**) lower reaches, COD, and flow; (**f**) lower reaches, NH$_3$-N, and flow (COD: chemical oxygen demand; NH$_3$-N: ammonia nitrogen).

3.3. Pollutant Load and Management

The study area was divided into 22 sub-basins according to elevation data and catchment area. Among them, the 22nd sub-basin is the catchment area of the main stream of Tangwang River and others flow into the mainstream. To control the pollutant load into the main stream, it is necessary to know the pollutant load into each sub-basin. According to the results of ECM simulation combined with the investigation of human pollution sources, the pollutant output of each sub-basin for 2015 was estimated (Figure 7). The pollutant load in the main stream area was the highest, followed by sub-basins 16 and 21, which are the focus of pollution prevention and control. The pollutant load in several sub-basins in the upper reaches was found to be relatively low. Overall, the basin showed a trend of an increasing load from the upstream to downstream areas.

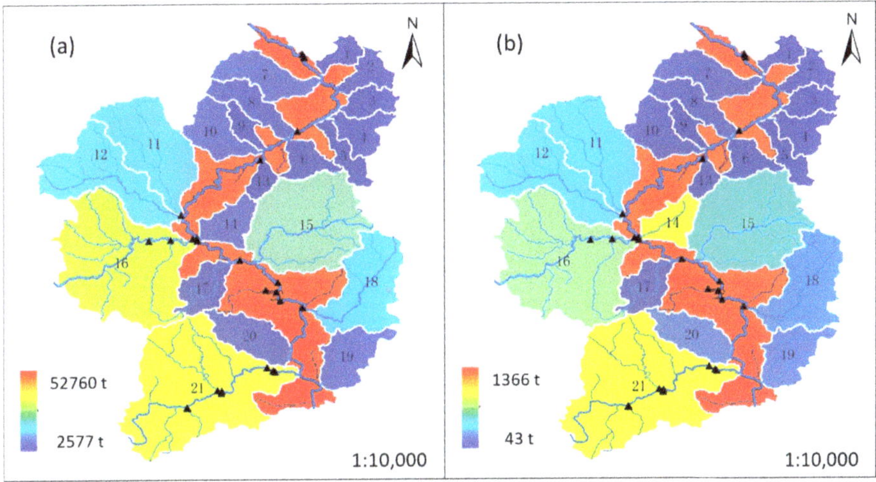

Figure 7. Spatial distribution of pollutant load in 2015. (**a**) Chemical oxygen demand; (**b**) ammonia nitrogen.

According to GB/T50594-2010, the Tangwang River Basin was divided into nine water function zones, and the pollutant output of the water function zones in 2015 was simulated (Figure 8). The simulated values of COD ranged from 6.2×10^4 to 0.21×10^4 t. The largest pollutant output was a, followed by h, b and c, with outputs of 5.0, 4.5, and 4.3×10^4 t; the NH$_3$-N simulation results ranged from 0.15 to 0.0086×10^4 t. According to the sewage discharge limit of each water function zone for the planning year 2020, a considerable amount of pollutant output should be reduced. However, a large part of the pollutant output in several water function zones is caused by background pollutants. Therefore, the regional sewage discharge limit should be reconsidered with the regional background pollutants in mind.

The spatial distribution of the NH$_3$-N pollutant output was different from that of COD. The main reason for this difference is that the discharge amounts of COD and NH$_3$-N in each water functional area were different. The output of COD was higher upstream than downstream, while the output of NH$_3$-N showed the opposite pattern. This shows that the influence on sewage discharge of NH$_3$-N was significantly higher than that of COD in the middle and downstream areas.

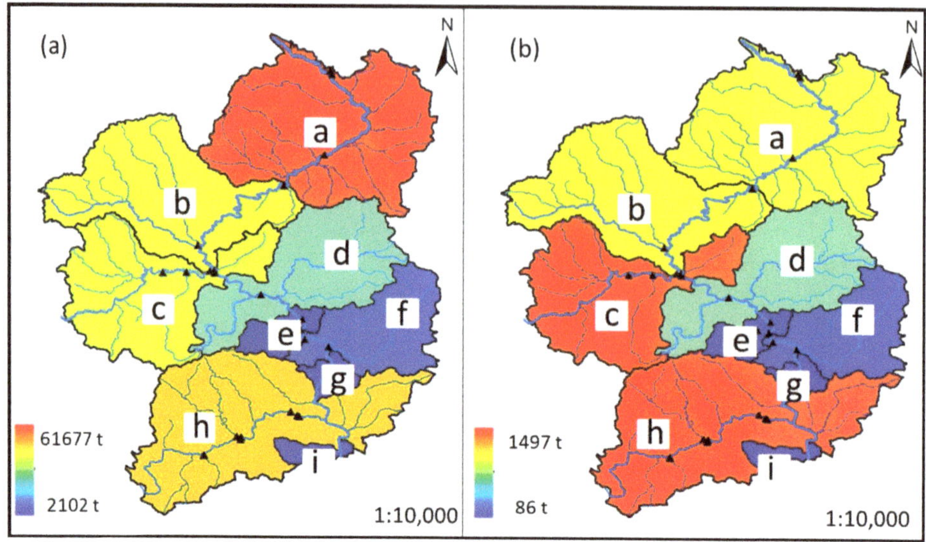

Figure 8. Pollutant output distribution in water function zones in 2015. (**a**) Chemical oxygen demand; (**b**) ammonia nitrogen.

In this paper, we suggest two methods for determining pollutant output. In the first method, the output of background pollutants (from forests and grasslands) is not taken into account. The regional pollutant output is calculated according to the output of human pollution sources and farmland pollutants. The second method is to design a sewage discharge limit for the planning year considering the background pollution output and to provide an invariable value for the yield coefficient of background pollutants. The purpose of this method is to control the output of background pollutants (disturbance from human activities such as logging and collection of resources from forests in mountains increases the loss of background pollutants). The area more representative of the background should be selected as the area for calculating the yield coefficient of forest and grassland.

We selected the first method as an example to account for the regional pollution output in 2015; forest and grassland as the source of background pollutants were not considered. The calculation results of pollutant output and sewage discharge limit are shown in Figure 9. A considerable amount of pollutant output should be reduced in water function zones *b*, *c*, *d*, and *h* in order to meet the emission requirements of the planning year, for which farmland pollutant output is the main target of reduction in *b*, *d*, and *h*, whereas sewage discharge from human production activities needs to be controlled in *c*. Figure 9b shows that the NH$_3$-N outputs of *b*, *c*, *d*, *f* and *h* significantly exceed the sewage discharge limit of the planning year. The output from farmland and sewage discharge are the key objects of pollution prevention and control in these function zones. Compared with COD output, human sewage discharge of NH$_3$-N accounts for a large proportion in water function zones that exceed the discharge limit, which also reflects the relationship between flow and pollutant concentration (Figure 6). In the upstream reaches, the concentration of non-point source pollutants entering the river is proportional to flow (Figure 6a,b), and for the middle and lower reaches of the river, the concentration of pollutants entering the river is not positively correlated with the flow because it is greatly affected by human sewage discharge. By revising the sewage discharge limit, the control object and control index with respect to the water function zoning in the study area needs to be clear. This method overcomes the interference of background pollutants in determining the discharge capacity limit and is convenient for regional water quality management.

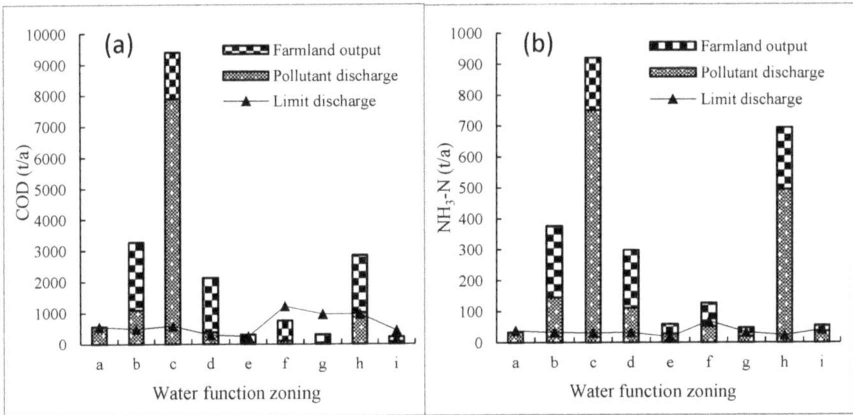

Figure 9. Total output of pollutants and sewage discharge limit of water function zones. (**a**) Chemical oxygen demand; (**b**) ammonia nitrogen.

3.4. Assessment Method Considering Background Value

For water quality monitoring, three values should be considered: "baseline value": self-produced pollutants of the river ecosystem; "background value": the impact of the natural environment on water quality; "pollution value": the impact of human activities on water quality. In order to objectively evaluate the impact of human activities on water quality, the influence of background value and baseline value should be considered on the basis of current water quality assessment methods. In other words, the background value should be subtracted from the monitoring value and then evaluated according to GB 3838-2002. Two methods are proposed to determine the background value (Figure 10). One is to set up monitoring stations in the upper reaches or areas without human activities, through which the baseline value is determined according to the concentration mean in the dry season to ensure that the water is not affected by surface recharge and exogenous pollutants. The monitoring value during the wet season contains the baseline value and the background value, through which the background value can be calculated. The second method is to consider the impact of rainfall on the output of background pollutants during the wet season. The total output of background pollutants is estimated by ECM for an entire year, and then the total output for each month is distributed according to the proportion of monthly rainfall to annual rainfall. Based on the monthly average flow of the monitoring station, the background concentration for the month is calculated, and then the monitoring value is revised.

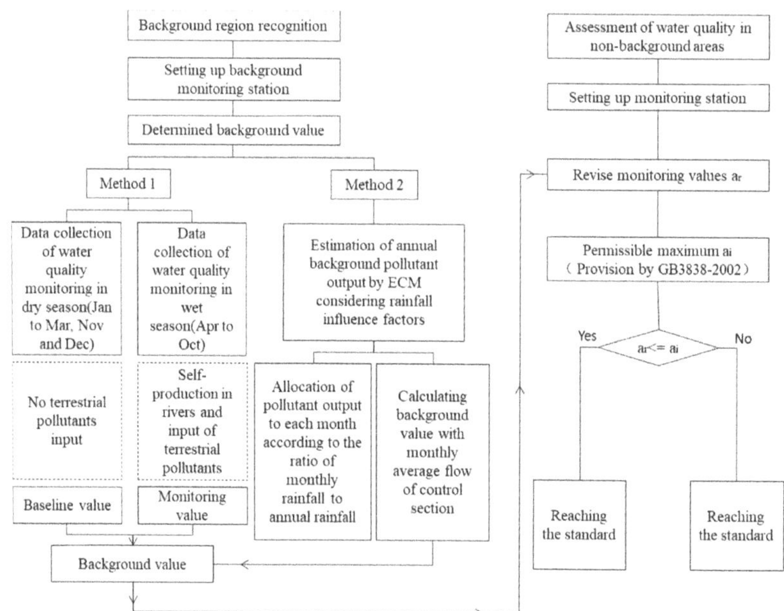

Figure 10. Flow chart of assessment methods considering background value.

4. Conclusions

By comparing the social, economic, and natural environments of 31 provinces in China and considering the results of this comparison in combination with the water quality standards for water function zones, it was clear that the surface water quality of Heilongjiang Province was seriously affected by background values (COD and NH_3-N); this was affecting water quality management and assessment in Heilongjiang Province. To manage and assess water quality reasonably and objectively in areas seriously affected by background values, the Tangwang River Basin in Heilongjiang Province was selected as the study area and an improved ECM was applied to quantify background pollutants. Firstly, model parameters in the ECM were determined by coupling the improved ECM with a mechanistic model—pollutant yield coefficients of forest and grassland (COD: 28.1 t/km^{2}a, NH_3-N: 0.47 t/km^{2}a) and farmland (COD: 15.8 t/km^{2}a, NH_3-N: 1.47 t/km^{2}a) and their validities were verified using water quality monitoring data with high pollutant concentrations from background monitoring stations (1–19). The results of the correlation analysis show that there is a significant positive correlation between pollutant concentration and flow in the upstream areas, indicating that runoff is the main factor driving background pollutant amounts. However, the correlation between the middle and lower reaches of the basin is not significant or even negative because of the impact of human sewage. Based on the 22 sub-basins and nine water function zones in the basin, the spatial distribution characteristics of pollutant output in the study area were analysed. Then, the regional sewage discharge limit scheme was re-approved, and water quality evaluation methods are proposed after considering background pollutants. The revised scheme effectively avoids the influence of background values and objectively reflects the impact of human activities on water quality. The results of this study provide a scientific basis for improving China's water quality management and assessment system. It is suggested that the relevant departments should consider regional differences when formulating schemes and standards.

This study did not fully demonstrate the background characteristic of the upstream area when determining the yield coefficient of the background pollution source (forest and grassland). It is suggested that the background characteristics of the simulated area be fully considered in the application of the ECM to study the background values of water quality. Due to limitations in monitoring data, the proposed water quality evaluation method was not verified via practical application, but the method is reasonable.

Author Contributions: W.P. provided overall guidance; X.D. and S.Z. analysed the data; L.Y. contributed analysis tools; M.D. wrote the paper.

Funding: This work was jointly supported by the IWHR Research and Development Support Program (Grant Nos. WE0145B052017); Beijing Natural Fund Program (J150005).

Acknowledgments: Yichun Water Environment Monitoring Centre and Heilongjiang Hydrological Bureau provide data.

Conflicts of Interest: The authors declare no conflicts of interest.

References

1. Butler, J.R.A.; Wong, G.Y.; Metcalfe, D.J.; Honzák, M.; Pert, P.L.; Rao, N.; van Grieken, M.E.; Lawson, T.; Bruce, C.; Kroon, F.J.; et al. An analysis of trade-offs between multiple ecosystem services and stakeholders linked to land use and water quality management in the Great Barrier Reef, Australia. *Agric. Ecosyst. Environ.* **2013**, *180*, 176–191. [CrossRef]

2. Yan, D.H.; Wang, H.; Li, H.H.; Wang, G.; Qin, T.L.; Wang, D.Y.; Wang, L.H. Quantitative analysis on the environmental impact of large-scale water transfer project on water resource area in a changing environment. *Hydrol. Earth Syst. Sci.* **2012**, *16*, 2685–2702. [CrossRef]

3. Hassan, N.E.; Ali, S.A.M. Screening and analysis of water quality of Zea River in Kurdistan region, Iraq. *Int. J. Adv. Appl. Sci.* **2016**, *3*, 61–67. [CrossRef]

4. Fan, X.; Cui, B.; Zhao, H.; Zhang, Z.; Zhang, H. Assessment of river water quality in Pearl River Delta using multivariate statistical techniques. *Proc. Environ. Sci.* **2010**, *2*, 1220–1234. [CrossRef]

5. Cassanego, M.B.B.; Droste, A. Assessing the spatial pattern of a river water quality in southern Brazil by multivariate analysis of biological and chemical indicators. *Braz. J. Biol.* **2016**, *77*, 118–126. [CrossRef] [PubMed]

6. Cheng, X.; Chen, L.; Sun, R.; Jing, Y. An improved export coefficient model to estimate non-point source phosphorus pollution risks under complex precipitation and terrain conditions. *Environ. Sci. Pollut. Res. Int.* **2018**, *25*, 20946–20955. [CrossRef]

7. Ding, X.; Shen, Z.; Hong, Q.; Yang, Z.; Wu, X.; Liu, R. Development and test of the Export Coefficient Model in the Upper Reach of the Yangtze River. *J. Hydrol.* **2010**, *383*, 233–244. [CrossRef]

8. Lai, Y.C.; Yang, C.P.; Hsieh, C.Y.; Wu, C.Y.; Kao, C.M. Evaluation of non-point source pollution and river water quality using a multimedia two-model system. *J. Hydrol.* **2011**, *409*, 583–595. [CrossRef]

9. Yang, S.; Dong, G.; Zheng, D.; Xiao, H.; Gao, Y.; Lang, Y. Coupling Xinanjiang model and SWAT to simulate agricultural non-point source pollution in Songtao watershed of Hainan, China. *Ecol. Modell.* **2011**, *222*, 3701–3717. [CrossRef]

10. Ye, Y.; Jia, K. A water quality assessment method based on sparse autoencoder. In Proceedings of the IEEE International Conference on Signal Processing, Communications and Computing, Ningbo, China, 19–22 September 2015.

11. Liu, S.; Lou, S.; Kuang, C.; Huang, W.; Chen, W.; Zhang, J.; Zhong, G. Water quality assessment by pollution-index method in the coastal waters of Hebei Province in western Bohai Sea, China. *Mar. Pollut. Bull.* **2011**, *62*, 2220–2229. [CrossRef]

12. Jähnig, S.C.; Cai, Q. River water quality assessment in selected Yangtze tributaries: Background and method development. *J. Earth Sci.* **2010**, *21*, 876–881. [CrossRef]

13. Hinman, J.J. American standards for quality of water. *J. Chem. Technol. Biotechnol.* **1921**, *40*, R325–R327. [CrossRef]

14. New York State Department of Environmental Conservation, Part 701 Classifications-Surface Waters and Ground Waters. Available online: https://docplayer.net/. (accessed on 1 December 2018).

15. Lumb, A.; Halliwell, D.; Sharma, T. Application of CCME water quality index to monitor water quality: A Case of the Mackenzie River Basin, Canada. *Environ. Monit. Assess.* **2006**, *113*, 411–429. [CrossRef] [PubMed]

16. Rosemond, S.D.; Duro, D.C.; Dubé, M. Comparative analysis of regional water quality in Canada using the Water Quality Index. *Environ. Monit. Assess.* **2009**, *156*, 223–240. [CrossRef]

17. Campos, M.L.; Pierangeli, M.A.P.; Guilherme, L.R.G.; Marques, J.J.; Curi, N. Baseline Concentration of Heavy Metals in Brazilian Latosols. *Commun. Soil Sci. Plant Anal.* **2003**, *34*, 547–557. [CrossRef]

18. Reimann, C.; Caritat, P.D. Distinguishing between natural and anthropogenic sources for elements in the environment: Regional geochemical surveys versus enrichment factors. *Sci. Total Environ.* **2005**, *337*, 91–107. [CrossRef] [PubMed]

19. Cobelo-Garcia, A.; Prego, R. Heavy metal sedimentary record in a Galician Ria (NW Spain): Background values and recent contamination. *Mar. Pollut. Bull.* **2003**, *46*, 1253–1262. [CrossRef]

20. Rodríguez, J.G.; Tueros, I.; Borja, A.; Belzunce, M.J.; Franco, J.; Solaun, O.; Valencia, V.; Zuazo, A. Maximum likelihood mixture estimation to determine metal background values in estuarine and coastal sediments within the European Water Framework Directive. *Sci. Total Environ.* **2006**, *370*, 278–293. [CrossRef]

21. Li, J.; Wu, Y. Historical changes of soil metal background values in select areas of China. *Water Air Soil Pollut.* **1991**, *57–58*, 755–761. [CrossRef]

22. Nsouli, B.; Darwish, T.; Thomas, J.P.; Zahraman, K.; Roumié, M. Ni Cu Zn and Pb background values determination in representative Lebanese soil using the thick target PIXE technique. *Nucl. Instrum. Methods Phys. Res. B* **2003**, *219*, 181–186. [CrossRef]

23. Ward, R.C. Development and use of water quality criteria and standards in the United States. *Reg. Environ. Chang.* **2001**, *2*, 66–72. [CrossRef]

24. Chen, J.; He, D.; Zhang, Y. Is COD a suitable parameter to evaluate the water pollution in the Yellow river? *Environ. Chem.* **2003**, *22*, 611–614. (In Chinese)

25. Chen, J.; Zhang, Y.; Yu, T.; He, D. A study on dissolution and bio-degradation of organic matter in sediments from the Yellow River. *Acta Sci. Circumstantiae* **2004**, *24*, 1–5. (In Chinese)

26. Chen, J.; Zhang, Y.; Yu, T.; He, D. Influences of the suspended matter on the water quality parameters including COD, Potassium Permanganate Index and BOD 5 in the Yellow River, China. *Acta Sci. Circumstantiae* **2004**, *24*, 369–375. (In Chinese)

27. Chen, J.; Zhang, Y.; Yu, T.; He, D. Problem and solution in assessing the oxygen-demanding organic matters of the Yellow River, China. *Acta Sci. Circumstantiae* **2005**, *25*, 279–284. (In Chinese)

28. Jacinthe, P.A.; Lal, R.; Owens, L.B.; Hothem, D.L. Transport of labile carbon in runoff as affected by land use and rainfall characteristics. *Soil Tillage Res.* **2004**, *77*, 111–123. [CrossRef]

29. Jordán, A.; Martínez-Zavala, L. Soil loss and runoff rates on unpaved forest roads in southern Spain after simulated rainfall. *For. Ecol. Manag.* **2008**, *255*, 919. [CrossRef]

30. Du, Z.; Cai, Y.; Wang, X.; Yan, Y.; Lu, X.; Liu, S. Research progress on the effects of soil freeze-thaw on plant physiology and ecology. *Chin. J. Eco-Agric.* **2014**, *22*, 1–9. [CrossRef]

31. Zhu, J.; He, X.; Wu, F.; Yang, W.; Tan, B. Decomposition of *Abies faxoniana* litter varies with freeze–thaw stages and altitudes in subalpine/alpine forests of southwest China. *Scand. J. For. Res.* **2012**, *27*, 11. [CrossRef]

32. Wu, F.; Yang, W.; Zhang, J.; Deng, R. Litter decomposition in two subalpine forests during the freeze–thaw season. *Can. J. For. Res.* **2010**, *36*, 135–140. [CrossRef]

33. Liu, R.M.; Yang, Z.F.; Shen, Z.Y.; Yu, S.L.; Ding, X.W.; Wu, X.; Liu, F. Estimating Nonpoint Source Pollution in the Upper Yangtze River Using the Export Coefficient Model, Remote Sensing, and Geographical Information System. *J. Hydraul. Eng.* **2009**, *135*, 698–704. [CrossRef]

34. Wu, L.; Li, P.; Ma, X.Y. Estimating nonpoint source pollution load using four modified export coefficient models in a large easily eroded watershed of the loess hilly–gully region, China. *Environ. Earth Sci.* **2016**, *75*, 1056. [CrossRef]

35. Wu, Q.; Liu, M.; Wang, X.; Di, L.; Kang, L.; Lin, L. Assessing the water environmental capacity of pollution consumption in Jiulong River Basin. In Proceedings of the Fourth International Conference on Agro-geoinformatics, Istanbul, Turkey, 20–24 July 2015.

36. Zhang, R.; Qian, X.; Yuan, X.; Ye, R.; Xia, B.; Wang, Y. Simulation of Water Environmental Capacity and Pollution Load Reduction Using QUAL2K for Water Environmental Management. *Int. J. Environ. Res. Public Health* **2012**, *9*, 4504–4521. [CrossRef]

37. Chen, Y.; Yuan, Q.; Han, F.; Zhou, L.; Hong, S. Estimation of Non-point Source Pollution Load of Yangtze Watershed Based on Improved Export Coefficient Model. *J. Geomat.* **2017**, *42*, 96–99. (In Chinese)
38. Zhang, C.; Liu, Z.; Zhang, G. A Study of the Agricultural Non-point Sources of Ammonia Nitrogen Load in Henan Province Based on the Export Coefficient Method. *China Rural Water Hydropower* **2017**, *10*, 35–39. (In Chinese)

Article

Multi-Objective Optimization for Selecting and Siting the Cost-Effective BMPs by Coupling Revised GWLF Model and NSGAII Algorithm

Zuoda Qi [1], Gelin Kang [1], Xiaojin Wu [1], Yuting Sun [2] and Yuqiu Wang [1,*]

[1] Tianjin Key Laboratory of Environmental Technology for Complex Trans-Media Pollution, College of Environmental Science and Engineering, Nankai University, Tianjin 300350, China; 1120170165@mail.nankai.edu.cn (Z.Q.); 1120180182@mail.nankai.edu.cn (G.K.); 2120190529@mail.nankai.edu.cn (X.W.)

[2] Khoury College of Computer Sciences, Northeastern University, San Jose, CA 95138, USA; sun.yut@husky.neu.edu

* Correspondence: yqwang@nankai.edu.cn; Tel.: +86-022-8535-8068

Received: 11 December 2019; Accepted: 10 January 2020; Published: 15 January 2020

Abstract: Best management practices (BMPs) are an effective way to control water pollution. However, identification of the optimal distribution and cost-effect of BMPs provides a great challenge for watershed policy makers. In this paper, a semi-distributed, low-data, and robust watershed model, the Revised Generalized Watershed Loading Function (RGWLF), is improved by adding the pollutant attenuation process in the river channel and a bank filter strips reduction function. Three types of pollution control measures—point source wastewater treatment, bank filter strips, and converting farmland to forest—are considered, and the cost of each measure is determined. Furthermore, the RGWLF watershed model is coupled with a widely recognized multi-objective optimization algorithm, the non-dominated sorting genetic algorithm II (NSGAII), the combination of which is applied in the Luanhe watershed to search for spatial BMPs for dissolved nitrogen (DisN). Fifty scenarios were finally selected from numerous possibilities and the results indicate that, at a minimum cost of 9.09×10^7 yuan, the DisN load is 3.1×10^7 kg and, at a maximum cost of 1.77×10^8 yuan, the total dissolved nitrogen load is 1.31×10^7 kg; with the no-measures scenario, the DisN load is 4.05×10^7 kg. This BMP optimization model system could assist decision-makers in determining a scientifically comprehensive plan to realize cost-effective goals for the watershed.

Keywords: BMPs; Revised GWLF; optimization; NSGAII; water quality

1. Introduction

Water pollution has received increasing attention, and many countries have increased their investments into water pollution control and water resources protection [1]. Designing scientific, reasonable, and efficient management measures to control or reduce pollutants at the watershed scale has become one of the most challenging problems for policy researchers and decision makers [2]. Many policies for the selection of best management practice (BMPs) have been created and applied to specific cases all around the world. For example, the United States and Europe have developed the corresponding Total Maximum Daily Loads and European Water Framework Directive for such purposes [3,4]. The implementation of these plans has provided a sufficient theoretical basis for subsequent watershed governance research [5].

BMPs are the most effective measure for controlling watershed pollution, including vegetative filter strips, land-use transformation, reducing the amount of fertilizer, terraces, and so on [6]. In general, BMP implementation plans should consist of a combination of maximum pollution reduction and minimal financial costs, due to limited budgets [7]. To our best knowledge, there are three optimization techniques which achieve the purpose. The first is setting a fixed number of scenarios manually and then calculating the corresponding pollutant production and cost separately [8,9]. By comparing the results of a limited number of different BMP scenarios, the final solution can be picked out. This method is straightforward and easy to implement, but may cause biased results as it depends on the experience of managers. Thus, the solution may not be the most cost-effective at the watershed scale [10]. The second is aggregating the environmental goals and economic factors into a single compromised objective function (e.g., a genetic algorithm or TaBu search algorithm) [11,12]. Through coupling the watershed model and optimization algorithm, only one optimal solution can be searched [13]. Compared with the first method, this method is more objective but usually takes more time, due to the necessary model runtime for each population per generation. The last technique is the coupling of a multi-objective optimization algorithm and a distributed watershed model to search over a set of solutions. This technique is similar to the second one, but it is able to provide a range of different trade-off BMPs among two or more conflicting objective functions. Due to its comprehensiveness and accuracy, it has been widely used in recent years [14].

The non-dominated sorting genetic algorithms II (NSGAII) is the most popular method for multi-objective optimization, whose ultimate goal is to find "Pareto-optimal" solutions, which is a modified version based on the genetic algorithm [15]. For example: Maringanti et al. utilized the Soil and Water Assessment Tool (SWAT) model and NGSAII to analyze the funding input under different combinations of fertilization reduction ratios, riparian filter belt widths, and other agricultural management measures in a tributary of the Mississippi River [16,17]; Ahmadi et al. combined the SWAT model and the NSGAII algorithm to evaluate the prevention and treatment effects of Atrazine through various evaluation indicators and non-point source management measures in the Eagle Creek Watershed in Indiana, USA [18]; and Geng et al. coupled the SWAT model and NSGAII algorithm to calculate the relationship between the amount of nutrient reduction and the required funding in the Chaohe River Watershed upstream of the Miyun Reservoir in China [19]. However, almost all studies in this category have focused on non-point source pollution BMPs and ignored point source BMP measures. It can be inferred, therefore, that the description of point source measures is not easy in the management practices of the watershed model.

Besides the selection of optimization technique, effective watershed management requires an understanding of the fundamental hydrologic and physicochemical processes in the watershed system, which are non-linear, dynamical, and complex [20]. Therefore, the applicability of the basin watershed model is very essential. A number of comprehensive watershed models have been developed to simulate hydrology and water quality in basins, and previous studies have demonstrated that some watershed models are well-behaved for the selection and targeted placement of BMPs (e.g., SWAT and AnnAGNPS) [21,22]. However, the watershed models in most previous studies have a high demand for data, and are difficult to apply in some areas where there is a lack of data [23]. The Revised Generalized Watershed Loading Function (RGWLF) is an improved semi-distributed hydrological model based on the Generalized Watershed Loading Function (GWLF) model, which has favorable stability, robustness, and less data requirements [24].

Given the above considerations, in this study, we incorporated RGWLF and NSGAII to identify a set of optimal BMPs based on both point and non-point source pollution control practices. Three tasks were completed to accomplish this research target: (1) adding the nutrient channel routing algorithms into the RGWLF then calibrating and verifying the parameters of the model; (2) determining the specific point source and non-point source management measures based on the established model; and (3) coupling the RGWLF model and NSGAII optimization algorithm based on a parameter sensitivity analysis to identify the optimal spatial allocation of BMPs for dissolved nitrogen.

2. Materials and Methods

2.1. Study Area and Data Sources

The Luanhe River watershed, which is located in North China (Figure 1), was selected in this study. It is a major component of one of the nine main river watersheds in China: the Haihe River watershed. It has been listed by the Chinese government as an important ecological conservation area in the Beijing–Tianjin–Hebei region. At the same time, its downstream reservoir is an important source of drinking water. The study watershed covers an area of about 30,000 km^2. According to the 2010 national Land Cover Data set, the watershed consists of 39.2% forest, 33.3% grassland, 21.6% agricultural area, and 1% water bodies. The climate is dominated by temperate semiarid monsoon climate. From 2000–2014, the annual average temperature for this area was 5.7 °C and the mean annual precipitation was 422 mm. Most of the precipitation was concentrated between April and August.

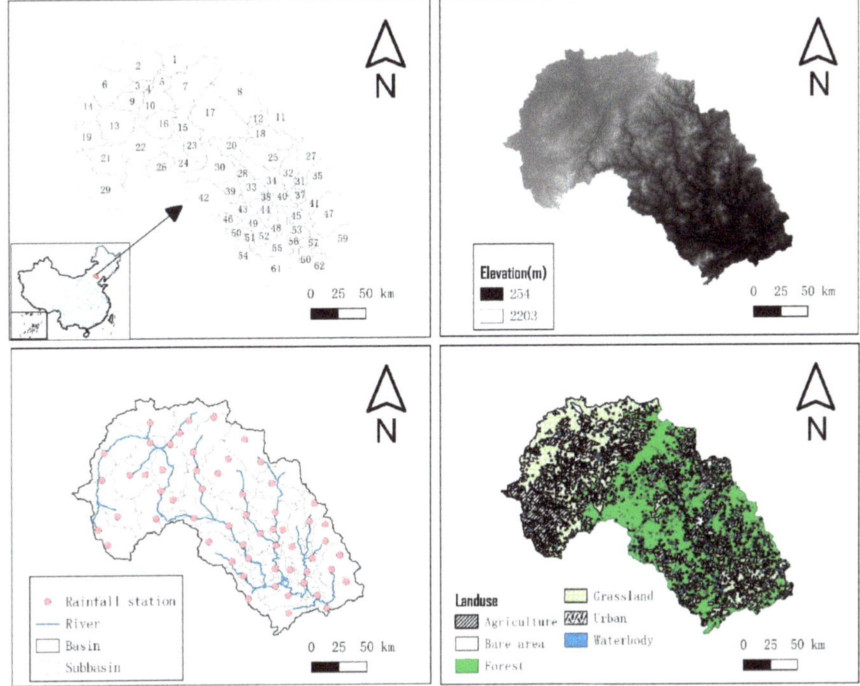

Figure 1. Location, sub-basins, land-use distribution, and elevation of the Luanhe watershed.

QGIS3 (https://qgis.org/en/site/index.html) and TauDEM (http://hydrology.usu.edu/taudem) were used to divide the Luanhe River Watershed into 62 sub-basins. Climate data were obtained from the Annual Hydrological Report P. R. China for precipitation data and the China Meteorological Data Service Center for temperature data. Thirty-meter resolution digital elevation models (DEMs) were downloaded from the Geospatial Data Cloud (http://www.gscloud.cn). Furthermore, a land-use map (in vector format) was provided by the National Earth System Science Data Center (http://www.geodata.cn). Observed flow data were gathered from the Annual Hydrological Report P. R. China. The hydrological monitoring station and the water quality monitoring station are located in sub-basin 62, which is at the outlet of the whole watershed. The above-mentioned data for the model setup are summarized in Table 1.

Table 1. Watershed Model input data used in the study.

Data Type	Data Description	Source	Time
Weather	Rainfall stations with the daily precipitation;	Annual Hydrological Report P.R. China	2005–2014
	Temperature stations with the daily average temperature	China Meteorological Data Service Center (http://cdc.cma.gov.cn/en)	2005–2014
DEM	Digital elevation model (30 m × 30 m)	Geospatial Data Cloud (http://www.gscloud.cn/)	2009
Land-use	Shapefile	Institute of Geographic Sciences and Natural Resources Research, CAS	2010
Hydrology	Streamflow/Monthly	Annual Hydrological Report P.R. China	2006–2014
Water Quality	Dissolved Nitrogen/Monthly	Chinese Academy of Environmental Planning	2006–2014
Point source	Annual discharge	Chinese Academy of Environmental Planning	2005–2014

2.2. The Watershed Model

The RGWLF model is a semi-distributed simulation model, which includes sub-basin calculation and the channel routing process, in contrast to the original model (GWLF) [25]. Detailed improvements and verification can be referred to in the author's previous article [24]. However, the original study did not include a description of the pollutant transport process in the channel. In this study, we added nutrient channel routing algorithms and made several references to the equations of the Sparrow models. Its main assumption is that contaminant flux along the stream satisfies a first-order decay process and the fraction of contaminant removed over a given stream distance is estimated as an exponential function of a first-order reaction rate coefficient and the cumulative water time of travel over this distance [26]:

$$NtrStore_{i,t} = NtrStore_{i-1,t} + UpNtr_{i,t} + LandNtr_{i,t} + PntNtr_{i,t}, \tag{1}$$

$$NtrOut_{i,t} = NtrStore_{i,t} \times FlowOut_{i,t}/FlowStore_{i,t}, \tag{2}$$

$$NtrOut_{i,t}' = NtrOut_{i,t} \times \exp(\theta_i \cdot TravelTime_{i,t}), \tag{3}$$

where

$NtrStore_{i,t}$ represents the amount of nutrient load in reach i at day t;
$UpNtr_{i,t}$ represents the amount of nutrient load from upstream in reach i at day t;
$LandNtr_{i,t}$ represents the amount of nutrient load from local land area in reach i at day t;
$PntNtr_{i,t}$ represents the amount of nutrient load from a point source in reach i at day t;
$NtrOut_{i,t}$ represents the amount of nutrient load to an outflow before attenuation in reach i at day t;
$NtrOut_{i,t}'$ represents the amount of nutrient load to an outflow after attenuation in reach i at day t;
θ_i represent the nutrient attenuation exponent in reach i; and
$TravelTime_{i,t}$ represents the flow travel time in reach i at day t.

2.3. BMPs and Costs

Three management practices were selected in this study: point source wastewater treatment, bank filter strips, and converting farmland to forest. The cost information for each practice is summarized in Table 2, which was based on published data and reports for this region.

According to the characteristics of point source wastewater in the study area, the sequencing batch reactor (SBR) treatment process was selected as the treatment method. This process is suitable for treating starch plant wastewater [27,28]. The cost of wastewater treatment consists of two parts: the construction of a wastewater treatment plant and wastewater treatment per unit volume:

$$Cost_{pns} = \sum_{i=1}^{n} \left[2.9178 \times Q_i^{0.9427} + \sum_{k=1}^{T} \left(q_{i,t} \times C_e \right) \right], \tag{4}$$

where $2.978 \times Q_i^{0.9427}$ represents the construction costs of the sewage treatment plant [29], where Q_i $(m^3 \cdot day^{-1})$ represents the scale of sewage treatment in sub-basin i; T represents the total number of days during the simulation; $q_{i,k}$ (m^3) represents the actual treatment flow of the sewage treatment plant in sub-basin i at day t; and Ce $(yuan/m^3)$ represents the costs of treating wastewater per cubic meter.

The cost of land-use conversion refers to the policy documents of the Chinese Ministry of Finance on returning farmland to forests. The bank filter strip trapping efficiency for nutrients is calculated using the following equations:

$$\text{trap}_{\text{surface}} = 0.367 \times \left(\text{width}_{\text{filtstrip}}\right)^{0.2967}, \tag{5}$$

$$\text{trap}_{\text{subsurf}} = 0.01 \times \left(2.1661 \times \text{width}_{\text{filtstrip}} - 5.1302\right), \tag{6}$$

where $\text{trap}_{\text{surface}}$ is the fraction of the constituent loading trapped by the filter strip, $\text{trap}_{\text{subsurf}}$ is the fraction of the subsurface flow constituent loading trapped by the filter strip, and $\text{width}_{\text{filtstrip}}$ is the width of the filter strip (m).

Table 2. Cost information and type of practices in the optimization.

Practice	Type	Sub-Basin	Cost
Wastewater treatment [1]	0%, 10%, 20%, 30%, 40%, 50%	8, 9, 34, 53–55	1.42 yuan/m^3
Converting farmland to forest	0%, 10%, 20%, 30%	All	5250 yuan/ha
Filter strips	0, 5, 10, 20, 30 m	All	2.83 yuan/m^2

[1] Sewage treatment plant scale is designed as the corresponding ratio of maximum wastewater discharge from the sub-basin during the simulation.

2.4. Multi-Objective Functions and NSGAII Optimization Processes

NSGAII generates offspring using a specific type of crossover and mutation and selects the next generation according to non-dominated sorting and crowding distance comparison. Figure 2 illustrates the genetic encoding of the various measure's structures for arrangement of conservation practices in the optimization algorithm. The sub-basins delineated by RGWLF and the configurations of management practices in each sub-basin form the basic chromosome units. The length of each chromosome is equal to the total number of sub-basins.

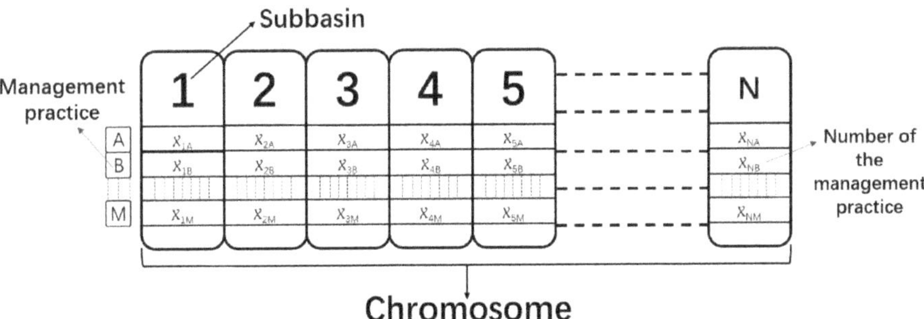

Figure 2. Gene string (chromosomes) for best management practices (BMPs) optimization in the watershed.

The analytical flow chart for this research is outlined in Figure 3. The watershed model was created to provide the nutrient load; the cost of management practices will be calculated by referring to the policy literature and actual surveys. To obtain the most cost-effective set of BMPs, the operation must satisfy two objective functions—the minimization of net total cost and the lowest dissolved nitrogen load—which are expressed by the equations:

$$min = \sum_{i=1}^{n} Cost_{pns,i} + Cost_{lu,i} + Cost_{strip,i}, \tag{7}$$

$$min = \sum_{i=1}^{n} DisN_{i,BMPs}, \tag{8}$$

where $Cost_{pns,i}$, $Cost_{lu,i}$, and $Cost_{strip,I}$ are the costs of wastewater treatment plant, converting land-use, and bank filter strisp in sub-basin i; and $DisN_{i,BMPs}$ is the total dissolved nitrogen load in sub-basin i.

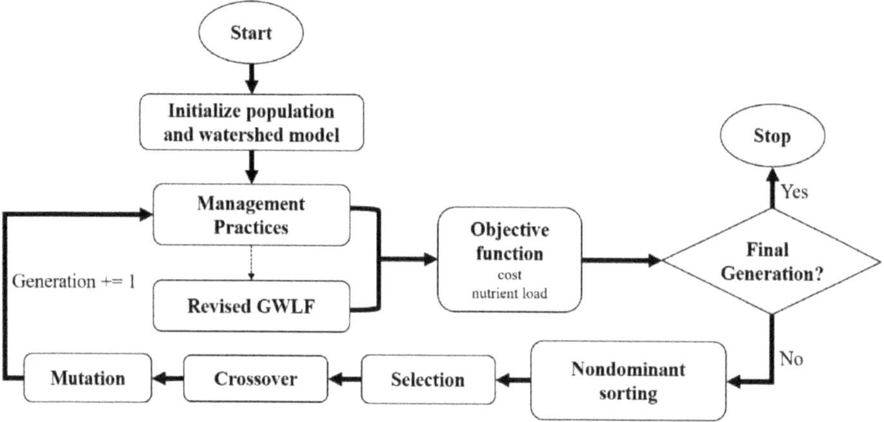

Figure 3. The optimization process for BMP selection and placement.

The optimization process consists of several procedures, as follows:

(1) Initialize the population and read the watershed model input data. Then, simulate the baseline scenario.

(2) For the two objective functions, obtain the pollutant load of the basin model and the net cost of each management measure combination.

(3) Through a series of processes, including non-dominated sorting, calculation of crowded distance, selection, crossover, and mutation, NSGAII obtains the Pareto-optimal result set of the current generation.

(4) Repeat the second process, determine whether it is the last generation, and then perform the third process.

3. Results and Discussion

3.1. RGWLF Model Calibration and Validation

Nine years of monthly records of observed streamflow and dissolved nitrogen data were used for model calibration and verification at the outlet of the watershed. The simulation period was from 2005–2014, in which the first year (2005) was used as a warming-up period, the data from 2006–2011 were used for the calibration process, and the rest were used for model validation.

The generalized likelihood uncertainty estimation (GLUE), a frequently used Bayesian parameter estimation method, was used for calibration analysis in terms of both hydrology and pollutants for the watershed model [30,31]. Table 3 lists the uniform prior distribution range and best value for each calibrated parameter after 10,000 iterations. The Nash–Sutcliffe efficiency (NES) and the coefficient of determination (R^2) were selected as the simulation evaluation criteria [32]. NSE can range from minus infinity to 1 and an efficiency of 1 indicates a perfect performance. R^2 ranges from 0 to 1, with higher values indicating a better fit for the model.

Table 3. Parameters selected for the calibration.

Parameters Name	Initial	Range	Calibration
Recession coefficient	0.008	[0, 0.015]	0.00782
Seepage coefficient	0.01	[0, 0.02]	0.00671
Recession threshold	10	[0, 20]	17.784
Seepage threshold	10	[0, 20]	6.497
Leakage coefficient	0.06	[0, 0.08]	0.0436
CN2 [1]	-	[−0.15, 0.15]	−0.0418
Agriculture	75	-	-
Forest	35	-	-
Grass	45	-	-
Urban	95	-	-
Dissolved nitrogen Concentration (mg/L)	-	-	-
Agriculture	4.0	[2.0, 6.0]	4.66
Forest	0.2	[0.1, 0.3]	0.12
Grass	2.0	[1.0, 3.0]	1.76
During fertilizer (mg/L) *Agriculture*	10.0	[7.0, 15.0]	12.47
Underground dissolved nitrogen concentration	0.1	[0.01, 0.5]	0.24
Dissolved nitrogen attenuation exponent	0.0004	[0.0, 0.0008]	0.000233

[1] Relative changes apply to the CN2 range.

As shown in Figure 4, during the calibration period, the NES was 0.93 for streamflow and 0.68 for DisN; moreover, the R^2 values were 0.93 and 0.68 for the simulated streamflow and DisN, respectively. At the same time, the model validation for the streamflow R^2 and NES were 0.77 and 0.78, respectively. Similar to streamflow, DisN had a lower validation value of 0.60 for R^2 and 0.58 for NES. Both streamflow and DisN simulations by the model had a good performance on a monthly time scale, which indicates robustness of the RGWLF model, according to previous research [33].

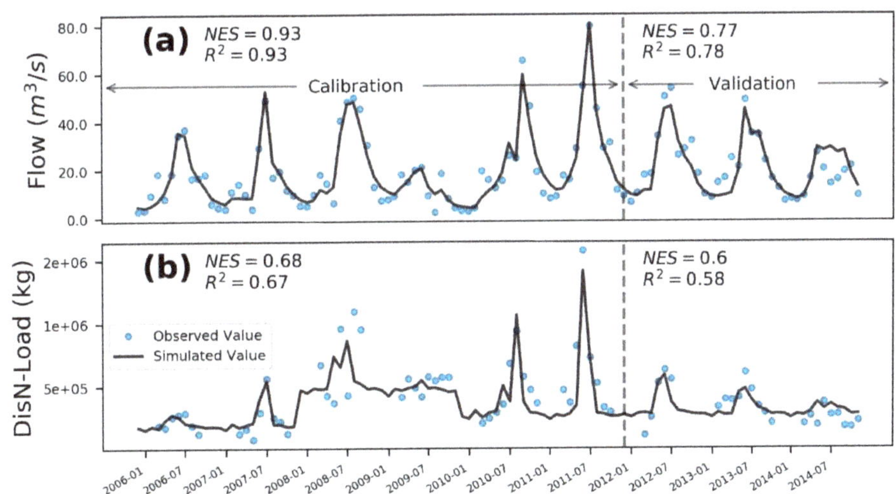

Figure 4. Time-series for the entire period (2006–2014) of observed and simulated monthly streamflow (**a**) and dissolved nitrogen (**b**).

3.2. Sensitivity Analysis of NSGAII Operational Parameters

To ensure accuracy of the NSGAII and optimization operational efficiency, sensitivity analyses were performed for the four key parameters, including population size, generations, crossover, and mutation probability, by the one-at-a-time sensitivity analysis method [16,34]. Table 4 lists the default and per-change values for the four NSGAII parameters.

Table 4. Parameters selected for sensitivity analysis of non-dominated sorting genetic algorithms II (NSGAII).

Order	Population Size	Generations	Crossover Probability	Mutation Probability
1	30	50	0.1	0.001
2	40	80	0.3	0.005
3	50	100	0.5	0.01
4	80	200	0.7	0.03
5	100	500	-	0.05
6	200	-	-	0.08
default	80	100	0.5	0.01
optimal	50	200	0.5	0.03

Figure 5a–d illustrate the Pareto-optimal fronts under per-change for the four key parameters. It is apparent in Figure 5a that the improvement in the Pareto-optimal fronts was remarkable as the population size increased from 30 to 50, but there was little gap for Pareto-optimal fronts when the population size was increased from 80 to 200. The main reason for this case is possibly that a population size of 50 had enough convergence chance for the solution space's freedom of the whole optimization system. Furthermore, a value of 50 would considerably reduce the computation time, compared to the default values of 80, 100 and 200.

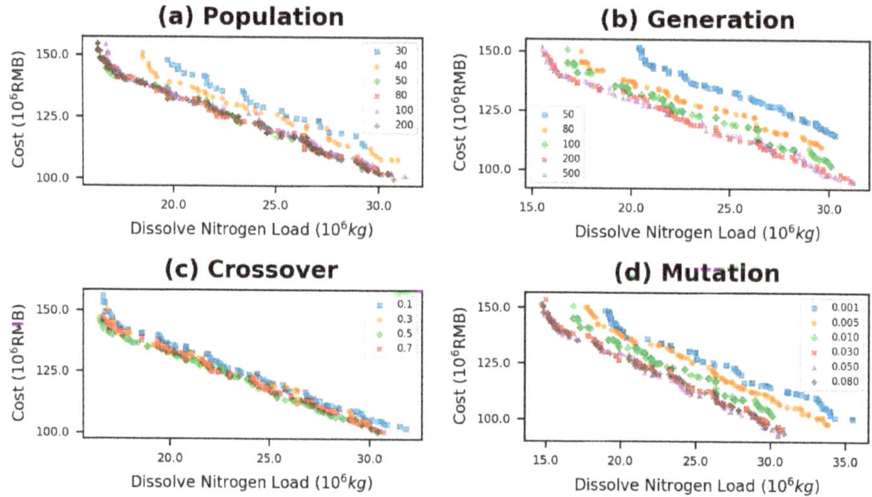

Figure 5. Pareto-optimal fronts for the sensitivity analysis of NSGAII.

As shown in Figure 5b, the number of generations also had an appreciable impact on the Pareto-optimal fronts. A significant improvement in the Pareto-optimal front was noticed, with the number of generations increasing from 50 to 200. In stark contrast, increasing from 200 to 500 did not bring any improvement for the Pareto-optimal fronts.

Unlike the above two parameters, the Pareto-optimal front was not sensitive to changes in the crossover probability, which is displayed in Figure 5c. The Pareto-optimal front had an inconspicuous shift approaching the origin as the crossover probability increasing from 0.1 to 0.5. When the crossover probability reached a value of 0.7, the Pareto-optimal front was farther away from the origin.

The trend of the Pareto-optimal front, as affected by the change of Mutation probability, is presented in Figure 5d. What stands out most in this figure is that mutation probability is a sensitive parameter, especially in the range from 0.001 to 0.03, where the Pareto-optimal front approached the origin with an increase of parameter value. However, there was no benefit for the Pareto-optimal front as the parameter increased from 0.03 to 0.08.

Considering all of the above analyses, the values 50, 200, 0.5 and 0.03 were used as population size, generations, crossover, and mutation probability, respectively, for the further optimization processes.

3.3. Optimization Result and Cost-Effectiveness Analysis

The baseline scenario for this watershed was no management practice and 4.05×10^7 kg for the total dissolved nitrogen load over nine years.

Figure 6 provides the final optimization progress with two objectives, minimizing the net cost and dissolved nitrogen load by scattering four Pareto-optimal fronts belonging to four generations (1, 50, 100 and 200). The overall trend for the four scatter diagrams was that the dissolved nitrogen decreased as the net cost increased. Each new generation's chromosomes were initialized by random numbers without no optimization process (such as non-dominance, selection, crossover, and so on), so it was regarded as the 0th generation and not shown in the figure. In the first generation, the distribution of points was more concentrated, indicating that the solutions were less differentiated. As the number of generations increased, the solution became more and more scalable, and came closer to the origin. For the final generation, the most widespread solution set, which was closest to the origin, can be seen. Thus, the best quality and broadest decision supports were supplied to the policy researchers and decision-makers.

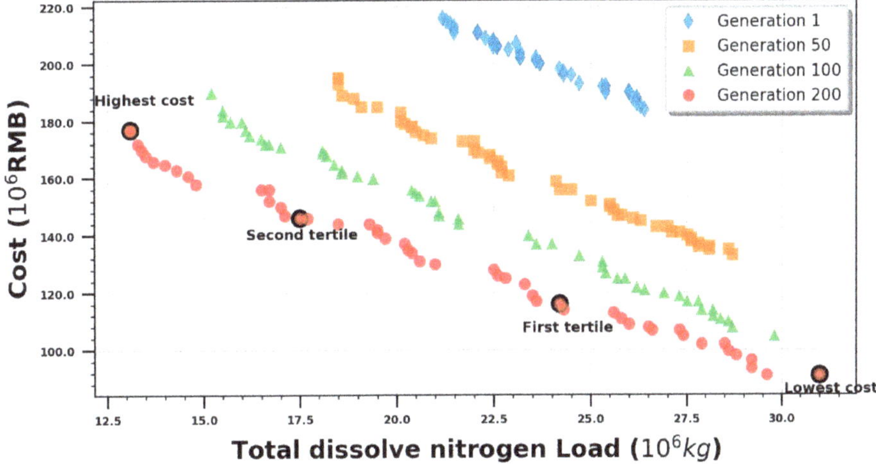

Figure 6. Pareto-optimal fronts for total dissolved nitrogen load and cost.

The set of solutions was divided into three segments with four endpoints, which were: the lowest cost, the lower tertile cost point, the higher tertile cost point, and the highest cost. The values of cost, dissolved nitrogen load, and spatial distribution of corresponding management measures for the four scenarios are shown in Figure 7. For the lowest cost scenario (Figure 7a), the scale of treatment of polluted water plants is small and the measures are mainly bank filter strips and converting farmland to forest. As cost increased, the intensity and scale of management practices also increased, which is illustrated in Figure 7b,c. At the point of maximum cost (Figure 7d), the load of dissolved nitrogen reached its minimum value, where the designed daily treatment capacity of the sewage plant was extremely high and there was no conversion of farmland to forests in some upstream sub-basins. The majority of management practices are concentrated in middle and lower regions of the watershed.

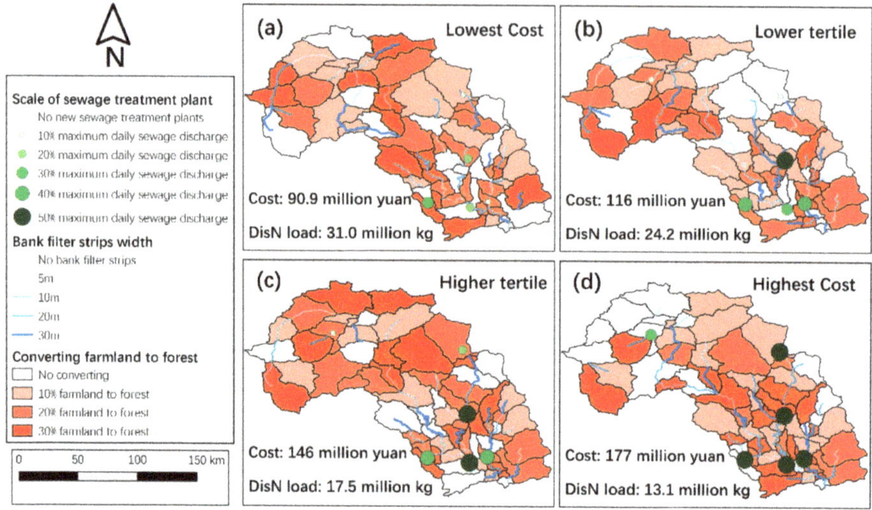

Figure 7. The types and locations of optimal BMPs for different costs, as provided in Figure 6.

4. Conclusions

In this study, we supplemented the RGWLF model with the addition of a pollutant attenuation process in the river channel and a bank filter strips reduction function, based on our previous research. Meanwhile, taking into account the regional pollution source composition of the watershed studied and the characteristics of the models used, three types of pollution control measures—point source wastewater treatment, bank filter strips, and converting farmland to forest—were considered, and the cost of each measure was determined. Furthermore, the optimization algorithm NGSAII was linked with the RGWFL watershed model and the implemented measures to search for a Pareto-optimal set of BMPs. Before the final optimization calculation, sensitivity analyses for the four key parameters of NGSAII were performed. According to the results of the sensitivity analysis, the entire coupled model system supplied 50 solutions which could provide managers with many pollution reduction options. In the end, depending on the cost, we chose four gradient solutions and demonstrated the geospatial distribution of their different management measure characteristics and briefly analyzed the differences and features of the four solutions.

The research results show that, with almost no increase in data requirements, dissolved nitrogen had an excellent simulation performance, expanding the spatial scope of pollutant simulation for the GWLF. Moreover, due to its robustness and semi-distribution, the RGWLF model was able to be coupled with NSGAII. The entire linkage system had good performance in the optimization process and provided a range of watershed implementation measures for DisN reduction and minimizing cost, which is a worthy reference for policy researchers and decision-makers to realize their watershed management goals. However, limited by the available data, funding, and researchers' capabilities, we did not consider other target indicators, such as sediment, phosphorus, and so on, which could be carried out in future research.

Author Contributions: Y.W. conceived and designed the framework; Z.Q. and G.K. wrote the paper; Z.Q. and X.W. conducted software and data analysis; Z.Q. and Y.S. Visualization; Y.W. contributed to improving the article. All authors have read and agreed to the published version of the manuscript.

Funding: This research was funded by Major Science and Technology Program for Water Pollution Control and Treatment, grant number 2017ZX07301-001-06.

Acknowledgments: We wish to thank the Chinese Academy of Environmental Planning for providing the hydrology quality data. We would also like to acknowledge the National Science Data Share Project Data Sharing Infrastructure of Earth System Science (China) for the data support.

Conflicts of Interest: The authors declare no conflict of interest.

References

1. Gleick, P.H. Global freshwater resources: Soft-path solutions for the 21st century. *Science* **2003**, *302*, 1524–1528. [CrossRef]
2. Fleifle, A.; Saavedra, O.; Yoshimura, C.; Elzeir, M.; Tawfik, A. Optimization of integrated water quality management for agricultural efficiency and environmental conservation. *Environ. Sci. Pollut. Res.* **2014**, *21*, 8095–8111. [CrossRef]
3. Borah, D.K.; Bera, M. Watershed-scale hydrologic and nonpoint-source pollution models: Review of applications. *Trans. ASAE* **2004**, *47*, 789–803. [CrossRef]
4. Hering, D.; Borja, A.; Carstensen, J.; Carvalho, L.; Elliott, M.; Feld, C.K.; Heiskanen, A.-S.; Johnson, R.K.; Moe, J.; Pont, D.; et al. The European Water Framework Directive at the age of 10: A critical review of the achievements with recommendations for the future. *Sci. Total Environ.* **2010**, *408*, 4007–4019. [CrossRef]
5. Qu, J.; Fan, M. The current state of water quality and technology development for water pollution control in China. *Crit. Rev. Environ. Sci. Technol.* **2010**, *40*, 519–560. [CrossRef]
6. Xie, H.; Chen, L.; Shen, Z. Assessment of agricultural best management practices using models: Current issues and future perspectives. *Water* **2015**, *7*, 1088–1108. [CrossRef]
7. Balana, B.B.; Vinten, A.; Slee, B. A review on cost-effectiveness analysis of agri-environmental measures related to the EU WFD: Key issues, methods, and applications. *Ecol. Econ.* **2011**, *70*, 1021–1031. [CrossRef]
8. Guo, H.; Hu, Q.; Jiang, T. Annual and seasonal streamflow responses to climate and land-cover changes in the Poyang Lake basin, China. *J. Hydrol.* **2008**, *355*, 106–122. [CrossRef]
9. Hundecha, Y.; Bardossy, A. Modeling of the effect of land use changes on the runoff generation of a river basin through parameter regionalization of a watershed model. *J. Hydrol.* **2004**, *292*, 281–295. [CrossRef]
10. Deb, K. Multi-objective genetic algorithms: Problem difficulties and construction of test problems. *Evol. Comput.* **1999**, *7*, 205–230. [CrossRef] [PubMed]
11. Kaini, P.; Artita, K.; Nicklow, J.W. Optimizing structural best management practices using SWAT and genetic algorithm to improve water quality goals. *Water Resour. Manag.* **2012**, *26*, 1827–1845. [CrossRef]
12. Qi, H.; Altinakar, M.S.; Vieira, D.A.N.; Alidaee, B. Application of Tabu search algorithm with a coupled AnnAGNPS-CCHE1D model to optimize agricultural land use. *J. Am. Water Resour. Assoc.* **2008**, *44*, 866–878. [CrossRef]
13. Muleta, M.K.; Nicklow, J.W. Decision support for watershed management using evolutionary algorithms. *J. Water Resour. Plan. Manag.* **2005**, *131*, 35–44. [CrossRef]
14. Qiu, J.; Shen, Z.; Huang, M.; Zhang, X. Exploring effective best management practices in the Miyun reservoir watershed, China. *Ecol. Eng.* **2018**, *123*, 30–42. [CrossRef]
15. Deb, K.; Pratap, A.; Agarwal, S.; Meyarivan, T. A fast and elitist multiobjective genetic algorithm: NSGA-II. *IEEE Trans. Evol. Comput.* **2002**, *6*, 182–197. [CrossRef]
16. Maringanti, C.; Chaubey, I.; Arabi, M.; Engel, B. Application of a multi-objective optimization method to provide least cost alternatives for NPS pollution control. *Environ. Manag.* **2011**, *48*, 448–461. [CrossRef]
17. Maringanti, C.; Chaubey, I.; Popp, J. Development of a multiobjective optimization tool for the selection and placement of best management practices for nonpoint source pollution control. *Water Resour. Res.* **2009**, *45*. [CrossRef]
18. Ahmadi, M.; Arabi, M.; Hoag, D.L.; Engel, B.A. A mixed discrete-continuous variable multiobjective genetic algorithm for targeted implementation of nonpoint source pollution control practices. *Water Resour. Res.* **2013**, *49*, 8344–8356. [CrossRef]
19. Geng, R.; Yin, P.; Sharpley, A.N. A coupled model system to optimize the best management practices for nonpoint source pollution control. *J. Clean. Prod.* **2019**, *220*, 581–592. [CrossRef]

20. Wellen, C.; Kamran-Disfani, A.-R.; Arhonditsis, G.B. Evaluation of the current state of distributed watershed nutrient water quality modeling. *Environ. Sci. Technol.* **2015**, *49*, 3278–3290. [CrossRef]

21. Arabi, M.; Govindaraju, R.S.; Hantush, M.M. Cost-effective allocation of watershed management practices using a genetic algorithm. *Water Resour. Res.* **2006**, *42*. [CrossRef]

22. Qi, H.; Altinakar, M.S. Vegetation buffer strips design using an optimization approach for non-point source pollutant control of an agricultural watershed. *Water Resour. Manag.* **2011**, *25*, 565–578. [CrossRef]

23. Qi, Z.; Kang, G.; Chu, C.; Qiu, Y.; Xu, Z.; Wang, Y. Comparison of SWAT and GWLF model simulation performance in humid south and semi-arid north of China. *Water* **2017**, *9*, 567. [CrossRef]

24. Qi, Z.; Kang, G.; Shen, M.; Wang, Y.; Chu, C. The improvement in GWLF model simulation performance in watershed hydrology by changing the transport framework. *Water Resour. Manag.* **2019**, *33*, 923–937. [CrossRef]

25. Haith, D.A.; Shoemaker, L.L. Generalized watershed loading functions for stream flow nutrients. *JAWRA J. Am. Water Resour. Assoc.* **1987**, *23*, 471–478. [CrossRef]

26. Schwarz, G.; Hoos, A.B.; Alexander, R.; Smith, R. The SPARROW surface water-quality model—Theory, application and user documentation. In *U.S. Geological Survey Techniques and Methods*; Series number 6-B3; U.S. Geological Survey: Reston, VA, USA, 2006; 248p.

27. Shi, H.; Wang, L. Treatment of the corn starch wastewater by UASB-SBR process. *Shanxi Archit.* **2003**, *29*, 71–72. (In Chinese)

28. Yu, H. UASB-SBR Technical Processing Starch Wastewater. Master's Thesis, Changan University, Xi'an, China, 2007. (In Chinese).

29. Liu, J.; Zheng, X.; Gao, C.; Chen, L. Study on area, operating and construction costs of urban wastewater treatment plants. *Chin. J. Environ. Eng.* **2010**, *4*, 2522–2526. (In Chinese)

30. Beven, K.; Binley, A. The future of distributed models: Model calibration and uncertainty prediction. *Hydrol. Process.* **1992**, *6*, 279–298. [CrossRef]

31. Beven, K.; Smith, P.; Freer, J. Comment on "Hydrological forecasting uncertainty assessment: Incoherence of the GLUE methodology" by Pietro Mantovan and Ezio Todini. *J. Hydrol.* **2007**, *338*, 315–318. [CrossRef]

32. Nash, J.E.; Sutcliffe, J.V. River flow forecasting through conceptual models part I—A discussion of principles. *J. Hydrol.* **1970**, *10*, 282–290. [CrossRef]

33. Moriasi, D.N.; Arnold, J.G.; Van Liew, M.W.; Bingner, R.L.; Harmel, R.D.; Veith, T.L. Model evaluation guidelines for systematic quantification of accuracy in watershed simulations. *Trans. ASABE* **2007**, *50*, 885–900. [CrossRef]

34. Hamby, D. A comparison of sensitivity analysis techniques. *Health Phys.* **1995**, *68*, 195–204. [CrossRef] [PubMed]

Article

Coupling Coordination Assessment on Sponge City Construction and Its Spatial Pattern in Henan Province, China

Kun Wang [1], Lijun Zhang [1,*], Lulu Zhang [2] and Shujuan Cheng [1]

[1] Key Laboratory of Geospatial Technology for Middle and Lower Yellow River Regions, College of Environment and Planning, Henan University, Kaifeng 475004, China; wangkun@vip.henu.edu.cn (K.W.); shujuancheng@henu.edu.cn (S.C.)

[2] Luoyang No.4 Vocational High School, Luoyang Service Outsourcing College, Luoyang 471023, China; zll@vip.henu.edu.cn

* Correspondence: zlj7happy@vip.henu.edu.cn

Received: 15 October 2020; Accepted: 7 December 2020; Published: 11 December 2020

Abstract: Coordinating the "green" and "gray" infrastructure construction and the socioeconomic development is essential to sponge city construction. Most previous research has investigated the structural and non-structural approach for urban water management, such as operational practice, engineered measures, technical solutions, or planning management. However, there is a shortage of strategic management approaches to identify pilot sponge cities, which is essential to cities in developing countries under huge financial pressures. Hence, this paper proposed a coupling coordination evaluation index system to assess the coordination degree between economic development and infrastructure construction in Henan Province in central China. Then, the paper analyzed the differences of the coordination level and its spatial statistical pattern of the coupled and coordinated development of sponge city construction in Henan Province. The results show that: (1) from the perspective of comprehensive level, the problems of inadequate and unbalanced development of infrastructure construction and economic development level are prominent; (2) from the perspective of coordinated development level, the level of coupling and coordination development in Henan Province increased during the sample period, but the level of coupling and coordination development in each region was small; (3) from the perspective of relative development, Zhengzhou City is lagging behind in infrastructure, indicating that economic growth is faster than infrastructure construction, and other regions are lagging economic development, indicating that infrastructure construction is faster than economic growth; and (4) from the spatial statistical analysis, there is spatial positive correlation, that is, the area with high coupling degree of infrastructure construction and economic development level tends to be significantly concentrated in space. Studies have shown that Henan Province should focus on strengthening the construction of "green" infrastructure and increasing the infiltration of the underlying surface to counter the precipitation in urban areas in extreme climates.

Keywords: sponge city; coupling coordination degree; spatial pattern

---

1. Introduction

In recent years, the acceleration of urbanization in China has caused to the change of underlying surface. Due to the increasement of extreme climate and the backwardness of urban rainwater management system, the increasing of surface runoff is the inevitable consequence of frequent urban flooding. The water problem has become a common urban problem in China, posing a serious threat to the safety of life and property of urban residents, and has resulted in a huge impact on the environment and social economy [1]. For instance, on 21 July 2012, a serious waterlogging disaster occurred in Beijing, causing 79 deaths and 11.64 billion CNY in economic losses [2]. The 2015 Environmental Status Report conducted water quality assessments of groundwater monitoring sites in 202 cities. Among them, 9.1% had good water quality, and 61.3% had poor water quality [3]. Other studies have also shown that life-saving threats and economic losses caused by urban floods are on the rise [4]. In order to solve the urban water disaster and water environment problems, in 2013, China's central urbanization work conference officially proposed the construction of the sponge city [2].

Sponge city means that the city can be like a "sponge". It has good "elasticity" in adapting to environmental changes and coping with natural disasters. When it rains, it absorbs water, stores water, seeps water, and cleans water. When necessary, it releases stored water [5]. In addition to improving the absorption and storage of rainwater, the sponge city combines "green" infrastructure to enhance ecological functions, enhance urban aesthetic value, and create additional comfort spaces [6]. In addition, the construction of China's sponge city also draws on the experience of some developed countries in urban rainwater drainage facilities, such as the United States' Low Impact Development (LID) [7]; the UK's Sustainable Urban Drainage System (SuDs) [8]; Australia's Water Sensitive Urban Design (WSUD) [9]; and New Zealand's Low Impact Urban Development Design (LIDUD) [10]. Chinese scholars have made a lot of progress in the exploration of sponge city construction, combined with local favorable conditions to shape the experience of surrounding landscape system and rainwater system management, and extract the surrounding landscape (river and lake) from the macro level by using modern technology (Geographic Information System, Remote Sensing, etc.). The system, from the microscopic level to the landscape design (green roof, etc.), combined with water landscape reconstruction, can achieve the sustainable development of human-water relationship [2,11–14].

However, in the face of unprecedented climate change and urbanization, it is more important to strengthen the elasticity of urban infrastructure [15–17]. But most of the research is limited to describing the flood recovery capacity of urban drainage systems [18,19]. Changes in climatic conditions, such as increased precipitation intensity, changes in precipitation patterns, and more extreme weather events, have caused urban drainage systems to be frequently hit by heavy rains [20,21]; urbanization has increased the concentration of population and economic activity, plus the burden of existing urban social development systems on sewage, urban runoff and water pollutant types [22]. A sustainable urban infrastructure system should be effective and adaptable in an uncertain future, which will contribute to the flexibility of green and grey infrastructure in the future, and to carry out sustainable development assessments of urban drainage systems [23]. Numerous studies have evaluated the role of different types of green infrastructure in stormwater management and carbon emission control and compared the performance of grey infrastructure, which is considered to be mitigating and adapting to climate change and urbanization. Interferences contribute to the effectiveness of sustainable development [23–25], but grey infrastructure is also a necessary condition for dealing with extreme rains [26]. The mutual development and application of grey infrastructure and green infrastructure has been used in many countries to mitigate urban floods [27,28]. Green infrastructure is an important measure in many common stormwater management strategies, such as low-impact development, best management practices, and water-sensitive urban design [29]. The study of China's sponge cities also evaluated the effects of green infrastructure [30] and evaluated the performance of low-impact development practices as a means of reducing water flooding in urban small watersheds [31]. However, the investment and income mechanism of Sponge City is currently in the exploration stage [32], and the proportion of gray infrastructure and green infrastructure investment is also rarely considered. Therefore, this requires urban managers to plan the rainwater management infrastructure from an economic perspective and an adaptation path [33].

Sponge city construction, as a comprehensive ecological project, should be evaluated from the aspects of cost, social, and economic benefits, hydrology, and water quality. Lack of adequate financial support and effective market incentives is one of the main obstacles to sustainable storm water management [34]. To solve this problem, the United States introduced the Stormwater Utility Fees

(SUF) in the 1970s, as user fees; SUF has increasingly been used by local governments as an alternative source of income for implementing sustainable rainwater projects [35]. In China, some researchers are also exploring revenue sources for sustainable stormwater management, and the research results show that 76% of the respondents agreed to pay for life-cycle maintenance of sponge city facilities, and the median amount of willingness to pay was 16.57 CNY (2.53 USD) per month [36]. In addition, under the guidance of ecological civilization construction, the green infrastructure in China's sponge city construction should be incorporated into the ecosystem service value, consider cities as a coupled nature-human system, for such a system, different urban ecosystem structures, such as lakes, wetlands, rivers, and parks, are an integral part, providing important services for the landscape, including water conservation and runoff regulation, and also helping urban residents to support social welfare [37]. The recycling of rainwater for the gray infrastructure of sponge city construction provides services for the community and green infrastructure. On the one hand, it promotes the efficiency of recycled water use in the community. On the other hand, it supports the water demand for green infrastructure.

However, these studies are still obviously insufficient for the construction of sponge cities. Chinese scholars rarely conduct research on economic development level and infrastructure, and the construction of sponge city is a comprehensive process, establishing a perfect "gray" infrastructure and "green" infrastructure system, actively promoting the fact that the pilot work of sponge city construction is the key to solving the problem of domestic embarrassment. To this end, this paper attempts to analyze the coupling and coordination degree of infrastructure and economic development in 18 prefecture-level cities in Henan Province, as well as provide a new research idea for the construction of the sponge city and a theoretical basis for the future development of the sponge city.

2. Materials and Methods

2.1. Study Area

Henan Province is located in the transition zone between central and eastern China, the southern region and the northern region. It has a large span from north to south. From the west to the east, the Yellow River runs through. There are many rivers in Henan Province, which provide a surrounding landscape (rivers and lakes) for the construction of the sponge city. Wetland and green space provide good carriers for the construction of sponge city. Henan Province has diverse terrain, most of which is dominated by mountains and plains. The "green" infrastructure construction in different regions is also very different. Most of Henan's climate is dominated by temperate monsoon climate, and some regions are transition from subtropical monsoon climate to temperate monsoon climate. In the zone, affected by the monsoon climate, the impact of the formation of the mountain microclimate is more obvious, and the differences between the regions are greater. Therefore, the construction of sponge cities in different regions varies from region to region; Henan Province has more latitudes, and most of the region belongs to the "northern region (dividing line between subtropical and temperate monsoons climate, northern region belongs to the temperate monsoon climate)"; however, Xinyang and Nanyang cities in the southern part of Henan province have subtropical climates, because they are near the Qinling Mountains-Huaihe River line (the boundary between temperate and subtropical monsoon climates). The difference between the south and the north of the south is obvious. The problems faced by various regions in Henan Province are similar to those faced by China at this stage of development. They are called "the epitome" of China (Figure 1); the construction of sponge cities in Henan Province can be studied. The commonality of the sponge city provides practical experience for the construction of China's sponge city. The research area is the smallest scale of 18 prefecture-level cities in Henan Province. The differences in climate, precipitation, and urban development between the regions are more obvious, but precipitation is mainly concentrated in the summer and autumn, more often with short-term heavy rain, annual average the temperature is about 16 degrees Celsius, and the annual precipitation is about 650 mm (Figure 2). In addition, most of the terrain in the southeast of Henan province is plain, with monsoon climate in summer and short-term precipitation, the precipitation is relatively large, and the terrain is low and flat, with slow surface runoff and infiltration, where it is easy to cause waterlogging. Due to the influence of climate and topography, the precipitation in the study area decreases from southeast to northwest. However, in recent years, due to unreasonable development and utilization, as well as the impact of extreme climate, some rivers and lakes are seriously deficient in water. In the summer and autumn, the urban drainage capacity is insufficient, and the urban shackles occur, which also hinders the development of the city to a certain extent.

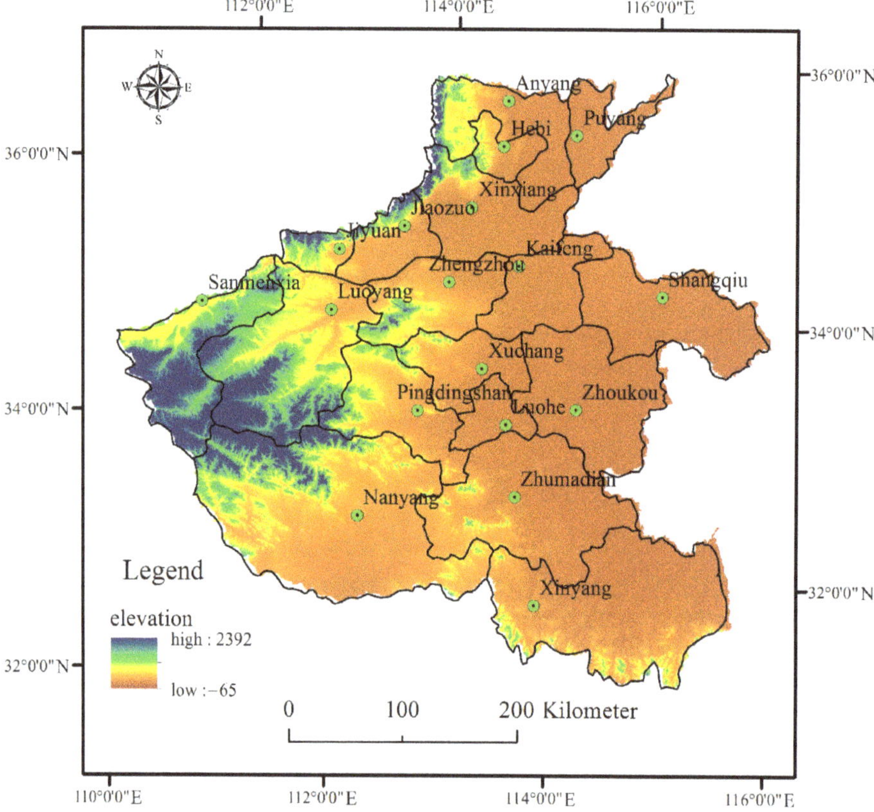

Figure 1. Research area.

2.2. Data Source

The construction of the sponge city not only requires the support of science and technology to change the landscape of the underlying surface but also requires a large amount of funds to build a "gray" infrastructure system and a "green" infrastructure system. The use of "green" and "gray" infrastructure can enhance urban resilience, and "green" infrastructure can provide greater adaptability and resistance to the unpredictability of future climate forecasts. In the face of unsustainable urban drainage practices, a good strategy is to integrate green and gray infrastructure into a cyclic utilization control system, which will bring more advantages and reduce problems, while improving existing elasticity of urban drainage systems [23]. In view of this, this study selected the road area, drainage pipe length, road length, sewage treatment rate, and number of bridges in the "gray" infrastructure system; the green area in the "green" infrastructure system, the per capita park green area, the park area; and the main indicators, such as water reuse rate and green coverage area, economic development level of GDP, fixed investment in garden green space, fixed drainage investment, urban population, and urbanization rate, to be analyzed (Table 1). The missing data of some indicators were simulated by linear interpolation. In this paper, the missing indicators are the fixed investment in Kaifeng garden green space and drainage in 2016, as well as the fixed investment in Luohe garden green space and Xinxiang Drainage in 2014. In this paper, linear interpolation is used, that is, a function method that is used to calculate an unknown quantity between two endpoints of a line, because the interpolation accuracy on nodes can be guaranteed to be higher, and it is more convenient than other interpolation methods, such as parabolic interpolation. The data mainly came from China Urban Construction Statistical Yearbook 2012–2016 and Henan Statistical Yearbook 2013–2017.

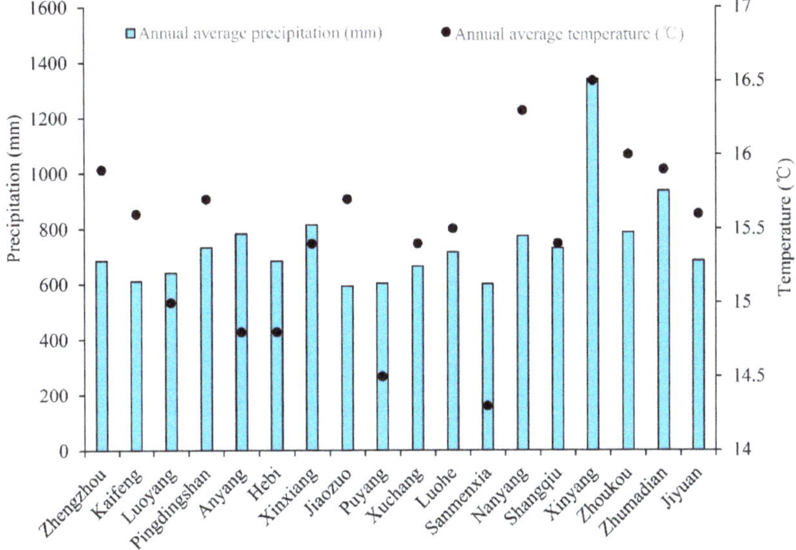

Figure 2. Annual average temperature and annual precipitation in the study area.

Table 1. Coupling coordination evaluation index system for sponge city construction.

Target Layer	System Layer	Indicator Layer	Index
Coupling Coordination Evaluation Index of Sponge City Construction in Research Area (A)	"gray" infrastructure construction (B$_1$)	Road area (10,000 square meters) X$_1$	Negative
		Drainage pipe length (km) X$_2$	Positive
		Road length (km) X$_4$	Negative
		Sewage treatment rate (%) X$_5$	Positive
		Number of bridges (seat) X$_8$	Positive
	"Green" infrastructure construction (B$_2$)	Green area (hectare) X$_3$	Positive
		Per capita park green area (m^2) X$_6$	Positive
		Park area (hectare) X$_7$	Positive
		Green coverage area (hectare) X$_{10}$	Positive
	The level of economic development (C$_1$)	GDP(Billion) X$_{11}$	Positive
		Garden green land fixed investment (ten thousand yuan) X$_{12}$	Positive
		Drainage fixed investment (ten thousand yuan) X$_{13}$	Positive
		Urban population (10,000 people) X$_{14}$	Positive
		Built-up area (square kilometers) X$_{15}$	Negative
		Urbanization rate (%) X$_{16}$	Positive
		The actual investment in the construction of municipal public facilities is in place (10,000 yuan) X$_{17}$	Positive

2.3. Coupling Coordination Mechanism

Coupling refers to the phenomenon that two or more systems interact and influence each other, while coupling degree is used to describe the degree of mutual influence of multiple systems. In general, the degree and quality of coupling action, using coordination to judge. Coordination is a benign embodiment based on coupling effect, and the coordination degree is used to measure the coordination degree of multiple systems in the coupling process. Both have a connection already and have distinction. The degree of coupling reflects the degree of system interaction, and the degree of coordination reflects the degree of coupling coordination [38].

The construction of sponge city is a composite system composed of green infrastructure, gray infrastructure, and economic development level, which interact and form a symbiotic coupling relationship. In China, with ecological civilization, green infrastructure is the leading driving force for the construction and development of sponge cities, and this driving force can improve the material structure of the underlying surface and increase the infiltration of rainwater, which, to a certain extent, is conducive to dealing with urban waterlogging caused by extreme climate. In addition, the construction of green infrastructure can provide ecosystem services value to surrounding communities. The construction of grey infrastructure restricts the improvement of the underlying surface but promotes the management of rainwater, domestic water, and sewage, which is conducive to the economical utilization of water resources and provides water demand for green infrastructure. The level of economic development is the material guarantee for the construction of green infrastructure and gray infrastructure, and it can provide financial support for the construction of green infrastructure and gray infrastructure through the investment effect (PPP, Public-Private Partnership model or willingness to pay) and the value of ecosystem services. Therefore, the three form a multiple correlation interaction coupling effect, which has both positive and negative effects. The sustainable development of sponge city can be realized only when the three cooperate and coordinate with each other.

Green Infrastructure construction in a sponge city references LID (low-impact development), an ecologically-based planning and engineering design approach to managing stormwater runoff and stormwater treatment technologies. In the practice of SuDS (Sensitive urban drainage system design) in Western Europe, SuDS sustainable stormwater management measures mitigate and adapt to climate change through carbon sequestration and urban cooling, with multiple ecological and environmental benefits, based on such a concept, as much as possible to restore the natural and pre-development drainage system. The construction of grey infrastructure in a sponge city is mainly used for sewage treatment and watercourse pipe network construction. In stormwater management, management systems link non-structural approaches to structural deployment for pollution prevention and drainage. Basically, similar to Best Management Practices (BMPs) in the United States and Canada. WSUD (Water-Sensitive Urban Design) is mainly to protect and strengthen the natural water system of urban development, better integrate rainwater into the landscape, minimize impermeable water surface, reduce the peak flow brought by urban development, and protect water quality, reduce the development costs of drainage infrastructure, while adding value at the same time, and all urban water systems should be better integrated in urban design [29]. And sponge city is a systematic framework with methods to improve urban water problems. Therefore, based on the coupling coordination mechanism, this paper provides a basis for decision-making and management of sponge city construction.

2.4. Evaluation Index System

The sponge city construction evaluation is a diagnostic analysis of the drainage capacity of an area and the environmental impact of the underlying surface. The sponge city construction evaluation is based on the construction of the infrastructure system and economic development level model. Among them, gray infrastructure is a traditional municipal infrastructure dominated by single-function municipal engineering, consisting of roads, bridges, pipelines, and other networks that ensure the proper functioning of the industrial economy [39]. However, the construction of road infrastructure changes the underlying surface structure, reduces the infiltration amount of surface runoff, and is also the infrastructure that has a great impact on urban residents' travel. In general, drainage pipeline network is under road infrastructure. In the construction of sponge cities, gray infrastructure facilities provide municipal entities and residents with municipal infrastructure services, such as flood protection, stormwater drainage, and wastewater treatment [39]. The "gray" infrastructure system reflects the traditional drainage system and the foundation of a good operation of the city. Green infrastructure refers to a green space network consisting of natural areas and other interconnected open spaces, including natural areas, public protected areas, and productive land with conservation values; it also represents a protected open system network that protects the value of natural resources and maintains the survival functions of humans, animals, and plants [40,41]. The development of green infrastructure is the result of joint promotion of parks, park systems, open spaces, greenways, ecological networks, biological corridors, and storm-water management [39]. The "green" infrastructure system reflects the sustainable development model of improving the underlying surface and is the basis of a good ecological environment of the city. The level of economic development reflects a city's strong support for the construction of sponge cities. Based on these evaluation methods and models, combined with the individual indicators that can reflect the construction of sponge cities in Henan Province, the index system for sponge city construction was initially determined (Table 1).

2.5. Research Methods

2.5.1. Determination of Entropy Weight

Entropy was first introduced into the information theory by Shannon. It has been widely used in engineering, social, and economic fields. It hypothesizes that the quantity or quality of information is an important factor in determining the reliability and accuracy of decision-making [42]. Entropy is often used to measure the amount of useful information provided by the dataset itself and is therefore considered to be a suitable indicator for use in various evaluation cases. Weights can be determined based on the data itself, thereby reducing decision bias and increasing the objectivity of the decision process [43]. The entropy weight method is also to calculate the comprehensive index by the size of the selected index information. The index weight is determined by the judgment matrix composed of the evaluation indicators. Since the evaluation system has positive and negative indicators, the sample matrix needs to be dimensionless [44]. The main indicators selected in this paper are studied through positive and negative indicators.

(1)   Data standardization processing

Assume that there are n research objects (mainly cities) in the study area, including m evaluation indicators, and the definition P is the original data matrix, expressed as:

$$P_{nm} = \begin{bmatrix} P_{11} & P_{12} & P_{13} & \cdot & \cdot & \cdot & \cdot & \cdot & P_{1m} \\ & \cdot & & & & & & & \cdot \\ & \cdot & & & & & & & \cdot \\ & \cdot & & & & & & & \cdot \\ & \cdot & & & & & & & \cdot \\ & \cdot & & & & & & & \cdot \\ P_{n1} & P_{n2} & P_{n3} & \cdot & \cdot & \cdot & \cdot & \cdot & P_{nm} \end{bmatrix} \quad (1)$$

Among them, P_{nm} is expressed as the m item of the n city (m = 1, 2, 3, 4, 5, 6$\cdots\cdots$; n = 1, 2, 3, 4, 5, 6, 7, 8 9, 10, 11$\cdots\cdots$). Because the dimension of each index coefficient is not uniform, the index coefficient is standardized by the method of extreme difference. The main indicators selected in this paper are studied by the forward index and the reverse index. When the positive index is larger than the index value, the better the index. The normalization method is: $Y_{nm} = (X_{nm} - minx_m)/(maxx_m - minx_m)$;

the inverse index, that is, shows that, the smaller the index value, the better the index, and the normalization method is: $Y_{nm} = (\text{maxx}_m - X_{nm})/(\text{maxx}_m - \text{minx}_m)$. In the normalized method formula, X_{nm} is expressed as a specific value, minx_m is represented as the minimum value of the m index, and maxx_m is expressed as the maximum value of the m index. The value is between $Y_{nm} \in [0, 1]$ after normalization [45,46].

(2)　Determination of indicator weight

In information theory, the larger the entropy value, the smaller the difference between the values of the evaluation indicators, and the smaller the weight of the index; the smaller the entropy value is, and vice versa [46], to calculate the information entropy of each index, assuming that E_m represents the m. The information entropy under the indicator is calculated as:

$$E_m = -k \sum_{n=1}^{h} f_{nm} \ln f_{nm} \tag{2}$$

where $k = 1/\ln h$, $f_{nm} = Y_{nm}/\sum_{n=1}^{h} y_{nm}$, if $f_{nm} = 0$, then define

$$\lim_{f_{nm} \to 0} f_{nm} \ln f_{nm} = 0 \tag{3}$$

Among them, the m indicator of the n city of Y_{nm} is a specific value.

According to the calculation formula of information entropy, the information entropy $E_1, E_2, \ldots,$ E_m of each index is calculated. The weight of each index is calculated by information entropy. W_m is the entropy weight of the m evaluation index, and then the weight of the index is calculated. The method is [45,46]:

$$W_m = \frac{1 - E_m}{\sum_{n=1}^{h} E_m} \tag{4}$$

where $W_m \in [0, 1]$, $\sum_{n=1}^{h} W_m = 1$.

2.5.2. Coupling Coordinated Development Model

(1)　Comprehensive evaluation model

The comprehensive evaluation model is used to measure the level of infrastructure and economic development. The calculation method is:

$$S = \sum_{i=1}^{n} (W_m \times Y_k) \tag{5}$$

Among them, S represents the comprehensive index of infrastructure construction or economic development; W_m represents the weight of each index within the system; Y_k represents the evaluation value of each indicator.

(2)　Coupling degree model

Coupling refers to the phenomenon in which two or more systems or forms of motion interact and interact with each other through some means. This paper establishes a coupling model of infrastructure and economic development. The calculation method is:

$$C = \left\{ \frac{(S_1 \times S_2)}{\left[\frac{(S_1 + S_2)}{2} \right]} \right\}^2 \tag{6}$$

Among them, C is the coupling degree between infrastructure construction and economic development, and the value range is [0, 1]. The larger C, the stronger the interaction between infrastructure construction and economic development; S1 and S2 are infrastructure construction and economy, respectively. The comprehensive index of development, k as the adjustment factor, in practice, should make k ≥ 2; this paper takes k = 2.

(3) Coupling coordination degree model

The coupling degree model can only indicate the existence of interaction between systems and cannot reflect the level of coupling coordination between systems. Therefore, this paper further constructs a coupling coordination model of infrastructure construction and economic development. The calculation method is:

$$\begin{cases} D = \sqrt{C \times T} \\ T = \alpha \times S_1 + \beta \times S_2 \end{cases} \tag{7}$$

where D is the coupling coordination degree; T is the inter-system comprehensive coordination index; α and β are undetermined coefficients, and $\alpha + \beta = 1$. This paper assumes that infrastructure construction and economic development interact, so take $\alpha = \beta = 0.5$.

Based on the relevant classification criteria proposed in the existing research [47,48], the median segmentation method is used to divide the D value into four stages. The classification criteria are shown in Table 2.

Table 2. Coordination level of coupling coordination.

Coupling Coordination	Coordination Level	S1 > S2	S2 > S1
$0 < D \le 0.3$	Low coupling coordination phase	Economic development lags behind	Infrastructure lag
$0.3 < D \le 0.5$	Moderate coupling coordination phase	Economic development lags behind	Infrastructure lag
$0.5 < D \le 0.8$	Highly coupled coordination phase	Economic development lags behind	Infrastructure lag
$0.8 < D \le 1$	Extreme coupling coordination phase	Economic development lags behind	Infrastructure lag

2.5.3. Spatial Statistical Methods

This paper introduces Moran's I to analyze the imbalance and spatial autocorrelation of the coupling and development of infrastructure construction and economic development between adjacent regions. The calculation formula of Moran's I is as follows:

$$I = \frac{\sum_{i=1}^{n} \sum_{j=1}^{n} w_{ij}\left(Y_i - \overline{Y}\right)\left(Y_j - \overline{Y}\right)}{S^2 \sum_{i=1}^{n} \sum_{j=1}^{n} W_{ij}} \tag{8}$$

$$I_i = \frac{\left(Y_i - \overline{Y}\right)}{S^2} \sum_{j=1}^{n} w_{ij}\left(Y_j - \overline{Y}\right) \tag{9}$$

where I represents the overall degree of correlation between regions, $S^2 = \frac{1}{n}\sum_{i=1}^{n}\left(Y_i - \overline{Y}\right)^2$; $\overline{Y} = \frac{1}{n}\sum_{i=1}^{n} Y_i$; Y_i represents the degree of coupling coordination of the i region; n is the number of regions, and w_{ij} represents the element of the spatial weight matrix W. I_i indicates the degree of correlation between the coordination degree of the i region and the surrounding area, and the local spatial features are displayed by using the Moran scattergram.

3. Results

3.1. Analysis of Coordination Degree between Regions

According to the coupling evaluation model of infrastructure construction and economic development, the comprehensive index of infrastructure construction (S1) and the comprehensive index of economic development (S2), and the coupling coordination degree (D) of the two are calculated. This paper will be in the process of empirical analysis. Infrastructure construction and economic development are regarded as two subsystems of equal importance. Therefore, the undetermined coefficients of the two are all 0.5, that is, $\alpha = \beta = 0.5$. Therefore, the comprehensive harmonic index of the two is $T = 0.5 \times S\_1 + 0.5 \times S\_2$, combined with the coupling coordination degree model for calculation; the obtained empirical results are shown in Table 3.

Table 3. Calculation results of the coupling degree of regional infrastructure construction and economic development.

Region	Coupling Coordination Degree in 2013				Coupling Coordination Degree in 2017			
	S1	S2	D	Coordination Level	S1	S2	D	Coordination Level
Zhengzhou City	0.623386	0.825219	0.846899	Extreme coordination	0.669786	0.816205	0.859872	Extreme coordination
Kaifeng City	0.322134	0.273311	0.54472	Highly coordinated	0.372635	0.27094	0.563689	Highly coordinated
Luoyang City	0.465014	0.334487	0.628002	Highly coordinated	0.502356	0.337348	0.641612	Highly coordinated
Pingdingshan City	0.41014	0.332901	0.607872	Highly coordinated	0.434414	0.291147	0.596354	Highly coordinated
Anyang City	0.434528	0.270151	0.585337	Highly coordinated	0.43343	0.2893	0.595068	Highly coordinated
Hebi City	0.404502	0.262699	0.570946	Highly coordinated	0.436042	0.287656	0.595114	Highly coordinated
Xinxiang City	0.40586	0.270464	0.575601	Highly coordinated	0.35694	0.278649	0.561582	Highly coordinated
Jiaozuo City	0.43791	0.30888	0.606448	Highly coordinated	0.500011	0.310613	0.627769	Highly coordinated
Puyang City	0.42392	0.246459	0.568535	Highly coordinated	0.455427	0.239409	0.574632	Highly coordinated
Xuchang City	0.445089	0.269307	0.588401	Highly coordinated	0.460815	0.336353	0.627452	Highly coordinated
Luohe City	0.447582	0.25596	0.581783	Highly coordinated	0.465954	0.288366	0.605441	Highly coordinated
Sanmenxia City	0.451144	0.271231	0.591444	Highly coordinated	0.439332	0.28157	0.593055	Highly coordinated
Nanyang City	0.600006	0.329184	0.666651	Highly coordinated	0.429884	0.345018	0.620581	Highly coordinated
Shangqiu City	0.32222	0.279107	0.547622	Highly coordinated	0.360593	0.328569	0.586693	Highly coordinated
Xinyang City	0.467129	0.324367	0.623905	Highly coordinated	0.381195	0.266289	0.564449	Highly coordinated
Zhoukou City	0.344228	0.225794	0.528007	Highly coordinated	0.392165	0.222019	0.543206	Highly coordinated
Zhumadian City	0.384815	0.299556	0.582683	Highly coordinated	0.482803	0.315774	0.624866	Highly coordinated
Jiyuan City	0.436586	0.338317	0.619938	Highly coordinated	0.462085	0.294883	0.607565	Highly coordinated

Comparing the coupling and coordination of infrastructure and economic development in each city in 2013 and 2017, we can find that the average system coupling coordination degree of 18 cities in Henan Province in 2016 is 0.6105, which is slightly higher than the average of 0.6036 in 2013. The infrastructure construction of each city and overall average level of economic development level coupling coordination is in the low and medium coupling coordination stage. Among them, the coordination degree of Zhengzhou City, Luoyang City, Jiaozuo City, Xuchang City, Nanyang City, and Zhumadian City is higher than the average level of Henan Province. The coupling coordination degree of most regions is rising continuously. Among them, the coupling coordination degree of Pingdingshan City, Nanyang City, and Jiyuan City shows a downward trend.

In view of the different development speeds of different regions, there are obvious differences in the coupling and coordination degree between infrastructure construction and economic development in various regions. Figure 3 shows that the overall evolution of regional infrastructure construction and economic development level coordination can be divided into two types in 2013–2017: the first category is the area where the coupling coordination degree is between 0.8 and 1, namely Zhengzhou City. Zhengzhou City has been at a relatively high level of coupling and coordination in 2013–2017, indicating that the infrastructure construction and economic development level are at an effective coupling development stage. In recent years, Zhengzhou Zheng Dong New Area has been built and developed according to the concepts and ideas of sustainable "ecological city", "metabolic city", and "sponge city". Among them, the completed water area covers 18 square kilometers, and the green area covers 39 square kilometers, and the green coverage rate in the urban core area is nearly 50%. Drawing on the experience of modern urban construction in the west, the sustainable water cycle is realized through tracking and integrating low impact development (LID) and rainwater utilization. According to the climate characteristics of the Zhengzhou Zheng Dong New Area, the PP module storage device is used to collect and purify rainwater for green irrigation, creating a sustainable green landscape [49]. The construction of Zhengzhou Zheng Dong New Area is inseparable from the strong economic support. The second category is the area where the coupling coordination degree is between 0.5 and 0.8. Although these areas are in a highly coupled and coordinated development stage, the infrastructure construction and economic development level are relatively insufficient, and the infrastructure construction and economic development level are relatively weak. The two have not yet formed a benign interactive coupling development model, and there is still much room for improvement in the coordinated and coordinated development. Government departments in various regions should formulate corresponding infrastructure construction and economic development level strategies according to local actual conditions, as well as upgrade infrastructure and economic development as soon as possible.

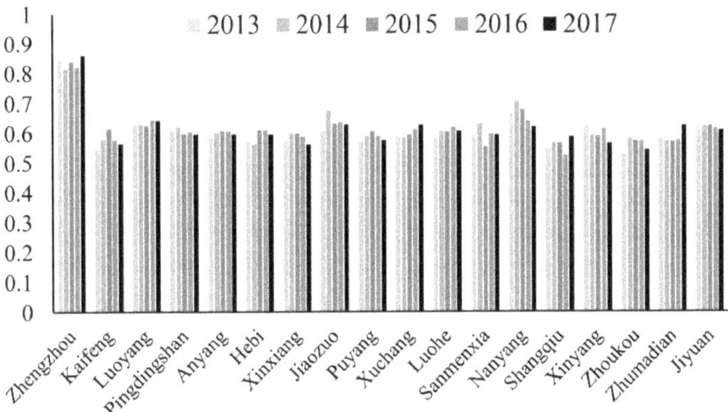

Figure 3. Evolution of the coupling degree of regional infrastructure construction and economic development level.

3.2. Evolution of Spatial Pattern of Coordinated Development

3.2.1. Evolution of Regional Differences

In order to better analyze the spatial pattern and dynamic evolution of the coupling and coordinated development of inter-regional infrastructure construction and economic development level, this paper uses 2014 and 2017 as time nodes, combined with Table 3, through ArcGIS 10.2 software, respectively, for an inter-regional 2014, 2017 spatial visualization of the system coupling coordination level for the year (Figure 4).

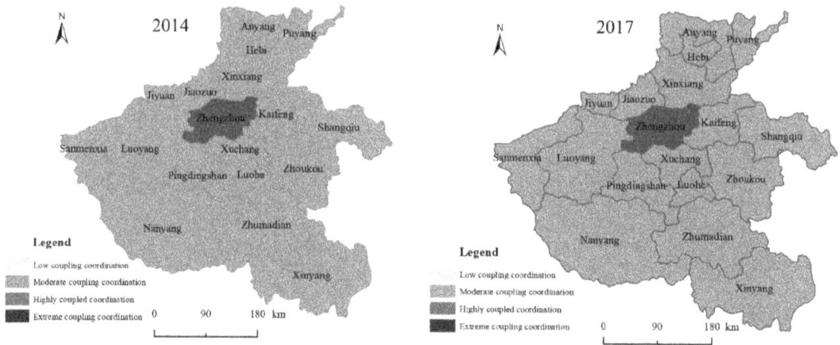

Figure 4. Spatial pattern evolution of the coupling degree of infrastructure construction and economic development in 18 cities in Henan Province.

Combined with Table 3 and Figure 4, it can be seen that the spatial pattern of infrastructure construction and economic development level is obviously different. The coupling and coordination distribution of Henan Province is basically consistent with the spatial location of economic development, and the degree of system coupling coordination is relatively high in regions with relatively developed economy in areas with low levels of economic development. During the study period, the coordination degree of Zhengzhou City, Nanyang City, Luoyang City, and Xuchang City has been maintained at a high level, while the coupling coordination degree of Hebi City, Zhumadian City, Zhoukou City, and Shangqiu City is in a rising trend state. The degree of economic development of the region is closely related. With the development of the economy, the investment in infrastructure construction is also increasing; the coupling coordination degree of Puyang City, Xinxiang City, Kaifeng City, and Pingdingshan City is declining, which is related to the development of the region and policy. Overall, the infrastructure construction and economic development level of Henan Province presents a spatial pattern of "high west and low east". However, the classification shows that the coupling degree between infrastructure construction and economic development level in Henan Province has not changed, but each prefecture-level city fluctuates in the grade interval.

3.2.2. Statistical Analysis of Local Space

Through the global Moran's I statistic, it can be seen that the coupling and coordinated development of infrastructure construction and economic development level in Henan Province has significant spatial agglomeration, as well as the correlation and concentration between regions, the spatial correlation with the surrounding areas, the degree of spatial difference. The distribution of the spatial pattern, on whether there is heterogencity, in this paper, is analyzed by Moran scatter plot (Figure 5).

It can be observed from Figure 5 that the level of coupling and coordination between infrastructure construction and economic development level in most regions in the past five years is in a stable upward and downward fluctuation state. Compared with 2014, the first quadrant of 2017 Moran's I is reduced, and the fourth quadrant is increased. And the Moran's I index is greater than 0, indicating that there is spatial positive correlation, that is, the area where the coupling degree of infrastructure construction and economic development level is higher (or lower) tends to be significantly concentrated in space, and the correlation is stronger; if Moran's I = 0, it means that the space is not correlated, and the distribution is in a random state. Table 4 reports the relative development degree and relative development type. The results show that Henan Province is generally synchronous development, Zhengzhou City is lagging behind infrastructure, indicating that economic growth is faster than infrastructure construction; other regions are economic development. The lag type indicates that the economic growth is slower than the infrastructure construction, and the construction of infrastructure is more advanced than consumption. Therefore, to promote economic growth, it is necessary to increase investment in land resource conservation and intensiveness in order to realize infrastructure construction and economy. Coordinated development levels and simultaneous development are important ways to achieve coordinated and coordinated development.

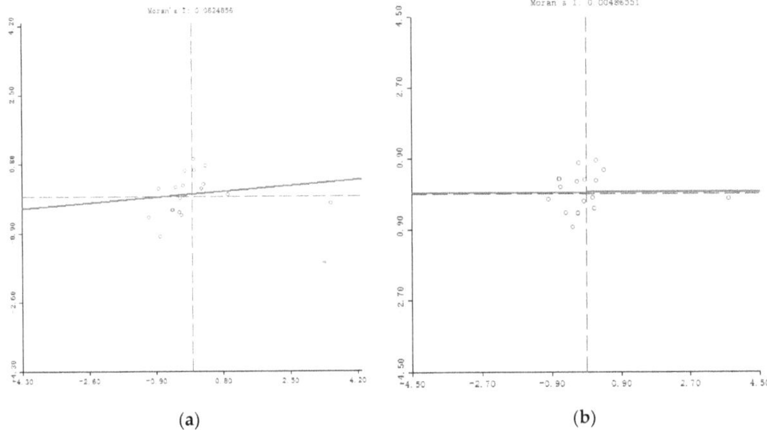

(a)   (b)

Figure 5. Moran scatter plot of the coupling degree of infrastructure construction and economic development level in Henan Province. (**a**): Moran's I index in 2014. (**b**) Moran's I index in 2017.

Table 4. Types of relative development of infrastructure construction and economic development level in 2013–2017.

Region	Infrastructure Construction Index	Economic Development Index	Relative Development	Coupling Coordination	Relative Development Type
Zhengzhou City	0.633578115	0.774055169	0.822082425	0.8364244	Infrastructure lag
Kaifeng City	0.396918457	0.279733272	1.42961227	0.5761015	Lag in economic development
Luoyang City	0.483916433	0.333859849	1.449215794	0.6338987	Lag in economic development
Pingdingshan City	0.430533325	0.313778336	1.383605691	0.6057436	Lag in economic development
Anyang City	0.441702908	0.291403433	1.517672072	0.598873	Lag in economic development
Hebi City	0.429776229	0.282747714	1.518187027	0.5899238	Lag in economic development
Xinxiang City	0.399777152	0.293814553	1.365102599	0.5850529	Lag in economic development
Jiaozuo City	0.474586257	0.34826711	1.396870733	0.6358566	Lag in economic development
Puyang City	0.456965847	0.25831484	1.774071396	0.5858622	Lag in economic development
Xuchang City	0.450395427	0.291419111	1.557051753	0.6014466	Lag in economic development
Luohe City	0.466232093	0.284559279	1.641295553	0.603342	Lag in economic development
Sanmenxia City	0.42761795	0.292404637	1.466660645	0.5935525	Lag in economic development
Nanyang City	0.530638566	0.366931643	1.461614198	0.6622117	Lag in economic development
Shangqiu City	0.329256604	0.296774658	1.109013939	0.5584797	Lag in economic development
Xinyang City	0.43525355	0.292081501	1.493643174	0.5966539	Lag in economic development
Zhoukou City	0.405418306	0.243993113	1.673831676	0.5598729	Lag in economic development
Zhumadian City	0.419954597	0.280990821	1.501264572	0.5854403	Lag in economic development
Jiyuan City	0.450710664	0.324022045	1.396122648	0.6179675	Lag in economic development

4. Conclusions and Recommendations

Based on the coupling coordination degree model, relative development degree model, and spatial statistical analysis, this research work studied the difference of horizontal and spatial statistical analysis of the coupled and coordinated development of sponge city construction in Henan Province. The results show that: (1) From the perspective of comprehensive level, the problems of inadequate and unbalanced development of infrastructure construction and economic development level are prominent. (2) From the perspective of coordinated development level, the level of coupling and coordination development in Henan Province increased during the sample period, but the level of coupling and coordination development in each region was small. (3) From the perspective of relative development, Zhengzhou City is lagging behind in infrastructure, indicating that economic growth is faster than infrastructure construction; other regions are lagging economic development, indicating that economic growth is faster than infrastructure construction. Slowly, (4) from the spatial statistical analysis, the Moran's I index is greater than 0, indicating that there is spatial positive correlation, that is, the area with high coupling degree of infrastructure construction and economic development level tends to be significantly concentrated in space.

According to the above conclusions, due to the different natural foundations, economic reserves, location advantages, historical background, social influence, and policy conditions of various regions, the coordinated development of sponge city construction in Henan Province requires differentiated regional infrastructure construction and economy development policy. In areas with better economic development, it is necessary to increase the proportion of investment in "green" infrastructure and "gray" infrastructure, as well as appropriately increase the proportion of investment in "green" infrastructure; in areas with insufficient economic development level, it is necessary to adapt to local conditions. In the built-up area, the proportion of "green" infrastructure and "gray" infrastructure will be created to reduce the frequent flooding in the city and to continuously improve the renewal and construction of the drainage system. At the same time, sponge buildings and residential areas should be promoted, and measures, such as roof greening, rainwater storage, collection and utilization, and micro-topography, should be taken according to local conditions to improve the rainwater storage and retention capacity of buildings and residential areas. Rainwater collection and recycling, on the one hand, can provide water for the vegetation of green infrastructure, and, on the other hand, they can provide the use of reclaimed water for the community. The construction of green infrastructure in sponge cities increases the vegetation cover and water area of cities, effectively weakens the urban heat island effect, and thus affects the precipitation process. Therefore, the construction of green infrastructure in sponge cities can conserve water resources, regulate runoff, purify water quality, save water resources, improve the carrying capacity of regional water resources, and enhance the capacity of natural water storage and drainage. Sustainable sponge city construction needs to coordinate the coordination and development of green infrastructure, grey infrastructure, and economic development.

Author Contributions: Conceptualization, L.Z. (Lijun Zhang); methodology, K.W. and L.Z. (Lijun Zhang); software, K.W.; validation, K.W., L.Z. (Lulu Zhang); Formal Analysis, K.W.; Investigation, K.W., S.C., and L.Z. (Lulu Zhang); Resources, L.Z. (Lijun Zhang); writing—original draft preparation, K.W.; writing—review & editing, K.W. and L.Z. (Lijun Zhang); visualization, K.W.; project administration, L.Z. (Lijun Zhang). All authors have read and agreed to the published version of the manuscript.

Funding: The present study is supported by the Key Scientific Research Projects of Institutions of Higher Learning in Henan Province in 2017 (Grant No. 17A170006).

Conflicts of Interest: The authors declare no conflict of interest.

References

1. Liu, H.; Jia, Y.; Niu, C. "Sponge city" concept helps solve China's urban water problems. *Environ. Earth Sci.* **2017**, *76*, 473–479. [CrossRef]
2. Xia, J.; Zhang, Y.; Xiong, L.; He, S.; Wang, L.; Yu, Z. Opportunities and challenges of the Sponge City construction related to urban water issues in China. *Sci. China Earth Sci.* **2017**, *60*, 652–658. [CrossRef]

3. MEP. *Report on the State of the Environment in China*; Environmental Information Centre, Ministry of Environmental Protection of the Peoples Republic of China: Beijing, China, 2015.

4. Li, H.; Ding, L.; Ren, M.; Li, C.; Wang, H. Sponge City Construction in China: A Survey of the Challenges and Opportunities. *Water* **2017**, *9*, 594. [CrossRef]

5. Shao, W.; Zhang, H.; Liu, J.; Yang, G.; Chen, X.; Yang, Z.; Huang, H. Data Integration and its Application in the Sponge City Construction of China. *Procedia Eng.* **2016**, *154*, 779–786. [CrossRef]

6. Chan, F.K.S.; Griffiths, J.A.; Higgitt, D.; Xu, S.; Zhu, F.; Tang, Y.-T.; Thorne, C.R. "Sponge City" in China—A breakthrough of planning and flood risk management in the urban context. *Land Use Policy* **2018**, *76*, 772–778. [CrossRef]

7. Pyke, C.; Warren, M.P.; Johnson, T.; Lagro, J., Jr.; Scharfenberg, J.; Groth, P.; Freed, R.; Schroeer, W.; Main, E. Assessment of low impact development for managing stormwater with changing precipitation due to climate change. *Landsc. Urban Plann.* **2011**, *103*, 166–173. [CrossRef]

8. Griffiths, J. Sustainable urban drainage. In *Encyclopedia of Sustainable Technologies*; Abraham, M., Ed.; Elsevier: Amsterdam, The Netherlands, 2017.

9. Morison, P.J.; Brown, R.R. Understanding the nature of publics and local policy commitment to Water Sensitive Urban Design. *Landsc. Urban Plann.* **2011**, *99*, 83–92. [CrossRef]

10. Voyde, E.; Fassman, E.; Simcock, R. Hydrology of an extensive living roof under sub-tropical climate conditions in Auckland, New Zealand. *J. Hydrol.* **2010**, *394*, 384–395. [CrossRef]

11. Hui, L.; Zihao, W.; Xifeng, L. Ecological Sensitivity Analysis of Sponge City Planning Area Based on GIS. *Land Resour. Guide* **2017**, *14*, 49–52.

12. Yu, K.; Li, D.; Yuan, H.; Fu, W.; Qiao, Q.; Wang, S. Theory and Practice of "Sponge City". *City Plan. Rev.* **2015**, *39*, 26–36.

13. Wang, Y.C.; Gao, J. Seattle stormwater management practice inspired for the development of sponge city. *J. Hum. Settl.* **2017**, *6*, 65–70.

14. Baoxing, Q. The Connotation, Ways and Prospects of Sponge City (LID). *Water Supply Drain.* **2015**, *51*, 1–7.

15. McDaniels, T.; Chang, S.; Cole, D.; Mikawoz, J.; Longstaff, H. Fostering resilien to extreme events within infrastructure systems: Characterizing decision contexts for mitigation and adaptation. *Glob. Environ. Chang.* **2008**, *18*, 310–318. [CrossRef]

16. Ouyang, M.; Duenas-Osorio, L.; Min, X. A three-stage resilience analysis framework for urban infrastructure systems. *Struct. Saf.* **2012**, *36*, 23–31. [CrossRef]

17. Hossain, F.; Arnold, J.; Beighley, E.; Brown, C.; Burian, S.; Chen, J.; Tidwell, V.; Wegner, D. Local-to-regional landscape drivers of extreme weather and climate: Implications for water infrastructure resilience. *J. Hydrol. Eng.* **2015**, *20*, 2005–2015. [CrossRef]

18. Mugume, S.N.; Gomez, D.E.; Fu, G.; Farmani, R.; Butler, D. A global analysis approach for investigating structural resilience in urban drainage systems. *Water Res.* **2015**, *81*, 15–26. [CrossRef] [PubMed]

19. Mugume, S.N.; Diao, K.; Astaraie-Imani, M.; Fu, G.; Farmani, R.; Butler, D. Enhancing resilience in urban water systems for future cities. *Water Sci. Technol. Water Supply* **2015**, *15*, 1343–1352. [CrossRef]

20. Willems, P.; Arnbjerg-Nielsen, K.; Olsson, J.; Nguyen, V.T.V. Impact assessment on urban rainfall extremes and urban drainage: Methods and shortcomings. *Atmos. Res.* **2012**, *103*, 106–118. [CrossRef]

21. Karamouz, M.; Hosseinpour, A.; Nazif, S. Improvement of urban drainage system performance under climate change impact:case study. *J. Hydrol. Eng.* **2011**, *16*, 395–412. [CrossRef]

22. Huong, H.T.L.; Pathirana, A. Urbanization and climate change impacts on future urban flooding in Can Tho City, Vietnam. *Hydrol. Earth Syst. Sci.* **2013**, *17*, 379–394. [CrossRef]

23. Dong, X.; Guo, H.; Zeng, S.Y. Enhancing future resilience in urban drainage system: Green versus grey infrastructure. *Water Res.* **2017**, *124*, 280–289. [CrossRef] [PubMed]

24. Liu, W.; Chen, W.; Peng, C. Assessing the effectiveness of green infrastructures on urban flooding reduction: A community scale study. *Ecol. Model.* **2014**, *29*, 6–14. [CrossRef]

25. Liu, W.; Chen, W.; Peng, C. Influences of setting sizes and combination of green infrastructures on community's stormwater runoff reduction. *Ecol. Model.* **2015**, *318*, 236–244. [CrossRef]

26. Kamil, P.; Daniel, S.; Sabina, K. The temporal variability of a rainfall synthetic hyetograph for the dimensioning of stormwater retention tanks in small urban catchments. *J. Hydrol.* **2017**, *549*, 501–511.

27. Eckart, K.; Mcphee, Z.; Bolisetti, T. Performance and implementation of low impact development—A review. *Sci. Total Environ.* **2017**, *608*, 413–432. [CrossRef]

28. Kong, F.; Ban, Y.; Yin, H.; James, P.; Dronova, I. Modeling stormwater management at the city district level in response to changes in land use and low impact development. *Environ. Model. Softw.* **2017**, *95*, 132–142. [CrossRef]
29. Fletcher, T.D.; Shuster, W.; Hunt, W.F.; Ashley, R.; Butler, D.; Arthur, S.; Trowsdale, S.; Barraud, S.; Semadeni-Davies, A.; Bertrand-Krajewski, J.-L.; et al. SUDS, LID, BMPs, WSUD and more—The evolution and application of terminology surrounding urban drainage. *Urban Water J.* **2016**, *12*, 525–542. [CrossRef]
30. Mei, C.; Liu, J.H.; Wang, H.; Yang, Z.Y.; Ding, X.Y.; Shao, W.W. Integrated assessments of green infrastructure for flood mitigation to support robust decision-making for sponge city construction in an urbanized watershed. *Sci. Total Environ.* **2018**, *639*, 1394–1407. [CrossRef]
31. Hu, M.; Sayama, T.; Zhang, X.; Tanaka, K.; Takara, K.; Yang, H. Evaluation of low impact development approach for mitigating flood inundation at a watershed scale in China. *J. Environ. Manag.* **2017**, *193*, 430–438. [CrossRef]
32. Jia, H.; Yu, S.L.; Qin, H. Low impact development and sponge city construction for urban stormwater management. *Front. Environ. Sci. Eng.* **2017**, *11*, 20. [CrossRef]
33. Manocha, N.; Babovic, V. Development and valuation of adaptation pathways for storm water management infrastructure. *Environ. Sci. Policy* **2017**, *77*, 86–97. [CrossRef]
34. Yike, W.; Qianhu, C. Study on Sustainable Stormwater Management Policies from the International Perspective: Based on the Comparison of the United States, the United Kingdom and China. *Urban Plan. Int.* **2020**, 1–15. [CrossRef]
35. Zhao, J.; Fonseca, C.; Zeerak, R. Stormwater Utility Fees and Credits: A Funding Strategy for Sustainability. *Sustainability* **2019**, *11*, 1913. [CrossRef]
36. Ding, L.; Ren, X.; Gu, R.; Che, Y. Implementation of the "sponge city" development plan in China: An evaluation of public willingness to pay for the life-cycle maintenance of its facilities. *Cities* **2019**, *93*, 13–30. [CrossRef]
37. Ma, Y.; Jiang, Y.; Swallow, S. China's sponge city development for urban water resilience and sustainability: A policy discussion. *Sci. Total Environ.* **2020**, *729*, 139078. [CrossRef] [PubMed]
38. Xiong, J.X.; Wang, W.H.; He, S.H.; Yin, Y.; Tang, C.F. Spatio-temporal Pattern and Influencing Factor of Coupling Coordination of Tourism Urbanization System in the Dongting Lake Region. *Sci. Geogr. Sin.* **2020**, *40*, 1532–1542.
39. Sun, Y.; Deng, L.; Pan, S.-Y.; Chiang, P.-C.; Sable, S.S.; Shah, K.J. Integration of Green and Gray Infrastructures for Sponge City: Water and Energy Nexus. *Water Energy* **2020**, *3*, 29–40. [CrossRef]
40. Woznicki, S.A.; Hondula, K.L.; Jarnagin, S.T. Effectiveness of landscape-based green infrastructure for stormwater management in suburban catchments. *Hydrol. Process* **2018**, *32*, 2346–2361. [CrossRef]
41. Alves, A.; Gersonius, B.; Sanchez, A.; Vojinovic, Z.; Kapelan, Z. Multi-criteria Approach for Selection of Green and Grey Infrastructure to Reduce Flood Risk and Increase CO-benefits. *Water Resour. Manag.* **2018**, *32*, 2505–2522. [CrossRef]
42. Shannon, C.E. A mathematical theory of communication. *ACM SIGMOBILE Mobile. Comput. Commun. Rev.* **2001**, *5*, 3–55. [CrossRef]
43. Wu, S.S.; Fu, Y.; Shen, H.; Liu, F. Using ranked weights and Shannon entropy to modify regional sustainable society index. *Sustain. Cities Soc.* **2018**, *41*, 443–448. [CrossRef]
44. Jiaqi, Z.; Yijin, W.; Yong, G.; Chenghao, W.; Kung, H. Comprehensive evaluation of ecological security in poor areas based on grey relational model—A case study of enshi poor area. *Geogr. Res.* **2014**, *33*, 1457–1466.
45. Zheng, S.; Fu, Y.; Lai, K.K.; Liang, L. An improvement to multipie criteria ABC inventory classification using Shannon entropy. *J. Syst. Sci. Complex.* **2017**, *30*, 857–865. [CrossRef]
46. Wu, J.; Sun, J.; Liang, L. DEA cross-efficiency aggregation method based upon Shannon entropy. *Int. J. Prod. Res.* **2012**, *50*, 6726–6736. [CrossRef]
47. Li, M.; Fengjun, J.; Yi, L. Analysis on the coupling pattern and industrial structure of China's economy and environmental pollution. *J. Geogr.* **2012**, *67*, 1299–1307.
48. Xia, W.; Zhaolin, G.; Jinyuan, L.; Hu, M. Multi-level gray method for evaluation of tourism resources development potential—Taking Laozi mountain scenic spot as an example. *Geogr. Res.* **2007**, *3*, 625–635.

49. Zheng Dong New District. Available online: http://www.zhengdong.gov.cn/sitesources/zhengdong/page_pc/index.html (accessed on 25 November 2020).

Publisher's Note: MDPI stays neutral with regard to jurisdictional claims in published maps and institutional affiliations.

Article

Delving into the Divisive Waters of River Basin Planning in Bolivia: A Case Study in the Cochabamba Valley

Nilo Lima-Quispe [1], Cláudia Coleoni [1], Wilford Rincón [2], Zulema Gutierrez [3], Freddy Zubieta [3], Sergio Nuñez [3], Jorge Iriarte [4], Cecilia Saldías [3], David Purkey [1], Marisa Escobar [5] and Héctor Angarita [1,*]

[1] Latin America Center, Stockholm Environment Institute, Calle 71 No. 11-10, 110231 Bogotá, Colombia; nilo.lima@sei.org (N.L.-Q.); claudia.coleoni@sei.org (C.C.); david.purkey@sei.org (D.P.)
[2] Independent Consultant, 170001 Manizales, Colombia; wrinconarango@gmail.com
[3] Independent Consultant, 2500 Cochabamba, Bolivia; zlgutierrez@hotmail.com (Z.G.); z_freddy@hotmail.com (F.Z.); sertau93@gmail.com (S.N.); cecisaldias@gmail.com (C.S.)
[4] Independent Consultant, 703030 Sucre, Bolivia; iriartejorge@gmail.com
[5] Stockholm Environment Institute, Davis, CA 95616, USA; marisa.escobar@sei.org
* Correspondence: hector.angarita@sei.org

Abstract: River basin planning in Bolivia is a relatively new endeavor that is primed for innovation and learning. One important learning opportunity relates to connecting watershed planning to processes within other planning units (e.g., municipalities) that have water management implications. A second opportunity relates to integrating watershed management, with a focus on land-based interventions, and water resources management, with a focus on the use and control of surface and groundwater resources. Bolivia's River Basin Policy and its primary planning instrument, the River Basin Master Plan (PDC in Spanish), provide the relevant innovation and learning context. Official guidance related to PDC development lacks explicit instructions related to the use of analytical tools, the definition of spatially and temporally dis-aggregated indicators to evaluate specific watershed and water management interventions, and a description of the exact way stakeholders engage in the evaluation process. This paper describes an effort to adapt the tenets of a novel planning support practice, Robust Decision Support (RDS), to the official guidelines of PDC development. The work enabled stakeholders to discern positive and negative interactions among water management interventions related to overall system performance, hydrologic risk management, and ecosystem functions; use indicators across varying spatial and temporal reference frames; and identify management strategies to improve outcomes and mitigate cross-regional or inter-sectorial conflicts.

Keywords: water resources systems; participatory modeling; river basin planning; watershed management; water scarcity; water conflicts; Robust Decision Support; WEAP; Integrated Water Resources Management; Bolivia

Citation: Lima-Quispe, N.; Coleoni, C.; Rincón, W.; Gutierrez, Z.; Zubieta, F.; Nuñez, S.; Iriarte, J.; Saldías, C.; Purkey, D.; Escobar, M.; et al. Delving into the Divisive Waters of River Basin Planning in Bolivia: A Case Study in the Cochabamba Valley. *Water* **2021**, *13*, 190. https://doi.org/10.3390/w13020190

Received: 31 October 2020
Accepted: 5 January 2021
Published: 14 January 2021

1. Introduction

Water resources managers worldwide face high levels of natural and human-induced hydrologic variability accompanied by climate change projections [1] suggesting increased risks of water scarcity [2]. Defining future changes in hydrologic variability is highly uncertain, hindering the prediction of extreme events such as floods and droughts [3]. In addition to hydrologic variability, water managers deal with growing long-term demands for water from rapidly expanding urban areas and increased consumption across sectors such as agriculture and energy [4]. The intensification of human–water interactions reaffirms the need for 'good governance' to improve water management [5,6]. As the UN World Water Development Report [7] stated, "the world's water crisis is one of water governance, essentially caused by the ways we mismanage water". Under scenarios of deep uncertainty, water governance approaches should support water-related decision making [1]. The recognition of the interconnected nature of the biophysical and socioeconomic factors that converge in water management has resulted in the progressive adoption of integrative governance approaches, such as the Integrated Water Resources Management (IWRM) [8,9]. The IWRM framework has been widely adopted as a template for water governance [10,11] and was recently included as an implementation target for the Sustainable Development Goal (SDG) 6 'Water and Sanitation' [12]. Numerous countries, including Bolivia, have attempted to adapt IWRM to their contexts and realities in the hope of improving outcomes in river basins characterized by scarcity and conflict. At its core, IWRM relies on placing biophysical and technical knowledge that is comprehensively developed at the scale of a river basin planning unit at the center of stakeholder interactions designed to identify integrated and coherent river basin scale solutions [8,9]. However, efforts to evaluate the river basin scale performance of specific interventions does not necessarily align with an individual stakeholder's more parochial objectives and interests. Nonetheless, within existing water policy frameworks, such as the European Water Framework Directive and many others, the river basin is commonly defined as the predominant spatial domain for water management, superseding other territorial and administrative boundaries [13,14]. The river basin unit also drives water management governance, generating new regulatory frameworks and institutional arrangements such as river basin organizations to promote the participation of multiple water users and civil society [15–17]. However, the river basin perspective may lead to potential conflicts or ambiguities with other existing levels of territorial governance more connected to individual stakeholder interests or frames of reference, such as municipalities or targeted water management entities (e.g., water utilities, irrigation districts). Although river basins certainly do define useful 'natural' boundaries [18,19], river basin management is also a result of 'political' processes and choices based on values and preferences often defined at sub-watershed scales, based on administrative boundaries [20,21].

Although the IWRM framework is often used interchangeably with the concept of water governance, Lautze et al. [22] warn that "setting pre-determined goals or outcomes associated with IWRM circumscribes a major role of water governance—that of determining goals". In practice, the attempt to uniformly apply IWRM principles leads to 'poor' rather than 'good' water governance since local conditions, preferences, and values would remain largely misrepresented [22]. Ensuring engaged and effective participation from across sub-basin jurisdictions is key to address collective action dilemmas arising from diverse and potentially conflicting users' interests over Common Pool Resources such as a shared river basin [23]. Human–water interactions are often characterized by mismanagement (e.g., depleted, overused water) and ecosystem impairment [24], threating the water commons. For instance, in river basins, collective action dilemmas can lead to unsustainable water use and insufficient or unsafe water supply for downstream users [25]. This is surely contrary to the interests of some sub-basin stakeholder interests. A common approach toward representing sub-basin interests within a comprehensive river basin scale planning process involves developing analytical tools sufficiently disaggregated to capture the diversity of conditions and interest within the river basin planning unit. Many approaches and case studies pertaining to the use of models to simulate water resources system are available in the literature, which are addressed and discussed by Mashaly and Fernald [26].

Mirchi et al. [27] define three modeling approaches in support of water resources planning and decision making: predictive simulation, integrated descriptive models, and participatory models. The first case predicts the future behavior of a particular subsystem, such as hydrological conditions [27], based on output from a calibrated and validated historical model [26]. The second case adopts a more holistic approach allowing for feedback between two or more disparate subsystems such as hydrological, social, ecological, economic, and political based on historical patterns of interaction [27]. The third approach promotes the participation of decision makers and stakeholders in the modeling process [26,27] where they can express their interests [28]. This approach allows for the performance evaluation of interventions in the face of changing climate or evolving societal prerogatives [29–32]. This requires an understanding of the relationships between climate, water resources, and user expectations, which can be addressed using integrated hydrological/water management modeling tools [33] that quantify potential impacts at varying spatial and temporal scales [34]. For instance, the Water Evaluation and Planning system (WEAP) [35] has been applied in several river basins worldwide as a simulation model for decision support while considering multiple objectives [36–38]. The Sacramento Water Allocation Model, known as SacWAM, is an example of a WEAP-based hydrologic and system model that illustrates the complex water system operation, besides showing how its water would flow if there were no dams, diversions, or infrastructure [39]. The use of modeling tools allows water resources managers and stakeholders to identify the operational, institutional, or infrastructure interventions that meet the disparate goals set out for a water system [4]. Although participatory frameworks for decision making under uncertainty have been increasingly applied to water resources management [40,41], major challenges remain, such as dealing with trade-offs between multiple goals and addressing various sources of uncertainty in contested political processes [4]. There are many barriers to the implementation of participatory decision support approaches—limited capacities in local institutions, associated costs, implementation time, the challenge of designing an adequate methodological framework, and the sustainability of the process in the long term [28]. In addition to these barriers, the knowledge and methods to tackle uncertainties should be considered in learning and decision-making processes [42].

If the goal is to avoid collective action dilemmas, the analytical tools used to evaluate potential water management interventions must, beyond basin-level consideration, align with river basin sub-jurisdictions and sectorial interest [15]. The complexity of this challenge is multi-fold:

- there must be a clear connection between the model scope and disaggregated user concerns, interests, and goals [26];
- there must be an appropriate level of granularity, i.e., fixed-scale models are often either too large to allow analysis of small-scale issues that affect communities and ecosystems, or too small to address connections with the river basin-scale drivers or imperatives [43];
- there must be a common language and a channel of communication between the model and the diverse stakeholders [44];
- the model must be credible, i.e., modeling as a representation of reality can be very subjective unless guided by stakeholders' perceptions [45].

The Robust Decision Support (RDS) framework proposed by Purkey et al. [46] consists of an iterative bottom–up process with active stakeholder participation where decisions are supported by the use of water resources models, which are accompanied by a strong process of local capacity building. Case studies for RDS include urban water management in the metropolitan region of La Paz/El Alto, Bolivia [47], evaluation of climate change impacts in a large basin in northern Patagonia, Argentina [36], and implementation of an IWRM planning process in the Yuba River basin, California [46,47]. This paper presents innovations in the RDS participatory framework as a contribution to Bolivia's National Watershed Policy, specifically in the formulation of the Rocha River Basin Master Plan (PDC in Spanish), which is located in the Cochabamba Valley, Bolivia. This effort took place within Bolivia's unique historical and political context related to water management. Our methodological approach consists of combining the RDS framework for water resources management [46] to extend Bolivia's guiding framework for the formulation of river basin master plans. We address the following questions: How does the water resources system model respond to water-related decision-making processes and institutional governance design at a range of scales within a river basin? How does the water resources system model contribute to the development of effective water planning instruments?

2. Materials and Methods

2.1. Study Area

The Rocha River Basin is located in the department of Cochabamba in the Plurinational State of Bolivia and contains all or part of 25 rural and urban municipalities (Figure 1). The basin has an area of approximately 3699 km^2 and a population of almost 1,300,000 people (13% of the country). From the hydrological point of view, it is comprised of three sub-basins: Rocha, Maylanco, and Sulty (also known as Valle Alto). The Rocha and Maylanco sub-basins—for decades predominantly agricultural areas—are rapidly urbanizing, including cities such as Cochabamba, Sacaba, Colcapirhua, Vinto, Tiquipaya, and Sipe Sipe, which make up the greater Cochabamba metropolitan area. In contrast, the Sulty sub-basin is mainly rural, with agriculture as its primary economic activity. Climatically, 80% of the annual precipitation is concentrated between the months of December and March (rainy season), 2% between the months of May and August (dry season), and the remaining precipitation in the months of the transition seasons (April, September to November) [48]. Annual rainfall varies between 300 and 900 mm, wetter to the north and the east, where most of the water supply reservoirs and current and potential sources of inter-basin transfer are located (Figure 1).

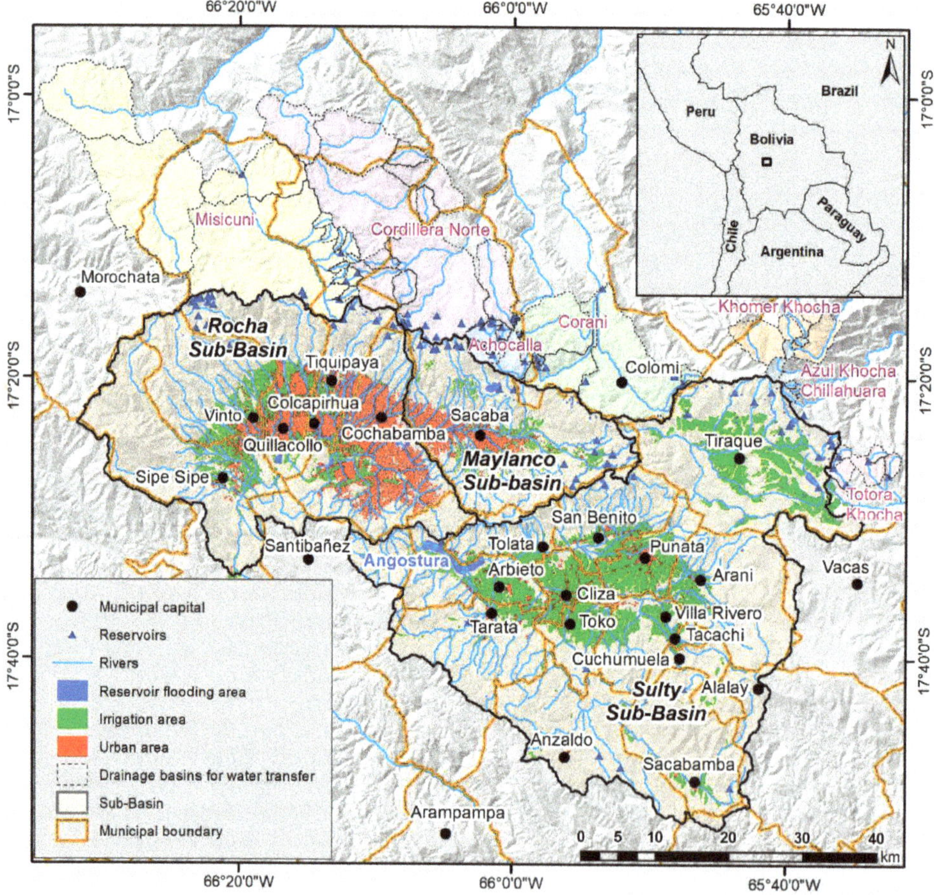

Figure 1. The Rocha River Basin and its geographical environment related to the urban area, agriculture, and the different hydrographic elements.

Challenging climatic conditions are likely to worsen under projected climate conditions. Previous studies have developed climate change scenarios on a basin scale for the 2020–2050 horizon, based on Intergovernmental Panel on Climate Change (IPCC)-5th Phase of the Coupled Model Intercomparison Project (CMIP5) climate models prioritized based on their capacity to capture long-term historical (observed) climate variability attributes in the period 1981–2005 [48]. According to the Institut Pierre-Simon Laplace Climate Model 5A Low-Resolution (IPSL-CM5A-LR) [49] model for Representative Concentration Pathway 8.5 (RCP8.5) [50], the annual precipitation could be reduced by 7%. In some months of the rainy season (February and March), precipitation could be reduced by up to 30%. In the remaining months of the rainy season (December and January), it could increase by up to 11%. The dry season could be even drier with reductions of up to 77%. In the transition season, particularly between September and November, reductions in precipitation of between 10 and 20% are also expected. The average temperature could increase by 1.1 °C.

Given these challenging climatic conditions, it is not surprising that the Rocha River has historically been affected by water supply problems. Limited water availability due to the prevailing semi-arid climate as well as long-standing conflicts over access, governance, and environmental degradation contribute to the basin's water-related challenges. Cochabamba, Bolivia's third largest city, has experienced conflicts over the expansion of water access, particularly in the rural areas, highlighting the complexity of rural–urban hydrosocial relations [51] where established water users confront emerging water use communities. The Cochabamba Water War (*la Guerra del Agua*) emerged as a conflict between a centralized, foreign and private water company and urban residents over water tariffs that increased by as much as 200% [52,53], internationally recognized as a "David versus Goliath success" [53,54]. Empowered by the success of the Water War, the peri-urban water committees (*comités de agua*) moved toward more decentralized, small-scale, and autonomous water management led by community-owned supply systems [52]. However, this shift led to large disparities in water access in many extensive areas of the basin where small-scale options based on the water availability in the immediate environment were no longer compatible with the current levels of water demands. In response, new centralized water transfer projects from neighboring basins [51] with favorable quantity and quality conditions emerged as options [55]. The construction and expansion of some water transfers is currently underway; yet, local water managers also seek short-term solutions [52] such as drilling wells to extract water from aquifers. In addition to these problems, the region is exposed to hazards such as floods [56], landslides, debris flow, and droughts. The quality of surface water [57] and groundwater is degraded by the direct discharge of domestic and industrial wastewater, which is later re-used in irrigation [58–60]. In addition, the absence of effective land planning policies has been causing unregulated urban growth to the detriment of agricultural areas, aquifer recharge zones, and national parks [61]. This is hardly the setting for successful river basin-scale decision making, more so in the absence of sufficient comprehensive basin-scale information.

2.2. Bolivia's Basin Plan in the Context of the River Rocha Basin Planning

Nonetheless, similar to other countries, Bolivia has progressively adopted the river basin as the spatial domain for water management. However, it is worth noting that Bolivia's original Integrated River Basin Management (IRBM) approach adopted more of a terrestrial focus rather than a focus on water resources. Small-scale actions such as soil conservation, forest conservation, afforestation and reforestation, flood control, and bank protection civil works were the focus. For instance, the 1991 Cochabamba's Integrated River Basin Program (*Programa de Manejo Integral de Cuenca*, PROMIC) [62] was a national IRBM reference in the implementation of projects aimed at reducing local damage from flood events in prioritized river basins. This created an expectation that large-scale river basin plans would justify small-scale, community-level interventions.

This expectation aligned with customary practices in the Andean region of Bolivia known as customs and habits (*usos y costumbres*) that establish a relationship between water management and existing local governance in communal territories [63,64]. In this context, "water belongs to the territory and the territory belongs to the community" [63,65]. However, by the 2000s, Bolivia expanded the initial terrestrial-oriented scope to incorporate the social, environmental, and sectoral dimensions in water resources management [66]. The combination of the IWRM and IRBM approaches attempt to represent both the socioeconomic dimension and the natural resources conditions of the river basin, broadening the national concept of water governance beyond strictly local imperatives. However, the IWRM and IRBM approaches may not offer an appropriate representation of water management for complex community-managed water supply systems [67] upon which large-scale water resources management intervention are superimposed.

Bolivia's Ministry of the Environment and Water (MMAyA for its Spanish initials) has developed the conceptual framework and the national policy for IWRM and IRBM through the National Basin Plan (PNC in Spanish), which was promulgated in 2006. The PNC framework seeks to deliver solutions to integrated land and water-related problems with intervention justified within a PDC. The PDC is a planning instrument aimed at establishing intergovernmental and intersectoral coordination to develop water resource governance [68]. An early effort at PDC development occurred within the Rocha River Basin before systematic learning from other implementation experiences became available [69]. The Cochabamba Departmental Government originally formulated strongly IRBM-oriented and less so IWRM-oriented guidelines for the Rocha River Basin in 2014 [70], hampering the implementation of the proposed plan, as there was little buy-in among the 25 basin municipalities to implement large-scale water management interventions. In the Cochabamba Valley, competing sectoral and regional (i.e., upstream vs. downstream) interests such as household water users, rural communities, and civic organizations in urban areas make intersectoral coordination extremely difficult to achieve. As a result, the MMAyA and the Cochabamba Departmental Government decided to update the planning instrument by formulating a package of medium- and long-term actions based on three main factors:

- climate change considerations;
- the development of analytical tools;
- the broad participation of key stakeholders in the basin.

2.3. Proposed Approach

The RDS framework developed by Purkey et al. [23] and Bolivia's guiding framework of the PDCs [11] provided our methodological approach for the formulation of a river basin master plan supported by participatory water resources systems modeling (Figure 2). The RDS framework has two phases: (i) preparation and formulation and (ii) evaluation and agreement. The first phase has six steps to identify the current and future vulnerability of the system. The second phase is a three-step participatory-driven process for the assessment of different management options leading to the identification of robust actions (actions that can satisfy disparate objectives under the assumed uncertainties). An essential step of this framework is the formulation of the problem using the XLRM matrix [71], where (X) stands for the uncertainties, (L) stands for the management options, (R) stands for the analytical tools that relate the (X) and (L), leading to performance measures, and (M) is used to evaluate the potential options. Uncertainties (X) are generally not contentious, i.e., all interest groups can agree that climate change or demographic growth are uncertainties that have the potential to impact outcomes related to water management. In contrast, preferences related to sectoral or strategic actions (L) are contentious, as stakeholders can oppose strategies offered by others. Metrics of performance (M) are identified for each sector to evaluate the outcome of each strategy identified, the stakeholder's preferred strategy, as well as those offered by others. These metrics are independent of any strategy, enabling a sector-specific strategy to improve outcomes (M) defined by another sector.

This is the basis of trade-off analysis and compromise, which serves for the formulation of a participatory-driven analysis in the decision process.

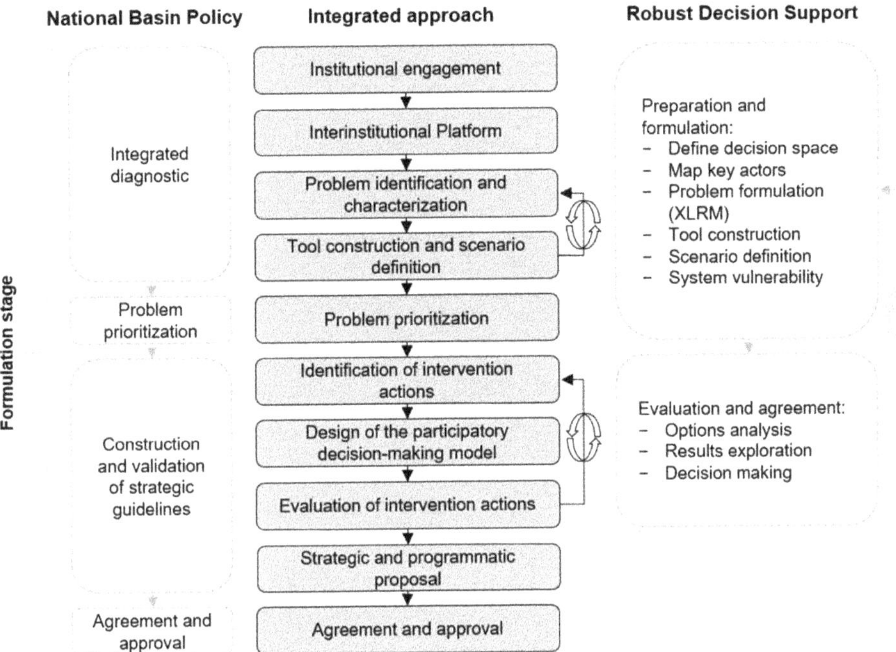

Figure 2. Framework to develop a Basin Master Plan based on model-driven participatory-based decisions. The new approach (center panel) allows grounding and implementation of the National Basin Policy guidelines (left panel) [68] and the robust decision support approach (right panel) [46].

The guiding framework of the PDC contemplates three stages: formulation, implementation, and evaluation and monitoring. Each of these stages has a step-by-step process with general guidelines (Figure 2). All stages rely on public participation of the basin's institutions and actors as an essential element to promote environmental governance. The formulation stage comprises an integral participatory appraisal that allows for the identification and prioritization of the main objectives of basin and water management, albeit without the benefit of detailed modeling and analysis. This stage also considers the identification, construction, and validation of actions to achieve the proposed objectives. Stakeholder participation is a key element for the PDC guidelines and the RDS framework. The institutional approach and the mapping of key actors are considered at the beginning of the process for both cases.

The comparison of PNC guidelines with the RDS framework reveals several gaps within the PDC guidelines. These gaps reduce the capacity of the PDC guidelines to inform water management in Bolivia's river basins, including:

- The prioritization of problems is done only based on the historical time horizon without considering potential future changes generated by uncertain factors (X) such as climate change, land-use change, population growth, etc.
- The application of water resources system models is limited to generating water balance data between supply and demand for historical conditions, and there is no use for modeling at this stage of evaluating trends or strategic actions.

- The strategic actions lack quantitative performance indicators or measures (M) to help forecast progress toward medium- and long-term goals and objectives, to which models have a great potential to contribute effectively.

However, the RDS general template to approach decision processes requires adapting to the local particulars of water governance and institutional design. To use RDS effectively to support the development of a new Rocha River basin PDC, we oriented the framework toward the development of a participatory decision-making process. A novel component of the RDS process was the introduction of 'hard coupled' decision interfaces to the WEAP model to support participatory forums. This allowed diverse institutions, interests, and organizations—that benefit or are affected by the decisions of the basin's intervention—to interactively explore the medium- and long-term implications of water management options, recognize disparities between their own and others' frames of reference, and share ideas to build a Master Plan. The implemented framework, shown in Figure 2, is detailed in the list below:

1. A mapping of key actors to identify the main institutional, sectoral, and civil society actors in basin and management problems.
2. The institutional engagement which, within the framework of the PDC, means the creation of the Inter-Institutional Platform as a space for the participation of the basin's stakeholders. This process is made up of the following instances: (i) the Board of Directors as the highest decision-making body, which is integrated by executive authorities from the national, departmental, and municipal levels; (ii) the Technical Council with representation from public institutions, non-governmental organization (NGOs), research institutes, and international cooperation initiatives; (iii) the Social Council as a space for dialogue and negotiation between the different social sectors such as water users, indigenous authorities, and civic organizations; and (iv) the Basin Management Unit (UGC) as the operational body for the implementation of the plan [11].
3. The identification of problems in the basin includes (i) participatory workshops, where problems must be addressed at appropriate levels of granularity, from the local to the regional scale and (ii) relevant indicators to assess those problems. This identification should be accompanied by the collection of existing information and fieldwork to characterize the problems and to construct a comprehensive database. Based on the participatory space and relevant information, a list of the uncertainties or uncertain factors (X) that define the problems can be consolidated.
4. The development of models of water resources systems that fully respond to the problems identified and have the capacity and flexibility to incorporate different future uncertainties (X) in the evaluation of intervention actions (L). Given the participatory approach, this step requires the involvement of key stakeholders in providing additional information, feedback, and local expert opinion to validate models, and possible trajectories of uncertain factors.
5. The prioritization of problems through participatory workshops, where the problems initially identified are presented, but this time characterized with model output information for both historical and future conditions. Stakeholders can discuss and agree on the priorities and problems that require intervention in the short, medium, and long term based on quantitative and qualitative information; they can also define specific objectives for each prioritized problem.
6. The identification of intervention actions (L) to achieve the specific objectives set. Since the PDC must connect to other planning instruments, it is important to start with the inventory of existing sectoral intervention actions and plans in the different territorial entities, mainly those in a pre-investment state. This will allow an analysis of how these actions contribute to the specific objectives and the need to identify additional actions for which participatory spaces can be generated. This should be accompanied by a characterization of the actions with basic technical information, the

scope, reference costs, and social and environmental conflicts that may make their implementation unfeasible.

7. The design of the Participatory Decision-Making Model (PDM) (R) as a facilitation-oriented tool for the construction of agreements between different actors, recognizing that there are coordination needs and that decision making is a process of negotiation between the different interconnected interests, which are not always coherent with each other. The PDM is made up of information, models of water resources systems, participation mechanisms, users, and use cases. The characteristics of the PDM are described in detail below:

    a. It allows the consideration of action proposals from the different territorial levels. It provides an overview of the coherent actions' prioritization and implementation, recognizing the present biophysical, sectoral, and territorial interconnections that make the basin an indivisible planning unit.

    b. The indicators (M) that consider a system of hierarchical analysis units, which allows a nested measurement between working scales, and where the basin domain is represented by sub-units of different scales, starting from indicators in basic modeling units (i.e., micro-basin, irrigation zone, urban demand unit).

    c. The specific objectives for the prioritized problems must be articulated with the general principles established in the national policy and planning instruments such as the PDC.

    d. Their application is not limited only to the formulation stage but also to the other stages such as monitoring or follow-up. In the formulation stage, it allows a comparative and participatory analysis to be made over time, making it possible to differentiate between medium- and long-term horizons. In the monitoring and follow-up stage, it will have the capacity to review the fulfillment of goals and potential strategic adjustments and redirection.

    e. The type and scope of intervention actions that become relevant in the performance of objectives and indicators at the intervention scales of a basin plan. Through the PDM, it is possible to evaluate the multiple effects (positive or negative) of the interaction of the various interventions.

    f. With respect to both frameworks, the set of quantitative indicators that can operate at various scales and the use of an interactive decision panel connected to a water resources system model were important innovations. Co-designed with stakeholders, these innovations create a common language and interactive feedback of model runs to specific requirements.

8. The intervention actions identified in previous steps should be implemented in the PDM to evaluate performance using multiscale indicators. This can begin with a participatory and iterative exercise in evaluating individual actions in the medium and long term. Interactions between the actors and the PDM, and discussions of the different interests should lead to the formulation of intervention packages for different time horizons.

9. The action packages should be grouped into strategic lines and a coherent and plausible programmatic implementation framework with clearly established goals and potential funding options leading to the development of a formulation document to be approved by the stakeholders. Within the PDC framework, this refers to the Inter-Institutional Platform.

At the implementation stage, intervention actions can be reviewed and adjusted as new redirections are needed. The water resources system model can be updated with new information to help improve its performance. To make this effective, it is necessary to build local capacity in the management of all the tools that constitute the PDM.

3. Implementation of the Proposed Approach

The results of the implementation of the proposed approach in the formulation of the Rocha River PDC are presented below.

3.1. Strategic Modeling of the Basin's Water Resources

The MMAyA and the Development Bank of Latin America (CAF) developed a previous instance of the Rocha River basin model using WEAP [48]. The modeling process included the following in the collection and systematization of available data: participatory co-development of the model, fieldwork, definition of scenarios, and discussions about the benefits of using models in support of river basin planning. The WEAP model of the basin includes hydrology, water demand and supply, rules of operation, use rights, and water quality in terms of organic contamination of the main rivers. The model has a monthly time step for a historical horizon of 1980–2015 and a prospective period of 2020–2050. Hydrology was implemented using WEAP's Soil Moisture Model (SMM) [35], which is a one-dimensional model based on the notion of water transfer between two buckets: an upper bucket representing the root zone and a lower bucket representing deep storage or regional aquifers (when applicable). These two buckets represent the dynamics between evapotranspiration, surface runoff, interflow, and percolation for each basic modeling unit (catchments). The model allows dividing the basin in catchments in a semi-distributed way. The SMM is forced with climate data such as precipitation, temperature, relative humidity, wind speed, and insolation. Climate data from the Bolivian Water Balance [72] were used as input. The SMM is also used to calculate irrigation demands in agricultural areas, which are obtained as deficits of water required to maintain soil moisture within desired boundaries following crop type and calendars and other technical restrictions such as distribution or application efficiency and water allocation priority.

The catchments were delimited based on the location of reservoirs, water use points, transfer basins, wastewater discharge, calibration points, and topographic transition zones between mountain and valley areas (Figure 3). To parameterize the hydrological model, vegetation coverage, slope, and geology were considered. The vegetation cover was classified with a Landsat 8 image (January 2018) for the following categories: agriculture, forest, dispersed vegetation, temporary flood areas, urban area, scrubland, badland, and water bodies. The slope was obtained from a HydroSHEDS digital elevation model (DEM) [73] for the following ranges: sloping (<10%), strongly sloping (10–15%), moderately steep (15–30%), steep (30–60%), and very steep (>60%). Using the 1:100,000 scale geological map from Bolivia's Geological Mining Service, it was possible to differentiate the Quaternary deposits. We used this information to identify the zones of recharge and the three aquifers (Sacaba Valley, Cochabamba Valley, and Alto Valley) with high probability of groundwater occurrence.

For the modeling, the 48 existing reservoirs were considered (Figure 3), of which 90% have a storage capacity of less than 1 Mm^3 and are used for small-scale irrigation. In the rainy season, storage is prioritized, and in the transition and dry seasons, water volumes are released for irrigation according to the agricultural calendar. The reservoirs for human consumption are operated according to water demand. Information regarding physical characteristics and location was collected from the national inventory [74] and potential investment project database.

Household water demand was represented in WEAP as a function of population, per capita consumption, and average losses reported by the water distribution systems of each municipality. In the Rocha and Maylanco sub-basins, there are two types of service providers: Water and Sanitation Service Providers (EPSA) and Small-scale Local Operators (OLPE) [75]. The EPSA is an entity that depends on the municipal government and whose service area is limited to the urban area, and it is regulated by the corresponding authorities. The OLPE are autonomous community organizations or cooperatives that provide water service to their areas of immediate territorial occupation. For the modeling, a distinction has been made between the coverage of EPSA and OLPE. A total of 26 water demand nodes have been represented (Figure 3). Water quality in the main section of the Rocha River was represented in the model as the organic contamination constituents that determine biological oxygen demand (BOD) using the Streeter–Phelps model [76] and water temperature of the river. The modeled river section and assessment points are shown in Figure 3. The assessment points are located downstream of the wastewater treatment plant (WWTP) discharge.

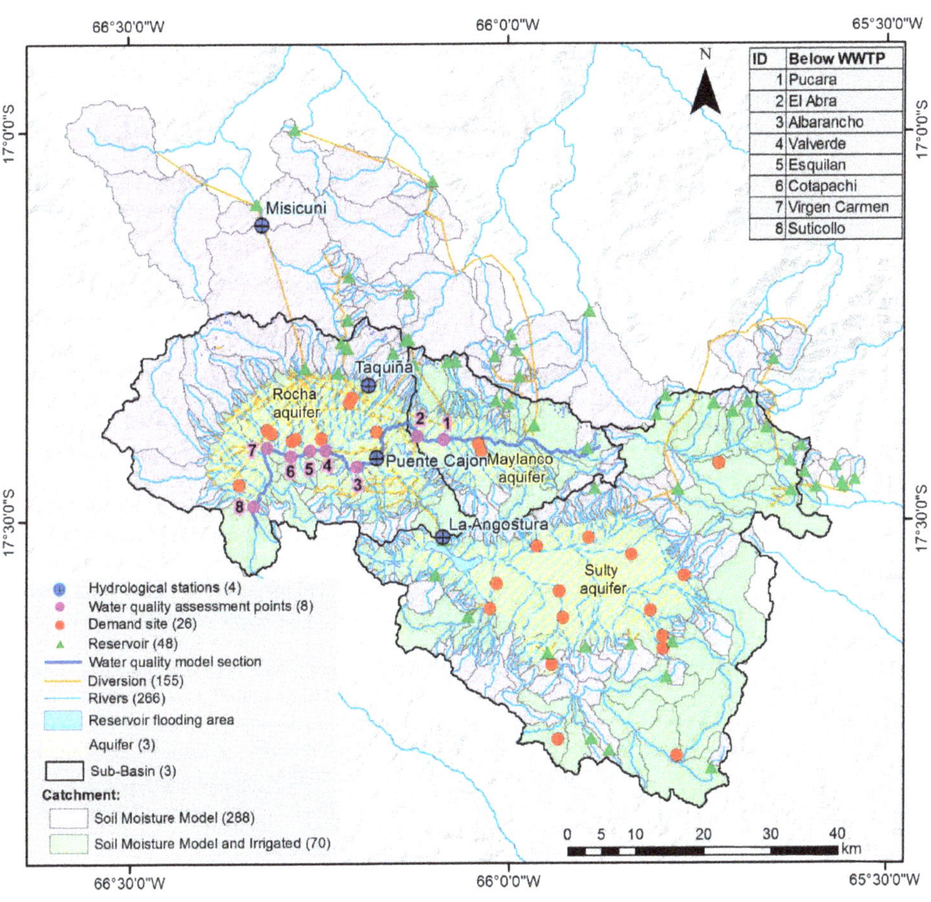

Figure 3. Key components of the Rocha River basin WEAP model. The numbers in parentheses in the legend refer to the number of objects simulated in WEAP.

To determine the irrigation demand and allocation, we calculated the water balance using the Food and Agriculture Organization of the United Nations (FAO) approach, which is based on reference potential evapotranspiration and crop coefficients (Kc) [77]. We defined the irrigation area with the digitalization of high spatial resolution images. We also assimilated information from previous studies to characterize the crop schedule, agricultural calendar, and extent and type of irrigation systems [78–80]. The water balance was implemented at an irrigation zone scale that can include one or several irrigation systems. The catchments that have irrigation are shown in Figure 3. Irrigated zones can cover more than one catchment; for this reason, the total number of modeled zones is 74, while the number of irrigated catchments is 70.

For the model calibration, we used the flows measured in Misicuni, Taquiña, and Puente Cajón stations, and the storage levels in La Angostura reservoir (Figure 3). In the drainage area of these stations, the flows are modified by the extractions and storage in reservoirs. Therefore, model calibration is not only limited to optimizing SMM parameters but also to the operation rules and extraction volume. The initial parameters were defined

panel, they could select the intervention actions, uncertainties, water allocation, and the planning horizon (medium—2025, and long term—2040). The selected decisions are sent to the WEAP model through a Microsoft VBA script and then, the model is run. The stakeholders were able to visualize the performance of the intervention measures by means of multi-scale indicators quantified at a range of scales from the basin, sub-basins, micro-basins, aquifers, irrigation zones, urban demand units, and river reach. The platform allowed for the visualization of these indicators at the precise scale of interest of each actor. The indicators used are shown in Table 3 where each has the unit of measurement, basic unit of spatial analysis, the upper-level spatial scale, and the function of aggregation. For example, for drinking water, the basic unit of spatial analysis is the urban demand unit, the next level is the sub-basin, and the subsequent level is the basin. The aggregation function depends on the unit of measurement; in the case of volume, the function is the sum. A cost-efficiency indicator was also considered. Multiscale indicators enabled a diverse audience of stakeholders to explore the positive and negative interactions of intervention actions, identify disparities in action performance across scales, and interactively compare different actions that help identify and mitigate emerging regional or sectoral conflicts.

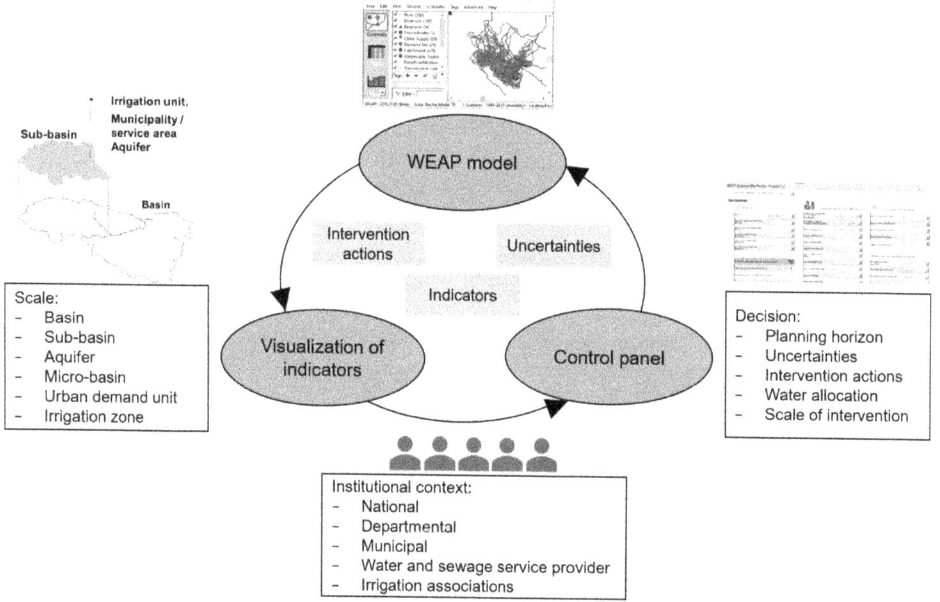

Figure 4. Conceptual design of the participatory decision-making model for the Rocha River Basin.

Table 3. Multiscale indicators (M) to evaluate the performance of intervention actions.

Indicator Name	Unit	Basic Unit of Spatial Analysis [1]	Aggregation Function	Aggregation Scales [2]
Water demand	m^3/year	Urban demand unit (26)	Sum	Basin (1), sub-basin (3)
Coverage	People	Urban demand unit	Sum	Basin, sub-basin
Coverage (driest condition)	People	Urban demand unit	Sum	Basin, sub-basin
Unmet demand	m^3/year	Urban demand unit	Sum	Basin, sub-basin
losses	%	Urban demand unit	Weighted average	Basin, sub-basin

Table 3. Cont.

Indicator Name	Unit	Basic Unit of Spatial Analysis [1]	Aggregation Function	Aggregation Scales [2]
Agricultural areas under optimal irrigation	Ha	Irrigation zone (74)	Sum	Basin, sub-basin
Unmet demand	m^3/year	Irrigation zone	Sum	Basin, sub-basin
Recharge	m^3/year	Aquifer (3)	-	-
Withdrawal	m^3/year	Aquifer	-	-
Use rate (withdrawal/recharge)	%	Aquifer	-	-
BOD	mg/L	Urban river reaches (9)	Weighted average	Rocha River (1)
Flow	L/s	Urban river reaches	Weighted average	Rocha River Outlet
Transfer volume	Mm^3	Sub-basin	Sum	Basin
Cost-effectiveness	$/$m^3$/people, $/$m^3$/ha	-	-	-

[1] The numbers in parentheses refer to the quantity of basic units. There are 26 urban demand units, 74 irrigation zones, 3 aquifers, and 9 river reaches. [2] The numbers in parentheses indicate that there is a basin, 3 sub-basins, and a main river.

3.5. Evaluation of Intervention Actions

First, each intervention was evaluated individually using the participatory decision-making model considering efficiency indicators for different uncertainty scenarios and their performance in the medium (2025) and long term (2040). Second, stakeholders were divided into two groups that through an iterative process identified and prioritized action packages for the time horizons considered and the collection of stakeholder-specific goals (Figure 5). Public institutions such as the MMAyA and different operational institutions of the departmental government (Departmental Watershed Service, and Directorate of Water Management and Basic Services) participated in this process. These institutions work on pre-investment and investment planning related to water (Figure 5). During the group sessions, participants offered four main insights:

- Given the scale and heterogeneity of conflicts over water access, projects currently in development, such as the Misicuni reservoir and water transfer, can provide important—but insufficient—gains toward the plan objectives, with several areas of the basin with high vulnerabilities requiring specific targeted actions.
- Most actions will require additional actions with long-term commitments; mid-term actions must focus on creating the enabling conditions such as pre-investment studies that can last several years accompanied with the search for additional funding sources.
- The solution to water scarcity problems requires large water transfer actions, which may take decades to implement.
- Short-term, efficiency-oriented actions can provide gains toward the required solutions; however, a strategic long-term plan is required to allow a coherent and compatible implementation of short-term and long-term actions. Hence, short-term actions may enable the implementation of middle- and long-term actions and prevent future contradictions among other strategic projects.

During the exercise, not only technical and financial indicators were analyzed but also potential social and environmental conflicts that may make implementation unfeasible. Aspects such as water distribution and allocation priorities were also addressed.

3.6. Proposal and Approval of the Strategic and Programmatic Framework

The prioritized actions were organized and structured in lines of action and strategic lines that focused on institutional, sectorial, and community interventions, always considering the principle of the basin approach. The PDC gives action plans that must be pursued in the medium and long term to make substantial progress in the well-being of the inhabitants and their way of life. It also provides an integrated framework to strengthen financing processes by coordinating the investments that the territorial authorities make in water management issues, which are currently carried out in a fragmented manner and generate environmental conflicts. Coordination allows for the expansion of the scope of benefits through the coherent use of resources. The PDC document was agreed to by all key stakeholders and approved by the Technical Council. Moreover, the PDC has been declared a departmental law (*Decreto Departamental* No 4544, 18th of September 2020) [84]. A legal framing means the PDC actions are now binding and must be fulfilled by the sub-basin jurisdictions, hence leveraging financial resources for action implementation and increasing the PDC's visibility across planning scales. Table 4 shows the plan in detail, which has five strategic lines and 82 interventions. The total investment for the period 2020–2040 amounts to USD 1.5 billion; for the current population, this is equivalent to approximately USD 58/person/year. Most of the total investment cost of the plan (98%) corresponds to the strategic line of water management, which has an implementation horizon of 20 years. For the first five years (planning horizon of the PDC), priority was given to the construction and improvement of the WWTP, the construction of water conduction infrastructure of the Misicuni, starting with detailed pre-investment studies for new water transfers (Cordillera Norte and Khomer Khocha), water demand management, and preservation of the drainage area of current and potential transfer basins (water reserve zone). The 2026–2040 horizon of the plan contemplates the construction and putting in operation of the new water transfers, the extension of the water distribution network and sanitary sewerage, and the new WWTP. With this package of actions in the medium term (2025), progress will be made in improving the quality of water in the Rocha River and reducing unsatisfied demand in the Rocha and Maylanco sub-basins. The solution to water scarcity problems in the entire basin depends mostly on new water transfers, which will require decades to implement. The other strategic lines have an implementation horizon in accordance with the times established in the national basin policy. These lines seek to improve the sustainable management of micro-basins, institutional strengthening for water resources management, improving information and knowledge, and water culture. The baseline and plan indicators agreed upon for the 2020 and 2040 horizons are described in detail below. The baseline is the current system of water resource management projected for different future time horizons.

Figure 5. Photograph of the workshop for the evaluation of medium- (2025) and long-term (2040) intervention actions using the participatory decision-making model.

Table 4. Summary of the Rocha River Basin master plan agreed with the Inter-Institutional Platform.

Strategic Line	Number of Actions	Horizon	Key Actions	Investment (USD) [1]
Water management	37	2025	- Infrastructure for the conduction and distribution of drinking water and irrigation (Misicuni). - Construction of WWTP in the Rocha sub-basin and expansion of sewerage coverage. - Pre-investment studies of strategic actions for new water transfer. - Water demand management actions in urban areas and irrigation areas. - Adoption of strategic micro-basins as water reserves to prevent the emergence of land-use conflicts.	532.8 M
		2026–2040	- Construction and operation of new water transfer infrastructure. - Expansion of water distribution network coverage in urban areas and irrigation. - Extension of sanitary sewerage coverage and new WWTPs in the Sulty sub-basin.	958.4 M
Integral and sustainable management of sub-basin and micro-basin	19	2025	- Mitigation of soil erosion, landslides, loss of vegetation cover, and flood risk. - Protection of aquifer recharge areas and management of saline soils. - Restoration of environmental functions of water regulation through afforestation, reforestation, and the recovery of native grasslands. - Incorporate the watershed approach in territorial planning and management of hydrological risks on a basin scale through preventive and mitigating actions.	20.8 M
Information and knowledge	8	2025	- Aquifer and river water quality monitoring, and construction of water quality laboratory in Sulty sub-basin. Detailed studies of groundwater modeling, feasibility of implementing WWTPs on a small scale, environmental performance of the industrial sector, and ecological flows. - Sustainability of participatory decision-making model.	1.8 M
Institutional strengthening	10	2025	- The operational consolidation of the Inter-Institutional Platform and strengthening and formation of Strategic Committees (e.g., groundwater). - Assigning normative hierarchy to the plan through a departmental law or ministerial resolution. - The promotion of the plan to generate knowledge and institutional commitment - Institutional strengthening of territorial entities in water management. - Identification of institutional mechanisms for conflict resolution.	2.4 M
Water culture	8	2025	- Promote environmental sustainability before institutions and citizens through actions such as art, competitive funds, exchange of experiences, spaces for reflection and recreation, academic research, and design of tourist routes.	3.3 M

[1] Conversion rate: 1 USD = 6.96 BOB (reference year 2019).

3.6.1. Baseline Indicators

Projections until 2025 and 2040 indicate that the basin would have a population of 1,518,502 and 2,228,632 people, respectively. The irrigable area is approximately 40,002 ha, which was assumed to be constant for future time horizons. Performance indicators up to 2025 (Table 5) indicate that the current supply system could supply only 66% of the population and optimally irrigate 19,443 ha. The annual volume of unsatisfied demand in the basin would be 340.3 Mm^3. The water quality in the Rocha River modeled by BOD would be 153 mg/L, which is approximately four times higher than the limits established in national legislation. The Sulty sub-basin would have an unsustainable exploitation of groundwater. By 2040, these indicators could worsen (see Table 5); for example, the unsatisfied demand could reach 362.7 Mm^3. The current supply system would only have the capacity to supply 59% of the population, and irrigation conditions would remain similar to those encountered in the medium term.

3.6.2. Indicators of the Agreed Plan

In this paper, we present the aggregated results at the basin and sub-basin scale; however, in the PDM, the results can be visualized at the basic service area unit scale of modeling. Table 5 shows the expected level of improvement in the main performance indicators up to 2025 (medium term). The plan expects to reduce the annual unmet demand in the basin by 64.9 Mm^3. The population benefited by the package of actions is 403,647 people. In irrigation, it is expected that the deficit will be reduced by 58.1 Mm^3 and the optimal irrigated area will increase by 1768 ha. In water quality, a reduction of BOD by 101 mg/L is expected—that is, a reduction of more than 70% of the current levels of contamination, and at the same time, an increase of approximately 20% in the flow during the low water season to improve the environmental functionality and the assimilation capacity of the Rocha River.

Table 5 shows the indicators expected in the long term (2040). The package of actions will reduce unsatisfied water demand by 227 Mm^3 in the basin, thus increasing access to safe water by 834,049 people and increasing the area under optimal irrigation by 10,497 ha. The increase in the supply of surface water will make it possible to achieve the sustainable use of all the aquifers in the basin—that is, by exploiting them within the limits of their natural recharge. In terms of water quality, a reduction in BOD of 125 mg/L is expected as well as an increase in the flow of the River Rocha in the months of low water by 359 L/s.

In the Maylanco sub-basin, indicators presented negative changes up until the year 2025. For example, drinking water coverage reduced by 3759 people, and unmet demand increased by 0.4 Mm^3 (Table 5). This is mainly due to the expansion of the water distribution system from Misicuni to the municipalities of the Rocha sub-basin. In the baseline scenario, the Maylanco sub-basin would be benefitted from higher water volumes; however, the proposed plan will reduce the water volumes given the expansion of the water distribution system. These changes are also reflected in the increase of the groundwater use rate by 52%. In the Pucara and Abra sections (Maylanco sub-basin), there is a small BOD increase due to the reduction in volumes delivered from Misicuni, which also reduces the river's self-purification capacity.

Table 5. Performance of the proposed actions (2025 and 2040).

Component	Indicator	Unit	Spatial Domain		Base Line 2025	Base Line 2040	Agreed Plan 2025	Agreed Plan 2040	Expected Change 2025	Expected Change 2040
Drinking water	Coverage	People	Sub-basin	Sulty	66,733	87,171	66,733	148,232	0	61,061
			Sub-basin	Rocha	715,953	853,199	1,123,359	1,612,808	407,406	759,609
				Maylanco	222,569	372,068	218,810	385,448	−3759	13,380
			Basin		1,005,254	1,312,438	1,408,901	2,146,487	403,647	834,049
	Unmet demand	Mm³	Sub-basin	Sulty	3.3	5.8	3.3	1.4	0.0	−4.4
				Rocha	9.6	29.3	2.5	1.9	−7.1	−27.4
				Maylanco	0.0	1.3	0.4	0.0	0.4	−1.3
			Basin		12.99	36.41	6.24	3.34	−6.75	−33.07
Irrigation	Optimal irrigation area	ha	Sub-basin	Sulty	12,042	12,045	12,743	18,795	701	6750
				Rocha	6411	6467	7315	8960	903	2493
				Maylanco	990	986	1154	2240	164	1254
			Basin		19,443	19,499	21,211	29,996	1768	10,497
	Unmet demand	Mm³	Sub-basin	Sulty	227.0	227.0	198.7	98.9	−28.3	−128.1
				Rocha	81.2	80.1	57.7	32.5	−23.5	−47.7
				Maylanco	19.1	19.2	12.8	1.1	−6.3	−18.1
			Basin		327.3	326.3	269.2	132.4	−58.1	−193.8
Water quality	BOD, Flow	mg/L, L/s	River reach	Pucara	38 / 213	53 / 266	39 / 210	49 / 282	1 / −3	−4 / 16
				Abra	59 / 437	73 / 613	61 / 422	66 / 660	3 / −15	−7 / 47
				Albarancho	123 / 277	138 / 469	46 / 483	49 / 876	−76 / 207	−89 / 407
				Valverde	105 / 277	117 / 469	64 / 483	56 / 876	−41 / 207	−61 / 407
				Esquilan	196 / 606	207 / 840	60 / 581	53 / 1003	−136 / −25	−154 / 164
				Cotapachi	198 / 614	209 / 848	61 / 800	55 / 1334	−137 / 186	−155 / 486
				Virgen Carmen	268 / 710	312 / 944	57 / 872	53 / 1485	−211 / 161	−258 / 541
				Suticollo	185 / 568	220 / 802	36 / 753	35 / 1446	−149 / 184	−184 / 644
			Rocha river		153 / 502	177 / 703	52 / 613	52 / 1062	−101 / 111	−125 / 359
Groundwater	Use rate	%	Aquifer	Sulty	105	105	102	65	−3	−40
				Rocha	64	82	55	54	−9	−28
				Maylanco	43	87	95	0	52	−87

Cost-effectiveness analysis was part of the performance indicators used in the process of evaluating intervention actions. Some sector interventions identified before the formulation of the PDC presented unfavorable values in terms of cost-effectiveness, especially water transfers such as Khomer Khocha. The intervention assessment process explored options to make interventions feasible within the framework of the basin planning. The multipurpose approach of the actions has been an option that allowed obtaining more encouraging indicators of economic efficiency. Table 6 shows the cost-effectiveness indicators of the agreed plan for the different time horizons. In the case of drinking water, the highest value is related to the implementation of the Misicuni additional water transfers. In irrigation, the indicators are well within the Bolivian viability threshold of 10,000 USD/ha. The highest value in the horizon 2035 is related to the implementation of the Khomer Khocha water transfer that benefits mainly irrigation in the Sulty.

Table 6. Cost-effectiveness indicators of the agreed plan for water supply actions for drinking water and irrigation.

Horizon	Drinking Water		Irrigation	
	USD/People	Thousands USD/Mm³	USD/ha	Thousands USD/Mm³
2025	6	377	2370	72
2030	20	782	1764	82
2035	6	179	4507	246
2040	0.2	6	858	46

4. Discussion

4.1. How Does the Water Resources System Model Respond to Water-Related Decision-Making Processes and Institutional Governance Design at a Range of Scales within a River Basin?

In the Rocha River Basin, all available water is in use, which means that there is no free water for new uses [85], and management decisions are fundamentally a redistribution between existing users. According to the modeling, the demand for water is potentially three times the natural availability of the basin; hence, there is extensive unsafe reuse and a growing dependence on external sources. The implementation or intervention actions that consider expanding the use of the basin's supply sources generate conflicts. In this context, the formulation of the basin plan faced many challenges. In the decision-making process, analytical modeling tools helped to recognize, quantify, and differentiate the local and regional impacts of water reallocation. Our multiscale approach allowed the diverse frames of reference of basin stakeholders such as small OLPES, individual irrigation zones, and municipal jurisdictions to identify their interdependence with other sector-specific goals or regions, including those operating at different scales, and engage in productive negotiations.

An example to illustrate this is the Siches Reservoir (Table 7), which in the Sulty sub-basin generates an incremental irrigation area of 797 ha, but in the Rocha sub-basin, the optimal irrigation area is reduced by 222 ha. The drainage area of this potential reservoir is one of the main tributaries to the La Angostura reservoir, which in turn supplies irrigation areas in the Rocha sub-basin. To compensate for the conflicts identified, iterative exercises (as shown in Figure 3) were carried out. These exercises contributed to the construction of a package of actions to achieve the full spectrum of objectives and goals (Table 7). In addition, it was possible to incorporate climate change into decision making with the model. In the process of building the action set, stakeholders were able to visualize results of how climate change in the long term could lead to problems in the reliability of water supply systems. This is how actions to build resilience in water supply systems were assessed. Based on this exercise, the new Cordillera Norte water transfer action was formulated as part of the plan.

Table 7. Performance in irrigation of the intervention action of Siches Reservoir.

Sub-Basin	Optimal Irrigation Area [ha]		
	Base Line	**Intervention Action**	**Change**
Sulty	12,042	12,839	797
Rocha	6411	6189	−222
Maylanco	990	990	0
Basin	19,443	20,018	575

Likewise, the results of the model showed that water-use efficiency actions and expansion of current supply sources in the basin, while providing important gains in performance, cannot compensate for existing conflicts on their own or fulfill the medium- and long-term PDC main goals, such as universal and equitable access to safe water. Therefore, solutions to scarcity problems will also require several strategic actions that increase the volume of water transfers. Due to the multipurpose nature and regional scope, it will also allow substantial progress in the sustainable use of groundwater by supplementing current wells with transferred surface water. By improving the environmental conditions of the rivers, it will also provide greater capacity for assimilation by increasing the flow in months of low water from the return flows.

The analytical tool for decision-making in the basin makes it possible to co-develop strategies in participatory spaces with the interested parties. The use of models in conflict management is recognized in the scientific literature [86,87]. However, formalizing their use in decision-making within the framework of the basin plan had many challenges related to limited data, the complexity of traditional water management, the spatial and temporal scale of water supply systems, and some resistance to the use of models. At the beginning of the process, the main challenge was the lack of credibility for the use of the model in formulating the plan, as stakeholders indicated that the available data were insufficient to develop the watershed WEAP model. In addition, incorporating the model into planning meant adjusting the PDC's methodological guidelines, which also generated resistance in certain government institutions. To address these challenges, partnerships were formed with local universities to maximize the use of available data and advances, intensive fieldwork to collect irrigation data, discussion meetings with local experts, and workshops to share the inputs and results. The participatory modeling approach generated opportunities to iterate the process while interacting with local experts. By involving stakeholders early in the process, we were able to make adjustments and continuous improvement in the formulation of the river basin plan. As mentioned by Loucks et al. [45], the credibility of a model is very subjective, and the modeling, besides trying to represent reality, is also an attempt to formalize and guide those perceptions.

It is important to recognize that there are conflicts around intervention actions that cannot be negotiated through the support of models because they require arrangements in traditional, sectoral, institutional, and legal use rights. For example, water transfer actions generate conflicts in the territory of new uses [55]. Based on experiences of the Misicuni water transfer, this type of action generates the displacement of rural community settlements, the loss of productive land, and the drying up of wetlands and springs [51] with unjust compensation [88,89]. With these precedents, the transfer actions proposed in the watershed plan could face similar conflicts.

4.2. How Does the Water Resources System Model Contribute to the Development of Effective Water Planning Instruments?

Formulating and implementing river basin master plans in Bolivia is a learning process built on experiences from strategic basins of varied geographic contexts. For instance, the first master plan of the Rocha River Basin showed progress in elaborating the diagnostic study and the strategic guidelines. It also made progress in the implementation of IRBM actions in priority micro-basins and flood mitigation infrastructure. Its water management strategy focused on water access and use. However, it was up to sectorial plans (e.g., the

Water Master Plan) to identify the actions needed to fulfill the master plan. The initial strategy lacked quantitative indicators and goals in its implementation horizon, which would have enabled monitoring and follow-up processes as well as connections with other plans at the national scale. For instance, the Bolivian Economic and Social Development Plan could have been a potential connection, since 'achieving universal access to water' is also one of its objectives. Since the formulation of the first basin plan, the Rocha and Maylanco sub-basins made significant progress with their Water and Sanitation Master Plan. However, the proposed actions were not aligned with other water uses, such as irrigation, since the Water and Sanitation Master Plan focused on water for human consumption. Lastly, the Sulty sub-basin made progress in managing financial sources for the preparation of the Water Master Plan.

The updating of the Basin Plan under the proposed approach was an opportunity to coordinate these sectoral interventions and the actions promoted from the different territorial entities that are not necessarily connected to objectives at the regional- and basin-level. In addition, the updated plan has a sequence of short-, medium-, and long-term actions, thus recognizing the potential effects of climate change. Its strategic and multisectoral approach made it possible to identify a wide range of co-benefits and compensation measures. The flexibility of the analytical decision-making tool will allow monitoring and follow-up as the plan is implemented, and if necessary, the goals could be adjusted according to new orientations in policy or new stakeholder priorities.

5. Conclusions

Our integrated approach lays the foundation to apply the RDS framework in water resources management, building upon and expanding on Bolivia's basin policy. The formulation of the River Rocha PDC led to formalizing the use of modeling tools in the context of river basin planning in Bolivia. However, we faced many challenges arising from the planning process to identify strategic actions, particularly those related to expanding the use of the basin's supply sources. We connected sector- and territorial-based actions across the river basin while engaging municipalities and the regional and national governments in the planning process. Our analytical modeling tools helped to identify and quantify the trade-offs between local and regional impacts that a new water use could generate, hence facilitating the decision-making process. Once we identified positive and negative effects of the proposed actions, the participants could negotiate and propose compensation measures within the PDC. We formulated a PDC considering short-, medium-, and long-term climate change scenarios, setting out measurable goals for each time horizon and establishing implementation phases for the proposed actions. Our iterative RDS process allowed for re-arrangements or changes within the water resources system modeling according to the stakeholders' inputs and needs.

The PDC covers a wide range of topics—from flood control measures to determining a financial plan for the implementation of strategic actions. As previously mentioned, the analytical tools are key to the river basin planning process. However, not all territorial-based actions are integrated in the water resources system model. Further research is needed to integrate territorial-based actions and water resources modeling for participatory decision-making process, hence minimizing potential conflicts in planning instruments such as the PDC. In addition, water resources system models alone may not recognize all water-related conflicts. Water resources modeling tools may be limited in their capacity to propose sound socioeconomic alternatives and/or compensations to people directly or indirectly affected by water-related projects. Although our water resources system model accommodated water volumes according to each implemented action in the river basin, our work has not directly addressed legal and institutional conflicts behind those actions. In addition to indicating the water availability through a detailed water budget, anticipating conflicts that might emerge in the strategy implementation phase could help with conflict management under scenarios of uncertainty. Moreover, the Andean region of Bolivia widely adopts *usos y costumbres* to ensure water rights in communal territories.

The Cochabamba Valley is no exception to these historical customary practices. As properly observed by Hendriks [65], "water belongs to the territory and the territory belongs to the community". Further research is needed to intentionally intersect planning at the river basin level with water management in communal territories, addressing any mismatching policies that may escalate potentially divisive water governance approaches.

The RDS approach applied in the Bolivian context may facilitate IWRM implementation at the river basin scale by providing both rigorous water resources system modeling and effective stakeholder participation. Our proposed framework creates an opportunity for stakeholders to engage in the water management process and highlights the need of ensuring participatory processes to legitimate planning instruments for the river basin. We are comfortable that the approach taken addresses the general challenges facing IWRM and the specific context of water management in Bolivia. The key was allowing for cross-scale evaluation of the performance of different actions and a direct integration of watershed and water resources management interventions.

Author Contributions: Conceptualization, N.L.-Q., C.C. and H.A.; methodology, N.L.-Q., W.R., Z.G. and H.A.; software, N.L.-Q. and H.A.; validation, N.L.-Q., C.C. and H.A.; formal analysis, N.L.-Q., Z.G., F.Z., S.N., J.I. and H.A.; investigation, N.L.-Q., W.R., Z.G., F.Z., S.N., J.I., C.S. and H.A.; resources, N.L.-Q., D.P., M.E. and H.A.; data curation, N.L.-Q., F.Z., S.N., J.I. and C.S.; writing—original draft preparation, N.L.-Q., C.C., D.P. and H.A.; writing—review and editing, C.C., D.P. and H.A.; visualization, N.L.-Q., J.I. and H.A.; supervision, H.A.; project administration, D.P.; funding acquisition, D.P. and M.E. All authors have read and agreed to the published version of the manuscript.

Funding: The "Pilot Program for Climate Resilience—Ministry of Environment and Water, Bolivia, grant number 01/2018" funded the formulation of the River Rocha PDC and the "Stockholm Environment Institute's Regional Engagement Funds" financed the preparation of the manuscript.

Institutional Review Board Statement: Not applicable.

Informed Consent Statement: Not applicable.

Data Availability Statement: Not applicable.

Acknowledgments: We gratefully acknowledge the support of the Pilot Program for Climate Resilience—Ministry of Environment and Water (PPCR-MMAyA), the World Bank office in La Paz, and the *Secretaría Departamental de los Derechos de la Madre Tierra* (including the *Servicio Departamental de Cuencas* and the *Dirección de Gestión de Agua y Servicios Básicos*). We would also like to acknowledge the support of some SEI consultants and the PPCR supervision team who assisted in the implementation of the proposed approach.

Conflicts of Interest: The authors declare no conflict of interest.

References

1. Lemos, M.C.; Kirchhoff, C. Climate Information and Water Management. In *The Oxford Handbook of Water Politics and Policy*; Conca, K., Weinthal, E., Eds.; Oxford University Press: Oxford, UK, 2016; Volume 1.
2. Hoegh-Guldberg, O.; Jacob, D.; Taylor, M.; Bindi, S.; Brown, I.; Camilloni, A.; Diedhiou, A.; Djalante, R.; Ebi, K.; Engelbrecht, J.; et al. *Impacts of 1.5 °C Global Warming on Natural and Human Systems*; Global Warming of 1.5 °C. An IPCC Special Report on the impacts of global warming of 1.5 °C above pre-industrial levels and related global greenhouse gas emission pathways, in the context of strengthening the global response to the threat of climate change, sustainable development, and efforts to eradicate poverty; IPCC: Geneva, Switzerland, 2018.
3. Hall, J.W.; Grey, D.; Garrick, D.; Fung, F.; Brown, C.; Dadson, S.J.; Sadoff, C.W. Coping with the curse of freshwater variability. *Science* **2014**, *346*, 429–430. [CrossRef] [PubMed]
4. Hall, J.W.; Borgomeo, E.; Mortazavi-Naeini, M.; Wheeler, K. Water Resource System Modelling and Decision Analysis. In *Water Science, Policy, and Management*; Dadson, S.J., Garrick, D.E., Penning-Rowsell, E.C., Hall, J.W., Hope, R., Hughes, J., Eds.; John Wiley & Sons: Hoboken, NJ, USA, 2019; pp. 257–273. ISBN 978-1-119-52060-3.
5. Pahl-Wostl, C. A conceptual framework for analysing adaptive capacity and multi-level learning processes in resource governance regimes. *Glob. Environ. Chang.* **2009**, *19*, 354–365. [CrossRef]
6. Jiménez, A.; Saikia, P.; Giné, R.; Avello, P.; Leten, J.; Liss Lymer, B.; Schneider, K.; Ward, R. Unpacking Water Governance: A Framework for Practitioners. *Water* **2020**, *12*, 827. [CrossRef]

7. Unatide Nations World Water Assessment Programme. *Water for People, Water for Life*; The United Nations World Water Development Report: Executive Summary; UNESCO Publishing: Paris, France, 2003.

8. Lubell, M.; Balazs, C. Integrated Water Resources Management. In *The Oxford Handbook of Water Politics and Policy*; Conca, K., Weinthal, E., Eds.; Oxford University Press: Oxford, UK, 2016; Volume 1.

9. *Global Water Partnership Integrated Water Resources Management*; TAC Background Papers; Global Water Partnership: Stockholm, Sweden, 2000.

10. Rahaman, M.M.; Varis, O. Integrated water resources management: Evolution, prospects and future challenges. *Sustain. Sci. Pract. Policy* **2005**, *1*, 15–21. [CrossRef]

11. Transforming our World: The 2030 Agenda for Sustainable Development. Available online: https://sdgs.un.org/2030agenda (accessed on 10 October 2020).

12. United Nations. *Sustainable Development Goal 6 Synthesis Report 2018 on Water and Sanitation*; United Nations Publications: New York, NY, USA, 2018.

13. Lautze, J.; de Silva, S.; Giordano, M.; Sanford, L. Putting the cart before the horse: Water governance and IWRM. *Nat. Resour. Forum* **2011**, *35*, 1–8. [CrossRef]

14. Neto, S.; Camkin, J.; Fenemor, A.; Tan, P.-L.; Baptista, J.M.; Ribeiro, M.; Schulze, R.; Stuart-Hill, S.; Spray, C.; Elfithri, R. OECD Principles on Water Governance in practice: An assessment of existing frameworks in Europe, Asia-Pacific, Africa and South America. *Water Int.* **2018**, *43*, 60–89. [CrossRef]

15. Molle, F. River Basin Management and Development. In *International Encyclopedia of Geography: People, the Earth, Environment and Technology*; Richardson, D., Castree, N., Goodchild, M.F., Kobayashi, A., Liu, W., Marston, R.A., Eds.; John Wiley & Sons: Oxford, UK, 2017; pp. 1–12. ISBN 978-0-470-65963-2.

16. Kauffman, G.J. Governance, Policy, and Economics of Intergovernmental River Basin Management. *Water Resour. Manag.* **2015**, *29*, 5689–5712. [CrossRef]

17. Gerlak, A.K.; Schmeier, S. River Basin Organizations and the Governance of Transboundary Watercourses. In *The Oxford Handbook of Water Politics and Policy*; Conca, K., Weinthal, E., Eds.; Oxford University Press: Oxford, UK, 2016; Volume 1.

18. Colon, M.; Richard, S.; Roche, P.-A. The evolution of water governance in France from the 1960s: Disputes as major drivers for radical changes within a consensual framework. *Water Int.* **2018**, *43*, 109–132. [CrossRef]

19. Newson, M. *Land, Water and Development: Sustainable and Adaptive Management of Rivers*; Routledge: Abingdon, UK, 2008; ISBN 1-134-11190-8.

20. Pereira, H.C. *Policy and Practice in the Management of Tropical Watersheds*; Routledge: Abingdon, UK, 2019; ISBN 1-00-030613-5.

21. Blomquist, W.; Schlager, E. Political Pitfalls of Integrated Watershed Management. *Soc. Nat. Resour.* **2005**, *18*, 101–117. [CrossRef]

22. Warner, J.; Wester, P.; Bolding, A. Going with the flow: River basins as the natural units for water management? *Water Policy* **2008**, *10*, 121–138. [CrossRef]

23. Poteete, A.R.; Janssen, M.; Ostrom, E. *Working Together: Collective Action, the Commons, and Multiple Methods in PRACTICE*; Princeton University Press: Princeton, NJ, USA, 2010; ISBN 978-0-691-14603-4.

24. Vörösmarty, C.J.; Pahl-Wostl, C.; Bunn, S.E.; Lawford, R. Global water, the anthropocene and the transformation of a science. *Curr. Opin. Environ. Sustain.* **2013**, *5*, 539–550. [CrossRef]

25. Baldwin, E.; McCord, P.; Dell'Angelo, J.; Evans, T. Collective action in a polycentric water governance system. *Environ. Policy Gov.* **2018**, *28*, 212–222. [CrossRef]

26. Mashaly, A.F.; Fernald, A.G. Identifying Capabilities and Potentials of System Dynamics in Hydrology and Water Resources as a Promising Modeling Approach for Water Management. *Water* **2020**, *12*, 1432. [CrossRef]

27. Mirchi, A.; Madani, K.; Watkins, D., Jr.; Ahmad, S. Synthesis of System Dynamics Tools for Holistic Conceptualization of Water Resources Problems. *Water Resour. Manag.* **2012**, *26*, 2421–2442. [CrossRef]

28. Halbe, J.; Pahl-Wostl, C.; Adamowski, J. A methodological framework to support the initiation, design and institutionalization of participatory modeling processes in water resources management. *J. Hydrol.* **2018**, *556*, 701–716. [CrossRef]

29. Freas, K.; Bailey, B.; Munévar, A.; Butler, S. Incorporating climate change in water planning. *J. AWWA* **2008**, *100*, 92–99. [CrossRef]

30. Anderson, J.; Chung, F.; Anderson, M.; Brekke, L.; Easton, D.; Ejeta, M.; Peterson, R.; Snyder, R. Progress on incorporating climate change into management of California's water resources. *Clim. Chang.* **2008**, *87*, 91–108. [CrossRef]

31. Groves, C.R.; Game, E.T.; Anderson, M.G.; Cross, M.; Enquist, C.; Ferdaña, Z.; Girvetz, E.; Gondor, A.; Hall, K.R.; Higgins, J.; et al. Incorporating climate change into systematic conservation planning. *Biodivers. Conserv.* **2012**, *21*, 1651–1671. [CrossRef]

32. Bonelli, S.; Vicuña, S.; Meza, F.J.; Gironás, J.; Barton, J. Incorporating climate change adaptation strategies in urban water supply planning: The case of central Chile. *J. Water Clim. Chang.* **2014**, *5*, 357–376. [CrossRef]

33. Leavesley, G.H. Modeling the Effects of Climate Change on Water resources—A Review. In *Assessing the Impacts of Climate Change on Natural Resource Systems*; Frederick, K.D., Rosenberg, N.J., Eds.; Springer: Dordrecht, The Netherlands, 1994; pp. 159–177. ISBN 978-94-011-0207-0.

34. Abbaspour, K.C.; Faramarzi, M.; Ghasemi, S.S.; Yang, H. Assessing the impact of climate change on water resources in Iran. *Water Resour. Res.* **2009**, *45*. [CrossRef]

35. Yates, D.; Sieber, J.; Purkey, D.; Huber-Lee, A. WEAP21—A demand-, priority-, and preference-driven water planning model: Part 1: Model characteristics. *Water Int.* **2005**, *30*, 487–500. [CrossRef]

36. Forni, L.; Escobar, M.; Cello, P.; Marizza, M.; Nadal, G.; Girardin, L.; Losano, F.; Bucciarelli, L.; Young, C.; Purkey, D. Navigating the Water-Energy Governance Landscape and Climate Change Adaptation Strategies in the Northern Patagonia Region of Argentina. *Water* **2018**, *10*, 794. [CrossRef]

37. Alamanos, A.; Latinopoulos, D.; Xenarios, S.; Tziatzios, G.; Mylopoulos, N.; Loukas, A. Combining hydro-economic and water quality modeling for optimal management of a degraded watershed. *J. Hydroinform.* **2019**, *21*, 1118–1129. [CrossRef]

38. Hadded, R.; Nouiri, I.; Alshihabi, O.; Maßmann, J.; Huber, M.; Laghouane, A.; Yahiaoui, H.; Tarhouni, J. A Decision Support System to Manage the Groundwater of the Zeuss Koutine Aquifer Using the WEAP-MODFLOW Framework. *Water Resour. Manag.* **2013**, *27*, 1981–2000. [CrossRef]

39. Lempert, R.J.; Groves, D.G.; Popper, S.W.; Bankes, S.C. A General, Analytic Method for Generating Robust Strategies and Narrative Scenarios. *Manag. Sci.* **2006**, *52*, 514–528. [CrossRef]

40. Simpson, M.; James, R.; Hall, J.W.; Borgomeo, E.; Ives, M.C.; Almeida, S.; Kingsborough, A.; Economou, T.; Stephenson, D.; Wagener, T. Decision Analysis for Management of Natural Hazards. *Annu. Rev. Environ. Resour.* **2016**, *41*, 489–516. [CrossRef]

41. Pahl-Wostl, C.; Sendzimir, J.; Jeffrey, P.; Aerts, J.; Berkamp, G.; Cross, K. Managing Change toward Adaptive Water Management through Social Learning. *Ecol. Soc.* **2007**, *12*. [CrossRef]

42. Vorosmarty, C.J.; Hoekstra, A.Y.; Bunn, S.E.; Conway, D.; Gupta, J. Fresh water goes global. *Science* **2015**, *349*, 478–479. [CrossRef]

43. Liu, Y.; Gupta, H.; Springer, E.; Wagener, T. Linking science with environmental decision making: Experiences from an integrated modeling approach to supporting sustainable water resources management. *Environ. Model. Softw.* **2008**, *23*, 846–858. [CrossRef]

44. Loucks, D.P.; Kindler, J.; Fedra, K. Interactive Water Resources Modeling and Model Use: An Overview. *Water Resour. Res.* **1985**, *21*, 95–102. [CrossRef]

45. Halbe, J.; Pahl-Wostl, C.; Lange, M.A.; Velonis, C. Governance of transitions towards sustainable development—The water–energy–food nexus in Cyprus. *Water Int.* **2015**, *40*, 877–894. [CrossRef]

46. Halbe, J.; Adamowski, J.; Bennett, E.M.; Pahl-Wostl, C.; Farahbakhsh, K. Functional organization analysis for the design of sustainable engineering systems. *Ecol. Eng.* **2014**, *73*, 80–91. [CrossRef]

47. Purkey, D.R.; Escobar Arias, M.I.; Mehta, V.K.; Forni, L.; Depsky, N.J.; Yates, D.N.; Stevenson, W.N. A Philosophical Justification for a Novel Analysis-Supported, Stakeholder-Driven Participatory Process for Water Resources Planning and Decision Making. *Water* **2018**, *10*, 1009. [CrossRef]

48. Forni, L.G.; Galaitsi, S.E.; Mehta, V.K.; Escobar Arias, M.I.; Purkey, D.R.; Depsky, N.J.; Lima, N.A. Exploring scientific information for policy making under deep uncertainty. *Environ. Model. Softw.* **2016**, *86*, 232–247. [CrossRef]

49. MMAyA. *Modelación Estratégica de la Cuenca del Río Rocha*; Ministerio de Medio Ambiente y Agua con apoyo financiero de la CAF: La Paz, Bolivia, 2018.

50. Boucher, O.; Servonnat, J.; Albright, A.L.; Aumont, O.; Balkanski, Y.; Bastrikov, V.; Bekki, S.; Bonnet, R.; Bony, S.; Bopp, L.; et al. Presentation and Evaluation of the IPSL-CM6A-LR Climate Model. *J. Adv. Model. Earth Syst.* **2020**, *12*, e2019MS002010. [CrossRef]

51. Schwalm, C.R.; Glendon, S.; Duffy, P.B. RCP8.5 tracks cumulative CO_2 emissions. *Proc. Natl. Acad. Sci. USA* **2020**, *117*, 19656–19657. [CrossRef]

52. Hoogendam, P. Hydrosocial territories in the context of diverse and changing ruralities: The case of Cochabamba's drinking water provision over time. *Water Int.* **2019**, *44*, 129–147. [CrossRef]

53. Marston, A. The scale of informality: Community-run water systems in peri-urban Cochabamba, Bolivia. *Water Altern.* **2014**, *7*, 72–88.

54. Bakker, K. The ambiguity of community: Debating alternatives to private-sector provision of urban water supply. *Water Altern.* **2008**, *1*, 236–252.

55. Assies, W. David versus Goliath in Cochabamba: Water rights, neoliberalism, and the revival of social protest in Bolivia. *Lat. Am. Perspect.* **2003**, *30*, 14–36. [CrossRef]

56. Rocha López, R.; Boelens, R.; Vos, J.; Rap, E. Hydrosocial territories in dispute: Flows of water and power in an interbasin transfer project in Bolivia. *Water Altern.* **2019**, *12*, 267–284.

57. Myrland, J. *Two-Dimensional Hydraulic Modeling for Flood Assessment of the Rio Rocha, Cochabamba, Bolivia*; Uppsala University: Uppsala, Sweden, 2014.

58. Jacobson, S.; Sekizovic, I. Physico-Chemical Evaluation of the Water Quality in Rocha River: A Qualitative and Comparative Analysis Including Aspects of Social and Environmental Factors. Master's Thesis, Lund University, Lund, Sweden, 2019.

59. Zabalaga, J.; Amy, G.; von Münch, E. Evaluation of agricultural reuse practices and relevant guidelines for the Alba Rancho WWTP (primary and secondary facultative ponds) in Cochabamba, Bolivia. *Water Sci. Technol.* **2007**, *55*, 469–475. [CrossRef] [PubMed]

60. Cossio, C.; Perez-Mercado, L.F.; Norrman, J.; Dalahmeh, S.; Vinnerås, B.; Mercado, A.; McConville, J. Impact of treatment plant management on human health and ecological risks from wastewater irrigation in developing countries—Case studies from Cochabamba, Bolivia. *Int. J. Environ. Health Res.* **2019**, 1–19. [CrossRef] [PubMed]

61. Perez-Mercado, L.F.; Lalander, C.; Joel, A.; Ottoson, J.; Iriarte, M.; Oporto, C.; Vinnerås, B. Pathogens in crop production systems irrigated with low-quality water in Bolivia. *J. Water Health* **2018**, *16*, 980–990. [CrossRef] [PubMed]

62. Sharifi, M.A.; van den Toorn, W.; Rico, A.; Emmanuel, M. Application of GIS and multicriteria evaluation in locating sustainable boundary between the tunari National Park and Cochabamba City (Bolivia). *J. Multi Criteria Decis. Anal.* **2002**, *11*, 151–164. [CrossRef]

63. *Programa Manejo Integral de Cuencas El Manejo Integral de Cuencas en el Desarrollo Local, un Proceso de Construcción y Aprendizaje*; Fundación AGRECOL Andes: Cochabamba, Bolivia, 2004.

64. Cossío, V.; Wilk, J. A paradigm confronting reality: The river basin approach and local water management spaces in the Pucara Basin, Bolivia. *Water Altern.* **2017**, *10*, 181–194.

65. Bustamante, R.; Vega, D. *Normas Indígenas y Consuetudinarias Sobre la Gestión del Agua en Bolivia*; Wageningen University: Wageningen, The Nederlands, 2006; ISBN 90-8585-056-8.

66. Hendriks, J. Legislacion de Aguas y Gestión de Sistemas Hídricos en Países de la Región Andina. In *Derechos Colectivos y Políticas Hídricas en la Región Andina*; Urteaga, P., Boelens, R., Eds.; IEP Instituto de Estudios Peruanos: Lima, Peru, 2006; pp. 47–111.

67. Ministerio del Agua. *Plan Nacional de Cuencas PNC: Marco Conceptual y Estratégico*; Ministerio del Agua y Viceministerio de Cuencas y Recursos Hídricos: La Paz, Bolivia, 2007; p. 40.

68. Saldías, C.; Boelens, R.; Wegerich, K.; Speelman, S. Losing the watershed focus: A look at complex community-managed irrigation systems in Bolivia. *Water Int.* **2012**, *37*, 744–759. [CrossRef]

69. Ministerio de Medio Ambiente y Agua. *Marco Orientador para la Formulación de Planes Directores de Cuencas (PDC)*; Viceministerio de Recursos Hídricos y Riego: La Paz, Bolivia, 2014.

70. Ministerio de Medio Ambiente y Agua. *Programa Plurianual de Gestión Integrada de Recursos Hídricos y Manejo Integral de Cuencas 2017–2020*; Viceministerio de Recursos Hídricos y Riego: La Paz, Bolivia, 2017.

71. Servicio Departamental de Cuencas. *Elaboración del diagnóstico de situación de la gestión, uso y manejo del agua de la cuenca del río Rocha*; Servicio Departamental de Cuencas: Cochabamba, Bolivia, 2013.

72. Lempert, R.J.; Popper, S.W.; Bankes, S.C. *Shaping the Next One Hundred Years: New Methods for Quantitative, Long-Term Policy Analysis*; RAND: Santa Monica, CA, USA, 2003; ISBN 0-8330-3275-5.

73. Ministerio de Medio Ambiente y Agua. *Balance Hídrico Superficial de Bolivia*; Ministerio de Medio Ambiente y Agua: La Paz, Bolivia, 2018.

74. Lehner, B.; Verdin, K.; Jarvis, A. New Global Hydrography Derived From Spaceborne Elevation Data. *Eos Trans. Am. Geophys. Union* **2008**, *89*, 93–94. [CrossRef]

75. Programa de Desarrollo Agropecuario Sustentable. *Inventario Nacional de Presas de Bolivia 2010*; Programa de Desarrollo Agropecuario Sustentable: Cochabamba, Bolivia, 2010; ISBN 978-99954-774-3-1.

76. Ministerio de Medio Ambiente y Agua. *Plan Maestro Metropolitano de Agua y Saneamiento de Cochabamba Bolivia*; Ministerio de Medio Ambiente y Agua: Cochabamba, Bolivia, 2014.

77. Chapra, S.C. *Surface Water Quality Modeling*; McGraw-Hill: New York, NY, USA, 1997.

78. Allen, R.G.; Pereira, L.S.; Raes, D.; Smith, M. Crop evapotranspiration-Guidelines for computing crop water requirements-FAO Irrigation and drainage paper 56. *FAO Rome* **1998**, *300*, D05109.

79. *Diseño Conceptual del Componente Riego del Proyecto Misicuni*; Ministerio de Medio Ambiente y Agua; C3B Consultora Boliviana Beccar Bottega Ltda: Cochabamba, Bolivia, 2016.

80. *Balance Hídrico de la Producción Agrícola en el Valle Central de Cochabamba*; Programa de Enseñanza e Investigación en Riego Andino y de los Valles; Centro de Levantamientos Aeroespaciales y Aplicaciones de SIG para el Desarrollo Sostenible de los Recursos Naturales: Cochabamba, Bolivia, 1999.

81. Ministerio de Medio Ambiente y Agua. *Inventario Nacional de Sistemas de Riego 2012*; Viceministerio de Recursos Hídricos y Riego: La Paz, Bolivia, 2012.

82. Ministerio de Medio Ambiente y Agua. *Guía Metodológica para la Elaboración de Balances Hídricos Superficiales*; Viceministerio de Recursos Hídricos y Riego: La Paz, Bolivia, 2016.

83. Moriasi, D.; Gitau, M.; Pai, N.; Daggupati, P. Hydrologic and Water Quality Models: Performance Measures and Evaluation Criteria. *Trans. ASABE* **2015**, *58*, 1763–1785. [CrossRef]

84. Yates, D.; Gangopadhyay, S.; Rajagopalan, B.; Strzepek, K. A technique for generating regional climate scenarios using a nearest-neighbor algorithm. *Water Resour. Res.* **2003**, *39*, 1199. [CrossRef]

85. *Decreto Departamental N° 4544*; Gobierno Autónomo Departamental de Cochabamba: Cochabamba, Bolivia, 2020; p. 3.

86. Rocha López, R.; Hoogendam, P.; Vos, J.; Boelens, R. Transforming hydrosocial territories and changing languages of water rights legitimation: Irrigation development in Bolivia's Pucara watershed. *Geoforum* **2019**, *102*, 202–213. [CrossRef]

87. Lund, J.R.; Palmer, R.N. Water resource system modeling for conflict resolution. *Water Resour. Update* **1997**, *3*, 70–82.

88. Nandalal, K.D.W.; Simonovic, S.P. Resolving conflicts in water sharing: A systemic approach. *Water Resour. Res.* **2003**, *39*. [CrossRef]

89. Hoogendam, P.; Boelens, R. Dams and Damages. Conflicting Epistemological Frameworks and Interests Concerning "Compensation" for the Misicuni Project's Socio-Environmental Impacts in Cochabamba, Bolivia. *Water* **2019**, *11*, 408. [CrossRef]

Article

Risk Assessment of China's Water-Saving Contract Projects

Qian Li [1,2], **Ziheng Shangguan** [3,*], **Mark Yaolin Wang** [4,5], **Dengcai Yan** [3], **Ruizhi Zhai** [6] **and Chuanhao Wen** [1,7,*]

1   Research Center for Economy of Upper Reaches of the Yangtze River, Chongqing Technology and Business University, Chongqing 400067, China; liqian@ncwu.edu.cn
2   School of Water Conservancy, North China University of Water Resources and Electric Power, Zhengzhou 450011, China
3   School of Public Administration, Hohai University, Nanjing 211100, China; dengcaiofyan@163.com
4   School of Geography, The University of Melbourne, Melbourne, VIC 3010, Australia; myw@unimelb.edu.au
5   Asia Institute, The University of Melbourne, Melbourne, VIC 3010, Australia
6   School of Mechanics and Materials, Hohai University, Nanjing 211100, China; zrzhhu@hhu.edu.cn
7   School of Economics, Yunnan University, Kunming 650091, China
*   Correspondence: sgzh@hhu.edu.cn (Z.S.); chhwen1972@ynu.edu.cn (C.W.)

Received: 31 August 2020; Accepted: 24 September 2020; Published: 25 September 2020

Abstract: In order to alleviate the problem of water shortage, the Ministry of Water Resources of China proposed a Water-Saving Contract (WSC) project management model in 2014, which is similar to the Energy Performance Contract (EPC). In this context, this research aims to explore the applicability of China's WSC projects by risk assessment, and to help promote WSC projects in China. Different from traditional risk assessment, this paper takes into account the uncertainty of the EPC project's risks, and adopts the multielement connection degree set pair analysis to evaluate both the level and trend of the risks. The results show: (1) the overall risk of China's WSC projects is low, so WSC projects are very suitable for promotion in China. However, the overall risk shows a trend of decelerated ascent, which shows that there are some potential high-risk factors in China's WSC projects; (2) among the many risks of the WSC projects, audit risk, financing risk, and payment risk are at a high-risk level; market competition risk is at a medium-risk level; the remaining risks are at a low-risk level; (3) among the medium and high risks, audit risk, financing risk, and market competition risk have a trend of accelerated ascent, while payment risk has a trend of decelerated decline; in low risks, inflation risk has a trend of decelerated ascent, while the remaining risks have a trend of accelerated decline.

Keywords: risk assessment; water-saving; set pair analysis; China

---

1. Introduction

Since the reform and opening up, China's population has continued to grow. The process of industrialization and urbanization has accelerated, which has led to a gradual increase in water consumption. With global warming and the pollution of water resources caused by industrial development, the problem of insufficient regional water supply in China has become increasingly prominent. Insufficient water resource carrying capacity has become the main constraint in China's sustainable development [1]. Normally, the way to solve water shortage can be divided into the increasing water supply and saving water. For a long time, China has been relying on the construction of a transbasin water diversion project to solve the problem of insufficient water supply in the north, which is a typical supply-oriented solution while throttling has been relatively ignored [2]. Therefore, in order to achieve the purpose of saving water, the Ministry of Water Resources of China (MWRC) began to vigorously implement Water-Saving Contract (WSC) projects in 2016 [3]. At present, China's WSC project is still in the initial stage, and it faces many uncertain risks. In order to promote the development of the WSC projects in China, it is of great significance to assess the risks of the projects and formulate relevant policies to reduce them.

1.1. Profit Model of the WSC Project and Its Stakeholders

The WSC contains a specific water-saving target which is beneficial to the water user. By providing advanced and applicable water-saving technologies, the water-saving service operators carry out a technological transformation, establish long-term management mechanisms, and eventually pay the full cost by water-saving benefits, while water users can also share the benefits. This is a market-based water-saving management model [4]. The profit model is shown in Figure 1. The project involves three subjects, namely: the government, water-saving service operators, and water users. The government mainly assumes the role of policy guidance. Water-saving service operators provide services to water users through technological innovation to alleviate the pressure on water consumption. Water users are the ultimate beneficiaries. Here, we mainly discuss the risks encountered by water-saving service operators.

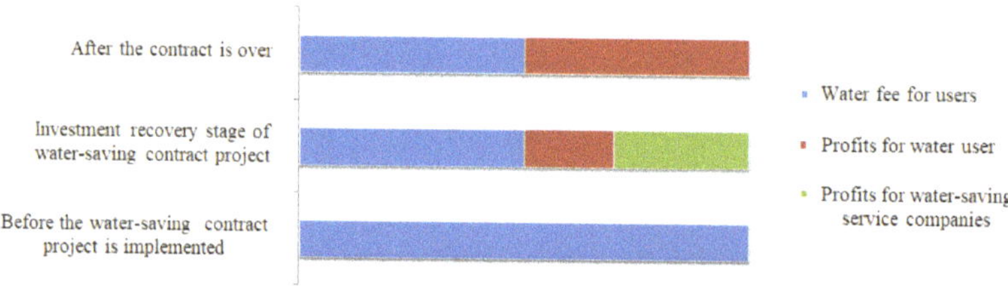

Figure 1. Profit model of water-saving contract projects.

1.2. Enlightenment of the Energy Performance Contract's (EPC) Risk Assessment to WSC

At present, there are few studies on the risk assessment of the WSC project, and there is no uniform analytical framework and research method for systematically assessing the risks. However, in essence, the WSC proposed by the MWRC is similar to the Energy Performance Contract (EPC); therefore, when analyzing the risks of China's WSC projects, the research results of EPC projects can be used for reference.

In the research of the EPC project, Mills et al. identified the inherent risks of EPC projects and divided them into five categories, namely, economic risks, environmental risks, technical risks, operational risks, and measurement and verification (M&V) risks [5]. On the basis of Mills et al.'s research, Lee et al. added financial risk, project design risk, and installation risk to the research of risk assessment in the EPC project, and identified the key risks of EPC projects through questionnaires. They believe that the key risks to energy service companies are possible payment defaults of hosts after installation, the uncertainty of baseline measurement, and the increase in installation costs in EPC projects [6]. Hu and Zhou further refined the risks of the EPC project in China. They considered that EPC project risks include political and legal risks, market risks, technical risks, management risks, financial risks, project quality risks, and customer risks [7]. Duan et al. constructed a life cycle analysis framework of EPC projects containing four stages, namely the contract signing stage, investment stage, implementation stage, and benefit-sharing stage [8]. Based on the framework proposed by Duan, Wu et al. used an improved analytic hierarchy process (AHP) to determine the weight of various risk indicators of the EPC project in China, and established a risk evaluation model using a fuzzy comprehensive evaluation method [9]; Huang et al. combined AHP and gray evaluation theory to construct a gray multilayer evaluation model of the EPC project's risks, and analyzed the high-risk factors of China's EPC project [10]. Garbuzova-Schlifter and Madlener systematically studied the common risk factors and causes of risk associated with EPC projects executed in three Russian sectors: (1) industrial; (2) housing and communal services; (3) public. They also conducted a quantitative assessment of risks based on an AHP approach, and proposed a widely applicable risk management framework for Russian EPC projects [11]. Valipour et al. divided EPC project risks in Iran into six categories: design risk, market risk, political risk, environmental risk, construction risk, and political risk, and assessed the level and occurrence probability of the EPC project's risks using Analytic Network Process (ANP) method. Their results indicate that political risk and design risk are the most significant types of risk in Iran's EPC projects [12].

1.3. The Risk Characteristics of the WSC Project and the Determination of Its Research Methods

The study of EPC projects provides effective analysis frameworks and research methods for the risk assessment of the WSC projects However, unlike EPC projects, WSC projects have the characteristics of the long project cycle, relatively low return on investment, and weak liquidity. These characteristics may cause the risk level of the WSC projects to change significantly during the whole life cycle of the project [4]. In addition, as WSC projects are still in their infancy in China, the level of the risk is prone to change under the influence of national development strategy, overall economic level, management technique, and market resource allocation [13]. Therefore, simply analyzing the level of risk can no longer meet the needs of risk assessment in the WSC project, and the trend of risk should also be analyzed.

In the previous studies on EPC projects, the trend of risk was hardly considered, and its uncertainty makes it difficult to assess. Set pair analysis (SPA) has strong adaptability in dealing with the interaction between certainty and uncertainty in the system [14]. Cui et al. built an evaluation model to quantitatively evaluate and diagnose the carrying capacity of regional water resources under uncertain conditions by applying set pair analysis [15]. Gao et al. put forward a model based on set pair analysis about information risk evaluation. This model can not only divide the extent of the information risk, but it describes the trend of the information risk. It could describe the information risk from static and dynamic [16]. Zheng et al. used the set pair analysis method to analyze the safety of the tailing pond. Through set pair analysis, the development trend of the safety status of the tailings pond can be judged [17].

Based on the studies above, this article intends to use a literature analytic method to determine the risks existing in the WSC project's life cycle which was proposed by Duan [8], and uses multielement connection number set pair analysis to evaluate the level and trend of the risks. Finally, suggestions are made based on the characteristics of various risks to control and reduce them.

2. Materials and Methods

2.1. Risk Identification in Water-Saving Contract Projects

This article divides the life cycle of the WSC project into four stages according to Duan et al. [8]. The four stages are the contract signing stage, investment stage, implementation stage, and benefit-sharing stage. Then, the risks in each stage of the WSC project were sorted out through a literature review, as shown in Table 1.

(1)   Contract signing stage

The contract signing stage includes the process of water audit, feasibility study, and contract signing. In the process of water consumption audit, if the water-saving service operator cannot accurately obtain the actual water consumption of the water user, the payback period will be lengthened [18]. The focus of the feasibility study is to evaluate and demonstrate the water-saving technique. If the technique obtained by water-saving service operators fails to reach the target, it will cause economic losses and waste of resources [19]. Li et al. consider that WSC projects have not yet formed a mature market-based management mechanism in China, which can easily cause malicious competition in the industry [20]. As a result, this article determines the risk evaluation indicators of the contract signing stage, such as: information risk, technical risk, and market competition risk.

(2) Investment stage

Sustainable funding is an important guarantee for the implementation of the project. In the investment stage, water-saving service operators need to finance to ensure the progress of the project, and its financing channel mainly comes from commercial banks [21]. At the same time, due to the long payback period of investment in water-saving contract projects, the bank's interest rate may increase during the project, which will also affect the company's financing costs and reduce the final profit of water-saving service operators [7]. Therefore, this article determines the risk indicators of the investment stage, such as: financing risk and interest rate risk.

(3) Implementation stage

The implementation stage includes engineering construction, equipment procurement, installation, and commissioning [26]. In this stage, construction safety directly affects the process and construction cycle of the WSC projects [22]. In addition, some external factors also increase the risk of the implementation stage, such as: policy changes [20], force majeure [23], and inflation [24]. Policy changes may directly affect the enthusiasm of water-saving service operators, so as to affect the final quality of the project; the risk of force majeure will directly lead to the termination of the project; the impact of inflation is the same as the increase in bank interest rates, which will reduce the final profit of the project. In summary, this article determines the risk indicators of the implementation stage as: construction risk, policy risk, force majeure risk, and inflation risk.

(4) Benefit-sharing stage

After the implementation stage is over, it enters the benefit-sharing stage. At this stage, water-saving service operators are responsible for project operation and equipment maintenance. At the same time, it recovers investment costs and obtains reasonable profits by sharing water-saving benefits. Li et al. pointed out that the depreciation rate of the equipment would greatly affect the operating costs, so they suggested strict maintenance of the equipment [20]. The payment default of the water user is also one of the important risks in the benefit-sharing stage. If water users have weak credit awareness or cannot reach the predetermined water consumption due to their own economic problems, the investment of the WSC projects will not be recovered [6]. The fluctuation of water prices will affect the payback period of the WSC projects [25]. As mentioned above, this article determines the risk indicators in the benefit-sharing stage as: facility depreciation risk, payment risk, and water price change risk.

Table 1. Risk index system of WSC project.

Stage	Indicator	Risk Consequences
Contract signing stage (CSS)	Audit risk (AR) [18]	Lengthen the payback period
	Technical risk (TR) [19]	The actual water-saving amount cannot meet the requirements of the contract
Investment stage (INS)	Market competition risk (MCR) [20]	Unfair competitive practice
	Financing risk (FR) [21]	Affect the progress of the project
	Interest rate risk (IRR) [7]	Reduce the final profit of the project
Implementation stage (IMS)	Construction risk (CR) [22]	Affects the process and construction cycle
	Policy risk (PR) [20]	Affect the enthusiasm of water-saving service operators
	Force majeure risk (FMR) [23]	Lead to the termination of the project
	Inflation risk (IR) [24]	Reduce the final profit of the project
Benefit-sharing stage (BSS)	Facility depreciation risk (FDR) [20]	Increase operating costs
	Payment risk (PMR) [6]	Investment in water-saving projects will not be recovered
	Water price change risk (PCR) [25]	Affect the payback period

2.2. Multielement Connection Degree Set Pair Analysis

2.2.1. Basic Theory of Set Pair Analysis

The set pair analysis (SPA), proposed by Zhao in 1989, is a modified uncertainty theory considering both certainties and uncertainties as an integrated certain–uncertain system and depicting the certainty and uncertainty systematically from three aspects as identity, discrepancy, and contrary [27]. In set pair analysis, the connection degree is usually expressed as follows:

$$\mu = a + bi + cj \tag{1}$$

where a is the identity degree, b the discrepancy degree, c the contradictory degree, $a + b + c = 1$ and $\forall a, b, c \in [0, 1]$; μ is the 3-element connection degree; i is the uncertainty coefficient of discrepancy, which has different values in $[-1, 1]$; j is the uncertainty coefficient of contradiction, which has value of -1.

2.2.2. Multielement Connection Degree and Partial Connection Degree

In Equation (1), bi is the measurement between identity degree a and contradictory degree c with uncertainty. This item could be often expanded in actual applications. The expanded equation is as follows [28]:

$$\mu = a + b_1 i_1 + b_2 i_2 + \cdots + b_{n-2} i_{n-2} - c \tag{2}$$

where μ in Equation (2) is the n-element connection degree; $a + b_1 + b_2 + \cdots + b_{n-2} + c = 1$; $\forall a, b_1, b_2 \cdots b_{n-2}, c \in [0, 1]$; $\forall i_1, i_2 \cdots i_{n-2} \in [-1, 1]$.

Similar to the concept of derivatives, the partial connection degree could be used to describe the development tendency of the connection degree. The first-order and second-order partial connection degree of the multielement connection degree could be described as follows [29]:

First-order partial connection degree:

$$\partial\mu = \partial a + i_1 \partial b_1 + i_2 \partial b_2 + \cdots + i_{n-2} \partial b_{n-2} \tag{3}$$

where $\partial a = \frac{a}{a+b_1}$, $\partial b_1 = \frac{b_1}{b_1+b_2}$, $\partial b_2 = \frac{b_2}{b_2+b_3}$, $\ldots$, $\partial b_{n-2} = \frac{b_{n-2}}{b_{n-2}+c}$. This equation describes the development trend from c to a. Its essence is the $n - 1$-element connection degree, which could be used to describe the development trend of Equation (2).

Second-order partial connection degree:

$$\partial^2\mu = \partial(\partial\mu) = \partial^2 a + i_1 \partial^2 b_1 + i_2 \partial^2 b_2 + \cdots + i_{n-3} \partial^2 b_{n-3} \tag{4}$$

Similar to Equation (3), Equation (4) is the $n - 2$-element connection degree which could use to describe the development trend of Equation (3), where $\partial^2 a = \frac{\partial a}{\partial a + \partial b_1}$, $\partial^2 b_1 = \frac{\partial b_1}{\partial b_1 + \partial b_2}$, $\partial^2 b_2 = \frac{\partial b_2}{\partial b_2 + \partial b_3}$, $\ldots$, $\partial^2 b_{n-3} = \frac{\partial b_{n-3}}{\partial b_{n-3} + \partial b_{n-2}}$.

In practical applications, system risks are often divided into five levels, namely: low, relatively low, medium, relatively high, and high risks. Therefore, the five-element connection degree is often used to analyze the level and trend of risk. Its expression is:

$$\mu = a + b_1 i_1 + b_2 i_2 + b_3 i_3 - c \tag{5}$$

If a is taken as the reference set and define it as low risk, then b_1 represents relatively low risk, b_2 medium risk, b_3 relatively high risk, and c high risk.

2.2.3. Set Pair Potential

When $c \neq 0$, the ratio a/c is the set pair potential [30], which is expressed as:

$$\text{Shi}(x) = a/c, \ (c \neq 0) \tag{6}$$

When $a/c > 1$, Shi (x) is at the same potential, which means that the risk is on the low side. When $a/c = 1$, Shi (x) is at the equal potential, which means the risk is at a medium size. When $a/c < 1$, Shi (x) is at opposite potential, which means that the risk is on the high side.

When using the five-element connection degree for risk analysis, the ratio a/c can show the situation of risk, but it fails to classify the level of risk that needs to be determined by the size of b, c, and d. According to the size of b, c, and d, the risks in the situation of the same potential and opposite potential can be divided into 65 levels (as shown in Appendix A). When Shi (x) is at the same potential, the higher the level (Level 1 is the highest level), the lower the risk; when Shi (H) is at the opposite potential, the higher the level, the higher the risk.

The concept of set pair potential can also be applied to first-order and second-order partial connection degree, which is expressed as:

$$\text{Shi}^1(x) = \frac{\partial a}{\partial b_{n-2}} \tag{7}$$

$$\text{Shi}^2(x) = \frac{\partial^2 a}{\partial^2 b_{n-3}} \tag{8}$$

where Shi^1 (x) is the set pair potential of first-order partial connection degree, and Shi^2 (x) is the set pair potential of second-order partial connection degree. Shi^1 (x) and Shi^2 (x) can be used to describe the trend of the risk, as shown in Appendix B.

2.3. Risk Assessment Process of Water-Saving Contract Project Based on Five-Element Connection Degree

(1)   Calculate the index weight by entropy method

In order to eliminate the subjectivity of experts in evaluating each risk, this article uses the entropy method to calculate the weight of risk indicators. The calculation steps are as follows:

-   Build the judgment matrix B, namely:

$$B = \begin{bmatrix} x_{11} & x_{12} & \cdots & x_{1n} \\ x_{21} & x_{22} & \cdots & x_{2n} \\ \vdots & \vdots & \ddots & \vdots \\ x_{m1} & x_{m2} & \cdots & x_{mn} \end{bmatrix} \tag{9}$$

In Equation (9), n is the number of risk assessment indicators, and m is the number of experts.

-   Calculate the entropy of the indicator j:

$$H_j = -\frac{1}{\ln(m)} \sum_{i=1}^{m} P_{ij} \ln(P_{ij}), \ (i = 1, 2, \cdots, m; j = 1, 2, \cdots, n) \tag{10}$$

$$P_{ij} = \frac{x_{ij}}{\sum_{j=1}^{n} x_{ij}} \tag{11}$$

- Calculate the weight of the indicator j:

$$\omega_j = \frac{1 - H_j}{n - \sum_{j=1}^{n} H_j} \tag{12}$$

where $0 \leq \omega_j \leq 1$ and $\sum_{j=1}^{n} \omega_j = 1$.

(2) Risk assessment based on the five-element connection degree

After determining the index weight ω, the calculation equation of the five-element connection degree can be obtained using Equation (13).

$$\mu = \omega \times R \times E^T = (\omega_1, \omega_2, \cdots, \omega_n)\begin{bmatrix} R_{11} & R_{12} & R_{13} & R_{14} & R_{15} \\ R_{21} & R_{22} & R_{23} & R_{24} & R_{25} \\ \vdots & \vdots & \vdots & \vdots & \vdots \\ R_{n1} & R_{n2} & R_{n3} & R_{n4} & R_{n5} \end{bmatrix}\begin{bmatrix} 1 \\ i \\ j \\ k \\ l \end{bmatrix}$$
$$= \sum_{i=1}^{n} \omega_r R_{r1} + \sum_{i=1}^{n} \omega_r R_{r2} i + \sum_{i=1}^{n} \omega_r R_{r3} j + \sum_{i=1}^{n} \omega_r R_{r4} k + \sum_{i=1}^{n} \omega_r R_{r5} l \tag{13}$$

where R is the occurrence probability matrix of the risk; E^T is the coefficient matrix of the five-element connection degree; $R_{ij} = N_{ij}/N$ ($i = 1, 2, \cdots, n$; $j = 1, 2, \cdots, 5$; $j = 1$ means low risk, $j = 2$ means relatively low risk, $j = 3$ means medium risk, $j = 4$ means relatively high risk, $j = 5$ means high risk). Among them, N_{ij} is the number of experts who determine the risk indicator i as risk level j, and N is the total number of experts. Finally, we can figure out $a = \sum_{i=1}^{n} \omega_r R_{r1}$, $b = \sum_{i=1}^{n} \omega_r R_{r2}$, $c = \sum_{i=1}^{n} \omega_r R_{r3}$, $d = \sum_{i=1}^{n} \omega_r R_{r4}$ and $e = \sum_{i=1}^{n} \omega_r R_{r5}$.

In the actual analysis, it is usually difficult to encounter equal potential. Therefore, this article ranks the risk level into 5 levels according to the five-element degree of the similar potential and the inverse potential, namely: high-risk level (Levels 1–26 of inverse potential), relatively high-risk level (Levels 27–52 of inverse potential), medium-risk level (Levels 53–65 of inverse potential, Levels 53–65 of the same potential), relatively low-risk level (Levels 27–52 of the same potential), and low-risk level (Levels 1–26 of the same potential). At the same time, use the set pair potential of the first-order partial connection degree and the second-order partial connection degree to analyze the trend of risk, which can be divided into 6 types: accelerated decline, decelerated decline, decelerated ascent, accelerated ascent, uniform decline, and uniform ascent.

2.4. Data Collection

This article adopts the entropy method to determine the weight of the risk indicator. In order to obtain x_{ij} in Equation (9), the evaluation of the expert i on the risk indicator j, this article adopts the interval classification method and requires 5 experts to score it. The scoring table is shown in Table 2. Five experts are from Hebei University of Engineering, North China University of Water Resources and Electric Power, Hohai University, Beijing Guotai Water-Saving Development Co., Ltd., and the Yellow River Conservancy Commission of the Ministry of Water Resources. Among them, Hebei University of Engineering and Beijing Guotai Water-Saving Development Co., Ltd. are the participants in China's first water-saving contract project; North China University of Water Resources and Electric Power and Hohai University are specialized universities for China's water conservancy and hydropower research; the Yellow River Conservancy Commission of the Ministry of Water Resources belongs to MWRC.

Table 2. The interval classification of the impact degree of the risk indicator.

Influence Level	Scoring Interval	Influence Degree
I	[0,0.2]	Low
II	[0.2,0.4]	Relatively low
III	[0.4,0.6]	Medium
IV	[0.6,0.8]	Relatively high
V	[0.8,1]	High

In order to obtain the risk occurrence probability of R_{ij} in Equation (13), this article uses the Likert five-point scale method to design the questionnaire. In the Likert five-point scale, 1 means low risk, 2 means relatively low risk, 3 means medium risk, 4 means relatively high risk, 5 means high risk. In this article, 300 questionnaires were distributed to staff members of water-saving service operators (including the technical manager, financial manager, procurement manager, and project manager) from 13 water-saving contract pilot projects, and 276 valid questionnaires were finally obtained. The descriptive statistics results are shown in Table 3

Table 3. Descriptive statistical results of the questionnaire.

Stage	Indicator	Mean	Std. Deviation	Skewness	Kurtosis
CSS	AR	3.036	1.416	−0.010	−1.318
	TR	2.587	1.423	0.284	−1.305
	MCR	3.337	1.078	−0.055	−0.858
INS	FR	3.007	1.404	0.154	−1.289
	IRR	2.638	1.506	0.222	−1.467
	CR	2.815	1.467	0.170	−1.385
IMS	PR	2.822	1.455	0.141	−1.331
	FMR	2.884	1.386	0.061	−1.214
	IR	2.449	1.370	0.363	−1.256
BSS	FDR	2.909	1.451	0.080	−1.290
	PMR	3.167	1.524	−0.110	−1.486
	PCR	2.786	1.394	0.063	−1.278

Note: Source: calculated by SPSS 19.

In this paper, Cronbach's Alpha is used for the reliability test. The results show that Cronbach's Alpha of the questionnaire is 0.912 (>0.9), indicating that the questionnaire has good reliability. The results are shown in Table 4

Table 4. Reliability Analysis of questionnaire on the risk level.

		N	%
Cases	Valid	276	100.0
	Excluded	0	0.0
	Total	276	100.0
	Cronbach's Alpha		N of items
	0.912		12

Note: Source: calculated by SPSS 19.

3. Results

The weight of each risk and their five- element connection degree are shown in Table 5.

Table 5. Calculation table of the five-element connection degree.

Stage	Weight	Indicator	Weight	Five-Element Connection Degree	Situation	Level
CSS	0.3876	AR	0.415	$0.1848 + 0.2174i + 0.1848j + 0.2029k + 0.2101l$	−	9
		TR	0.241	$0.3406 + 0.1630i + 0.1812j + 0.1993k + 0.1123l$	+	25
		MCR	0.344	$0.0290 + 0.2174i + 0.3043j + 0.2862k + 0.1630l$	−	61
		Total	1	$0.1686 + 0.2042i + 0.2250j + 0.2306k + 0.1703l$	−	47
INS	0.2155	FR	0.616	$0.1558 + 0.2790i + 0.1993j + 0.1341k + 0.2319l$	−	25
		IRR	0.384	$0.3732 + 0.1159i + 0.1522j + 0.2174k + 0.1377l$	+	25
		Total	1	$0.2393 + 0.2164i + 0.1812j + 0.1661k + 0.1957l$	+	3
IMS	0.2754	CR	0.248	$0.2609 + 0.2174i + 0.1486j + 0.1920k + 0.1812l$	+	7
		PR	0.237	$0.2681 + 0.1739i + 0.2065j + 0.1703k + 0.1812l$	+	21
		FMR	0.205	$0.2283 + 0.1739i + 0.2464j + 0.1884k + 0.1630l$	+	19
		IR	0.310	$0.3732 + 0.1667i + 0.1739j + 0.2101k + 0.0761l$	+	25
		Total	1	$0.2907 + 0.1824i + 0.1902j + 0.1917k + 0.1449l$	+	25
BSS	0.1215	FDR	0.306	$0.2464 + 0.1522i + 0.2536j + 0.1413k + 0.2065l$	+	21
		PMR	0.404	$0.1957 + 0.2029i + 0.1377j + 0.1667k + 0.2971l$	−	9
		PCR	0.290	$0.2754 + 0.1413i + 0.2355j + 0.2174k + 0.1304l$	+	19
		Total	1	$0.2343 + 0.1695i + 0.2015j + 0.1736k + 0.2210l$	+	21
Total				$0.2265 + 0.1982i + 0.2027j + 0.1931k + 0.1839l$	+	19

Note: "+" means same potential, "−" means opposite potential.

As shown in Table 5, $u_{CCS} = 0.1686 + 0.2042i + 0.2250j + 0.2306k + 0.1703l$, $shi(CCS) = 0.9900 < 1$. It is at Level 47 of the opposite potential, which means the risk in the contract signing stage is relatively high. In this stage, the audit risk is at Level 9 of the opposite potential, which is at a high-risk level; technical risk is at Level 25 of the same potential, which is at a low-risk level; the market competition risk is at Level 61 of the opposite potential, which is at a medium-risk level.

$u_{IS} = 0.2393 + 0.2164i + 0.1812j + 0.1661k + 0.1957l$, $shi(INS) = 1.2227 > 1$. It is at Level 3 of the same potential, which means the risk of this stage is low. In this stage, financing risk is at Level 25 of the opposite potential, which is at a high-risk level; technical risk is at Level 25 of the same potential, which is at a low-risk level.

$u_{OS} = 0.2907 + 0.1824i + 0.1902j + 0.1917k + 0.1449l$, $shi(IMS) = 2.0062 > 1$. It is at Level 25 of the same potential, which means the risk of this stage is low. In this stage, construction risk, policy risk, force majeure risk, and inflation risk are at Levels 7, 21, 19, and 25 of the same potential, respectively, which means they are all at low-risk levels.

$u_{BSS} = 0.2343 + 0.1695i + 0.2015j + 0.1736k + 0.2210l$, $shi(BSS) = 1.0601 > 1$. It is at Level 21 of the same potential, which means the risk of this stage is low. In this stage, facility depreciation risk and water price change risk are at Levels 21 and 19 of the same potential, respectively, which means they are at a low-risk level; payment risk is at Level 9 of the opposite potential, which is at a high-risk level.

$u_{WSC} = 0.2265 + 0.1982i + 0.2027j + 0.1931k + 0.1839l$, $shi(WSC) = 1.2316 > 1$. It is at Level 19 of the same potential, which means the overall risk of the WSC project is low. Therefore, WSC projects are suitable for development in China.

In each stage of the WSC project, the contract signing stage is at a relatively high level of risk; the investment stage, implementation stage, and benefit-sharing stage are at a low-risk level. Therefore, the contract signing stage is the focus of risk control in the WSC project. From the perspective of each risk, audit risk, financing risk, and payment risk are at a high-risk level, they are the primary concern in risk control; market competition risk is at a medium-risk level, it is the secondary concern in risk control; the remaining risks are at a low-risk level, but it does not mean that these risks can be ignored, because the risk level may be easily affected by factors such as national development strategy, overall economic level, management technology, and market resource allocation as the WSC project is still in its infancy in China. Therefore, it is necessary to analyze $Shi^1(x)$ and $Shi^2(x)$ of each risk, as shown in Table 6.

Table 6. Calculation of partial connection degree and trend analysis.

Stage	Indicator	First-Order Partial Connection Degree	Situation	Second-Order Partial Connection Degree	Situation
CSS	AR	$0.4595 + 0.5405i + 0.4767j + 0.4913k$	−	$0.4595 + 0.5314i + 0.4924j$	−
	TR	$0.6763 + 0.4736i + 0.4762j + 0.6396k$	+	$0.5882 + 0.4986i + 0.4268j$	+
	MCR	$0.1177 + 0.4167i + 0.5153j + 0.6371k$	−	$0.2202 + 0.4471i + 0.4472j$	-
	Total	$0.4522 + 0.4759i + 0.4938j + 0.5752k$	−	$0.4872 + 0.4907i + 0.4619j$	+
INS	FR	$0.3583 + 0.5833i + 0.5978j + 0.3664k$	−	$0.3805 + 0.4939i + 0.6200j$	−
	IRR	$0.7630 + 0.4323i + 0.4118j + 0.6122k$	+	$0.6383 + 0.5121i + 0.4021j$	+
	Total	$0.5230 + 0.5513i + 0.5333j + 0.4444k$	+	$0.4868 + 0.5083i + 0.5454j$	−
IMS	CR	$0.5455 + 0.5940i + 0.4363j + 0.5145k$	+	$0.4787 + 0.5765i + 0.4589j$	+
	PR	$0.6066 + 0.4572i + 0.5480j + 0.4845k$	+	$0.5702 + 0.4548i + 0.5308j$	+
	FMR	$0.5676 + 0.4138i + 0.5667j + 0.5361k$	+	$0.5784 + 0.4220i + 0.5139j$	+
	IR	$0.6912 + 0.4894i + 0.4529j + 0.7341k$	−	$0.5855 + 0.5194i + 0.3815j$	+
	Total	$0.6144 + 0.4896i + 0.4980j + 0.5696k$	+	$0.5565 + 0.4958i + 0.4665j$	+
BSS	FDR	$0.6182 + 0.3751i + 0.6422j + 0.4063k$	+	$0.6224 + 0.3687i + 0.6125j$	+
	PMR	$0.4910 + 0.5957i + 0.4524j + 0.3594k$	+	$0.4518 + 0.5684i + 0.5572j$	−
	PCR	$0.6609 + 0.3750i + 0.5200j + 0.6251k$	+	$0.6380 + 0.4190i + 0.4541j$	+
	Total	$0.5802 + 0.4569i + 0.5372j + 0.4399k$	+	$0.5595 + 0.4596i + 0.4541j$	+
Total		$0.5281 + 0.4920i + 0.5062j + 0.5321k$	−	$0.5159 + 0.4926i + 0.4889j$	+

Note: "+" means same potential, "−" means opposite potential.

As shown in Table 6, Shi^1 (CSS) is at the opposite potential while Shi^2 (CSS) is at the same potential, which means the risk of contract signing stage has a trend of decelerated ascent according to Appendix B. In this stage, audit risk and market competition risk have a trend of accelerated ascent, and technical risk has a trend of accelerated decline.

Shi^1 (INS) is at the same potential while Shi^2 (INS) is at the opposite potential, which means the risk of investment stage has a trend of decelerated decline. In this stage, financing risk has a trend of accelerated ascent, and interest rate risk has a trend of accelerated decline.

Shi^1 (IMS) and Shi^2 (IMS) are both at the same potential, which means the risk of implementation stage has a trend of accelerated decline. In this stage, construction risk, policy risk, and force majeure risk have a trend of accelerated decline, and inflation risk has a trend of decelerated ascent.

Shi^1 (BSS) and Shi^2 (BSS) are both at the same potential, which means the risk of the benefit-sharing stage has a trend of accelerated decline. In this stage, facility depreciation risk and water price change risk have a trend of accelerated decline, and payment risk has a trend of decelerated decline.

Shi^1 (WSC) is at the opposite potential while Shi^2 (WSC) is at the same potential, which means the overall risk of the WSC project has a trend of decelerated ascent. Although the overall risk of the WSC project is low as we concluded before, it shows a trend of decelerated ascent. This indicates that there are some potential high risks in WSC projects. From the perspective of each risk, audit risk and financing risk are not only at a high-risk level but also show a trend of accelerated ascent, so they are at the highest risk. Although the market competition risk is at a medium-risk level, it shows a trend of accelerated ascent, so it should also be considered as high risk. Payment risk is at a high-risk level, it shows a trend of decelerated decline; while inflation risk is at a low-risk level, it shows a trend of decelerated ascent. For these two risks, based on conservative principles, the former should still be treated as high risk, while the latter should be treated as medium risk. The remaining risks are all at a low-risk level and show a trend of accelerated decline, so their impact can be ignored under normal circumstances.

4. Discussion

In this section, we focus on discussing the risk level and risk trend of the WSC projects, and then put forward some policy recommendations for the high risks.

4.1. Audit Risk

Audit risk is not only at a high-risk level but also shows a trend of accelerated ascent. After interviewing the managers of water-saving service operators, it is concluded that the audit risk is at a high-risk level for the following two reasons: (1) For urban permanent residents, water-saving transformation can save costs in the long run. Therefore, these water users will deliberately over-report their water consumption in order to allow water-saving service operators to transform their water supply facilities, which will result in a longer payback period for water-saving service operators; (2) For rural water users who need agricultural irrigation, it is difficult to accurately assess their water consumption due to the influence of climate, environment and market demand, which leads to the uncertainty of investment payback period of water-saving service operators. Obviously, the payback period is the main factor that affects audit risk. The long payback period will increase the debt burden of the water-saving service operators [18]. Moreover, it will also increase the probability of other risks, resulting in some secondary risks. Therefore, audit risk shows a trend of accelerated ascent.

4.2. Financing Risk

As same as audit risk, financing risk is also at a high-risk level with a trend of accelerated ascent. In terms of financing risks, credit, mortgage, and loan mechanisms have not been established for WSC projects in the bank's financial system, which is the main reason for the high financing risk [31]. First of all, WSC projects are still in their infancy in China, and water-saving service operators have not yet obtained good credit ratings from the credit evaluation departments of financial institutions. Second, assets formed by WSC projects, such as equipment and contract receivables, can only be evaluated at the implementation stage and benefit-sharing stage. Banks and other financial institutions often do not recognize such assets or accept them as credit collateral during the investment stage. Thirdly, the technique and risk of the WSC projects are not well known by commercial banks, which greatly increases the cost of loan examination. According to the above reasons, the loan review of commercial banks will inevitably be stricter, and the requirements for loan guarantees will inevitably increase, making it more difficult for water-saving service operators to obtain financing. Li et al. pointed out that the water-saving service operators in China are generally small- and medium-sized enterprises. Unlike state-owned enterprises, these enterprises have difficulty obtaining financing as the market economy is not developed at a high level and the preferential policies are not strong enough [20]. In addition, commercial banks tend to lend to projects with short cycles and high returns [32]. Therefore, the majority of water-saving service operators will fall into a vicious circle of difficulty in obtaining loans, which leads financing risk to show a trend of accelerated ascent.

4.3. Market Competition Risk

Market competition risk is at a medium-risk level, but it has a trend of accelerated ascent, so it should also be regarded as a high-risk level. The high market competition risk of China's WSC projects is mainly due to the monopoly of local water-saving service operators and state-owned enterprises. First of all, no clear industry access standard has been established for WSC projects in China, and there is a lack of authoritative evaluation standard for water-saving efficiency and service level, which has led to irregular operation and market monopoly by some local water-saving service operators [33]. Second, state-owned enterprises have monopolized almost all large-scale WSC projects with their unique resource endowments (https://wsmc-china.com/home/main.html), these resource endowments include advanced technology, standardized management, high reliability, and state subsidies [34]. Local monopolies can be eliminated by regulating the market, but the monopoly of state-owned enterprises is difficult to change in China. As the current trend of "guojingmingtui" (the retreat of the private sector and advancement of state-owned enterprises) in China becomes more and more intense [35], the market competition risk caused by the monopoly of state-owned enterprises has a trend of accelerated ascent.

4.4. Payment Risk

Payment risk is at a high-risk level, it shows a trend of decelerated decline. Based on conservative principles, it should be treated as high risk. After interviewing the managers of water-saving service operators, it is concluded that the reasons for the high payment risk are customer default and business problems. There are two main cases of customer default: (1) Water users do not pay the water-saving benefits belonging to water-saving service operators; (2) As other water-saving operators gave more favorable terms, the water user breached the contract and resigned the contract with other water-saving service operators. The business problem refers to the customer's inability to reach the expected water consumption due to economic and demand pressures. In the final analysis, customer defaults are caused by customers' low moral standards and creditworthiness, and customer's business problems are caused by inaccurate assessments of customers' economic level and water consumption. Wang et al. believe that the application of blockchain and big data can help companies obtain more customer information, so that they can conduct a comprehensive evaluation of customers [36]. At present, blockchain and big data have begun to be applied in China in terms of information sharing, so the payment risk will be reduced in the long run [37].

4.5. Inflation Risk

Inflation risk is at a low-risk level, but it shows a trend of decelerated ascent, so it should be regarded as a medium level risk. Yadav et al. compared the inflation rates of China, the United States, and India. Their research shows that the inflation rates of China and the United States are basically the same and are at a relatively low level, while the inflation rate of India is much higher than that of China and the United States [38]. However, although the risk of inflation is very low in China, it has a steady upward trend, which is related to China's currency oversupply in recent years [39], and China's inflation rate in recent years has also proved this trend (as shown in Figure 2).

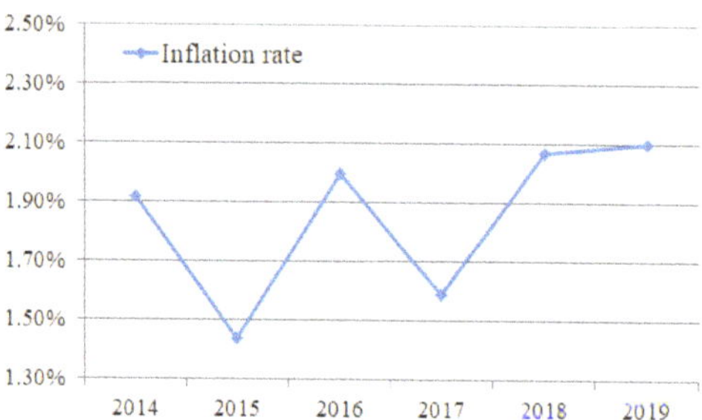

Figure 2. China's inflation rate (2014–2019).

4.6. Other Risks

The remaining risks are all at a low-risk level and show a trend of accelerated decline. Among these risks, technical risks, construction risks, and facility depreciation risks are technical and management risks. Such risks can be reduced significantly by the accumulation of project experience and the rapid development of technologies, so the risks show an accelerated decline.

Interest rate risk, policy risk, force majeure risk, and water price change risk are risks outside of technology and management, which are greatly affected by the fluctuations of the external environment.

In recent years, in order to promote economic development, China has continuously cut interest rates. The benchmark lending rate of China's central bank showed a continuous and large-scale decline (by 1.25%) from 22 November 2014 to 24 October 2015, and remained stable after 24 October 2015 [40], as shown in Figure 3. Therefore, the interest rate risk has a trend of accelerated decline.

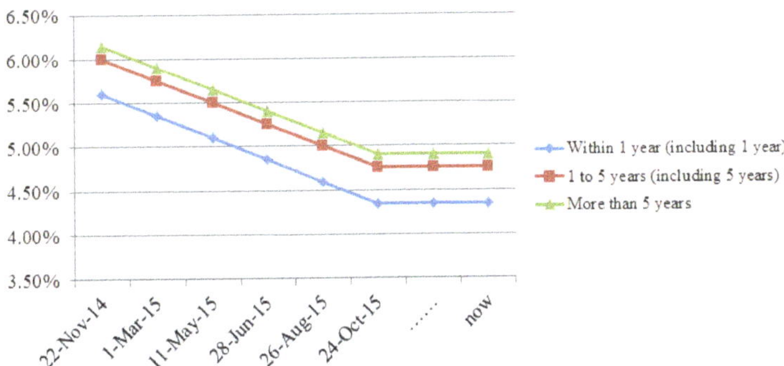

Figure 3. The benchmark lending rate of the People's Bank of China (2014–2020).

For the policy risk, although the preferential policies for water-saving contract projects are not enough at present [20], related policies are still being introduced [41]. Therefore, policy risk has a trend of accelerated decline.

We have counted the force majeure events encountered in 13 water-saving contract pilots, and the results showed that none of the projects were affected by force majeure. Therefore, it can be seen that the force majeure risk is low. In water-saving contract projects, except for the core technologies of water-saving, infrastructure construction takes up the majority of the projects. With the continuous accumulation of experience in China's infrastructure construction, the impact of force majeure on it has become smaller and smaller [42]. Therefore, Force majeure risk has a trend of accelerated decline.

China's water price has been at a low level due to the government's macrocontrol [43], and the overall water price in China has shown a downward trend since 2014 (http://www.h2o-china.com/price/). Therefore, the risk of water price changes has a trend of accelerated decline.

5. Conclusions

This paper uses the literature analysis method to determine the risks in the life cycle of the WSC project, and uses the multielement connection degree set pair analysis to evaluate the level and trend of the risks. The results show:

(1)   The overall risk of China's WSC projects is low, so water-saving contract projects are very suitable for promotion in China. However, the overall risk shows a trend of decelerated ascent, which shows that there are some potential high-risk factors in China's WSC projects.

(2)   Among the many risks of the WSC projects, audit risk, financing risk, and payment risk are at a high-risk level; market competition risk is at a medium-risk level; technical risk, interest rate risk, construction risk, policy risk, inflation risk, facility depreciation risk, and water price change risk are at low-risk level.

(3)   Among the medium and high risks, audit risk, financing risk, and market competition risk have a trend of accelerated ascent, while payment risk has a trend of decelerated decline; in low risks, inflation risk has a trend of decelerated ascent, while the remaining risks have a trend of accelerated decline.

In summary, audit risk, financing risk, market competition risk, payment risk, and inflation risk are the risks that should be focused on in water-saving contract projects. Since inflation risk can only be avoided through financial analysis, and is not controllable, this article proposes the following recommendations for audit risk, financing risk, market competition risk, and payment risk:

(1)   Audit risk and payment risk

Audit risk and payment risk can be reduced through effective third-party management mechanisms. Therefore, it is necessary to cultivate a group of qualified third-party management institutions for

WSC projects. These institutions not only supervise the performance of contracts by water users and water-saving service operators, but also coordinate and arbitrate contradictions between both parties.

(2) Market competition risk

China has not yet formed a standard market competition environment. In today's economic globalization, China should increase international cooperation, learn from the mature experience and models accumulated by other countries in EPC projects, and actively create a good market-oriented competition environment, so as to ensure the high quality and sustainable development of the WSC projects.

(3) Financing risk

The financing difficulty of small- and medium-sized enterprises is a universal problem, which does not only exist in WSC projects. In order to help water-saving service operators obtain financing, special funds for WSC projects in the banking system could be considered; Secondly, in order to solve the loan guarantee problem, water-saving service operators should be allowed to use the improved technology as collateral. Finally, insurance mechanisms can be introduced into ESC projects, that is, taking insurance premiums as a means of financing.

Author Contributions: Conceptualization, Q.L. and M.Y.W.; data curation, Z.S., D.Y., and R.Z.; investigation, Z.S. and R.Z.; methodology, Q.L. and Z.S.; software, Z.S.; supervision, M.Y.W., D.Y., and C.W.; writing—original draft preparation, Q.L. and Z.S.; writing—review and editing, M.Y.W. and C.W. All authors have read and agreed to the published version of the manuscript.

Funding: Australian Research Council: DP170104138; National Social Science Fund: 18BGL006; the Major Projects of the Key Research Base of Humanities and Social Sciences of the Ministry of Education: 18JJD790018.

Conflicts of Interest: The authors declare no conflict of interest.

Appendix A

Table A1. The rank of five-element connection degree of similar potential.

Ranking of Potential Situation $a>e$					
1	$a>b\,b>c\,c>d\,d>e$	23	$a>b\,b<c\,c=d\,d=e$	45	$a=b\,b<c\,c=d\,d>e$
2	$a>b\,b>c\,c>d\,d=e$	24	$a>b\,b<c\,c=d\,d<e$	46	$a=b\,b<c\,c<d\,d>e$
3	$a>b\,b>c\,c>d\,d<e$	25	$a>b\,b<c\,c<d\,d>e$	47	$a<b\,b>c\,c>d\,d>e$
4	$a>b\,b>c\,c=d\,d>e$	26	$a>b\,b<c\,c<d\,d=e$	48	$a<b\,b>c\,c>d\,d=e$
5	$a>b\,b>c\,c=d\,d=e$	27	$a>b\,b<c\,c<d\,d<e$	49	$a<b\,b>c\,c>d\,d<e$
6	$a>b\,b>c\,c=d\,d<e$	28	$a=b\,b>c\,c>d\,d>e$	50	$a<b\,b>c\,c=d\,d>e$
7	$a>b\,b>c\,c<d\,d>e$	29	$a=b\,b>c\,c>d\,d=e$	51	$a<b\,b>c\,c=d\,d=e$
8	$a>b\,b>c\,c<d\,d=e$	30	$a=b\,b>c\,c>d\,d<e$	52	$a<b\,b>c\,c=d\,d<e$
9	$a>b\,b>c\,c<d\,d<e$	31	$a=b\,b>c\,c=d\,d>e$	53	$a<b\,b>c\,c<d\,d>e$
10	$a>b\,b=c\,c>d\,d>e$	32	$a=b\,b>c\,c=d\,d=e$	54	$a<b\,b>c\,c<d\,d=e$
11	$a>b\,b=c\,c>d\,d=e$	33	$a=b\,b>c\,c=d\,d<e$	55	$a<b\,b>c\,c<d\,d<e$
12	$a>b\,b=c\,c>d\,d<e$	34	$a=b\,b>c\,c<d\,d>e$	56	$a<b\,b=c\,c>d\,d>e$
13	$a>b\,b=c\,c=d\,d>e$	35	$a=b\,b>c\,c<d\,d=e$	57	$a<b\,b=c\,c>d\,d=e$
14	$a>b\,b=c\,c=d\,d=e$	36	$a=b\,b>c\,c<d\,d<e$	58	$a<b\,b=c\,c>d\,d<e$
15	$a>b\,b=c\,c=d\,d<e$	37	$a=b\,b=c\,c>d\,d>e$	59	$a<b\,b=c\,c=d\,d>e$
16	$a>b\,b=c\,c<d\,d>e$	38	$a=b\,b=c\,c>d\,d=e$	60	$a<b\,b=c\,c<d\,d>e$
17	$a>b\,b=c\,c<d\,d=e$	39	$a=b\,b=c\,c>d\,d<e$	61	$a<b\,b<c\,c>d\,d>e$
18	$a>b\,b=c\,c<d\,d<e$	40	$a=b\,b=c\,c=d\,d>e$	62	$a<b\,b<c\,c>d\,d=e$
19	$a>b\,b<c\,c>d\,d>e$	41	$a=b\,b=c\,c<d\,d>e$	63	$a<b\,b<c\,c>d\,d<e$
20	$a>b\,b<c\,c>d\,d=e$	42	$a=b\,b<c\,c>d\,d>e$	64	$a<b\,b<c\,c=d\,d>e$
21	$a>b\,b<c\,c>d\,d<e$	43	$a=b\,b<c\,c>d\,d=e$	65	$a<b\,b<c\,c<d\,d>e$
22	$a>b\,b<c\,c=d\,d>e$	44	$a=b\,b<c\,c>d\,d<e$		

Table A2. The rank of five-element connection degree of inverse potential.

	Ranking of Inverse Potential $a<e$				
1	$a<b\,b<c\,c<d\,d<e$	23	$a=b\,b=c\,c>d\,d<e$	-	$a<b\,b=c\,c>d\,d=e$
2	$a=b\,b<c\,c<d\,d<e$	24	$a>b\,b=c\,c>d\,d<e$	46	$a<b\,b>c\,c>d\,d=e$
3	$a>b\,b<c\,c<d\,d<e$	25	$a<b\,b>c\,c>d\,d<e$	47	$a<b\,b<c\,c<d\,d>e$
4	$a<b\,b=c\,c<d\,d<e$	26	$a=b\,b>c\,c>d\,d<e$	48	$a=b\,b<c\,c<d\,d>e$
5	$a=b\,b=c\,c<d\,d<e$	27	$a>b\,b>c\,c>d\,d<e$	49	$a>b\,b<c\,c<d\,d>e$
6	$a>b\,b=c\,c<d\,d<e$	28	$a<b\,b<c\,c<d\,d=e$	50	$a<b\,b=c\,c<d\,d>e$
7	$a>b\,b>c\,c<d\,d<e$	29	$a=b\,b<c\,c<d\,d=e$	51	$a=b\,b=c\,c<d\,d>e$
8	$a=b\,b>c\,c<d\,d<e$	30	$a>b\,b<c\,c<d\,d=e$	52	$a>b\,b=c\,c<d\,d>e$
9	$a<b\,b>c\,c<d\,d<e$	31	$a<b\,b=c\,c<d\,d=e$	53	$a<b\,b>c\,c<d\,d>e$
10	$a<b\,b<c\,c=d\,d<e$	32	$a=b\,b=c\,c<d\,d=e$	54	$a=b\,b>c\,c<d\,d>e$
11	$a=b\,b<c\,c=d\,d<e$	33	$a>b\,b=c\,c<d\,d=e$	55	$a>b\,b>c\,c<d\,d>e$
12	$a>b\,b<c\,c=d\,d<e$	34	$a<b\,b>c\,c<d\,d=e$	56	$a<b\,b<c\,c=d\,d>e$
13	$a<b\,b=c\,c=d\,d<e$	35	$a=b\,b>c\,c<d\,d=e$	57	$a=b\,b<c\,c=d\,d>e$
14	$a=b\,b=c\,c=d\,d<e$	36	$a>b\,b>c\,c<d\,d=e$	58	$a>b\,b<c\,c=d\,d>e$
15	$a>b\,b=c\,c=d\,d<e$	37	$a<b\,b<c\,c=d\,d=e$	59	$a<b\,b=c\,c=d\,d>e$
16	$a<b\,b>c\,c=d\,d<e$	38	$a=b\,b<c\,c=d\,d=e$	60	$a<b\,b>c\,c=d\,d>e$
17	$a=b\,b>c\,c=d\,d<e$	39	$a>b\,b<c\,c=d\,d=e$	61	$a<b\,b<c\,c>d\,d>e$
18	$a>b\,b>c\,c=d\,d<e$	40	$a<b\,b=c\,c=d\,d=e$	62	$a=b\,b<c\,c>d\,d>e$
19	$a<b\,b<c\,c>d\,d<e$	41	$a<b\,b>c\,c=d\,d=e$	63	$a>b\,b<c\,c>d\,d>e$
20	$a=b\,b<c\,c>d\,d<e$	42	$a<b\,b<c\,c>d\,d=e$	64	$a<b\,b=c\,c>d\,d>e$
21	$a>b\,b<c\,c>d\,d<e$	43	$a=b\,b<c\,c>d\,d=e$	65	$a<b\,b>c\,c>d\,d>e$
22	$a<b\,b=c\,c>d\,d<e$	44	$a>b\,b<c\,c>d\,d=e$		

Appendix B

Table A3. Trend curve of the risk.

$Shi^1\,(x)$	Same potential	Same potential	Opposite potential
$Shi^2\,(x)$	Same potential	Opposite potential	Same potential
Change type	accelerated decline	decelerated decline	decelerated ascent
Change curve			

$Shi^1\,(x)$	Opposite potential	Same potential	Opposite potential
$Shi^2\,(x)$	Opposite potential	Equal potential	Equal potential
Change type	Accelerated ascent	Uniform decline	Uniform ascent
Change curve			

References

1. Su, X.; Kang, S. Research advances and key topics on optimal allocation of water resources based on ecosystem in the arid areas. *Trans. Chin. Soc. Agric. Eng.* **2005**, *21*, 167–172.
2. Rogers, S.; Chen, D.; Jiang, H.; Rutherfurd, I.; Wang, M.; Webber, M.; Crow-Miller, B.; Barnett, J.; Finlayson, B.; Jiang, M.; et al. An integrated assessment of China's South—North Water Transfer Project. *Geogr. Res.* **2020**, *58*, 49–63. [CrossRef]
3. Guo, H.; Chen, X.; Dong, Z.; Zhang, H. Construction method of water right trading based on water-saving management contract. *Water Resour. Prot.* **2019**, *35*, 33–38.
4. Guo, H.; Chen, X.; Liu, J.; Zhang, H.; Svensson, J. Joint analysis of water rights trading and water-saving management contracts in China. *Int. J. Water Resour. Dev.* **2020**, *36*, 716–737. [CrossRef]
5. Mills, E.; Kromer, S.; Weiss, G.; Mathew, P.A. From volatility to value: Analysing and managing financial and performance risk in energy savings projects. *Energy Policy* **2006**, *34*, 188–199. [CrossRef]
6. Lee, P.; Lam, P.T.I.; Lee, W.L. Risks in energy performance contracting (EPC) projects. *Energy Build.* **2015**, *92*, 116–127. [CrossRef]
7. Jinrong, H.; Enyi, Z. Engineering risk management planning in energy performance contracting in China. *Syst. Eng. Procedia* **2011**, *1*, 195–205. [CrossRef]
8. Duan, X.; Chen, F. Research on the Financing Risk of Contract Energy Management Project Based on the Whole Life Cycle. *Sci. Technol. Manag. Res.* **2018**, *38*, 235–243.
9. Wu, Z.; Dong, X.; Pi, G. Risk evaluation of China's petrochemical Energy Performance Contracting (EPC) projects: Taking the Ningxia Petrochemical Company as an example. *Nat. Gas Ind.* **2017**, *37*, 112–119.
10. Huang, Z.Y.; Zhang, Y.S.; Han, Y. Risk Identification and Comprehensive Evaluation in Energy Performance Contracting Projects in China. *J. Eng. Manag.* **2013**, *27*, 48–52.
11. Garbuzova-Schlifter, M.; Madlener, R. AHP-based risk analysis of energy performance contracting projects in Russia. *Energy Policy* **2016**, *97*, 559–581. [CrossRef]
12. RezaValipour, A.; HadiSarvari, N.; Noor, N.M.; Rashid, A.S.A. Analytic network process (ANP) to risk assessment of gas refinery EPC projects in Iran. *J. Appl. Sci. Res.* **2013**, *9*, 1359–1365.
13. Guo, L. Current situation and prospect analysis of contracted water-saving management in China. *China Water Resour.* **2016**, *15*, 18–21.
14. Zhao, K.Q. *Set Pair Analysis and its Preliminary Application*; Zhejiang Science and Technology Press: Hangzhou, China, 2000; pp. 1–200.
15. Cui, Y.; Feng, P.; Jin, J.; Liu, L. Water resources carrying capacity evaluation and diagnosis based on set pair analysis and improved the entropy weight method. *Entropy* **2018**, *20*, 359. [CrossRef]
16. Gao, H. Information Risk Evaluation and Application: Based on the Set Pair Analysis. In *Proceedings of the Fifth International Forum on Decision Sciences*; Springer: Singapore, 2018; pp. 285–294.
17. Zheng, X.; Xu, K.; Li, Q. A set pair analysis approach for dynamic risk assessment of tailings dam failure. In *Proceedings of the 11th International Mine Ventilation Congress*; Springer: Singapore, 2019; pp. 1024–1035.
18. Changshun, L.; Xian, C.; Jianhua, Q. Revelation of foreign water audit to building water-saving society in China. *China Water Resources* **2005**, *13*, 128–130.
19. Sorrell, S. The economics of energy service contracts. *Energy Policy* **2007**, *35*, 507–521. [CrossRef]
20. Li, R.; Leng, J. Risk of contract Water-saving management projects. *Water Sav. Irrig.* **2016**, *11*, 88–90.
21. Zhu, J.; Chen, L.; Zhao, Y.; Xie, J. Analysis of cooperative game of benefit distribution of contract for water saving in college based on modified Shapley value method. In *IOP Conference Series: Earth and Environmental Science, Proceedings of the 5th International Conference on Water Resource and Environment (WRE 2019), Macao, China, 16–19 July 2019*; IOP Publishing: Bristol, UK, 2019.
22. Seckler, D.; Sampath, R.K.; Raheja, S.K. An index for measuring the performance of irrigation management systems with an application. *JAWRA J. Am. Water Resour. Assoc.* **1988**, *24*, 855–860. [CrossRef]
23. Da-li, G. Energy service companies to improve energy efficiency in China: Barriers and removal measures. *Procedia Earth Planet. Sci.* **2009**, *1*, 1695–1704. [CrossRef]
24. Mizrachi, K. Force majeure in project finance: A comparative and practical analysis of risk allocation. *J. Struct. Financ.* **2006**, *12*, 76–97. [CrossRef]
25. Ellis, J. *Energy Service Companies (ESCOs) in Developing Countries*; International Institute for Sustainable Development: Winnipeg, MB, Canada, 2010.

26. Liu, D.; Yin, Q. Research on benefit distribution of contract water saving management based on modified Shapley model. *J. Econ. Water Resour.* **2016**, *34*, 53–58.

27. Aili, Z.K.X. Set Pair Theory–A New Theory Method of Non-Define and Its Applications. *Syst. Eng.* **1996**, *1*, 3.

28. Luan, W.; Lu, L.; Li, X.; Ma, C. Integrating Extended Fourier Amplitude Sensitivity Test and Set Pair Analysis for Sustainable Development Evaluation from the View of Uncertainty Analysis. *Sustainability* **2018**, *10*, 2435. [CrossRef]

29. Yang, G.; Gao, H. Uncertain risk assessment of knowledge management: Based on set pair analysis. *Sci. Program.* **2016**. [CrossRef]

30. Feng, D.; Xu, Y.; Jia, C. A Risk Assessment of PPP Projects Based on Five-element Connection Number. *J. Eng. Manag.* **2017**, *31*, 77–82.

31. Zeng, M.; Chen, C.; Duan, K.; Li, N.; OuYang, S. Life Cycle Risk Assessment for Energy Management Contract Project. *East China Electr. Power* **2012**, *40*, 1666–1670.

32. Guo, L.; Liu, L. Investment modes of water-saving management contract. *J. Econ. Water Resour.* **2017**, *35*, 45–48.

33. Yang, Z.; Zhao, J.; Wang, S. The development and problems of contract energy management in China. *Energy Conserv. Environ. Prot.* **2004**, *12*, 19–21.

34. Brødsgaard, K.E. Moving Ahead in China: State-Owned Enterprises and Elite Circulation. *China Int. J.* **2020**, *18*, 107–122.

35. Gu, T. The behavior of private entrepreneurs in an imperfect financial market. *Econ. Bull.* **2020**, *40*, 349–358.

36. Wang, Q.; Su, M.; Li, R. Is China the world's blockchain leader? Evidence, evolution and outlook of China's blockchain research. *J. Clean. Prod.* **2020**, *264*, 121742. [CrossRef]

37. Hayrutdinov, S.; Saeed, M.S.; Rajapov, A. Coordination of Supply Chain under Blockchain System-Based Product Lifecycle Information Sharing Effort. *J. Adv. Transp.* **2020**. [CrossRef]

38. Yadav, D.K.; Jameel, S. Pattern of GDP Growth Rate, Inflation, Interest Rate in Post Reform Period: A Comparative Analysis of India China and USA. Available online: https://papers.ssrn.com/sol3/papers.cfm?abstract_id=3643799 (accessed on 5 July 2020).

39. Zheng, Y. A Study on the Relationship between the Independence of the Central Bank and China's Inflation. *Open J. Soc. Sci.* **2020**, *8*, 263. [CrossRef]

40. Fang, J.; Lau, C.K.M.; Lu, Z.; Tan, Y.; Zhang, H. Bank performance in China: A Perspective from Bank efficiency, risk-taking and market competition. *Pac. Basin Financ. J.* **2019**, *56*, 290–309. [CrossRef]

41. Tang, Z. Refinement and strengthening of contract water-saving management fiscal and tax incentive policies. *Environ. Econ.* **2017**, *22*, 58–61.

42. Zhong, A.C.; Hu, B.R.; Wang, C.M.; Xue, D.W.; He, E.L. The Impact of Urbanization on urban agriculture: Evidence from China. *J. Clean. Prod.* **2020**, *276*, 122686. [CrossRef]

43. Li, J.; Lei, X.; Qiao, Y.; Kang, A.; Yan, P. The Water Status in China and an Adaptive Governance Frame for Water Management. *Int. J. Environ. Res. Public Health* **2020**, *17*, 2085. [CrossRef]

 *water*

Article

Are the Financial Markets Sensitive to Hydrological Risk? Evidence from the Bovespa

José Manuel Feria-Domínguez [1,*], Pilar Paneque [2] and Fanny de la Piedra [3]

1   Department of Financial Economics and Accounting, Pablo de Olavide University, Ctra. De Utrera, Km 1, 41013 Seville, Spain
2   Department of Geography, History and Philosophy, Pablo de Olavide University, Ctra. De Utrera, Km 1, 41013 Seville, Spain; ppansal@upo.es
3   Center of Postgraduates Studies, University Pablo de Olavide, Ctra. De Utrera, Km 1, 41013 Seville, Spain; fderiv@alumno.upo.es
*   Correspondence: jmferdom@upo.es; Tel.: +34-682-042-557

Received: 30 August 2020; Accepted: 19 October 2020; Published: 27 October 2020

Abstract: This research analyzes the BOVESPA stock market response to the worst drought occurred in the last 100 years in Brazil. For this purpose, we conducted a standard event study analysis in order to assess the financial response to such hydrological risk on a sample of seven Brazilian agri-food firms. We found statistically significant negative cumulative average abnormal returns (CAARs) around the drought official announcement for different event windows used. Particularly, the highest impact was obtained for the narrowest temporary window, five days around the event disclosure. Moreover, we also found the drought announcement affects even more negatively those companies that sell perishable products, five out of seven in our sample, versus those selling nonperishable ones by running a two-sample *t*-test on CAARs. This study brings awareness to the climate change impact into the emerging financial markets and the risk faced by shareholders when investing in the agri-food sector, not only in Brazil but also in other Latin American countries, due to the increasing probability to suffer from droughts.

Keywords: droughts; hydrological risk; agri-food sector; event study; financial markets; BOVESPA

---

1. Introduction

Climate change is one of the main risks we are facing worldwide. According to the Intergovernmental Panel on Climate Change [1] "Climate change refers to a change in the state of the climate that can be identified by changes in the mean and/or the variability of its properties and that persists for an extended period, typically decades or longer". It means a change of climate caused by human activity that modifies the composition of the global atmosphere and is observed over comparable time periods [2]. For the World Bank (WB, henceforth) and Inter−American Development Bank (IDB, henceforth), climate change includes changes in the variability of the weather, these are jointly represented as changes in the extremes of the weather, such as a greater number of rainfall events, bringing floods or droughts [3]. To reinforce this idea, the [1] has pointed out that changes are already occurring not only in terms of frequency and severity of the hydrological events, but also in their geographical location [4]. That is why the United Nations Environment Program [5] initiated an investigation to align the financial system to sustainable development needs. The changes observed are connected to growths in the areas affected by droughts, in the number of events causing floods, and in the duration and intensity of certain kinds of tropical storms. Drought-related disasters are also rising, with 3.5 times as many in the past decade, in comparison with 1970–1979. Moreover, the frequency of heat-related disasters has increased 10-fold compared with the 1970s [6].

In relative terms, drought is defined as a negative pluviometric anomaly, which is intense and of sufficient duration to produce social effects. Droughts are, therefore, natural in origin. The complexity of the phenomenon, the geographical and temporal limits of which are hard to determine, make it difficult to have a consensus on the definition [7–9]. In addition, droughts vary depending on the geographical location and the context [10]. The ambiguity of the term also gives rise to certain confusion and overlap between the concepts of drought (temporary and natural), water deficit (temporary and anthropic), and scarcity (permanent and anthropic) [11]. Moreover, we should also take into account an operational definition, used for the development of prevention and mitigation strategies, in order to identify the phenomenon in terms of its frequency, severity, and duration for a given return period [8]. The level of impact of a drought is also related to the ways in which societies manage water resources and in which develop strategies to cope with these events [9].

Within a well-defined spatial and temporal framework, [12] differentiate four categories of droughts: meteorological drought (decrease in rainfall), agricultural drought (reduction in soil humidity), hydrological drought (reduction in the availability of surface and underground water sources), and socioeconomic drought (reduction of water availability with regard to existing demands). Recently, this widely accepted classification has been considered insufficient as it is approached from a human-centered perspective. Thus, a new category named ecological drought has been added to emphasize the ecosystems and ecosystem services consequences of a drought event [13]. However, so far, interest in droughts is usually due to its socioeconomic aspects (impact, damage, loss), which are, in turn, tied to other complex concepts such as vulnerability and perception [14]. Ref. [15] stated that over the next 30 to 90 years, southern Africa, the United States, southern Europe, Brazil, and southeast Asia will suffer from an increasing drought severity. Particularly, it is highlighted that in a world warmer by 4 °C, precipitation is greatly reduced in the Caribbean, Central America, Central Brazil, and Patagonia, between 20% and 40%. It is expected that the drought conditions increase more than 20%. It is estimated that limiting warming to 2 °C will reduce considerably the risk of drought, a 1% increase in days with drought conditions in the Caribbean and a 9% increase in South America.

Some regions in Brazil, particularly the semi−arid north-east, are suffering from continuous droughts; producing devastation to some agricultural, livestock, and industrial producers. The intensification of extreme droughts in Brazil has promoted the interest among natural resource managers, farmers, development practitioners, researchers, and policymakers to understand the extent to which climate change will impact water resources, food production, incomes, and livelihoods [16]. However, the UK Parliamentary Office of Science and Technology and the Stern report [17] point out that one of the main challenges is "convincing" businesses that climate change is actually happening and will impact in their activities [18]. Most of the existing literature focuses on agriculture as the most sensitive economic sector to the correlated risks tied to natural disasters, but other sectors are also affected, such as the financial one. In practice, such catastrophic events have led some microcredit providers to ration credit [19] or to restructure loans before they have occurred.

In this regard, a useful tool to show the economic impact on different listed companies or sectors is through conducting an event study analysis. In this paper, we attempt to analyze the financial perceptions on hydrological risk, particularly caused by the recent, most severe drought registered in the past 50 years in Brazil. For this purpose, we examine the reaction of the Brazilian agri-food sector, listed in the BOVESPA [20] stock market.

The research is structured as follows. The next section, corresponding to Section 2, explains the data and sample. Section 3 explains the methodological approach. Section 4 develops the research design, Section 5 presents our main findings, and Section 6 summarizes the main conclusions.

2. Literature Review

Ref. [21] analyzed the impact of earthquake affecting the Umbria region in Central Italy (1997) on the tourism sector. By using an event study, they found a reduction of arrivals through the whole event window, particularly evident in the Assisi district where the "missing arrivals" exceeded by 50%

the value forecasted by the model for that period. Ref. [22] studied the impact of Hurricane Floyd on the market value of insurance firms, determining how financial markets reacted to changing news about the storm's characteristics through the insurer stock prices. They demonstrated that the market value of insurance firms is significantly affected by such catastrophic events; however, this effect is not constant, nor is it always negative on each day of the cycle. Ref. [23] also ran an event study to assess the impact of the tsunami in the financial markets by using the CAPM (Capital Asset Pricing Model); the analysis shows that stock markets were almost insensitive to this event. They highlighted a general increase in the long-term market risk of thirteen industrial portfolios in four countries that were directly affected. They conclude that there was no major wealth destruction for equity investors as a result of the Boxing Day tsunami, in line with the findings of [24]. Ref. [25,26], postulate that rare and extreme events influence financial markets' risk premiums. Finally, they believe that the observed effects are more likely to involve region-specific issues, as shown by [27,28]. Ref. [29] analyzed the impact of the oil spills disasters on the corporate reputation of a sample of oil and gas companies, listed on New York Stock Exchange (NYSE). By using an event study, they demonstrated that reputational risk arises depending on the investors' perception that the company may incur losses in the future as a result of a possible negligence concerning such environmental disasters. Ref. [30] examined the US market reaction to ISO 14001 certification announcements. They carried out an event study analysis on a sample of 140 announcements, in order to assess if the new deal between companies and environmental care could be particular important for shareholders, more specifically, in manufacturing, finding a negative impact on the stock returns. Moreover, they proved that the shareholder wealth was reduced due to these certifications announcements. Ref. [31] used an event study method to assess how the stocks of listed companies reacted before and after announcing their partnership with the Unites States Environmental Protection Agency (USEPA) Climate Leaders program. The results suggest that these firms' public announcements of joining the USEPA Climate Leaders partnership did not have a positive impact on stock returns. While this study demonstrated no immediate financial benefit from the practices implemented by these firms to reduce their greenhouse gas emissions, it may still bode well for long-term corporate earnings and attractiveness to investors, in line with the findings of [32]. Ref. [33] analyzed the stock market's reaction to information disclosure of environmental violation events (henceforth, EVEs). Using the methodology of event study, daily abnormal return (AR) and accumulative abnormal return (CAR) are calculated within different event windows in order to examine the extent to which the stock market responds to such EVEs. Their findings revealed that the average reduction in market value is much lower than the estimated changes in market value for similar events in other countries, demonstrating that the negative environmental events of Chinese listed companies currently have a weak impact on the stock market. The authors think it is due to the weak implementation effectiveness of the Chinese law system, making investors have a low expectation of the legal impact of such events in the future. This study raises the concerns of environmental conservation to the emerging economies and highlights the need to strengthen up government regulations as well as to improve the effectiveness of law implementation and penalty. To our knowledge, the literature on the effects of droughts on financial markets is almost nonexistent, so this paper contributes to enrich the existing literature of natural disasters (earthquake, tsunamis, hurricanes, floods, etc.).

3. Data and Sample

We have selected some of the companies listed in the industrial index (INDX) of the Brazilian stock market (BOVESPA). This index is formed by 150 companies; most of the important ones, in terms of market capitalization, belong to the agri-industrial and livestock sector and the segment of processed food, which are sensitive to the potential consequences of the drought events. The sample is presented in Table 1:

Table 1. Sample of agri-food companies.

Ticker	Company	Weight (%)	Main Activity
BRFS3	BRF SA	20	Operational holding: frozen foods: meat and derivatives.
JBSS3	JBS	7.56	Frozen foods: meat and derivatives.
CSAN3	COSAN	1.6	Sugar production, alcohol, ethanol, combustible distribution
MDIA3	M.DIASBRANCO	1.11	Production and sale of wheat flour, sponge cakes, margarine, snacks, etc.
MRFG3	MARFRIG	0.85	Production and distribution of meats and derivatives.
SMTO3	SAO MARTINHO	0.71	Production and sale of sugar and derivatives.
BEEF3	MINERVA	0.4	Production and sale of pork meat, poultry, etc.

The percentages represent the participation weight in the BOVESPA stock market index, being the BRF SA the most representative stock with 20% out of the total weight.

Brazil is the third-largest agricultural exporter in the world and one of the major producers of soybeans, sugar, orange juice, maize, cotton, chicken, meat, and pigs, representing approximately 6% of the country's GDP. Although the total agricultural land area has remained stable since the mid-seventies, production has increased by nearly 300% due to technological innovation. However, the severe drought (2015) that occurred in the north-east is considered one of the worst in the past 100 years [34], in terms of water availability. On 3 February 2015, the Food and Agriculture Organization's (FAO) Chairman, Ms. Jose Graziano da Silva, announced officially the drought declaration and its consequences for the agricultural sector (see Table 2).

Table 2. Event description. Source: LexisNexis® Academic.

Event	Date	Operational Loss	Event Description
FAOAnnouncement	3 February 2015	$4300 MM	Brazil risks a massive loss of crops of all kinds of agricultural products, which will generate an increase in prices in the next months.

In order to run the event study analysis, we have selected all companies belonging to the processed food segment in the industrial index of BOVESPA. The daily stock prices, the market value, and the index quotes have been obtained from Thomson Reuters Data Stream. A detailed description of the event has been collected from Reuters verified via LexisNexis® Academic.

4. Methodological Background

Originally introduced by [35,36] in two respective seminal papers, the use of event studies has become widespread in recent financial and accounting research. The event study methodology is based on the Efficient Market Hipothesis [37] that states that asset prices reflect all available information. If markets are efficient then new information is reflected quickly into market prices. Thus, stock prices trade at their fair value. There are three levels of market efficiency:

- Weak-form efficiency: stock prices reflect all historical information published.
- Semi-strong-form efficiency: stock prices fully reflect all past public information as well as recent public information available.
- Strong-form efficiency: stock prices reflect public and unpublicized or "insider" information.

Under the Efficient Market Hipothesis (EMH), any piece of information regarding the firm is supposed to be reflected in the stock prices. Conceptually, event study technique differentiates between the expected returns obtained in the case the event would not have occurred—normal returns, NR—and the returns arising from the event—abnormal returns, AR [29].

Thus, the abnormal return AR_{it} is calculated as the difference between the actual return R_{it} of a stock and the normal or expected return NR_{it}:

$$AR_{it} = R_{it} - NR_{it} \tag{1}$$

The normal returns are calculated following the market model [38] as a benchmark. Thus, the normal or expected returns of stock "i" on day "t" are estimated by ordinary least squares (OLS) as follows:

$$NR_{it} = \hat{\alpha}_i + \hat{\beta}_i R_{mt} \tag{2}$$

where

NR_{it} the normal return of a stock "i" is at time "t".

R_{mt} is the market index;

$\hat{\alpha}_i$ and $\hat{\beta}_i$ are the parameters estimated by OLS, using 250 trading days prior to the event window.

According to [39], and assuming there are N listed firms in the sample, we can draw a matrix of abnormal returns, denoted by Σ, as

$$\Sigma = \begin{pmatrix} AR_{1,T_1} & \cdots & AR_{N,T_1} \\ \vdots & \cdots & \vdots \\ AR_{1,0} & \cdots & AR_{N,0} \\ \vdots & \cdots & \vdots \\ AR_{1,T_2} & \cdots & AR_{N,T_2} \end{pmatrix} \tag{3}$$

Each column of this matrix represents a time series of abnormal returns for firm "i", whereas each row is a cross-section of abnormal returns for each day within the event window (T_1, T_2). In order to examine the stock reactions around events, each firm's return data could be analyzed separately. Such analysis is usually improved by averaging the information over the whole sample, given rise to the average abnormal return (AAR):

$$AAR = \frac{1}{N} \sum_{i=1}^{N} AR_{it} \tag{4}$$

Ref. [40] also suggest studying the cumulative abnormal returns (CAR), calculated within the event window, by aggregating AR_{it} from T_1 to T_2:

$$CAR_i = \sum_{t=T_1}^{T_2} AR_{it} \tag{5}$$

Finally, in event studies, CAR_i's usually aggregated over the sample cross-section, giving rise to the cumulative average abnormal returns ($CAAR$):

$$CAAR = \frac{1}{N} \sum_{i=1}^{N} CAR_i \tag{6}$$

The last step of the event study methodology is to test the statistical significance of the observed stock prices changes in order to assess they are not random. For this purpose, we carry out both parametric and nonparametric tests following [39–41]. Parametric test statistics are based on a standard *t*-test of difference between two means. Since nonparametric tests do not rely on the assumption of normality, they are more reliable than the parametric ones.

5. Research Design

For designing this research, we have defined three event windows, including the event date, $E1(-40,+40)$, $E2(-20,+20)$ and $E3(-5,+5)$, that is, from 5, 20, and 40 days, before and after, the drought announcement; in other words, 81, 41, and 21 days for each event window, respectively. Existing research shows that shorter event periods provide better and more accurate estimations of the impact of the disclosures on the stock prices since the likelihood of confounding factors not related to the event decrease. Later, we estimated the abnormal returns for each particular event window as well as the corresponding cumulative abnormal returns for the sample of agri-food firms.

The first step was to calculate both the sample daily stock returns and the Brazilian market index (BOVESPA) returns by applying natural logarithm [42,43]:

$$R_{it} = ln\left(\frac{P_{it}}{P_{it-1}}\right) \tag{7}$$

where

P_{it} is the price of the firm's stock i at day t.
P_{it-1} is the price of the firm's stock i the day before.

Once the observed returns for both the sample companies and the BOVESPA index were calculated, we estimated the corresponding expected returns by using ordinary least squares for 250 trading days, about a year of trading prior to the event window (see Equation (2)). Reaching this point, we calculated the stock's abnormal returns (AR_{it}) following the Equation (1).

In Table 3, we summarize the average of abnormal returns for each Brazilian company within the corresponding event window. We find much higher negative returns in the shorter event window $E(-5,+5)$ in comparison with the largest ones. In the last column of the table, we calculated the mean of the abnormal returns for the whole sample in each event window. Note that the mean for $E(-5,+5)$ is around double that for the $E(-20,+20)$ and $E(-40,+40)$, indicating the higher impact of the event in the very short term.

Table 3. Average abnormal returns (*AAR*) for Brazilian companies.

Event Window	BRFS3	JBSS3	CSAN3	MDIA3	MRFG3	SMTO3	BEEF3	Average
$E(-40,+40)$	−0.3%	0.28%	−0.09%	−0.03%	−0.50%	−0.13%	−0.38%	−0.14%
$E(-20,+20)$	−0.09%	0.17%	0.02%	−0.06%	−0.71%	−0.05%	−0.10%	−0.12%
$E(-5,+5)$	0.20%	−0.04%	−0.19%	−0.48%	−0.76%	−0.56%	−0.41%	−0.32%

Then, we drew the Σ matrix (see Equation (3)) for the whole sample, which summarizes the abnormal returns of the seven companies analyzed for each event window. Moreover, we have also calculated the average abnormal return (*AAR*), the cumulative abnormal returns (CAR), and the cumulative average abnormal returns (CAAR), following the Equations (4)–(6), respectively.

At this point, we ran a normality test by using Anderson–Darling (AD) test since it requires smaller samples than the Kolmogorov–Smirnov (KS) test to achieve sufficient statistical power [44]. As in our case seven companies were analyzed, the Anderson–Darling test is then more appropriate. Therefore, we have conducted the AD test on the average abnormal return (AAR) for the event windows $E(-40,+40)$, $E(-20,+20)$, and $E(-5,+5)$, and the normality assumption is not rejected, as the following Probability-Probability (P-P) plots illustrate (see Figure 1):

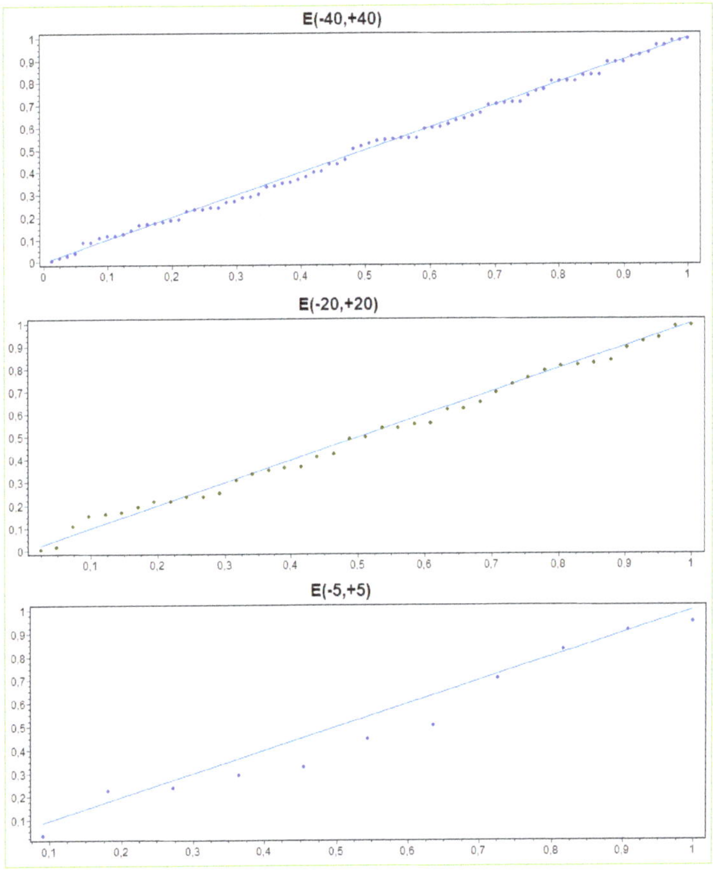

Figure 1. P-P plots for $E(-5,+5)$, $E(-20,+20)$, and $E(-40,+40)$.

After checking the normality assumption, it was time to test the significance of the cumulative abnormal returns. For this purpose, we first applied the parametric t-test [45–48]. The traditional t-test relies on the normality assumption of abnormal returns. The null hypothesis is that stock prices do not react to the drought announcement. Assuming that the abnormal returns are independent and identically distributed, the statistic follows a Student's distribution.

The use of nonparametric test provides a double-check of the robustness of the parametric test results [49]. In this sense, we used both the sign test [50–52] and the Wilcoxon test [53,54]. The sign test is a binomial test which calibrates if the frequency of abnormal positive residuals is equal to 50%. To apply this test, we calculated the proportion of values in the sample that shed no negative AR's under the null hypothesis. The null value is estimated as the average fraction of stocks with no negative AR's in the estimation period. If AR's are independent, under the null hypothesis, the number of positive abnormal return values follows a binomial distribution. The Wilcoxon test considers both the sign and the magnitude of abnormal returns. This test assumes that none of the absolute values are the same and each is non-zero. Under the null hypothesis, the probability of positive and negative abnormal returns is equal.

6. Finding and Results

As mentioned in the previous Section, a parametric test (*t*-test) was first conducted to study the reactions of the agri-food Brazilian companies to the draught announcement for the three event windows, $E(-40,+40)$, $E(-20,+20)$, and $E(-5,+5)$. The idea behind this is to test if the mean of the CAARs is equal (null hypothesis, H_0) or different to zero (H_1), that is

$$H_0 = E(CAAR) = 0 \tag{8}$$

$$H_1 = E(CAAR) \neq 0 \tag{9}$$

Table 4 summarizes the results obtained from the parametric *t*-test:

Table 4. *T*-test on cumulative average abnormal returns (CAARs).

Variable	Mean	t-Value	p-Value	95% CI
CAAR $(-5,+5)$	−0.01877	−3.22	0.009	$(-0.03174; -0.00580)$
CAAR $(-20,+20)$	−0.03529	−13.00	0.000	$(-0.04078; -0.02981)$
CAAR $(-40,+40)$	−0.0664	−15.58	0.000	$(-0.07488; -0.05791)$

Setting the specific confidence level (95%), we observe that the *p*-values are smaller than our choice of significance α (0.05), so we can reject the null hypothesis H_0, accepting H_1 for all the event windows. As a consequence, we find a statistically significant abnormal behavior produced by the draught announcement. In the following Figures 2–7, we illustrate those results of the *t*-test (95%) by using the corresponding histograms and P-P plots. It is easy to infer that the null hypothesis, represented by the dot H_0, is rejected for all the estimation windows, since the mean of CAARs is located far from it.

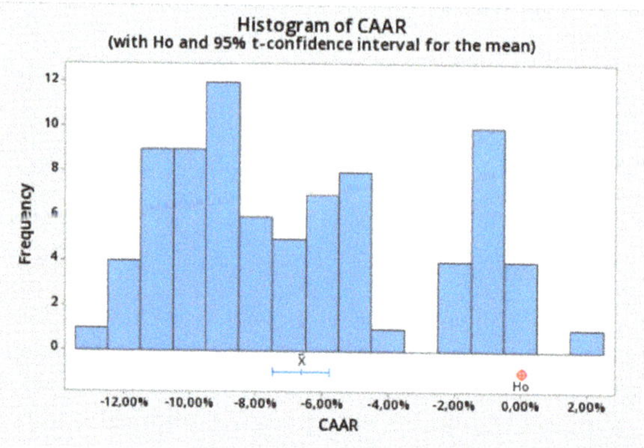

Figure 2. *t*-Test histogram of CAAR for $E(-40,+40)$.

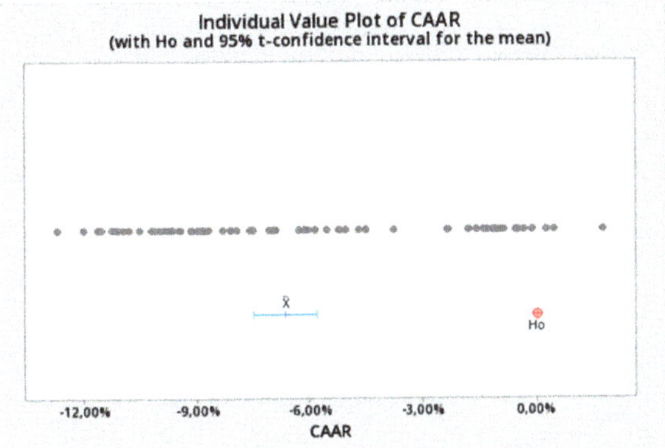

Figure 3. *t*-Test P-P plot of CAAR for $E(-40,+40)$.

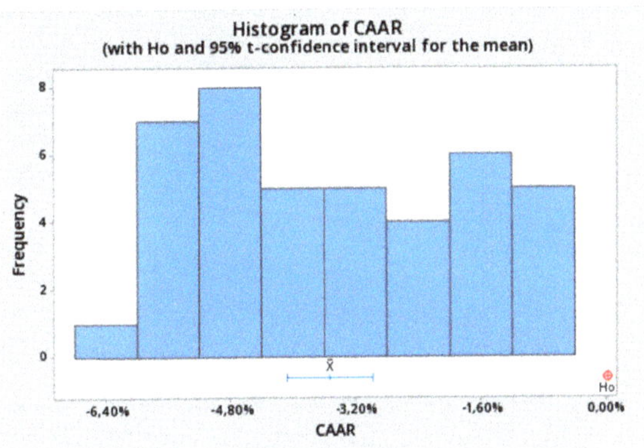

Figure 4. *t*-Test histogram of CAAR for $E(-20,+20)$.

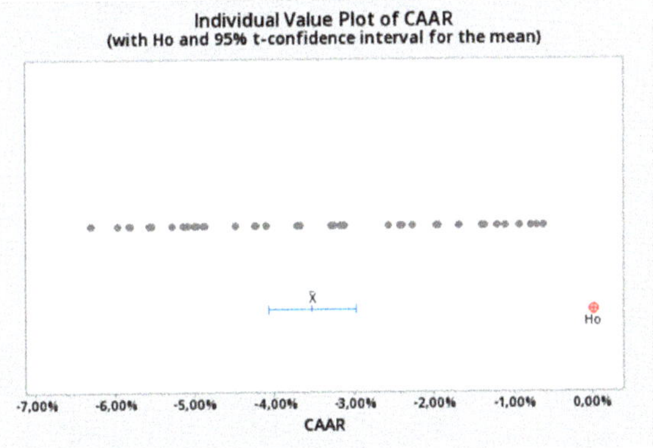

Figure 5. *t*-Test P-P plot of CAAR for $E(-20,+20)$.

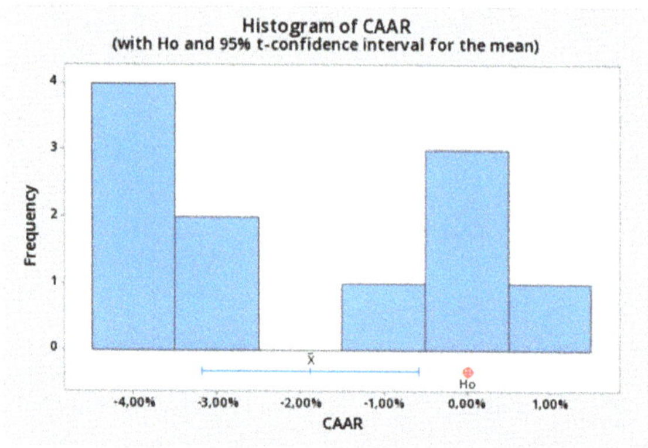

Figure 6. *t*-Test histogram of CAAR for $E(-5,+5)$.

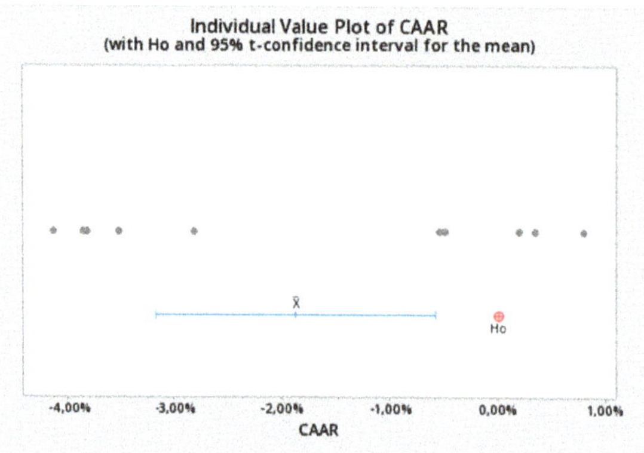

Figure 7. *t*-Test P-P plot of CAAR for $E(-5,+5)$.

After applying this parametric test, we also conducted two nonparametric tests in order to reinforce our research: The sign test and Wilcoxon test. The results of both tests are shown below, in Table 5 for Sign Test and Table 6 for Wilcoxon Test, respectively:

Table 5. Sign test on CAAR.

Variable	Median	Achieved Confidence	Confidence Interval		*p*-Value
			Lower	Upper	
CAAR (−5,+5)	−0.02808	93.46%	−0.03816	0.00194	0.027
		95.00%	−0.03819	0.00206	
		98.83%	−0.03852	0.00346	
CAAR (−20,+20)	−0.03683	94.04%	−0.04879	−0.02584	0.000
		95.00%	−0.0488	−0.02558	
		97.25%	−0.04883	−0.02437	
CAAR (−40,+40)	−0.07515	92.46%	−0.08821	−0.06133	0.000
		95.00%	−0.08841	−0.06047	
		95.45%	−0.08847	−0.06024	

Table 6. Wilcoxon test on CAAR.

Variable	Median	p-Value	Achieved Confidence	Confidence Interval	
				Lower	Upper
CAAR (−5,+5)	−0.0178	0.029	95.5%	−0.0347	−0.0015
CAAR (−20,+20)	−0.035	0.000	95.0%	−0.0414	−0.0297
CAAR (−40,+40)	−0.0677	0.000	95.0%	−0.0787	−0.0572

When running the sign test, since the statistic is discrete, the specified confidence level (95%) was not always achieved, so, in Table 5, three confidence intervals appear depending on the achievements. Since the p-value is smaller than the significance level α (0.05), we can conclude that the difference between the median and the hypothesized median of CAAR is statistically significant, thus rejecting the null hypothesis H_0.

Following the same reasoning that in the previous test, for the three event windows, the p-value is smaller than the significance level α (0.05), so we also reject the null hypothesis H_0; accepting that the median of CAARs differs from the hypothesized one.

Reaching this point, we split the original sample into two main subsamples, just to check if there was a potential different impact of the drought event between nonperishable (BRFS3 and JBSS3) and perishable firms (CSAN3, MDIA3, MRFG3, SMTO3, BEEF3). In the following charts (see Figures 8–10), we have drawn the corresponding CAARs for each subsample, as well as the full sample, to illustrate the existence of higher negative impact on the perishable dataset, independently of the event window used.

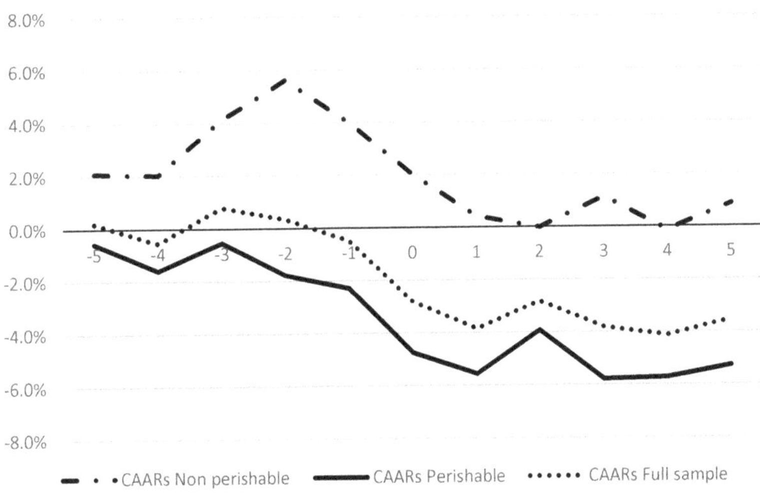

Figure 8. CAARs for the event window (−5,+5).

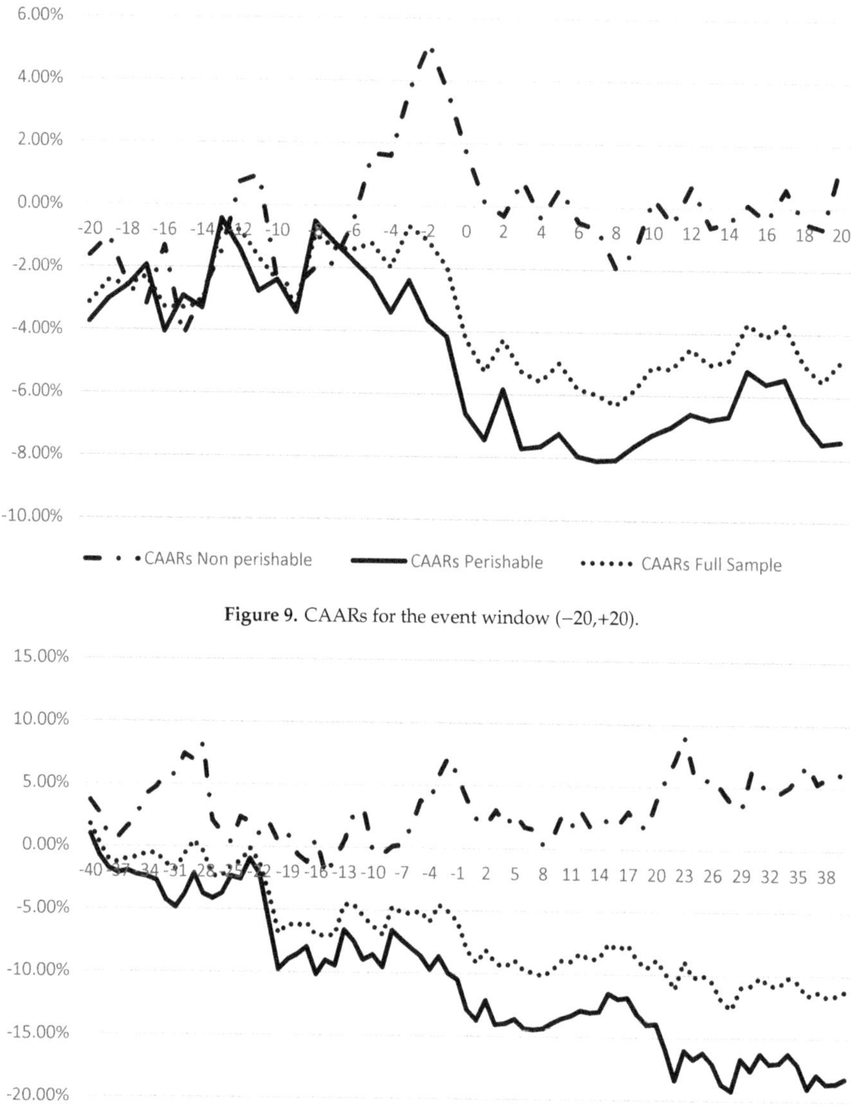

Figure 9. CAARs for the event window (−20,+20).

Figure 10. CAARs for the event window (−40,+40).

From a statistical point of view, in order to compare the different impact of the drought event in the two subsample distributions we focus on the comparison of means. The two-sample t−test [55] was then used to determine whether the two-sample means are equal or not. The null hypothesis is defined as H_0: $\mu_1 - \mu_2 = 0$, indicating that the difference between the means, in terms of CAARs, of the two subsamples (nonperishable versus perishable one) is equal to 0. On the contrary, the alternative hypothesis states H_1: $\mu_1 - \mu_1 \neq 0$. In Table 7, we summarize the results of the t-test.

Table 7. Summary of T-test on CAARs subsamples.

$E(-5,+5)$	N	Mean	StDev	SE Mean	t-Value	DF	*p*-Value
CAARs nonperishable	11	0.0201	0.0185	0.0056	6.42	19	0.000
CAARs perishable	11	−0.0344	0.0211	0.0064			
$E(-20,+20)$	N	Mean	StDev	SE Mean	t-Value	DF	*p*-Value
CAARs nonperishable	41	−0.0029	0.0189	0.003	9.42	75	0.000
CAARs perishable	41	−0.0482	0.0243	0.0038			
$E(-40,+40)$	N	Mean	StDev	SE Mean	t-Value	DF	*p*-Value
CAARs nonperishable	81	0.0315	0.0245	0.0027	20.17	109	0.000
CAARs perishable	81	−0.1056	0.0561	0.0062			

Since the *p*-value is 0.000, which is less than the significance level of 0.05, we can reject the null hypothesis and conclude that the CAARs of the two subsamples are statistically different for each of the event windows. Note that the mean of CAARs for the perishable subsample remains negative within the three event windows, whereas the mean of CAARs for the nonperishable subsample appears to be positive in two of the three event windows, that is, in $E(-5,+5)$ and $E(-40,+40)$. We can also infer this idea by using the following charts (see Figures 11–16):

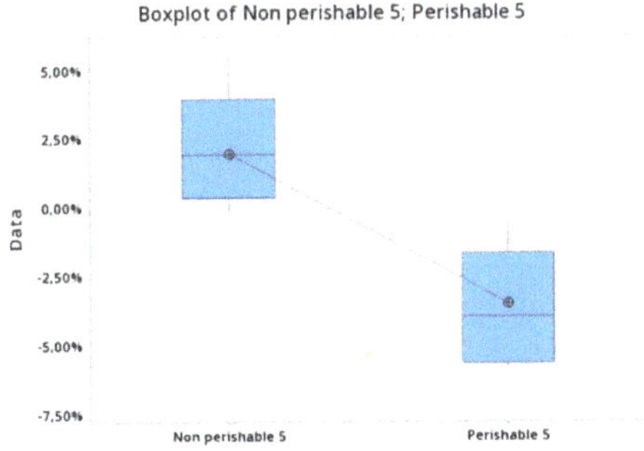

Figure 11. Individual plot CAARs event window $E(-5,+5)$.

Figure 12. Boxplot CAARs event window $E(-5,+5)$.

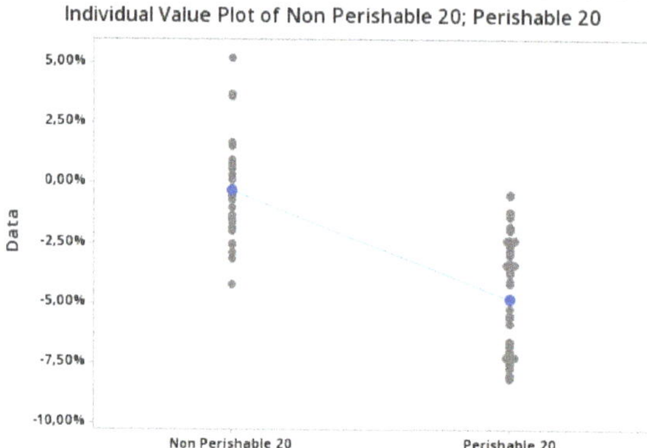

Figure 13. Individual plot CAARs event window $E(-20,+20)$.

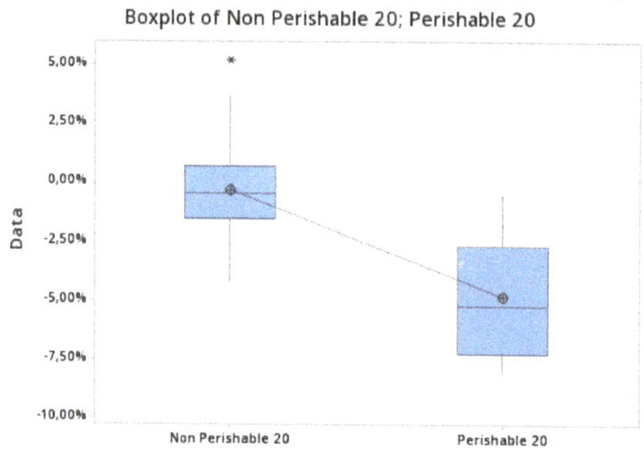

Figure 14. Boxplot CAARs event window $E(-20,+20)$.

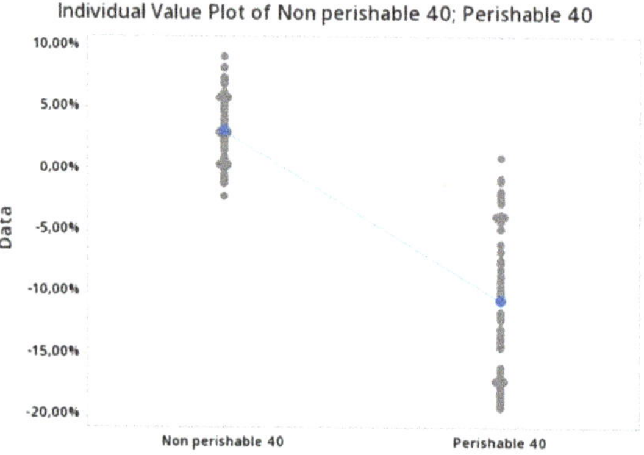

Figure 15. Individual plot CAARs event window $E(-40,+40)$.

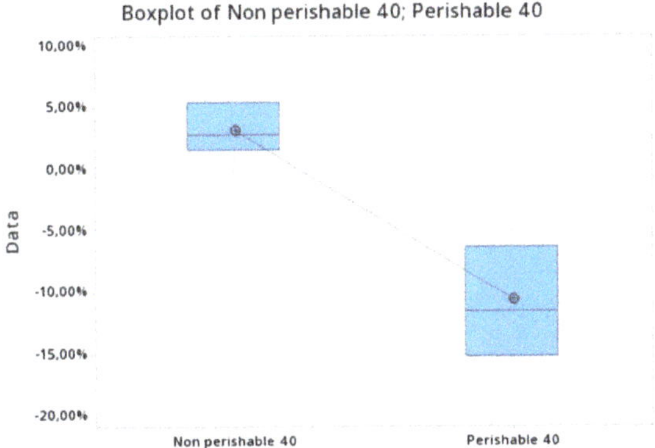

Figure 16. Boxplot CAARs event window $E(-40,+40)$.

7. Conclusions

In this paper, we examine how sensitive is the BOVESPA stock market to the worst drought occurred in the last 100 years in Brazil, particularly analyzing the impact of the official announcement on a sample of seven main agri-food listed companies. For this purpose, we carry out a standard short-term event study by using three event windows: ±40, ±20 and ±5 days. After applying both parametric and nonparametric tests, we confirm the existence of statistically significant negative cumulative abnormal returns around the drought official announcement within the three event windows. The highest impact was obtained for the narrowest window, that is, five days before and after the event date, around twice the decline observed for wider windows. Particularly, we observe daily abnormal returns under this mean in three of the companies analyzed, which operate in the frozen food sector, whereas the rest of firms register significant abnormal returns above this threshold. In other words, the impact of the drought announcement affects even more negatively those companies that sell perishable products versus nonperishable ones. To our knowledge, the literature on the effects of droughts on financial markets is almost nonexistent, so this paper contributes to enrich the existing literature of natural disasters (earthquake, tsunamis, hurricanes, floods, etc.). In this regard, we highlight the sensitivity of the BOVESPA stock market to the hydrological risk as well as the complexity of the consequences of climate change in terms of risk perceptions for investors. The results of this study point out the negative impact in terms of cumulative abnormal returns suffered by the agri-food firms as evidence to the next droughts that can be approximate and even intensified in the mid-term. So, shareholders and investors should be aware of the specific risk they face when investing in the agri-food sector, not only in Brazil but also in other Latin American countries, where there is a high degree of probability to see increasing drought severity and extent over the next 30 to 90 years. As with the majority of studies, the current study is subject to limitations since we exclusively focus on the short-term effects of the official drought announcement. In order to capture longer-term effects of the drought, alternative methodologies could be applied, such as the buy-and-hold abnormal return (BHAR) approach and the Jensen's alpha approach, so this limitation is seen as an opportunity to be addressed in a future research.

Author Contributions: Conceptualization, J.M.F.-D.; Data curation, F.d.l.P.; Formal analysis, P.P.; Investigation, J.M.F.-D. and P.P.; Software, J.M.F.-D.; Supervision, J.M.F.-D. and P.P. All authors have read and agreed to the published version of the manuscript.

Funding: This work has been funded by Andalusia's Regional Government (PAIDI-SEJ141).

Conflicts of Interest: The authors declare no conflict of interest. The founding sponsors had no role in the design of the study; in the collection, analyses, or interpretation of data; in the writing of the manuscript, and in the decision to publish the results.

References

1. Planton, S.A., III. Glossary. In *Climate Change 2013: The Physical Science Basis. Contribution of Working Group I to the Fifth Assessment Report of IPCC the Intergovernmental Panel on Climate Change*; Stocker, T.F., Qin, D., Plattner, G.K., Tignor, M.M., Allen, S.K., Boschung, J., Nauels, A., Xia, Y., Bex, V., Midgley, P.M., Eds.; Cambridge University Press: Cambridge, UK, 2013.
2. Nations, U. United Nations Framework Convention on Climate Change. Available online: http://unfccc.int/resource/docs/convkp/conveng.pdf (accessed on 27 August 2020).
3. INDES. Recursos Educativos Abiertos Fundamentos del Cambio Climático. Available online: https://cursos.iadb.org/es/indes-catalogo/fundamentos-del-cambio-clim-tico (accessed on 24 June 2019).
4. Parry, M.L.; Rosenzweig, C.; Iglesias, A.; Livermore, M.; Fischer, G. Effects of climate change on global food production under SRES emissions and socio-economic scenarios. *Glob. Environ. Chang.* **2004**, *14*, 53–67. [CrossRef]
5. UNEP. United Nations Environment Programme. Available online: https://www.unep.org/news-and-stories (accessed on 29 July 2019).
6. World Bank; Global Facility for Disaster Risk and Recovery (GFDRR). *Preventive Resettlement of Populations at Risk of Disaster: Experiences from Latin America*; World Bank: Washington, DC, USA, 2011; p. 121.
7. Wilhite, D.; Pulwarty, R. *Drought and Water Crises Integrating Science, Management, and Policy*, 1st ed.; CRC Press: Boca Raton, FL, USA, 2018; ISBN 9781315265551.
8. Mishra, A.K.; Singh, V.P. A review of drought concepts. *J. Hydrol.* **2010**, *391*, 202–216. [CrossRef]
9. Vargas, J.; Paneque, P. Methodology for the analysis of causes of drought vulnerability on the River Basin scale. *Nat. Hazards* **2017**, *89*, 609–621. [CrossRef]
10. Quiring, S.M. Developing Objective Operational Definitions for Monitoring Drought. *J. Appl. Meteorol. Climatol.* **2009**, *48*, 1217–1229. [CrossRef]
11. Paneque, P. Drought Management Strategies in Spain. *Water* **2015**, *7*, 6689–6701. [CrossRef]
12. Wilhite, D.A.; Glantz, M.H. Understanding: The Drought Phenomenon: The Role of Definitions. *Water Int.* **1985**, *10*, 111–120. [CrossRef]
13. Crausbay, S.D.; Ramirez, A.R.; Carter, S.L.; Cross, M.S.; Hall, K.R.; Bathke, D.J.; Betancourt, J.L.; Colt, S.; Cravens, A.E.; Dalton, M.S.; et al. Defining ecological drought for the 21st century. *Bull. Am. Meteorol. Soc.* **2017**, *98*, 2543–2550. [CrossRef]
14. Kallis, G. Droughts. *Annu. Rev. Environ. Resour.* **2008**, *33*, 85–118. [CrossRef]
15. World Bank. *Turn Down the Heat Why a 4 C Warmer World Must Be Avoided*; World Bank: Washington, DC, USA, 2012; ISBN 9781464800535.
16. Gutiérrez, A.P.A.; Engle, N.L.; De Nys, E.; Molejón, C.; Martins, E.S. Drought preparedness in Brazil. *Weather Clim. Extrem.* **2014**, *3*, 95–106. [CrossRef]
17. House of Commons Treasury Committee. *Climate Change and the Stern Review: The Implications for Treasury Policy*; House of Commons Treasury Committee: London, UK, 2007.
18. Bleda, M.; Shackley, S. The dynamics of belief in climate change and its risks in business organisations. *Ecol. Econ.* **2008**, *66*, 517–532. [CrossRef]
19. Collier, B.; Katchova, A.L.; Skees, J.R. Loan portfolio performance and El Niño, an intervention analysis. *Agric. Financ. Rev.* **2011**, *71*, 98–119. [CrossRef]
20. BOVESPA Mercado de Valores de Brasil. Available online: http://www.b3.com.br/en_us/ (accessed on 24 June 2019).
21. Mazzocchi, M.; Montini, A. Earthquake effects on tourism in central Italy. *Ann. Tour. Res.* **2001**, *28*, 1031–1046. [CrossRef]
22. Ewing, B.T.; Hein, S.E.; Kruse, J.B. Insurer Stock Price Responses to Hurricane Floyd: An Event Study Analysis Using Storm Characteristics. *Weather Forecast.* **2006**, *21*, 395–407. [CrossRef]
23. Ramiah, V. Effects of the Boxing Day tsunami on the world capital markets. *Rev. Quant. Financ. Account.* **2013**, *40*, 383–401. [CrossRef]
24. Roll, R. The International Crash of October 1987. *Financ. Anal. J.* **1988**, *44*, 19–35. [CrossRef]
25. Rietz, T.A. The equity risk premium a solution. *J. Monet. Econ.* **1988**, *22*, 117–131. [CrossRef]
26. Barro, R.J. Rare Disasters and Asset Markets in the Twentieth Century. *Q. J. Econ.* **2006**, *121*, 823–866. [CrossRef]

27. Bird, M.; Cowie, S.; Hawkes, A.; Horton, B.; Macgregor, C.; Ong, J.E.; Hwai, A.T.S.; Sa, T.T.; Yasin, Z. Indian Ocean Tsunamis: Environmental and Socio-Economic Impacts in Langkawi, Malaysia. *Geogr. J.* **2007**, *173*, 103–117. [CrossRef]

28. Bandara, J.S.; Naranpanawa, A. The Economic Effects of the Asian Tsunami on the 'Tear Drop in the Indian Ocean': A General Equilibrium Analysis. *South Asia Econ. J.* **2007**, *8*, 65–85. [CrossRef]

29. Feria-Domínguez, J.M.; Jiménez-Rodríguez, E.; Fdez-Galiano, I.M. Financial perceptions on oil spill disasters: Isolating corporate reputational risk. *Sustainability* **2016**, *8*, 1090. [CrossRef]

30. Antony, P.; de Pieter, J. The effect of ISO 14001 certification announcements on stock performance. *Int. J. Oper. Prod. Manag.* **2011**, *31*, 765–788. [CrossRef]

31. Keele, D.M.; DeHart, S. Partners of USEPA Climate Leaders: An Event Study on Stock Performance. *Bus. Strateg. Environ.* **2011**, *20*, 485–497. [CrossRef]

32. Mintzberg, H. Patterns in Strategy Formation. *Manag. Sci.* **1978**, *24*, 934–948. [CrossRef]

33. Xu, X.D.; Zeng, S.X.; Tam, C.M. Stock Market's Reaction to Disclosure of Environmental Violations: Evidence from China. *J. Bus. Ethics* **2012**, *107*, 227–237. [CrossRef]

34. Mueller, B.; Mueller, C. *The Economics of the Brazilian Model of Agricultural Development.*; The Global Development Institute: Manchester, UK, 2014.

35. Ball, R.; Brown, P. An Empirical Evaluation of Accounting Income Numbers. *J. Account. Res.* **1968**, *6*, 159–178. [CrossRef]

36. Fama, E.; Fisher, L.; Jensen, M.; Roll, R. The adjustment of stock prices to new information. *Int. Econ. Rev. (Philadelphia).* **1969**, *10*, 1–21. [CrossRef]

37. Fama, E.F. Efficient capital markets: A review oF theory and empirical work. *J. Financ.* **1970**, *25*, 383–417. [CrossRef]

38. Sharpe, W.F. A Simplified Model for Portfolio Analysis. *Manag. Sci.* **1963**, *9*, 277–293. [CrossRef]

39. de Jong, F. *Event Studies Methodology*; Lecture Notes Written for the Course Empirical Finance and Investment Cases; Academia.edu: San Francisco, CA, USA, 2007.

40. Fiordelisi, F.; Soana, M.-G.; Schwizer, P. The determinants of reputational risk in the banking sector. *J. Bank. Financ.* **2013**, *37*, 1359–1371. [CrossRef]

41. Cummins, J.D.; Lewis, C.M.; Wei, R. The market value impact of operational loss events for US banks and insurers. *J. Bank. Financ.* **2006**, *30*, 2605–2634. [CrossRef]

42. Maynes, E.; Rumsey, J. Conducting event studies with thinly traded stocks. *J. Bank. Financ.* **1993**, *17*, 145–157. [CrossRef]

43. Strong, N. Modelling abnormal returns: A review article. *J. Bus. Financ. Account.* **1992**, *19*, 533–553. [CrossRef]

44. Engmann, S.; Couisenau, D. Comparing distributions: The two-sample Anderson–Darling test as an alternative to the Kolmogorov–Smirnov test. *J. Appl. Quant. Methods* **2011**, *6*, 1–17.

45. Arnold, S. *Mathematical Statistics*; Prentice-Hall: Englewood Cliffs, NJ, USA, 1990; ISBN 0135610516, ISBN 9780135610510, ISBN 0135630991, ISBN 9780135630990.

46. Casella, G.; Berger, R.L. *Statistical Inference*; Wadsworth & Brooks/Cole: Pacific Grove, CA, USA, 1990.

47. Moore, D.S.; McCabe, G.P. *Introduction to the Practice of Statistics*; W.H. Freeman and Company: New York, NY, USA, 1993.

48. Srivastava, A.B.L. Effect of Non-Normality on the Power Function of t-Test. *Biometrika* **1958**, *45*, 421–430. [CrossRef]

49. Campbell, C.J.; Wasley, C.E. Measuring abnormal daily trading volume for samples of NYSE/ASE and NASDAQ securities using parametric and nonparametric test statistics. *Rev. Quant. Financ. Account.* **1996**, *6*, 309–326. [CrossRef]

50. Dixon, W.J.; Mood, A.M. The Statistical Sign Test. *J. Am. Stat. Assoc.* **1946**, *41*, 557–566. [CrossRef]

51. Sanger, G.C.; McConnell, J.J. Stock Exchange Listings, Firm Value, and Security Market Efficiency: The Impact of NASDAQ. *J. Financ. Quant. Anal.* **1986**, *21*, 1. [CrossRef]

52. Cowan, A.R.; Sergeant, A.M.A. Trading frequency and event study test specification. *J. Bank. Financ.* **1996**, *20*, 1731–1757. [CrossRef]

53. Wilcoxon, F. Individual Comparisons by Ranking Methods. *Biom. Bull.* **1945**, *1*, 80–83. [CrossRef]

54. Siegel, S. *Nonparametric Statistics for the Behavioral Sciences*; McGraw-Hill: New York, NY, USA, 1956; ISBN 0070856893, ISBN 9780070856899, ISBN 0070573484, ISBN 9780070573482.
55. Snedecor, G.; Cochran, W. *Statistical Methods*; Iowa State University Press: Ames, IA, USA, 1989.

Publisher's Note: MDPI stays neutral with regard to jurisdictional claims in published maps and institutional affiliations.

 *water*

Article

Research on the Impact of Tourism Development on the Sustainable Development of Reservoir Headwater Area Using China's Tingxi Reservoir as an Example

Chih-Chien Shen [1], **Chou-Fu Liang** [2,*], **Chin-Hsien Hsu** [3], **Jung-Hul Chien** [4] and **Hsiao-Hsien Lin** [3,*]

1. Institute of Physical Education and Health, Yulin Normal University, Yulin 537000, China; g169168@gmail.com
2. School of Environmental and Life Sciences, Nanning Normal University, Nanning 530001, China
3. Department of Leisure Industry Management, National Chin-Yi University of Technology, Taichung 41170, Taiwan; hsu6292000@yahoo.com.tw
4. Department of Social Work, Toko University, Chiayi 61249, Taiwan; chien0106@yahoo.com.tw
* Correspondence: Lcf66131@yahoo.com.tw (C.-F.L.); chrishome12001@yahoo.com.tw (H.-H.L.)

Received: 19 October 2020; Accepted: 21 November 2020; Published: 25 November 2020

Abstract: The purpose of this study was to understand the impact of tourism development on the sustainable development of Tingxi Reservoir. Based on tourism impact theory, 804 questionnaires were statistically validated and analyzed, followed by a semi-structured interview with five respondents, and finally examined by a multivariate verification method. The study found that not only did development fail to raise land and housing prices, develop leisure activities, improve medical facilities, and supplement police manpower, but it also increased consumer costs and environmental damage. There were also problems such as insufficient interpreters, parking and rest facilities, and ineffective management of communication channels, bicycle facilities, and tourist waste, which did not help youths to return to their hometowns. Furthermore, due to the disparities in the performance of leisure opportunities, medical and health care, spatial planning, and cultural development, there were different opinions among the stakeholders. Suggestions: (1) Satisfy the needs of different stakeholders; (2) Improve the environmental literacy of tourists and provide more garbage cans; (3) Develop additional scenic spots to divert tourists; (4) Stabilize prices and attract investment from enterprises; and (5) Increase the participation of residents in community development to supplement industrial manpower.

Keywords: multifunctional water source area; ecotourism, people with different stakeholders; balanced decision-making

1. Introduction

Tingxi Reservoir was built in 1956 as an artificial lake in Tong'an District, Xiamen City, Fujian Province, China. In addition to providing water storage and irrigation, the surrounding area features diverse natural ecology, rich catches, numerous cultural sites, Song dynasty porcelain, and historical relics [1], so the nearby villages are striving to build and develop tourism resources in order to improve development and realize the goal of revitalizing the village economy. The area has now become a lake and water resource area with a variety of functions including water storage, drinking water, flood control, power generation, irrigation, aquaculture, art, culture, archaeology, fishery, livestock, hot springs, and forestry resources [2], as shown in Figure 1. It is estimated that it attracts 777,800 visitors each year, and it has set a record of 1,013,300 visitors in seven days, generating US$34,146,000 in revenue [3], which shows the effectiveness of tourism development.

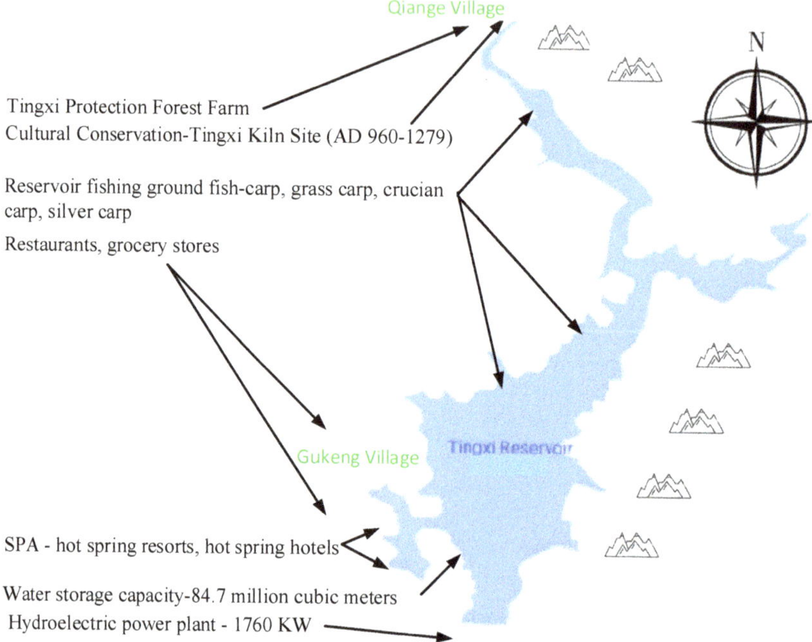

Figure 1. Distribution map of the tourism resources at Tingxi Reservoir.

However, policy decisions affect the direction and effectiveness of tourism development [4,5]. While local development can bring positive improvements in the living standards of the people and environmental sanitation, it can also have negative effects [6,7], destroying existing environments or cultures. In addition, tourism development can bring about changes. Overall, the impact of development can be explored from three levels: economic, social, and environmental [8,9].

The construction of the reservoir source area is not only to provide livelihood and industrial development, but also to bring economic improvement to the local village after proper planning. However, changes in the economic level may have positive and negative effects. Basically, we can get a comprehensive answer from the perspective of the cost of living, business development, economic infrastructure, village development, etc. [10,11]. However, researchers believe that economic development can promote business interaction, increase local tax revenue, and provide funds for improving local public facilities. Therefore, we focused on the cost of living, industrial construction, and community development as the main direction of investigation, and then carried out an in-depth exploration of the development status of local medical care, employment, wages, consumption, construction, industry, facilities, prices, concessions, sanitation, cultural and creative industries, rewards, leisure activities, community feedback, and policy coordination [10–13], so that we can understand the actual extent of changes caused by development to the economy.

The construction of a reservoir is tantamount to altering the appearance of an existing ecological and human habitat on a large scale. Although moderate destruction can create better living conditions and bring about social stability, the rights of the original villagers to survive will be sacrificed. According to previous studies, sustainable development should take into account the needs and expectations of the local villages and people. Therefore, the impact of the reservoir development on the society is relatively diversified, and we should be able to understand the needs of the local people by looking at tourism facilities, community development, living atmosphere, culture and customs, fire prevention and security [10,14], etc. However, the unique culture and customs of local villages provide an image of stability to the people, so that they can live in a pure and simple manner, stabilizing their living conditions and reducing the occurrence of security incidents. Furthermore, local development is based on the premise of improving the local community and people's lives to create a better living environment. Therefore, with the village development, living atmosphere as well as culture and security as the main directions of investigation, we then looked into issues such as local reputation, quality of services and activities, policy participation, tourism organization planning, tourism indicators, cultural and architectural characteristics, security maintenance, community construction, interaction among people, and youth returning to their hometowns [14–17], through which we believed that we could see how development has changed the villages.

Reservoir development changes the existing natural environment and ecology to achieve the original purpose of reservoir construction and improve people's livelihood and economy, but at the same time, it affects the entire society and creates a conflict between local social benefits and losses, but more importantly, it changes the original natural and ecological environment. Therefore, from the analysis of tourism facilities and natural ecology [10,11,14], it should be possible to get a full picture of the impact of development on the environment. However, development not only causes changes in the appearance of the water source area, but also in the current state of local human and ecological development. Furthermore, construction is a type of change and should be a change for the better, to better meet the needs of human beings and to make the existing natural landscape and ecology more sustainable. Therefore, from the perspective of conservation measures, leisure environment, tourism facilities, landscape and ecological environment, we can look into issues such as public transportation, parking and open space, Internet communication, monuments and buildings, residents' environmental awareness, visitors' environmental quality, garbage volume, forest and ecological habitat, motor vehicle fumes, water source, and air quality [14–21], which is helpful to understand the impact of development on the existing environment.

The current development of water sources is moving toward a multifunctional development and management concept based on sustainable development. However, change can be tangible or intangible, and with the accumulation of time [12,13], the most realistic feelings of long-time residents can be recorded with their eyes and body perceptions [10–15]. As a matter of fact, however, it is not objective to describe the merits of development decisions from a single perspective, although the residents' perceptions may present the most realistic picture. This is because development not only aims at preserving the original functions and maintaining the existing appearance, but also aims at making the original facilities or resources more valuable and satisfying more human needs [11,22]. After all, the number of natural, socio-cultural, and economic resources needed to be obtained, and the level of consideration are different for different positions and roles [14,23]. For existing residents, development is expected to improve their quality of life by improving economic conditions, sanitation, transportation, and medical facilities, but they do not want to destroy their original living style and environment [24]. For tourists, water source areas are one of the most attractive tourism resources due to their ecological diversity and the differences in the customs and cultures of existing villages [11], and fulfilling their psychological needs through tourism [12,25] will satisfy tourists and strengthen their tourism and consumption behaviors. It can be seen that both the residents and tourists expect the development of water resource areas to be effective, but they are afraid of the gap between the results and their expectations.

In order to strike a balance, the current tourism research and development should be considered from the perspective of both residents and tourists. As development is aimed at improving local conditions, and tourism development is aimed at attracting tourists and boosting the local economy [6], it takes time to prove the effectiveness of development, and policymakers need the assistance of residents in order to achieve success. Therefore, the most effective way to understand the development process and its effectiveness is to conduct a review with the residents as the subject [6,26]. However, development is the betterment of the existing predicament, while tourism development hopes to attract tourists to travel and spend, in order to achieve the purpose of boosting the economy, increasing people's income, raising revenue, and improving the quality of life of the villages and people [24]. Therefore, the more time tourists spend on in-depth tourism, the more the desire and opportunities for consumption can be stimulated. It shows that the effectiveness of tourism development should be based on the needs and feelings of residents and tourists [11,14].

Taken together, a certain degree of tourism development in Tingxi Reservoir can satisfy the needs of both residents and visitors, but it can also lead to rejection. By conducting a survey based on environmental issues such as the community environment, the villagers' environmental literacy, preservation of historical sites, awareness of ecological conservation, coordination of conservation policies, maintenance of tourist trails and bicycle lanes, tourism transportation planning, Wi-Fi network speed, bicycle rental, community modernization and scale, rest and parking space, experience of tourism activities, water and air quality, fumes from steam and locomotives, mountain slope development, etc. can be determined [14–20]. Social issues can be identified such as tourism visibility, quality of services and activities, content of community activities, friendly treatment of village culture, tourism indicators or descriptions, tourism and leisure facilities, human resources, DIY activities, hardware and software facilities, tourism environment and space quality, indigenous cultural traditions and historical relics, the image of tourism companies or organizations, the promotion of traditional cultural activities, the interaction mechanism between villagers and tourists, the number and popularity of traditional culture and characteristic industries, the management and safety maintenance of activities, the allocation of service or management personnel, the sense of travel security, the willingness to travel again, etc. [10,14–17]. Economic issues such as employment and entrepreneurship opportunities, tourism activity prices, tourism consumption costs, increased tourism facilities and local characteristic industrial products, tourism diversification, provision of explanatory guides, increase leisure and life consumption options, obtain promotion or priority use rights, and expand community tourism. Through the scale of development, the quality of public facilities construction and maintenance, the quality of medical and health services, the communication channels between the community and the government, the protection policies of the local tourism industry, the mechanisms and norms involved in formulating tourism development policies, the development of DIY or product portfolios, etc., it should be possible to obtain the feelings of both sides and to explore the most in-depth issues to obtain a balanced view for future decision-making [10,11,14,24].

Until now, however, the majority of reservoir and lake studies have been conducted from the perspective of residents [27–29], while fewer have been conducted from the perspective of tourists [30–32], and even fewer have been conducted from the perspective of both residents and tourists [33,34]. Most of the studies that focus on Tingxi Reservoir are about water quality and ecology, water flow, water quality, and power generation [35–37], and almost no scholars have focused on tourism issues. Therefore, the researchers believed that by examining the current tourism development of Tingxi Reservoir from the perspective of residents and tourists, the latest and most realistic answers could be obtained for the reference of local organizations and people. The purpose of this study was to: (1) understand residents' awareness of the current tourism development of Tingxi Reservoir; (2) explore tourists' awareness of the current tourism development of Tingxi Reservoir; and (3) analyze the difference in awareness between residents and tourists on the tourism development of Tingxi Reservoir.

2. Methods and Instruments

2.1. Study Framework and Hypotheses

The literature finds that most studies on reservoirs and lakes are conducted from the perspective of residents [27–29], while the research conducted from the perspective of tourists [30–32] is rarely conducted, from the perspective of residents and tourists. There is even less discussion [33,34]. Most of the research on Tingxi Reservoir involves water quality and ecology, water flow, water quality and power generation [35–37], and almost no scholars pay attention to tourism issues. Therefore, the researchers believe that the latest development results and dilemmas can be obtained by examining the tourism development status of Tingxi Reservoir from the perspective of residents and tourists.

We sought to collect the experiences of two different groups including residents and tourists on the tourism development of Tingxi Reservoir, point out the current problems, and make suggestions for improvement. The research methods and tools were determined according to the existing information of the case and the relevant literature on the reservoir [1,2,35,36] and the lake and other tourism development issues [3–34]. By applying tourism impact theory and combining the opinions of residents and visitors [10,11,14,16,21,38], we used methods of surveys, interviews, and observations to collect research information [39], and then by comparing and verifying the data [40], we used induction, organization, and analysis sequences to construct this paper [16], in order to obtain correct and reasonable information to revise the development plan of Tingxi Reservoir. As shown in Figure 2.

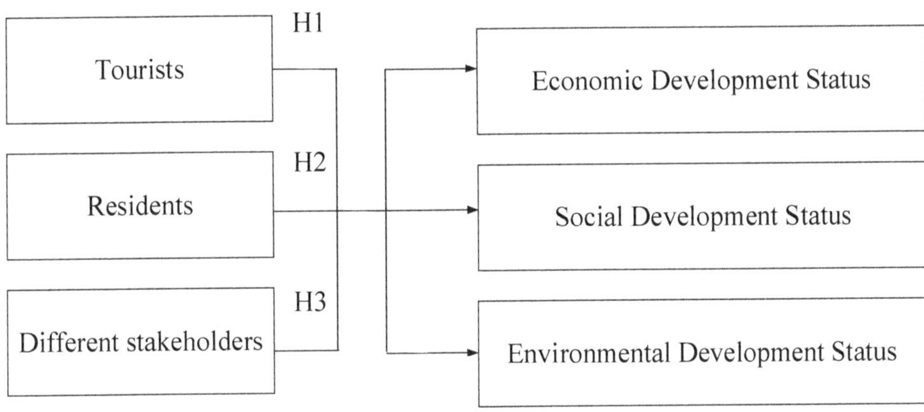

Figure 2. Study framework.

According to the research framework, we propose the following hypothesis:

Hypothesis 1. *There is consistency in the recognition of current economic development among different stakeholders.*

Hypothesis 2. *There is consistency in the recognition of current social development among different stakeholders.*

Hypothesis 3. *There is consistency in the recognition of the current status of environmental development among the different stakeholders.*

2.2. Study Procedure and Instruments

Tourism development needs to be considered in a multi-faceted way and the impact of tourism development is more complicated nowadays [6,10,12]. More detailed answers can be obtained through comprehensive surveys [10,11,14,24,28]. Therefore, the authors first reviewed the theories and literature on tourism development [3–32], interpreted the research results [38,39], compiled a 60-question questionnaire on the current status of tourism development, and then discussed with experts and scholars to complete the outline of the questionnaire, which was divided into two parts: background information and tourism development. In addition to the background information on issues such as identity, gender, age, and occupation, tourism development included 57 issues: economy (15), society (21) and environment (21), as shown in Table 1.

Table 1. Initial questionnaire issue preparation.

Level	Issue Content	Number of Questions
Background	Identity, gender, age	1–3
Economic	Increase employment opportunities Enhance in consumer prices for tourist activity content Enhance in tourism consumption expenditure costs Tourism development to increase tourism facilities and local characteristic industrial products More diversified tourism industry categories Provide interpretive guides during travel or services for the use of amusement facilities Increase the choice of tourist, leisure and life consumption opportunities Get opportunities for promotion or preferential use of local tourist facilities Enhance scale of community tourism development Public facilities and construction and maintenance quality promotion Tourist consumption health and service quality promotion Establish communication channels between the community and the government Establish protection policies for local tourism development, to prevent monopoly by consortia Mechanisms and norms that can participate in the formulation of tourism development policies Develop DIY or product portfolio or creative products to enhance tourists' willingness to spend	4–18
Society	Enhance tourism visibility Enhance the quality of tourism services and activities Enhance understanding of the content of community sightseeing activities More treasure the community environment of tourist destinations Sufficient introduction to local tourism related indicators or descriptions Provide more choices of local tourist and leisure facilities More young people are employed locally Increased opportunities to participate in local DIY activities Improving the establishment of hardware and software facilities in local communities Tourism environment and space quality enhance Indigenous cultural traditions and historical sites are preserved Good image of local tourism-related industry business or organizations Intensive promotion and development of traditional cultural activities To feel friendly and trust the local residents Nice interaction among tourists, and between tourists and residents Increased number and visibility of traditional culture and industries tourist activity management and safety maintenance quality enhance Sufficient number of fireman and police, security officer Have a sense of security during leisure travel Willing to travel here again	19–39

Table 1. *Cont.*

Level	Issue Content	Number of Questions
Surroundings	The community's natural environment is maintained cleaner The environmental literacy concept of the villagers is good, do not throw away trash Historic buildings are preserved Willing to attach importance to and participate in the conservation of local natural ecology Willing to cooperate with environmental protection announcement slogans or manuals, care for the environment The tourist trail has a complete appearance and smooth movement The bicycle lane facilities are well preserved and circulation is best Public transportation construction is helpful for personal travel round trip Wireless Wi-Fi online provides tourist and travel information inquiry It is convenient to rent bicycle facilities and location planning To feel smooth when moving in transportation or round trip Increase in the size of community construction area and the number of facilities Sufficient planning of parking and pavilion leisure facilities Tourist will affect the quality of the local natural environment Increase leisure and experience space in tourist tours The local water source is polluted Local air quality is nice Good quality of water for people's livelihood in tourist The development area of hillside vegetation and forest land in the village increased The artificial landscape area increased Emission intensive of fumes from car and motorcycle during tourism	40–60

Except for background information, all issues were measured on a 5-point Likert-type scale, with 1 indicating strongly disagree and 5 indicating strongly agree. The questionnaire was first compiled by referring to the relevant literature, then reviewed by three experts for content validity, and 50 questionnaires were distributed. The results were analyzed using SPSS 22.0 statistical software and then tested with statistical methods. When Kaiser–Meyer–Olkin (KMO) > 0.06 and the p-value of the Bartlett test was less than 0.01 ($p < 0.01$), the scale was suitable for continuous factor analysis [40]. A coefficient α greater than 0.60 indicates that the questionnaire has good reliability [41]. The results of the analysis are shown in Table 2.

Table 2. Analysis of the Tingxi Reservoir Tourism Development Perception Questionnaire.

Construct	Subfacet	Issues	Cronbach's α
Economic	People's livelihood price	1. Increase employment opportunities 2. Enhance in land and house prices 3. Enhance in expenditure costs	0.618–0.654
	Industry construction	1. Combination of local characteristic industries 2. Increase tourism industry 3. Increase interpretation facilities 4. Increase leisure opportunities 5. Preferential tourism facilities 6. Increase tourism construction	0.809–0.815
Society	Community development	1. Maintain complete public facilities 2. Enhance medical and health standards 3. Establish community communication channels 4. Develop protection policy settings 5. Participate in tourism policy planning 6. Develop creative products	0.774–0.783
	Village construction	1. Enhance tourism visibility 2. Improve the quality of tourism services 3. Participate in community tourism affairs 4. Actively clean up the community environment 5. Sufficient instructions for local tourism 6. Increased selection of leisure facilities	0.756–0.786
	Atmosphere of life	1. Youth return to their hometowns for development 2. Industry to contribute to local development 3. Improve the living environment 4. Improve the quality of tourist activities 5. Protect the indigenous culture	0.699–0.728
Environment	Cultural safety	1. Best image of foreign consortia 2. Development of traditional cultural activities 3. Tourists feel friendly 4. Best interaction among residents 5. Invest in the indigenous cultural industry 6. Enhance community self-government management 7. Sufficient fireman and police, security officer 8. Have a sense of security in life 9. Willing to travel again or buy a property in the local area	0.847–0.856
	Conservation measures	1. Clean community environment 2. Do not throw away trash by tourists 3. Complete preservation of historical sites 4. Participation in nature conservation 5. Public environment awareness of environmental literacy	0.748–0.783
	Leisure environment	1. Complete tourists trails 2. Perfect bicycle lane management 3. Public transportation facilitates tourism 4. Wi-Fi online coverage 5. Cheap bicycle rental 6. Complete transportation facilities	0.781–0.800
	Tourist facility	1. Increased facility construction area 2. Adequate parking and leisure facilities 3. Environmental quality affected by tourists 4. Adequate personal living space	0.675–0.715

Economy (15) had a KMO > 0.940, a Bartlett approximate χ^2 value of 2386.692, and a degree of freedom (df) of 105 with a significance of 0.000 ($p < 0.001$), making it suitable for factor analysis. Explained variances for the scale were 18.72%, 14.509%, and 14.491% for a total explained variance of 47.72%. All these were retained after factor analysis and taking into account the understanding of the actual state of economic development. The three areas were named: people's livelihood price (3), industrial construction (6), and community development (6). They contained a total of 15 questions and the three scales were 0.729, 0.838, and 0.810, respectively.

Society (21) had a KMO > 0.943, a Bartlett approximate χ^2 value of 3303.559, a degree of freedom (df) of 210, and a significance of 0.000 ($p < 0.001$), thus making it suitable for factor analysis. Explained variances of the scale were 15.187%, 15.094%, and 11.783%, for an overall explained variance of 42.064%. All these were retained after factor analysis and taking into account the understanding of the actual situation of social development. Three areas were named: village construction (6), atmosphere of life (5), and cultural safety (9). In total, they contained 21 questions with three scales of 0.807, 0.756, and 0.864, respectively.

The environment (21) had a KMO > 0.950, a Bartlett approximate χ^2 value of 3658.093, a degree of freedom (df) of 210, and a significance of 0.000 ($p < 0.001$), making it suitable for factor analysis. Explained variances in the scale were 16.495%, 14.142%, 9.245%, and 7.665%, respectively, for a total explained variance of 47.546%. All of these were retained after factor analysis and taking into account an understanding of the physical conditions of environmental development. The four areas were named: Conservation Measures (5), Leisure Environment (6), Tourism Facilities (4), and Landscape and Ecological Environment (6). They contained a total of 21 questions with three scales of 0.799, 0.817, 0.748, and 0.779, respectively.

Subsequently, the researcher went to the local area to conduct fieldwork in October 2019, but due to the impact of the COVID-19 epidemic, the questionnaire sample collection process and effectiveness were affected. The questionnaire was collected from November 2019 to May 2020. Initially, the questionnaires were collected on site by random sampling, and later, the questionnaires were collected through an online platform in a snowballing fashion, and a total of 804 valid questionnaires were collected. The data were collected and analyzed using the SPSS for Windows 22.0 statistical package in order to statistically check the reliability of the questionnaire issues and analyze the results using descriptive and t-test analysis. Interviews were subsequently conducted to supplement missing information. With the consent of the interviewees, a semi-structured design and open-ended interviews were conducted with five interviewees including tourism practitioners, residents, and academics who had experience in traveling to Tingxi Reservoir or had some knowledge of the current development of the reservoir. Researchers interacted by video, taking the results of the questionnaire analysis as the topic, in order to solicit their opinions on the results of the questionnaire. As shown in Table 3.

Table 3. Background information of the interviewees and outline of the interview.

Identity	Gender	Residence Time/Years of Work Experience	Identity	Gender	Residence Time/Years of Work Experience
The elderly	Male	25	Tourist guide	Male	40
The elderly	Female	30	Tourist guide	Female	25
Professor	Male	15			

Construct	Issues
Impact of tourism development	1. What impact does tourism development have on the economic, social, and environmental development of the community? 2. According to research and investigation, what causes the impact of economic, social and environmental issues?

After collecting the opinions of the interviewees, we recorded the interview content and then asked the interviewees to verify the accuracy of the recorded content. The information from the questionnaire was then integrated, and the results were analyzed, and the research paper was constructed with the order of induction, organization, and analysis [16]. The information from the questionnaire was then integrated, and the results were analyzed, and the research paper was constructed with the order of induction, organization, and analysis [16]. Finally, using the multivariate verification and analysis method and combining information from different research subjects, research theories and methods to examine multiple data from multiple viewpoints and to compare the results of various studies [38,40], accurate knowledge and meaning were obtained in order to examine the current situation of promoting tourism development in Tingxi Reservoir.

2.3. Study Scope and Limitations

The study aimed to investigate the current tourism development of Tingxi Reservoir by applying a mixed-method research approach using the reservoir as a location, surrounding villages as the range, and local residents and people who had traveled to the reservoir as subjects.

The initial phase of the study began in October 2019, but due to the extensive study area, the research team was unable to complete the sample collection immediately as there were limitations in manpower, resources, and funding during the study period, and the outbreak of COVID-19 in December 2019 further delayed the sample collection process, which took a total of seven months. Although the online questionnaire platform was adopted to collect the information, it was limited by the willingness of the respondents and their proficiency in using 3C products, which led to the shortcomings of the information collected by the researcher. Summing up the above explanations, it is unlikely that more comprehensive information can be obtained due to the limitations of the sample background. If this resulted in any discrepancy in the study, it will be taken into consideration for further study.

3. Analysis of Results

A total of 804 samples were obtained. Results showed that most of the sample subjects were residents (67.7%), while only 30.8% were tourists. Most of the samples were from females (75.9%), with males accounting for the least (24.1%). Most of the respondents were aged 21–30 (67.2%), followed by those under 20 years old (20.4%), and the least were aged 51 or above (1.7%).As shown in Table 4.

Table 4. Descriptive characteristics of the participants.

Identity			
Identity	**Percentage**	**Age**	**Percentage**
Residents	69.7%	Under 20	20.4%
Tourists	30.8%	21–30	67.2%
Gender	**Percentage**	31–40	6.7%
Male	24.1%	41–50	2.2%
Female	75.9%	51–60	1.7%
		Over 61	1.7%

3.1. Cognitive Analysis of Economic Development

The questionnaire was revised with reference to previous literature [3–32] where a score of 1 means strongly disagree and 5 means strongly agree. First, the perceptions of residents and tourists were investigated by statistical tests; next, the *t*-test was used to analyze the differences in perceptions among different stakeholders; then the respondents' perceptions were combined, and finally, a multivariate test was used to explore the differences [16,38,40].

The perceptions of the residents and tourists of the current economic development were analyzed, as shown in Table 5. Results showed that residents and visitors agreed only on increasing job opportunities and construction, land, and depressed prices, but disagreed on the rest.

Table 5. The perceptions of residents and tourists of the current economic development.

	Facets	Highest	M	Lowest	M
Residents	People's livelihood price	Increase employment opportunities	3.86	Land and housing prices are rising Enhanced expenditure costs	3.69
	Industry construction	Enhanced tourism construction	3.87	Increased leisure opportunities	3.68
	Community development	Participate in tourism policy planning	3.88	Enhanced medical and health standards	3.71
Tourists	People's livelihood price	Increase employment opportunities	4.04	Land and housing prices are rising Enhanced expenditure costs	3.74
	Industry construction	Enhanced tourism construction	3.96	Enhanced interpretation facilities	3.77
	Community development	Developing creative products	4.03	Establish community communication channels	3.78

The inference is that since the government has been actively improving the public infrastructure and utilizing the natural resources of the area to attract investment, it is hoped that this will bring business opportunities and create jobs. Although many villages have been preserved and are suitable for rural and eco-tourism activities, the hinterland of Tingxi Reservoir is large, the industries in the surrounding villages are highly similar, and the population is aging, resulting in poor overall development planning. As a result, residents believe that although there are more opportunities for employment, tourism construction, and participation in tourism development policies, the leisure options are low, the medical and health conditions are poor, and the consumption conditions have not improved.

The diverse nature of the local ecology has created abundant tourism resources and brought different travel experiences. With the existing resources and culture, the development of handicraft products and agricultural products could attract tourists to spend money. However, although the hinterland can be developed and the infrastructure upgraded, there is a shortage of manpower in the industry due to the obvious trend of aging. As a result, tourists believe that development can help increase job opportunities, develop creative products, increase leisure options, and maintain a low consumption level. Nevertheless, due to the sparse population, there is a lack of interpretive facilities, and communication channels in the scenic spots are poor.

In summary, residents and visitors agree that the government's interest in tourism development has resulted in improved local infrastructure, stimulated industrial operations, business opportunities for the villages, and increased employment opportunities, but perhaps imperfect overall development decisions have also resulted in low land prices and low consumer willingness in most areas. In addition, the aging population, shortage of manpower, lack of tourism facilities, and high similarity of industries have led to a divergence of views.

Further exploring the differences in the perceptions of the impact of economic development among different stakeholders, we found that there were significant differences ($p < 0.001$) in the issues of leisure opportunities and health care standards, indicating that different stakeholders had different views on the increase in leisure opportunities and the current development of health care standards, as shown in Table 6.

Table 6. An analysis of the differences in the cognition of the current economic development among different stakeholders.

	Issue	Residents		Tourists		T	*p*-Value
		M	SD	M	SD		
People's livelihood price	Increase in employment opportunities	3.86	0.841	4.04	0.724	4.792	0.029
	Land and housing prices are rising	3.69	0.939	3.74	0.809	7.901	0.005
	Enhanced expenditure costs	3.69	0.928	3.78	0.763	9.035	0.003
Industry construction	Local characteristic industry combination	3.71	0.946	3.85	0.797	9.692	0.002
	Enhanced tourism construction	3.75	0.854	3.91	0.788	6.143	0.014
	Enhanced interpretation facilities	3.72	0.774	3.77	0.840	0.079	0.779
	Increased leisure opportunities	3.68	0.963	3.96	0.747	23.521	0.000 *
	Tourist facility discounts	3.75	0.817	3.82	0.763	4.026	0.045
	Enhanced tourism construction	3.87	0.812	3.90	0.730	5.721	0.017
Community development	Complete maintenance of public facilities	3.78	0.794	3.91	0.742	3.777	0.053
	Enhanced medical and health standards	3.71	0.912	3.89	0.797	10.742	0.001 *
	Establish community communication channels	3.74	0.781	3.78	0.760	1.082	0.299
	Development and protection policy settings	3.75	0.830	3.86	0.773	3.131	0.078
	Participate in tourism policy planning	3.88	0.820	3.92	0.764	4.351	0.038
	Developing creative products	3.86	0.878	4.03	0.784	3.562	0.060

$* = p < 0.001$.

It is inferred that due to the ecological diversity of Tingxi Reservoir, the village is rich in tourism resources and agricultural specialties, but the similarity between existing tourism resources and industries is high, making it less attractive to tourists and unable to meet the changing tourism needs. In addition, due to the inconvenient transportation and the aging population, the local medical resources are only sufficient to meet the needs of tourists, and not residents. As a result, different stakeholders have different views on leisure opportunities and the current development of medical and health care facilities. As shown in Figure 3.

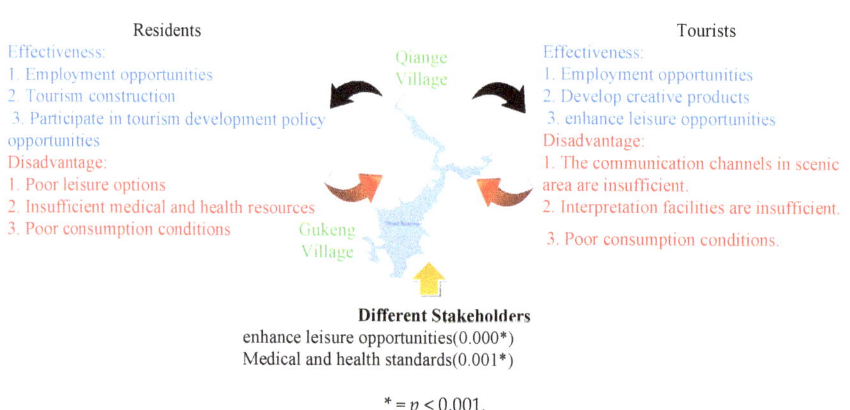

Figure 3. Perception analysis of the current economic development.

3.2. Cognitive Analysis of Social Development

The perceptions of residents and tourists on the impact of social development were first investigated separately, as shown in Table 7. The results showed that residents and tourists shared the same views on the residents' initiative to clean up the community environment, the youths' low willingness to return to their hometowns, and the shortage of police and fire safety personnel, but differed in all other aspects.

Table 7. The perceptions of the residents and tourists of the current social development.

	Facets	Highest	M	Lowest	M
Residents	Village construction	Participate in community tourism affairs	3.94	Increased choice of leisure facilities Actively organize the community environment	3.80
	Atmosphere of life	Improve the living environment	3.89	Youth back to the country development	3.73
	Culture safety	Feel safe in life	3.87	Sufficient fireman and police, security officer	3.68
Tourist	Village construction	Increased choice of leisure facilities	3.94	Actively sort out community environment	3.84
	Atmosphere of life	Industry gives back to local development	3.94	Youth back to the country	3.75
	Culture safety	Development of traditional cultural activities	3.93	Sufficient fireman and police, security officer	3.63

It is inferred that, since improving local conditions, enhancing safety and well-being, and meeting people's expectations are the development goals of the policy, combining the existing resources and residents' manpower to jointly promote tourism development, attract tourists to spend money, and promote the village economy is the expectation for the development of remote areas. However, the vast expanse of rural areas is not easy to develop and manage, resulting in a declining population, a lack of mobility for the elderly, and a shortage of human resources, creating a manpower gap in the industry, which affects investment intentions and reduces job opportunities. As a result, residents believe that the development will help improve their living environment, give them a sense of security, and increase their opportunities to participate in community tourism activities. However, residents are not willing to participate in community development, there are few leisure facilities to choose from, there is a lack of police and fire safety manpower, and young people are not willing to return to their hometowns for career development.

The natural ecological richness of the water source area makes the village very attractive to tourists. The government is willing to build tourism facilities and improve the local community environment to encourage investment from enterprises. However, in the early stages of development, job opportunities were scarce, forcing young people to look for jobs elsewhere. In addition, villagers rely on traditional industries to make a living, and the high workload, coupled with the aging of the population, makes it impossible to meet the demand for manpower for industrial development and community maintenance in villages or scenic areas. As a result, tourists believe that after development, the industry can give back to the local community and promote traditional cultural activities to increase the choice of recreational facilities, but the residents have a low awareness of community development, resulting in a low willingness of young people to return to their hometowns for development and a shortage of police and fire safety officers.

Further exploring the differences in perceptions of the impact of social development among different stakeholders revealed that there were significant differences ($p < 0.001$) on the issue of traditional cultural activities, indicating that different stakeholders had different views on the effectiveness of the development of traditional cultural activities, as shown in Table 8.

Table 8. Analysis of the differences in perceptions of social development status among different stakeholders.

Issue		Residents		Tourists		T	*p*-Value
		M	SD	M	SD		
Village construction	Enhance tourism visibility	3.92	0.862	3.91	0.823	3.113	0.078
	Improve the quality of tourism services	3.85	0.858	3.86	0.739	4.493	0.035
	Participate in community tourism affairs	3.94	0.827	3.87	0.743	1.903	0.169
	Actively sort out community environment	3.81	0.928	3.84	0.808	5.848	0.016
	Sufficient tourist pointer	3.84	0.909	3.79	0.766	4.175	0.042
	Increased choice of leisure facilities	3.81	0.833	3.94	0.761	4.393	0.037
Atmosphere of life	Youth back to the country	3.73	0.995	3.75	0.926	2.068	0.151
	Industry gives back to local development	3.79	0.865	3.94	0.752	5.079	0.025
	Improve the living environment	3.89	0.828	3.93	0.740	4.708	0.031
	Enhance the quality of sightseeing activities	3.80	0.909	3.92	0.793	7.156	0.008
	Protect the indigenous culture	3.88	0.845	3.91	0.811	0.654	0.419
Culture safety	best image of foreign business	3.75	0.890	3.76	0.741	7.019	0.008
	Development of traditional cultural activities	3.78	0.892	3.93	0.744	15.780	0.000 *
	Tourists feel friendly	3.79	0.828	3.79	0.779	2.352	0.126
	best interaction among residents	3.81	0.882	3.80	0.788	5.969	0.015
	Invest in unique cultural industries	3.82	0.878	3.88	0.720	9.088	0.003
	Community autonomy management rises	3.74	0.869	3.83	0.740	8.944	0.003
	Sufficient fireman and police, security officer	3.68	0.919	3.63	0.821	3.271	0.071
	Feel safe in life	3.87	0.812	3.80	0.739	2.932	0.088
	Willing to travel or buy a local	3.79	0.776	3.80	0.846	0.341	0.560

$* = p < 0.001.$

As a result of technological and civilizational advances and the evolution of lifestyles, it is inferred that traditional village culture and customs, which allow visitors to experience the characteristics of ancient civilizations, have become a special tourist image and attraction. However, the culture and customs of the villages have been with the inhabitants for their entire lives, and although they have a unique culture, they still need the construction and knowledge of modern civilization and technology to improve the quality of life of the inhabitants. This has led to different views on the effectiveness of traditional cultural activities among different stakeholders. As shown in Figure 4.

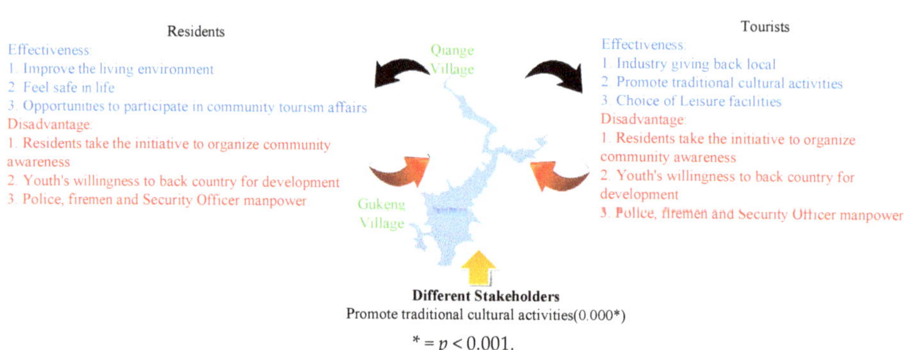

Figure 4. Perception analysis of current social development.

3.3. Cognitive Analysis of Environmental Development

The views of residents and tourists on the impact of environmental development were analyzed separately, as shown in Table 9, and it was found that residents and tourists only agreed on the wide coverage of a Wi-Fi network, good air quality, and not littering from the tourists, while the rest were different.

Table 9. Perception of the current state of environmental development by residents and visitors.

	Facets	Highest	M	Lowest	M
Residents	Conservation measures	Historic monuments are kept intact	3.95	Tourists do not throw away garbage	3.78
	leisure environment	Wide Wi-Fi online coverage	3.87	Bicycle rental is cheap	3.63
	Tourist facility	Environmental quality is affected by tourists	3.95	Sufficient personal living space	3.69
	Landscape and ecological environment	Best air quality	3.93	Pollution of water quality	3.63
Tourist	Conservation measures	Participate in nature conservation	3.97	Tourists do not throw away garbage	3.73
	leisure environment	Public transport helps travel Wide Wi-Fi online coverage	3.93	Perfect bicycle lane management	3.79
	Tourist facility	Sufficient personal living space	3.91	Adequate parking and leisure facilities	3.81
	Landscape and ecological environment	best air quality	3.90	Car and motorcycle oil fume pollution	3.67

It is inferred that the government's commitment to construction, improving the efficiency of the nation's Internet, advancing technology, civilization and quality of life as well as preserving the natural ecology and environment and protecting village culture and historical buildings has helped this area become a major tourist attraction. However, the influx of tourists attracts huge business opportunities, and local business owners try to inflate prices to seek huge profits; in addition, the hinterland is vast, construction funds are limited, and the existing parking facilities cannot meet the demand; moreover, the spending power of tourists has increased dramatically, but the quality is difficult to control, and the amount of garbage is large, which affects the environment. Therefore, residents believe that after development, historical monuments are preserved intact, Wi-Fi network coverage is wide, and the water and air quality is good, but the amount of garbage and environmental quality is affected by tourists, bicycle rental is expensive, and their living space is insufficient.

Since air quality, clear water sources, and a beautiful natural environment are the main appeal of Tingxi Reservoir, the government is willing to invest in upgrading Internet services and improving transportation. The residents are willing to work together to protect the natural ecology in order to attract tourists and improve the economic situation. However, crowds are booming, the amount of tourist waste is increasing, and the environmental literacy of tourists varies. In addition, due to the vast land area and the lack of funds for village construction, the existing bicycle facilities and parking space planning cannot meet the needs of a large number of tourists. As a result, tourists think that after development, there is a high level of participation in nature conservation, that public transportation facilitates tourism, the Wi-Fi network coverage is wide, personal living space is sufficient, there are few vehicles, air quality is good, but that tourists litter, bicycle lanes are not well managed, and parking rest facilities are insufficient.

The differences in perceptions of the impact of economic development among different stakeholders were found to be significant ($p < 0.001$), indicating that different stakeholders had different perceptions of the current status of living space planning and development, as shown in Table 10.

Table 10. Analysis of the differences in awareness of the current state of environmental development among different stakeholders.

Issue		Residents		Tourists		T	*p*-Value
		M	SD	M	SD		
Conservation measures	To clean community environment	3.91	0.792	3.88	0.769	0.163	0.687
	Tourists do not throw away garbage	3.78	0.869	3.73	0.757	3.736	0.054
	Historic monuments are kept intact	3.95	0.834	3.94	0.776	4.227	0.040
	Participate in nature conservation	3.92	0.786	3.97	0.760	1.512	0.220
	Public cognition of environmental literacy	3.87	0.772	3.92	0.746	0.387	0.534
Leisure environment	Complete tourist trail	3.86	0.854	3.84	0.762	1.814	0.179
	Perfect bicycle lane management	3.74	0.950	3.79	0.788	4.282	0.039
	Public transport helps travel	3.80	0.931	3.93	0.769	6.832	0.009
	Wide Wi-Fi online coverage	3.87	0.772	3.93	0.800	0.186	0.666
	Bicycle rental is cheap	3.63	0.882	3.88	0.800	4.883	0.028
	Perfect transportation line facilities	3.86	0.816	3.85	0.757	0.888	0.346
Tourist facility	Increased facility construction area	3.88	0.767	3.87	0.727	0.462	0.497
	Adequate parking and leisure facilities	3.74	0.869	3.81	0.834	2.084	0.150
	Environmental quality is affected by tourists	3.95	0.846	3.88	0.785	1.377	0.241
	Sufficient personal living space	3.69	0.834	3.91	0.730	10.397	0.001 *
Landscape and ecological environment	Pollution of water quality	3.63	0.893	3.77	0.766	7.078	0.008
	best air quality	3.93	0.807	3.90	0.758	0.592	0.442
	Vegetation forest land has been developed	3.81	0.858	3.81	0.767	3.656	0.057
	Destroy the original habitat	3.77	0.823	3.87	0.773	3.531	0.061
	Car and motorcycle oil fume pollution	3.82	0.817	3.67	0.838	0.167	0.683

$* = p < 0.001.$

It can be inferred that while residents expected the development to bring economic growth, increase their income, and improve their quality of life, the influx of tourists has taken over their living space and affected their daily routines. Although there are few scenic spots in the villages and not enough tourism infrastructure and industries to accommodate the huge number of tourists, visitors still look forward to a comfortable environment and space away from the hustle and bustle of the city, where they can relax both physically and mentally. Therefore, although the land is vast and the environment is spacious, the definition of living space varies according to the conditions and needs of individuals in different roles and positions. As a result, different stakeholders had different views on the current situation of adequate personal living space. As shown in Figure 5.

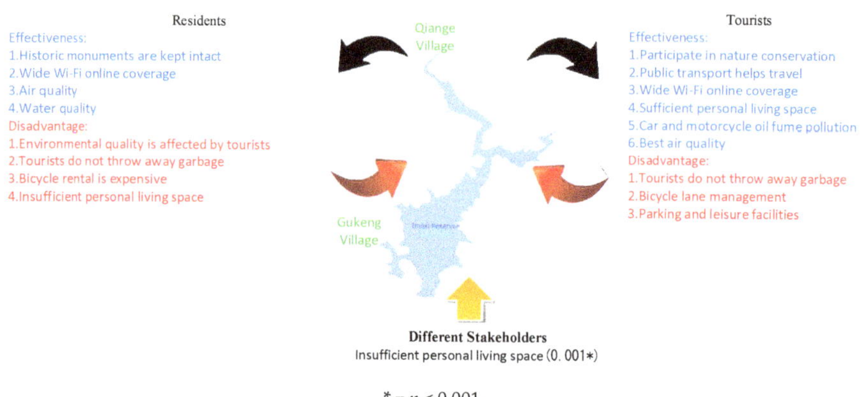

$* = p < 0.001.$

Figure 5. Perception analysis of current environmental developments.

4. Conclusions and Recommendations

4.1. Conclusions

Research believes that the current tourism development can help villages and residents around the reservoir to obtain employment opportunities and construction, increase the residents' willingness to participate in policies, improve the quality of life and the environment, protect cultural and historical relics, improve network information systems, and will not harm the air quality. However, it does not help increase the land and housing prices, develop leisure activities, and improve conditions for medical facilities, coupled with increased consumption costs, environmental damage caused by tourists, and insufficient police and firefighting personnel, which does not help attract young people to return to their homes.

Although development has enabled the region to obtain company resources, increase employment opportunities, develop creative products, increase leisure options, promote traditional cultural activities, increase people's awareness of ecological protection, improve transportation and network facilities, and maintain low consumption level, spacious area, good air quality, and ample living space, which ttract tourists to travel and consume, however, the shortage of interpreters, parking and entertainment facilities, poor communication channels, poor management of bicycle facilities and waste management as well as the low awareness of participation of residents have led to insufficient manpower and low consumption willingness of young tourists.

At present, different stakeholders have different opinions on the development effectiveness of leisure opportunities, medical care, traditional cultural activities and residential space planning.

4.2. Recommendations

4.2.1. The resident population

It is necessary to emphasize tourism development, invest in tourism industry planning, stabilize commodity prices, and bring in manpower to satisfy the tourism demand, in order to alleviate hardship and attract young people to return home.

4.2.2. The tourists

It is necessary to improve personal environmental literacy, carry out waste separation, reduce the production of tourist waste, and work together to protect the environment in order to preserve the beautiful scenic environment.

4.2.3. The government

Invest funds in public facilities, provide channels for tourists to lodge complaints, develop scenic spots to divert tourists, introduce enterprises and technologies to develop new industries, and train interpretive talents to provide in-depth tourism services.

4.2.4. Research suggestions

It is suggested that the follow-up study should understand the travel behavior of tourists, analyze the impact of development on the current leisure behavior of residents, investigate the tourism resource potential of the reservoir, and finally discuss the impact of development on neighboring villages to complete the relevant research information.

Author Contributions: C.-C.S. research information distribution and resource input. H.-H.L. conceived, designed and wrote this paper. C.-H.H. and J.-H.C. help check spelling. J.-H.C. helped check the grammar of the article. C.-F.L. data curation, writing-review & editing. All authors have read and agreed to the published version of the manuscript.

Funding: This research received no external funding.

Conflicts of Interest: The authors declare no conflict of interest.

References

1. Xiamen Water Resources Bureau. Tingxi Reservoir. 2020. Available online: http://sl.xm.gov.cn/wsbs/ (accessed on 25 September 2020).
2. Wu, J.-H. Analysis of the Forest Resources Status of Tingxi State-owned Forest Farm. *J. Green Sci. Technol.* **2016**, *9*, 105–108.
3. Harmonicare Institute. Tingxi Town, Tongan District, Xiamen. 2018. Available online: https://kknews.cc/zh-tw/travel/rbze2ko.html (accessed on 1 October 2020).
4. Kingdon, J.W. *Agendas, Alternatives, and Public Policies*; Addison-Wesley: Boston, MA, USA; Longman: New York, NY, USA, 1995.
5. Benner, M. From Overtourism to Sustainability: A Research Agenda for Qualitative Tourism Development in the Adriatic. *Munich Pers. RePEc Arch.* **2019**, 2213. Available online: https://Mpra.ub.uni-muenchen.de/92213 (accessed on 25 September 2020).
6. Ap, J.; Crompton, J.L. Developing and testing a tourism impact scale. *J. Travel Res.* **1998**, *37*, 120–130. [CrossRef]
7. Niñerola, A.; Sánchez-Rebull, M.V.; Hernández-Lara, A.B. Tourism research on sustainability: A bibliometric analysis. *Sustainability* **2019**, *11*, 1377. [CrossRef]
8. Johnson, J.D.; Snepenger, D.J.; Aki, S. Residents' perceptions of tourism development. *Ann. Tour. Res.* **1994**, *21*, 629–642. [CrossRef]
9. Haritha, W.D.; Atan, H.; Ali Khatib, S.M.; Azam, F.; Tham, J. Developing a Framework for Scrutinizing Strategic Green Orientation and Organizational Performance with Relevance to the Sustainability of Tourism Industry. *Eur. J. Soc. Sci. Stud.* **2019**, *4*, 1–18. [CrossRef]
10. Lin, H.H. The Study on Tourism Policy, Current Development Status, and Impact Perception of Sun Moon Lake. Doctoral Dissertation, DaYeh University, Institute of Environmental Engineering Department, Changhua, Taiwan, 2019.
11. Lin, H.-H.; Lee, S.-S.; Perng, Y.-S.; Yu, S.-T. Investigation about the Impact of Tourism Development on a Water Conservation Area in Taiwan. *Sustainability* **2018**, *10*, 2328. [CrossRef]
12. Lankford, S.V.; Howard, D.R. Developing a Tourism Impacts Attitude Scale. *Ann. Tour. Res.* **1994**, *21*, 121–139. [CrossRef]
13. Rothman, R. Residnets an dtransients: Community reaction to seasonal visitors. *J. Travel Res.* **1978**, *16*, 8–13. [CrossRef]
14. Hsu, C.-H.; Lin, H.-H.; Jhang, S. Sustainable Tourism Development in Protected Areas of Rivers and Water Sources: A Case Study of Jiuqu Stream in China. *Sustainability* **2020**, *12*, 5262. [CrossRef]
15. Husbands, W. Social status and perception of tourism in Zambia. *Ann. Tour. Res.* **1989**, *16*, 237–253. [CrossRef]
16. Gursoy, D.; Jurowski, C.; Uysal, M. Resident Attitudes: A structural odeling approach. *Ann. Tour. Res.* **2002**, *20*, 79–105. [CrossRef]
17. De Waal, F. *Are we Smart Enough to Know How Smart Animals Are?* W. W. Norton & Company: London, UK, 2017.
18. Plucinski, B.; Sun, Y.; Wang, S.Y.S.; Gillies, R.R.; Eklund, J.; Wang, C.C. Feasibility of Multi-Year Forecast for the Colorado River Water Supply: Time Series Modeling. *Water* **2019**, *11*, 2433. [CrossRef]
19. Yagoub, M.M.; AlSumaiti, T.S.; Ebrahim, L.; Ahmed, Y.; Abdulla, R. Pattern of Water Use at the United Arab Emirates University. *Water* **2019**, *11*, 2652. [CrossRef]
20. Nowak, B.; Lawniczak-Malińska, A.E. The Influence of Hydrometeorological Conditions on Changes in Littoral and Riparian Vegetation of a Meromictic Lake in the Last Half-Century. *Water* **2019**, *11*, 2651. [CrossRef]
21. Jurowski, C.; Uysal, M.; Williams, D.R. A theoretical analysis of host community resident reactions to tourism. *J. Travel Res.* **1997**, *36*, 3–11. [CrossRef]
22. Paulo, E.J.; Belmiro, V.P. Olfactory information from the path is relevant to the homing process of adult pigeons. *Behav. Ecol. Sociobiol.* **2017**, *72*, 5. [CrossRef]
23. Schreiber, R.L.; Diamond, A.W.; Peterson, R.T.; Cronkite, W. *Save the Birds*; Cambridge at the University Press: Lanham, MD, USA, 1987; p. 384, ISBN 0395511720.

24. Galeote, L.C.; Mestanza, J.G. Qualitative Impact Analysis of International Tourists and Residents' Perceptions of Málaga-Costa Del Sol Airport. *Sustainability* **2020**, *12*, 4725. [CrossRef]

25. Campón-Cerro, A.M.; Folgado-Fernández, J.A.; Hernández-Mogollón, J.M. Rural Destination Development Based on Olive Oil Tourism: The Impact of Residents' Community Attachment and Quality of Life on Their Support for Tourism Development. *Sustainability* **2017**, *9*, 1624. [CrossRef]

26. Widaningsih, T.T.; Diana, R.; Rahayunianto, A. Community Based Cultural Tourism Development Setu Babakan, Jakarta. *J. Environ. Manag. Tour.* **2020**, *11*, 486–495. [CrossRef]

27. Kim, S.; Kang, Y. Why do residents in an overtourism destination develop anti-tourist attitudes? An exploration of residents' experience through the lens of the community-based tourism. *Asia Pac. J. Tour. Res.* **2020**, *25*, 858–876. [CrossRef]

28. Zuo, L.; Zhang, J.; Zhang, R.J.; Zhang, Y.; Hu, M.; Zhuang, M.; Liu, W. The transition of soundscapes in tourist destinations from the perspective of residents' perceptions: A case study of the Lugu Lake scenic spot, Southwestern China. *Sustainability* **2020**, *12*, 1073. [CrossRef]

29. Liu, R. The state-led tourism development in Beijing's ecologically fragile periphery: Peasants' response and challenges. *Habitat Int.* **2020**, *96*, 102119. [CrossRef]

30. Sánchez-Martín, J.M.; Sánchez-Rivero, M.; Rengifo-Gallego, J.-I. Water as a tourist resource in Extremadura: Assessment of its attraction capacity and approximation to the tourist profile. *Sustainability* **2020**, *12*, 1659. [CrossRef]

31. Sánchez-Rivero, M.; Rodríguez-Rangel, M.; Fernández-Torres, Y. The Identification of Factors Determining the Probability of Practicing Inland Water Tourism Through Logistic Regression Models: The Case of Extremadura, Spain. *Water* **2020**, *12*, 1664. [CrossRef]

32. Furgała-Selezniow, G.; Jankun-Woźnicka, M.; Mika, M. Lake regions under human pressure in the context of socio-economic transition in Central-Eastern Europe: The case study of Olsztyn Lakeland, Poland. *Land Use Policy*. *Land Use Policy* **2020**, *90*, 104350. [CrossRef]

33. Yusuf, M.; Purwandani, I. Ecological politics of water: The ramifications of tourism development in Yogyakarta. *South East Asia Res.* **2020**, 1–17. [CrossRef]

34. Godovykh, M.; Ridderstaat, J. Health outcomes of tourism development: A longitudinal study of the impact of tourism arrivals on residents' health. *J. Destin. Mark. Manag.* **2020**, *17*, 100462. [CrossRef]

35. Chen, N.; Hong, H. Securing the Drinking Water Supply for the Growing Population of Xiamen City, PR China. *Case Study* **2016**, *29*, 334–346.

36. Kou, L.; Li, X.; Lin, J.; Kang, J. Simulation of urban water resources in Xiamen based on a WEAP model. *Water* **2018**, *10*, 732. [CrossRef]

37. De Souza, C.F.; Julia, M.A.; Pereira, S.; Benassi, F.; Filho, R.A.R. Trophic State Index (TSI) of the Reservoir of the Itaipu Binacional Hydroelectric Power Plant, Brazil. *Mod. Environ. Sci. Eng.* **2018**, *5*, 615–624. [CrossRef]

38. Corbin, J.; Strauss, A. *Basics of Qualitative Research: Grounded Theory Procedures and Techniques*; Sage: Newbury Park, CA, USA, 1990.

39. Janesick, V.J. The choreography of qualitative research design: Minuets, improvisations, and crystallization. In *Handbook of Qualitative Research*; Denzin, N.K., Lincoln, Y.S., Eds.; Sage: Thousand Oak, CA, USA, 2000; pp. 379–399.

40. Ajzen, I. The Theory of Planned Behavior. *Organ. Behav. Human Decis. Process.* **1991**, *50*, 179–211. [CrossRef]

41. Devellis, R.F. *Scale Development: Theory and Applications*; Sage: Newbury Park, CA, USA, 1991.

Publisher's Note: MDPI stays neutral with regard to jurisdictional claims in published maps and institutional affiliations.

 *water*

Case Report

Assessing Inter-Administrative Cooperation in Urban Public Services: A Case Study of River Municipalities in the Internal Border Area between Aragon and Catalonia (Spain)

Albert Santasusagna Riu [1,*], Ramon Galindo Caldés [2] and Joan Tort Donada [1]

[1] GRAM (Grup de Recerca Ambiental Mediterrània), Department of Geography, Universitat de Barcelona, Montalegre 6, 08001 Barcelona, Spain; jtort@ub.edu
[2] GADE (eGovernança: Administració i Democràcia Electrònica), Faculty of Law and Political Sciences, Universitat Oberta de Catalunya, Av. Carl Friedrich Gauss 5, 08860 Castelldefels, Spain; rgalindoca@uoc.edu
* Correspondence: asantasusagna@ub.edu

Received: 6 July 2020; Accepted: 5 September 2020; Published: 8 September 2020

Abstract: The proper management of urban public services (UPS) ensures that a territory functions efficiently, since it guarantees optimal waste disposal, water supply, and the maintenance of communication infrastructure, among other things. In areas of high urban density located close to metropolitan cities, UPS are usually provided properly and efficiently. However, in less populated territories, lying in the periphery, significant problems and deficiencies are often encountered, being most evident in rural areas located on the administrative limits of a state or region. This paper seeks to analyze the management of UPS in the internal border area between two Spanish regions, Aragon and Catalonia. A total of 72 stakeholders (mayors and town clerks) from 49 river municipalities were involved in this study that employs a quantitative methodology (questionnaire). The perception that there are deficiencies to correct and a clear will to reach agreements and establish cooperation mechanisms is detected in many of the municipalities in the border area. A clear need to cooperate is also apparent in a series of priority UPS, including the promotion of river tourism, town access roads, urban collective passenger transport, and environmental protection.

Keywords: urban public services; inter-administrative cooperation; border studies; internal borders; river municipalities

---

1. Introduction

Urban public services (UPS) can be defined as those activities that meet citizen needs through a physical system of the production, distribution, provision, and consumption of basic goods [1–3]. Many studies have influenced both the technical and economic importance afforded to UPS, which are fundamental for the operation of cities [4–7] and which include the provision of resources and the collection and disposal of waste [8,9], the distribution of energy and public lighting [10,11], and the maintenance of communication and transport infrastructure [12,13]. Among them, water management is a key element, being provided by means of various UPS: on the one hand, the supply and distribution of drinking water [14–16] and, on the other, the sewage collection system [17]. Moreover, a municipality's urban policy is typically dependent on a combination of public, private, and mixed UPS management. Indeed, while the ownership of the services remains public, there are many instances around the world where municipalities prefer to outsource these services and privatize their management to achieve greater efficiency [18,19].

The proper planning of UPS provides citizens with a better quality of living. Here, the relationship between the core and the periphery can take on particular importance. Thus, suburbs usually experience complex problems related to a UPS deficit [20–22], and rural areas with low population density likewise present problems of accessibility for UPS [23–26], typified by few transportation and mobility resources [27,28]; an intermittent water supply system [29,30]; and scarce or remote health facilities [31–33], schools [34,35], and police and fire stations [36,37]. Indeed, various studies propose an enhanced distribution of UPS in rural areas based on a location-allocation approach using Geographic Information System (GIS) techniques [38–41].

Policies to decentralize and improve accessibility to UPS are one of the challenges faced by governance at different territorial levels, including border areas [42–47]. Such policies, as noted above, are necessary in remote and rural areas. Furthermore, if these areas are located along national or regional peripheral borders (i.e., external or internal borders, respectively), the outlook may be even worse in the absence of both cordial relations and inter-administrative cooperation. Indeed, the "barrier effect" can result in a testing situation for the administrations involved [48–50], making cooperation between cross-border areas essential in such sectors as tourism [51–54], healthcare [55,56], and natural resource management [57–59], among others. The policy of supra-state entities—most notably, the European Union (EU)—has, in recent decades, worked in this direction—that is, the strengthening of cooperation between states and a curtailing of the adverse effects of the classic border [60–65].

Inter-administrative cooperation is more readily addressed in internal border areas that form part of the same state and which share common policies, such as UPS management. However, cooperation is closely linked to the state's internal policy of organization; for example, in most decentralized states the barrier effect of the state's internal borders tends to be more acute, as is the case in Spain [66,67]. The need for the cooperative, shared management of UPS becomes essential for proper spatial planning in territories located some distance from the metropolitan region and suffering marked socioeconomic deficiencies.

The aim of this paper is to describe the role and determine the performance of UPS management in a peripheral rural border area, and to explore and analyze the perceptions that stakeholders have of inter-administrative cooperation between the border regions in the same decentralized state. The border area we study here is differentiated at an administrative level between the Spanish regions or Autonomous Communities of Aragon and Catalonia, yet the territory shares a common physical environment (the basin of the river Ebro) and presents considerable potential for implementing common objectives centered on the management of their UPS. A further aim of our study is that its results might be taken into account by the corresponding administrations and practitioners so as to create the appropriate instruments to solve existing deficiencies and to achieve greater efficiency in the management of UPS.

In seeking to fulfil these aims, this paper (1) reports a quantitative study conducted in the internal border area between two Spanish regions (Aragon and Catalonia); (2) identifies and analyzes inter-administrative cooperation in the delivery of UPS using quantitative methods; and (3) proposes future research on the issues addressed.

2. Materials and Methods

The present study is framed in the context of a broader research project, focused on the analysis of different types of problem and conflict that have been generated in recent times (especially over the last four decades) in the internal border area (IBA) between three Spanish Autonomous Communities: Aragon, Catalonia, and the Valencian Community. The results of this research have been reported in a number of studies conducted at different scales and focusing on different themes [68–70]. These previous studies, based on the conducting of focus groups with public stakeholders (mayors and town clerks) in the territory analyzed, conclude that, for a significant majority of the problems considered, it is essential that cooperation be promoted between the autonomous administrations. The present article seeks to build on the findings of these earlier studies by employing a questionnaire as a valid and rigorous methodology for collecting information about stakeholder perceptions [71–75].

While some of our previous studies have focused on the Catalonia–Valencian Community IBA, the present study focuses solely on the Aragon–Catalonia IBA (ARCAT-IBA, Figure 1). This area, with a border extending some 360 km, forms part of the Ebro basin and is characterized by its tributaries that run from north to south (Noguera Ribagorçana, Cinca, Matarranya) and a sub-tributary (Algars), which serve as the boundary. Specifically, a total of 57 border municipalities make up the ARCAT-IBA (CM in Figure 1), with an additional 19 (SBM in Figure 1) which, due to their size and proximity, play a secondary role in the border dynamics.

A questionnaire for the ARCAT-IBA public stakeholders (mayors and town clerks) was created with five main objectives: (i) to determine their perception of the deficiencies in UPS management as a result of the different regulations being operated in Aragon and Catalonia, respectively (Q1 in Table 1); (ii) to identify the existence of any formal or informal mechanisms of cooperation being employed by the Catalan and Aragonese administrations (Q2 and Q3 in Table 1); (iii) to appreciate their willingness to strengthen inter-municipal cooperation in the management of UPS, so that the citizens of ARCAT-IBA municipalities might access these services regardless of their origin (Q4 and Q5 in Table 1); (iv) to identify instruments to correct the deficiencies detected (Q6 in Table 1); and, finally, (v) to determine their perception of deficiencies at higher administrative levels (Q7 in Table 1). The questionnaire was answered in person between January and June 2017, following focus group sessions analyzed in previous studies devoted to water management [70]. Although there is evidence of general problems affecting local government in Spain and UPS management (e.g., budget deficits and shortages of administrative personnel), the questionnaire focuses on the specific problems attributable to their condition as border municipalities. Note that Q5 refers to the UPS specifically listed in Spanish regulations [76].

Table 1. Questionnaire used in this study (source: authors).

Q1: During your term of office as mayor or town clerk, have you encountered situations in which the different regulations applied in Catalonia and Aragon have given rise to problems or difficulties of an administrative nature?

☐ Yes

☐ No

Q2: Has the Town Council participated in any collaborative projects with neighboring municipalities that belong to the other Autonomous Community?

☐ Yes

☐ No

Q3: Do you know of any joint cultural, social or political initiatives that have been taken between your municipality and the border municipalities that form part of another Autonomous Community?

☐ Yes

☐ No

Q4: Do you think that there should be more cooperation between your municipality and the border municipalities that form part of Catalonia/Aragon?

☐ Yes

☐ No

Q5: In which areas does cooperation (formal or informal) exist, and in which do you think cooperation would be a good idea?

Table 1. *Cont.*

UPS Competences	Cooperation	
	Cooperation Exists	There Should Be Cooperation
Waste collection		
Street cleaning		
Town access roads		
Paving and maintenance of public roadways		
Public libraries		
Selective waste collection		
Civil defence		
Social services		
Sports installations		
Urban collective passenger transport		
Environmental protection		
Municipal welfare (Administration)		
Specialist social services (children, elderly, etc.)		
Promotion of tourism (river tourism)		
Police		
Housing		
Healthcare		
Urban planning		
Museums		
Music conservatories		
Nursery schools		
Others		

Q6: Do you think legal mechanisms should be put in place to somehow reduce the "negative" (that is, unwanted, albeit legal) effects of the border?

- ☐ No, it is not necessary
- ☐ No, the agreements between the Autonomous Communities are sufficient
- ☐ Yes, the way the rules are applied should be modulated in certain municipalities
- ☐ Yes, the town councils should be able to activate/implement legal mechanisms
- ☐ Yes, a special regime should be created, i.e., that of the "border municipality"
- ☐ Yes, a consortium (partnership) should be created
- ☐ Yes, an inter-municipal association of municipalities (mancomunidad) should be created
- ☐ Yes. Other options

Q7: In your opinion, where do you think the different levels of public administration stand on the question of cooperation between the border municipalities?

Administration	Opposed	Indifferent	Favorable	Other
Catalan Government (Generalitat)				
Aragonese Government (Diputación General)				
Catalan provincial councils				
Aragonese provincial councils				
Catalan county councils (comarcas)				
Aragonese county councils (comarcas)				

The study carried out presents a series of specific methodological characteristics: (i) to facilitate comparison with our previous studies, the numbering given to the contiguous border municipalities (CM) is respected (11–70 in Figure 1); (ii) the second buffer municipalities (SBM) (AA-I in Figure 1) are those located adjacent to the CM and play a secondary role in the border dynamics, so are not included in this study; (iii) as in previous studies, some border municipalities (DM in Figure 1) have been discarded due to their secondary role in the border dynamics resulting from physical geographical barriers or the extent of their border area; (iv) 8 of the 57 Catalan and Aragonese municipalities did not participate (Table 2); (v) 72 stakeholders participated in our study by answering the questionnaire (38 mayors and 34 town clerks), representing 57.1% of the potential stakeholders (Table 1); (vi) in most municipalities, both stakeholders participated (mayor and town clerk), but in some only one of the two participated: mayors only (11, 12, 36, 42, 45, 54, 57, 60, and 64 in Figure 1) and town clerks only (15, 25, 32, 35, 38, 39, 52, and 53 in Figure 1).

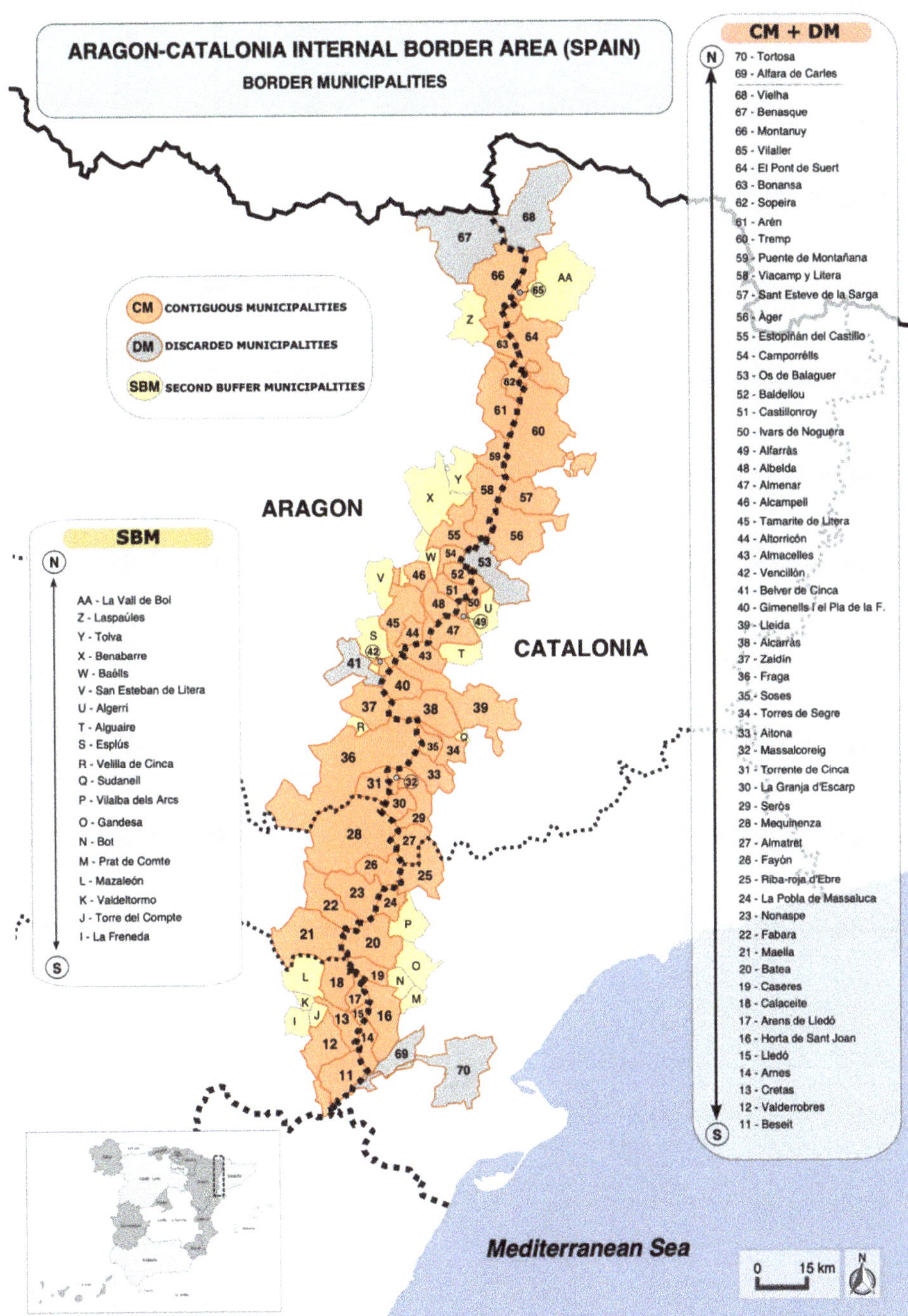

Figure 1. Border municipalities between Aragon and Catalonia (source: authors).

Table 2. Technical data of the questionnaires answered. Breakdown of the answers received in relation to the total number of potential stakeholders (source: authors).

	CAT						AR				Total	
	TAR		LED		TER		ZAR		HUE			
	N°	%	N°	%	N°	%	N°	%	N°	%	N°	%
M	4	66.7	10	50	6	100	2	40	16	80	38	56.7
TC	4	80	15	83.3	3	60	2	40	10	62.5	34	57.6
Total	8	72.7	25	65.8	9	81.8	4	40	26	72.2	72	57.1

CAT = Catalonia (Spanish Autonomous Community); AR = Aragon (Spanish Autonomous Community); TAR = Tarragona (Province of Catalonia); LED = Lleida (Province of Catalonia); TER = Teruel (Province of Aragon); ZAR = Zaragoza (Province of Aragon); HUE = Huesca (Province of Aragon); M = Mayors; TC = Town Clerks; N° = Number of responses; % = percentage of responses compared to potential responses.

3. Results

3.1. Perception of Deficiencies

Table 3 shows the responses to Q1, aimed at gauging the perception of possible deficiencies in UPS management. The results confirm that, in the ARCAT-IBA, there is a majority perception (75%) that the internal border is a problematic element from an administrative point of view that also affects the UPS management. This perception is shared on both sides of the border.

Table 3. Answers obtained in Q1 (Source: Authors).

Q1: During your term of office as mayor or town clerk, have you encountered situations in which the different regulations applied in Catalonia and Aragon have given rise to problems or difficulties of an administrative nature?						
Responses	CAT	%	AR	%	Total	%
Yes	25	75.7	29	74.4	54	75 *
No	8	24.3	10	25.6	18	25
Total	33	100	39	100	72	100

CAT = Catalonia (Spanish Autonomous Community). AR = Aragon (Spanish Autonomous Community). (*) = Outstanding result.

3.2. Existing Formal or Informal Cooperation Mechanisms

As can be seen in Table 4, most of the responses (75%) affirm that their municipality has not participated in the creation of cooperation mechanisms with the neighboring municipality on the other side of the border. Thus, although there may be specific cases, cooperation mechanisms do not proliferate between the border municipalities.

Table 4. Answers obtained in Q2 (source: authors).

Q2: Has the Town Council participated in any collaborative projects with neighboring municipalities that belong to the other Autonomous Community?						
Responses	CAT	%	AR	%	Total	%
Yes	8	25	10	25	18	25
No	24	75	30	75	54	75 *
Total	32	100	40	100	72	100

CAT = Catalonia (Spanish Autonomous Community). AR = Aragon (Spanish Autonomous Community). (*) = Outstanding result.

Table 5 shows an unequal response on the two sides of the ARCAT-IBA: while most of the public stakeholders in Catalonia (59.4%) state that there are no cultural, social, or political initiatives that have been taken between neighboring municipalities, more than half of the public stakeholders (52.5%) in Aragon affirm just the opposite.

Table 5. Answers obtained in Q3 (source: authors).

Q3: Do you know of any joint cultural, social, or political initiatives that have been taken between your municipality and the border municipalities that belong to another Autonomous Community?						
Responses	**CAT**	**%**	**AR**	**%**	**Total**	**%**
Yes	13	40.6	21	52.5 *	34	47.2
No	19	59.4 *	18	45	37	51.4
DK/NA	0	0	1	2.5	1	1.4
Total	32	100	40	100	72	100

CAT = Catalonia (Spanish Autonomous Community). AR = Aragon (Spanish Autonomous Community). (*) = Outstanding result.

Thus, from the results shown in Tables 4 and 5, it can be concluded that there is an uneven perception of cooperation between ARCAT-IBA municipalities, although most public stakeholders claim to be unaware of the existence of formal or informal mechanisms. Most of the border area municipalities between Catalonia and Aragon have a low population density, and their population centers are often physically separated by very large distances. In addition, there are sometimes very obvious surface differences in municipal area between both communities, as a result of a divergent evolution of the administrative division. This means that, on many occasions, the perception of interrelationships is different on both sides of the border area. In the discussion section, we examine this issue in greater detail.

3.3. Willingness to Strengthen Inter-Municipal Cooperation in UPS

As can be seen in Table 6, the vast majority of public stakeholders (86.1%) believe that a scenario of greater cooperation between the municipalities on both sides of the ARCAT-IBA would be positive.

Table 6. Answers obtained in Q4 (source: authors).

Q4: Do you think that there should be more cooperation between your municipality and the border municipalities that form part of Catalonia/Aragon?						
Responses	**CAT**	**%**	**AR**	**%**	**Total**	**%**
Yes	29	87.9 *	33	84.6	62	86.1 *
No	2	6.1	5	12.8	7	9.7
DK/NA	2	6.1	1	2.6	3	4.2
Total	33	100	39	100	72	100

CAT = Catalonia (Spanish Autonomous Community). AR = Aragon (Spanish Autonomous Community). (*) = Outstanding result.

Table 7 shows interesting data because of its specificity regarding local and supralocal UPS management competences based on Spanish regulations [76]. On the one hand, the results regarding possible cooperation that already exists (Q5A in Table 7) show that, for most municipalities, there are no mechanisms promoting joint cooperation in the management of UPS. However, it should be underlined that the perception of existing cooperation is not so great on the Catalan side, while in Aragon there is a greater awareness of specific agreements on competences such as healthcare and civil defence (on these differences in perception, see the specific discussion in Section 3.2 above).

Table 7. Answers obtained in Q5 (source: authors).

| Q5: In which areas does cooperation (formal or informal) exist, and in which do you think cooperation would be a good idea? ** | | | | | | | | | | | | |
| Local (L) and Supralocal (SL) UPS Competences | (Q5A) Cooperation Exists | | | | | | (Q5B) There Should Be Cooperation | | | | | |
	C	%	A	%	T	%	C	%	A	%	T	%
(L) Waste collection	3	10.0	3	7.7	6	8.7	6	20.0	13	33.3 *	19	27.5
(L) Street cleaning	2	6.7	2	5.1	4	5.8	4	13.3	6	15.4	10	14.5
(L) Town access roads	2	6.7	5	12.8	7	10.1	15	50.0 *	21	53.8 *	36	52.2 *
(L) Paving and maintenance of public roadways	2	6.7	4	10.3	6	8.7	11	36.7 *	13	33.3 *	24	34.8 *
(L) Public libraries	1	3.3	2	5.1	3	4.3	2	6.7	5	12.8	7	10.1
(SL) Selective waste collection	3	10.0	4	10.3	7	10.1	7	23.3	10	25.6	17	24.6
(L) Civil defence	1	3.3	9	23.1 *	10	14.5	14	46.7 *	17	43.6 *	31	44.9 *
(SL) Social services	0	0.0	5	12.8	5	7.2	7	23.3	15	38.5 *	22	31.9 *
(L) Sports installations	2	6.7	6	15.4	8	11.6	8	26.7	7	17.9	15	21.7
(L) Urban collective passenger transport	3	10.0	4	10.3	7	10.1	14	46.7 *	21	53.8 *	35	50.7 *
(SL) Environmental protection	1	3.3	3	7.7	4	5.8	17	56.7 *	18	46.2 *	35	50.7 *
(SL) Municipal welfare	1	3.3	2	5.1	3	4.3	5	16.7	6	15.4	11	15.9
(SL) Specialist social services	2	6.7	2	5.1	4	5.8	6	20.0	10	25.6	16	23.2
(L/SL) Promotion of river tourism	5	16.7	3	7.7	8	11.6	18	60.0 *	24	61.5 *	42	60.9 *
(L/SL) Police	2	6.7	2	5.1	4	5.8	14	46.7 *	12	30.8 *	26	37.7 *
(L) Housing	0	0.0	1	2.6	1	1.4	5	16.7	7	17.9	12	17.4
(SL) Healthcare	2	6.7	11	28.2 *	13	18.8	11	36.7 *	16	41.0 *	27	39.1 *
(L) Urban planning	0	0.0	0	0.0	0	0.0	12	40.0 *	8	20.5	20	29.0
(L) Museums	0	0.0	2	5.1	2	2.9	4	13.3	7	17.9	11	15.9
(SL) Music conservatories	2	6.7	3	7.7	5	7.2	3	10.0	6	15.4	9	13.0
(SL) Nursery schools	0	0.0	3	7.7	3	4.3	6	20.0	10	25.6	16	23.2
Others	0	0.0	0	0.0	0	0.0	2	6.7	1	2.5	3	4.3

C = Catalonia (Spanish Autonomous Community). A = Aragon (Spanish Autonomous Community). T = Total. (*) = Outstanding result (Q5A > 20%) (Q5B > 30%). (**) = Multiple choice question. Based on 69 responses (30 in the Catalan side and 39 in the Aragonese side). There are 3 blank responses.

On the other hand, there is a perception on both sides of the border that it would be positive to enter into agreements to meet common management objectives in the delivery of several UPS (>50% in Table 7): the promotion of river tourism (60.9%), town access roads (52.2%), urban collective passenger transport (50.7%), and environmental protection (50.7%). A moderate level of support is also recorded for cooperation in relation to the delivery of other UPS (>30% in Table 7), including civil defence (44.9%), healthcare (39.1%), police (37.7%), the paving and maintenance of public roadways (34.8%), and social services (31.9%). Interestingly, most of the UPS competences that achieve the greatest agreement between both sides of the border are local in nature.

Overall, Table 7 highlights a significant number of results in support of cooperation (both in the sense of recognizing its existence and in favor of its implementation) in the delivery of UPS. There is a clear local perception that cooperation is needed; if a percentage of 30% can be considered a significant indication in this regard, then obviously higher percentages cannot be ignored. In other words, there is a clear perception in the case of certain UPS (town access roads, environmental protection, and the promotion of river tourism) that cooperation mechanisms are essential to guarantee a cohesive and efficient management of public services. It should, however, be borne in mind that, in relation to certain competences, public stakeholders on one side of the border are more interested in cooperating than are those on the other side. For example, in Catalonia they are more interested in cooperating in urban planning (40% vs. 20.5%) and police (46.7% vs. 30.8%), while in Aragon there are calls for greater cooperation in social services (38.5% vs. 23.3%) and waste collection (33.3% vs. 20%).

3.4. Instruments to Correct Deficiencies in UPS Management

A wide variety of possible responses was presented to public stakeholders as administrative solutions to correct deficiencies in UPS management (Table 8). On the Catalan side, the creation of specific cooperation mechanisms in Spanish regulations (48.5%), the possibility of creating a special entity or "border municipality" in Spanish local regulations (42.4%), and the activation and implementation of cooperation mechanisms by the municipalities (39.4%) are seen as valid solutions. On the Aragonese side, only the creation of a special entity attracted a significant degree of agreement (46.2%), a solution that achieved the greatest support when considering stakeholders on both sides of the border (44.4%). In contrast, other solutions that are frequently adopted at the Spanish local level, such as consortiums (partnerships) or inter-municipal associations, are not seen here as effective solutions for correcting deficiencies.

Table 8. Answers obtained in Q6 (source: authors).

Q6: Do you think legal mechanisms should be put in place to somehow reduce the "negative" (that is, unwanted, albeit legal) effects of the border? **						
Responses	**CAT**	**%**	**AR**	**%**	**Total**	**%**
No, it is not necessary	1	3.0	7	17.9	8	11.1
No, the agreements between the Autonomous Communities are sufficient	1	3.0	3	7.7	4	5.6
Yes, the way the rules are applied should be modulated in certain municipalities	16	48.5 *	11	28.2	27	37.5 *
Yes, the Town Councils should be able to activate/implement legal mechanisms	13	39.4 *	10	25.6	23	31.9 *
Yes, a special entity should be created, i.e., that of the "border municipality"	14	42.4 *	18	46.2 *	32	44.4 *
Yes, a consortium (partnership) should be created	0	0	3	7.7	3	4.2
Yes, an inter-municipal association of municipalities (mancomunidad) should be created	3	9.1	5	12.8	8	11.1
Yes. Other options	3	9.1	0	0	3	4.2
DK/NA	0	0	2	5.1	2	2.8

CAT = Catalonia (Spanish Autonomous Community). AR = Aragon (Spanish Autonomous Community). (*) = Outstanding result (>30%). (**) = Multiple choice question. Based on 72 responses (33 on the Catalan side and 39 on the Aragonese side). Three responses were left blank.

3.5. Perception of UPS Deficiencies at Higher Administrative Levels

Table 9 shows the perceptions of public stakeholders regarding the stance taken by higher tiers of administration (district or comarcal, provincial and regional) on the question of cooperation between border municipalities. Most of the stakeholders do not perceive that the supralocal administrations have adopted a position contrary to cooperation, but only those on the Aragonese side consider that the Aragonese supralocal administrations (provincial and district, not regional) have been favorable in their stance. On the Catalan side, the general feeling is that the entire supralocal administration (be it district, provincial, or regional, regardless of which side of the border they are located) has been indifferent to cooperation. Thus, opinions are only shared with regards as to what is perceived as indifference on the part of the regional and provincial administrations to cooperation.

Table 9. Answers obtained in Q7 (source: authors).

Q7: In your opinion, where do you think the different levels of public administration stand on the question of cooperation between the border municipalities? *

Responses N° (%)	Opposed			Indifferent			Favourable			Other		
	C	A	T	C	A	T	C	A	T	C	A	T
Catalan Government (*Generalitat*) (R)	4 (12.9)	1 (3.2)	5 (8.1)	20 (64.5) *	22 (71) *	42 (67.7) *	5 (16.1)	7 (22.6)	12 (19.4)	2 (6.5)	1 (3.2)	3 (4.8)
Aragonese Government (*Diputación General*) (R)	6 (26.1)	4 (12.5)	10 (18.2)	14 (60.9) *	23 (71.9) *	37 (67.3) *	1 (4.3)	4 (12.5)	5 (9.1)	2 (8.7)	1 (3.1)	3 (5.5)
Catalan provincial councils (P)	2 (7.1)	0 (0)	2 (3.6)	17 (60.7) *	12 (42.9)	29 (51.8) *	8 (28.6)	15 (53.6) *	23 (41.1)	1 (3.6)	1 (3.6)	2 (3.6)
Aragonese provincial councils (P)	2 (8.7)	1 (2.9)	3 (5.3)	15 (65.2) *	10 (29.4)	25 (43.9)	5 (21.7)	22 (64.7) *	27 (47.4)	1 (4.3)	1 (2.9)	2 (3.5)
Catalan county councils (*comarcas*) (D)	2 (7.4)	0 (0)	2 (3.7)	17 (63.0) *	9 (33.3)	26 (48.1)	7 (25.9)	17 (63) *	24 (44.4)	1 (3.7)	1 (3.7)	2 (3.7)
Aragonese county councils (*comarcas*) (D)	1 (4.5)	0 (0)	1 (1.8)	14 (63.6) *	6 (17.6)	20 (35.7)	6 (27.3)	27 (79.4) *	33 (58.9) *	1 (4.5)	1 (2.9)	2 (3.6)

C = Catalonia (Spanish Autonomous Community). A = Aragon (Spanish Autonomous Community). (R) = Regional administration. (P) = Provincial administration. (D) = District or comarcal administration. (*) = Outstanding result (>50%). (**) = Multiple choice question. Based on 69 responses (32 on the Catalan side and 37 on Aragonese side). There are 3 blank responses.

4. Discussion and Conclusions

UPS management is essential for the well-being of the population, whose basic needs must be met through the provision of these services by different administrative levels (local and supralocal). Rural areas, located far from metropolitan urban centers, are more likely to suffer a lack of proper UPS management. Moreover, when rural areas are also peripheral border areas, this situation is likely to be exacerbated. Here, with the aim of analyzing the management of UPS in rural-border areas, we have carried out a quantitative study of the perception of public stakeholders (mayors and town clerks) in the case of the Spanish internal border area between Catalonia and Aragon (ARCAT-IBA), characterized by the river municipalities of the Ebro basin.

The perception of the existence of deficiencies in UPS management is shared on both sides of the border. Moreover, there is also a common perception that there are not enough cooperation mechanisms to correct these deficiencies. In fact, on both sides of the border, a significant percentage of stakeholders agree that the creation and implementation of cooperation mechanisms for UPS management would be a positive step forward.

The differences in perception regarding the degree of cooperation (existing or desirable) between the municipalities on both sides of the border cannot be considered significant in themselves. There are a number of factors of a geographical nature, linked to the heterogeneity of the whole border area (including the discontinuous distribution of urban settlements and the weak relationship between some municipalities), which condition this perception and which mean that, in many cases, the same situation or problem is interpreted differently on the two sides of the border. These divergences in perception (which can be considered inherent to the border territories, given their usual condition of "periphery" in relation to their respective "centres") could be better understood by conducting a detailed study of just a few municipalities and, in this way, leaving to one side the problems faced by the whole border area.

There is also a shared perception of the positive effects of the collaborative management of UPS for achieving common objectives for people on both sides of the border. This is particularly the case for both local and supralocal UPS competences, such as the promotion of river tourism, town access roads, urban collective passenger transport, and environmental protection, which Catalan and Aragonese public stakeholders alike feel would benefit from greater cooperation. There is also a moderate level of agreement that other competences, such as civil defence, healthcare, police, the paving and maintenance of public roadways, and social services, would benefit from cooperation. All these competences are basic for the social and economic development of peripheral rural border areas.

The promotion of cooperation mechanisms via the creation of a new special entity in the Spanish legal system (the "border municipality") could be way to achieve a satisfactory agreement between the two sides in the long term. This entity could usher in the establishment of different, specific, and more favorable regulations for the socioeconomic development of peripheral municipalities located on Spain's internal borders. In contrast, other more frequently employed formal solutions (i.e., agreements, consortiums, and commonwealths) do not, in many instances, result in a significant degree of cooperation, as they are usually designed for specific scenarios or to address specific problems.

Furthermore, the common perception is that the supralocal administration has been indifferent and distant (neither contrary nor favorable) in its stance to the mechanisms of cooperation. We conclude that the ARCAT-IBA is a territory that is favorable to cooperation in different competences that directly affect UPS management, and that local and supralocal public administrations should take into account this perception of stakeholders to achieve beneficial outcomes for both sides of the border.

Further quantitative research on the questions studied here is needed. The geolocation of UPS and associated statistical analyses aimed at creating efficient location-allocation models should help promote the willingness to cooperate that has been detected using the quantitative methods employed in this study. In addition, the need should be stressed for good decentralization policies and for the consideration of IBAs as a whole territory subject to the same deficiencies in UPS management and, hence, sharing the same common objectives. Finally, more research on possible cooperation mechanisms, including at the international level, should shed further light on the subject.

We would like to complete this study by highlighting the need also to undertake further research on the management of public services. First of all, because we start from the principle that public services should be implemented equally throughout a territory (whatever its scale) and that it is not admissible, from the point of view of the provision of these services, that a distinction be made by the administration between "central" territories, on the one hand, and "peripheral" territories on the other. From an academic point of view, it is important to highlight that border areas (whether at the regional or state scale) often tend to become peripheral spaces (that is, spaces where deficits accumulate and where the limitations of administrative action are accentuated) and that, in such circumstances, it is essential that public authorities seek to correct these situations of imbalance in order to guarantee equity and territorial cohesion. Secondly, our research illustrates, we believe, the rich scientific possibilities opened up by conducting research in the field and more specifically by entering into dialogue with the stakeholders involved in the situations analyzed. In the course of this study, we have been able to observe that, above and beyond the problems identified and the material difficulties that often exist to address them when taking a "top-down" approach, cooperation mechanisms (which, as a rule, operate from the "bottom-up") often offer practical and highly effective solutions that are worth careful consideration with a view to the future.

Author Contributions: A.S.R., R.G.C. and J.T.D. contributed equally to this paper. They conducted the analyses and wrote the paper together. All the authors have read and agreed to the published version of the manuscript. All authors have read and agreed to the published version of the manuscript.

Funding: This research was funded by the Escola d'Administració Pública de Catalunya (EAPC) Research Programme (JT089150) and Programme 2017SGR1344 (Grup de Recerca Ambiental Mediterrània) supported by the Generalitat de Catalunya. The authors also wish to acknowledge funding from CSO2015-6787-C6-4-P of the Ministry of Economy and Competitiveness of the Government of Spain and the postdoctoral scientific project concerted between the University of Barcelona and Societat General d'Aigües de Barcelona (Agbar).

Acknowledgments: We wish to thank Iain Robinson for reviewing the English manuscript and Roger Clavero for his advice on undertaking the cartography.

Conflicts of Interest: The authors declare no conflict of interest.

References

1. Kelleher, C.; Lowery, D. Tiebout Sorting and Selective Satisfaction with Urban Public Services: Testing the Variance Hypothesis. *Urban Aff. Rev.* **2002**, *37*, 420–431. [CrossRef]
2. Shan, X.; Yu, X. Citizen Assessment as Policy Tool of Urban Public Services: Empirical Evidence from Assessments of Urban Green Spaces in China. *Sustainability* **2014**, *6*, 7833–7849. [CrossRef]
3. Ouyang, W.; Wang, B.; Tian, L.; Niu, X. Spatial deprivation of urban public services in migrant enclaves under the context of a rapidly urbanizing China: An evaluation based on suburban Shanghai. *Cities* **2017**, *60 Pt B*, 436–445. [CrossRef]
4. Seetharam, K. Reforming Delivery of Urban Services in Developing Countries: Evidence from a Case Study in India. *Econ. Political Wkly.* **2007**, *33*, 3404–3413.
5. Alozie, N.; McNamara, C. Anglo and Latino Differences in Willingness to Pay for Urban Public Services. *Soc. Sci. Q.* **2008**, *89*, 406–427. [CrossRef]
6. Benito, B.; Bastida, F.; Guillamón, M. Urban Sprawl and the Cost of Public Services: An Evaluation of Spanish Local Governments. *Lex Localis J. Local Self Gov.* **2010**, *8*, 245–264. [CrossRef]
7. Fernández-Aracil, P.; Ortuño-Padilla, A. Costs of providing local public services and compact population in Spanish urbanized areas. *Land Use Policy* **2016**, *58*, 234–240. [CrossRef]
8. Adewole, A.T. Waste Management towards sustainable development in Nigeria: A case study of Lagos state. *Int. NGO J.* **2009**, *4*, 173–179.
9. Boex, J.; Malik, A.A. The Political Economy of Urban Governance in Asian Cities: Delivering Water, Sanitation and Solid Waste Management Services. In *New Urban Agenda in Asia-Pacific*; Dahiya, B., Das, A., Eds.; Springer: Singapore, 2020; pp. 301–329.

10. Ye, H.; He, X.; Song, Y.; Li, X.; Zhang, G.; Lin, T.; Xiao, L. A sustainable urban form: The challenges of compactness from the viewpoint of energy consumption and carbon emission. *Energy Build.* **2015**, *93*, 90–98. [CrossRef]

11. Garrido-Jiménez, F.J.; Magrinyà, F.; Del Moral-Ávila, M.C.; Rodríguez-García, G. The Relationship between Urban Morphology and Street Lighting Operating Costs: Evidence from Medium-sized Spanish Cities. *Appl. Spat. Anal. Policy* **2017**, *10*, 381–399. [CrossRef]

12. Hutabarat Lo, R. The City as a Mirror: Transport, Land Use and Social Change in Jakarta. *Urban Stud.* **2010**, *47*, 529–555. [CrossRef]

13. Zhang, C.; Xiao, G.; Liu, Y.; Yu, F. The relationship between organizational forms and the comprehensive effectiveness for public transport services in China? *Transp. Res. Part A Policy Pract.* **2018**, *118*, 783–802. [CrossRef]

14. Terhorst, P. 'Reclaiming public water': Changing sector policy through globalization from below. *Prog. Dev. Stud.* **2008**, *8*, 103–114. [CrossRef]

15. Kurian, M.; McCarney, P. *Peri-Urban Water and Sanitation Services: Policy, Planning and Method*; Springer: Dordrecht, The Netherlands, 2010.

16. Nealer, E.; Van Eeden, E. The essence of water services management according to surface water catchment regions. A case study of Delmas Municipality. *Adm. Publica* **2010**, *18*, 133–148.

17. Gulbenkian Think Tank on Water and the Future of Humanity. Integrated Urban Water Resources Management. In *Water and the Future of Humanity: Revisiting Water Security*; Gulbenkian Think Tank on Water and the Future of Humanity; Springer: Cham, Switzerland, 2014; pp. 109–132.

18. Argo, T.; Firman, T. To Privatize or not to Privatize? Reform of Urban Water Supply Services in Jabotabek, Indonesia. *Built Environ.* **2001**, *27*, 146–155.

19. Morgan, B. *Water on Tap: Rights and Regulation in the Transnational Governance of Urban Water Services*; Cambridge University Press: New York, NY, USA, 2011.

20. Firman, T. Urban development in Indonesia, 1990–2001: From the boom to the early reform era through the crisis. *Habitat Int.* **2002**, *26*, 229–249. [CrossRef]

21. Pérez Campuzano, E.; Tello, C.A.; Everitt, J.C. Spatial Segregation in a Tourist City: The Case of Puerto Vallarta, Mexico. *J. Lat. Am. Geogr.* **2014**, *13*, 87–112. [CrossRef]

22. Zhu, J.; Xu, Q.; Pan, Y.; Qiu, L.; Peng, Y.; Bao, H. Land-Acquisition and Resettlement (LAR) Conflicts: A Perspective of Spatial Injustice of Urban Public Resources Allocation. *Sustainability* **2018**, *10*, 884. [CrossRef]

23. Rahaman, K.R.; Salauddin, M. A spatial analysis on the provision of urban public services and their deficiencies: A study of some selected blocks in Khulna City, Bangladesh. *Theor. Empir. Res. Urban Manag.* **2009**, *4*, 120–132.

24. Qian, J. China's New Urbanisation Plan to Revamp Public Service Provision. *East Asian Policy* **2014**, *6*, 20–32. [CrossRef]

25. Brueckner, J.K.; Vall, S.V. Chapter 21-Cities in Developing Countries: Fueled by Rural-Urban Migration, Lacking in Tenure Security, and Short of Affordable Housing. In *Handbook of Regional and Urban Economics*; Duranton, G., Vernon Henderson, J., Strange, W.C., Eds.; Elsevier: Amsterdam, The Netherlands, 2015; pp. 1399–1455.

26. Qian, J. Improving Policy Design and Building Capacity in Local Experiments: Equalization of Public Service in China's Urban-rural Integration Pilot. *Public Adm. Dev.* **2017**, *37*, 51–64. [CrossRef]

27. Nutley, S. Indicators of transport and accessibility problems in rural Australia. *J. Transp. Geogr.* **2003**, *11*, 55–71. [CrossRef]

28. Dalkmann, H.; Hutfilter, S.; Vogelpohl, K.; Schnabel, P. Sustainable mobility in rural China. *J. Environ. Manag.* **2008**, *87*, 249–261. [CrossRef] [PubMed]

29. Mulwafu, W.; Chipeta, C.; Chavula, G.; Ferguson, A.; Nkhoma, B.G.; Chilima, G. Water demand management in Malawi: Problems and prospects for its promotion. *Phys. Chem. Earth Parts A/B/C* **2003**, *28*, 787–796. [CrossRef]

30. Olayiwola, L.M.; Adeleye, O.A. Rural Infrastructural Development in Nigeria: Between 1960 and 1990—Problems and Challenges. *J. Soc. Sci.* **2005**, *11*, 91–96. [CrossRef]

31. Laditka, J.N.; Laditka, S.B.; Probst, J.C. Health care access in rural areas: Evidence that hospitalization for ambulatory care-sensitive conditions in the United States may increase with the level of rurality. *Health Place* **2009**, *15*, 761–770. [CrossRef]

32. Weinhold, I.; Gurtner, S. Understanding shortages of sufficient health care in rural areas. *Health Policy* **2014**, *118*, 201–214. [CrossRef]

33. Douhit, N.; Kiv, S.; Dwolatzy, T.; Biswas, S. Exposing some important barriers to health care access in the rural USA. *Public Health* **2015**, *129*, 611–620. [CrossRef]

34. Koo, A.; Ming, H.; Tsang, B. The Doubly Disadvantaged: How Return Migrant Students Fail to Access and Deploy Capitals for Academic Success in Rural Schools. *Sociology* **2014**, *48*, 795–811. [CrossRef]

35. Nieto Masot, A.; Cárdenas Alonso, G. Research on the accessibility to health and education services in the rural areas in Extremadura. *Eur. Countrys.* **2015**, *1*, 57–67. [CrossRef]

36. Grant, W. The Provision of Fire Services in Rural Areas. *Public Policy Adm.* **2005**, *20*, 67–79. [CrossRef]

37. Espinoza, D.; Reed, D. Wireless technologies and policies for connecting rural areas in emerging countries: A case study in rural Peru. *Digit. Policy Regul. Gov.* **2018**, *20*, 479–511. [CrossRef]

38. Rahman, S.; Smith, D.K. Use of location-allocation models in health service development planning in developing nations. *Eur. J. Oper. Res.* **2000**, *123*, 437–452. [CrossRef]

39. Kitchen, H.; Slack, E. Providing public services in remote areas. In *Perspectives on Fiscal Federalism*; Bird, R.M., Vaillancourt, F., Eds.; The International Bank for Reconstruction and Development/The World Bank: Washington, DC, USA, 2006; pp. 123–139.

40. Tali, J.A.; Malik, M.M.; Divya, S.; Nusrath, A.; Mahalingam, B. Location-allocation model applied to urban public services: Spatial analysis of fire stations in Mysore urban area Karnataka, India. *Int. J. Adv. Res. Dev.* **2017**, *2*, 795–801.

41. Mindahun, W.; Asefa, B. Location Allocation Analysis for Urban Public Services Using GIS Techniques: A Case of Primary Schools in Yeka Sub-City, Addis Ababa, Ethiopia. *Am. J. Geogr. Inf. Syst.* **2019**, *8*, 26–38.

42. Paasi, A.; Prokkola, E. Territorial Dynamics, Cross-border Work and Everyday Life in the Finnish-Swedish Border Area. *Space Polity* **2008**, *12*, 13–29. [CrossRef]

43. Brandsen, T.; Van Hout, E. Co-management in public service networks. The organizational effects. *Public Manag. Rev.* **2006**, *8*, 537–549. [CrossRef]

44. Medeiros, E. Old vs recent cross-border cooperation: Portugal-Spain and Norway-Sweden. *Area* **2010**, *42*, 434–443. [CrossRef]

45. Badulescu, A.; Bucur, C.; Badulescu, D. Fostering Euroregional Cooperation in Public Services. Evidence from Bihor-Hadju-Bihar Euroregion. *Eur. J. Sci. Theol.* **2013**, *9*, 127–136.

46. Badulescu, D.; Badulescu, A.; Bucur, C. Considerations on the Effectiveness of Cross-Border Cooperation in Public Order and Civil Protection Services. The Case of the Romanian-Hungarian Border Area. *Lex Localis J. Local Self Gov.* **2015**, *13*, 559–578. [CrossRef]

47. Carter, C.L.; Post, A.E. Decentralization and Urban Governance in the Developing World. Experiences to Date and Avenues for Future Research. In *Decentralized Governance and Accountability: Academic Research and the Future of Donor Programming*; Rodden, J.A., Wibbels, E., Eds.; Cambridge University Press: Cambridge, UK, 2019; pp. 178–204.

48. Medeiros, E. Barrier effect and cross-border cooperation. The Sweden-Norway INTERREG–A territorial effects. *Finisterra* **2014**, *97*, 89–102. [CrossRef]

49. Medeiros, E. Territorial Impact Assessment and Cross-Border Cooperation. *Reg. Stud. Reg. Sci.* **2015**, *2*, 97–115. [CrossRef]

50. Braunerhielm, L.; Alfredsson Olsson, E.; Medeiros, E. The importance of Swedish-Norwegian border residents' perspectives for bottom-up cross-border planning strategies. *Nor. Geogr. Tidsskr. Nor. J. Geogr.* **2018**, *73*, 96–109. [CrossRef]

51. Tosun, C.; Timothy, D.J.; Parpairis, A.; Macdonald, D. Cross-Border Cooperation in Tourism Marketing Growth. *J. Travel Tour. Mark.* **2005**, *18*, 5–23. [CrossRef]

52. Weidenfield, A. Tourism and cross border regional innovation systems. *Ann. Tour. Res.* **2013**, *42*, 191–213. [CrossRef]

53. Studzieniecki, T.; Palmowski, T.; Korneevets, V. The system of cross-border tourism in the Polish-Russian borderland. *Procedia Econ. Financ.* **2016**, *39*, 545–552. [CrossRef]

54. Makkonen, T.; Williams, A.M.; Weidenfield, A.; Kaisto, V. Cross-border knowledge transfer and innovation in the European neighborhood: Tourism cooperation at the Finnish-Russian border. *Tour. Manag.* **2018**, *68*, 140–151. [CrossRef]

55. Wismar, M.; Palm, W.; Figueras, J.; Ernst, K.; Van Ginneken, E. *Cross-Border Health Care in the European Union. Mapping and Analyzing Practices and Policies*; World Health Organization/European Observatory on Health Systems and Policies: Copenhagen, Denmark, 2011.

56. Rudawska, I.; Fedorwoski, J.J. Cross-Border Care and Cooperation. *Econ. Sociol.* **2016**, *9*, 11–13. [CrossRef]

57. Hovik, S.; Vabo, S.I. Norwegian Local Councils as Democratic Meta-governors? A Study of Networks Established to Manage Cross-border Natural Resources. *Scand. Political Stud.* **2005**, *28*, 257–275. [CrossRef]

58. Crabb, P.; Dovers, S. Managing natural resources across jurisdictions: Lessons from the Australian Alps. *Australas. J. Environ. Manag.* **2007**, *14*, 210–219. [CrossRef]

59. Zabelina, I.A.; Klevakina, E.A. Environmental and Economic Aspects of Natural Resource Use and Problems of Cross-Border Cooperation in Regions of Siberia. *Probl. Econ. Transit.* **2012**, *55*, 39–48. [CrossRef]

60. Yoder, J.A. Bridging the European Union and Eastern Europe: Cross-border Cooperation and the Euroregions. *Reg. Fed. Stud.* **2003**, *13*, 90–106. [CrossRef]

61. Dimitrov, M.; Petrakos, G.; Totev, S.; Tsiapa, M. Cross-Border Cooperation in Southeastern Europe. *East. Eur. Econ.* **2003**, *41*, 5–25. [CrossRef]

62. Knippschild, R. Cross-Border Spatial Planning: Understanding, Designing and Managing Cooperation Processes in the German-Polish-Czech Borderland. *Eur. Plan. Stud.* **2011**, *19*, 629–645. [CrossRef]

63. Sousa, L. Understanding European Cross-border cooperation: A Framework for Analysis. *J. Eur. Integr.* **2013**, *35*, 669–687. [CrossRef]

64. Scott, J.W. Bordering, Border Politics and Cross-Border Cooperation in Europe. In *Neighbourhood Policy and the Construction of the European External Borders*; Celata, F., Coletti, R., Eds.; Springer: Cham, Switzerland, 2015; pp. 27–44.

65. Castanho, R.A.; Loures, L.; Cabezas, J.; Fernández-Pozo, L. Cross-Border Cooperation (CBC) in Southern Europe—An Iberian Case Study. The Eurocity Elvas-Badajoz. *Sustainability* **2017**, *9*, 360. [CrossRef]

66. Mansvelt Beck, J. Has the Basque borderland become more Basque after opening the Franco-Spanish border? *Natl. Identities* **2008**, *10*, 373–388. [CrossRef]

67. Gakdner, J.A.; Abad, A. Sustainable Decentralization: Power, Extraconstitutional Influence and Subnational Symmetry in the United States and Spain. *Am. J. Comp. Law* **2011**, *59*, 491–527. [CrossRef]

68. Tort Donada, J.; Galindo Caldés, R. *L'articulació Geogràfica i Jurídica dels Municipis Fronterers: Radiografia de la Cooperació en els Límits Autonòmics Entre Catalunya, Aragó i la Comunitat Valenciana*; Escola d'Administració Pública de Catalunya: Barcelona, Spain, 2018; (In Catalan). Available online: http://eapc.gencat.cat/web/.content/home/publicacions/col_leccio_estudis_de_recerca_digital/18_articulacio_geografia_juridica_municipis_fronterers/estudis_recerca_digitals_18.pdf (accessed on 18 June 2020).

69. Galindo Caldés, R.; Santasusagna Riu, A.; Tort Donada, J. La frontera como espacio de conflicto y como espacio de cooperación: La Ribagorza como paradigma. In *España: Geografías Para un Estado Posmoderno*; Farinós Dasí, J., Ojeda Rivera, J.F., Trillo Santamaría, J.M., Eds.; AGE/Geocrítica: Madrid/Barcelona, Spain; pp. 255–268, (In Spanish). Available online: http://www.ub.edu/geocrit/estadoposmoderno.pdf (accessed on 18 June 2020).

70. Santasusagna Riu, A.; Galindo Caldés, R.; Tort Donada, J. Furthering Internal Border Area Studies: An Analysis of Dysfunctions and Cooperation Mechanisms in the Water and River Management of Catalonia, Aragon and the Valencian Community (Spain). *Sustainability* **2019**, *11*, 4499. [CrossRef]

71. Lindsay, J.M. *Techniques in Human Geography*; Routledge: London, UK, 2006.

72. Mayoux, L. Quantitative, Qualitative or Participatory? Which Method, for What and When? In *Doing Development Research*; Desai, V., Potter, R., Eds.; SAGE Publications: London, UK, 2006; pp. 115–129.

73. Phellas, C.N.; Bloch, A.; Seale, C. Structured methods: Interviews, questionnaires and observation. In *Researching Society and Culture*; Seale, C., Ed.; SAGE Publications: London, UK, 2012; pp. 181–205.

74. Kitchin, R.; Tate, N. *Conducting Research in Human Geography*; Routledge: London, UK, 2013.

75. McGuirk, P.; O'Neill, P. Using questionnaires in qualitative human geography. In *Qualitative Research Methods in Human Geography*; Hay, I., Ed.; Oxford University Press: Don Mills, ON, Canada, 2016; pp. 246–273.

76. Ley 7/1985, de 2 de abril, Reguladora de las Bases del Régimen Local. Boletín Oficial del Estado (BOE) n° 80, 03/04/1985(In Spanish). Available online: https://www.boe.es/buscar/act.php?id=BOE-A-1985-5392 (accessed on 18 June 2020).